# Interprofessional E-Learning and Collaborative Work:
## Practices and Technologies

Adrian Bromage
*University of Birmingham, UK*

Lynn Clouder
*Coventry University, UK*

Jill Thistlethwaite
*University of Warwick, UK*

Frances Gordon
*Sheffield Hallam University (SHU), UK*

Information Science REFERENCE

**INFORMATION SCIENCE REFERENCE**
Hershey · New York

Director of Editorial Content: Kristin Klinger
Director of Book Publications: Julia Mosemann
Acquisitions Editor: Lindsay Johnston
Development Editor: Joel Gamon
Publishing Assistant: Jamie Snavely and Thomas Foley
Typesetter: Keith Glazewski
Production Editor: Jamie Snavely
Cover Design: Lisa Tosheff
Printed at: Yurchak Printing Inc.

Published in the United States of America by
Information Science Reference (an imprint of IGI Global)
701 E. Chocolate Avenue
Hershey PA 17033
Tel: 717-533-8845
Fax: 717-533-8661
E-mail: cust@igi-global.com
Web site: http://www.igi-global.com



Library of Congress Cataloging-in-Publication Data

Interprofessional e-learning and collaborative work : practices and technologies / Adrian Bromage ... [et al.], editors.
p. cm.
Includes bibliographical references and index.
Summary: "This book provides relevant theoretical frameworks and the latest case driven research findings to improve understanding of interprofessional possibilities through e-learning at the level of universities, networks and organizations, teams and work groups, information systems and at the level of individuals as actors in the networked environments"--Provided by publisher. ISBN 978-1-61520-889-0 -- ISBN 978-1-61520-890-6 (ebk.) 1. Internet in education. 2. Interdisciplinary approach in education. I. Bromage, Adrian, 1964-
LB1044.87.I59 2010
371.33'44678--dc22

2010007332

British Cataloguing in Publication Data
A Cataloguing in Publication record for this book is available from the British Library.

All work contributed to this book is new, previously-unpublished material. The views expressed in this book are those of the authors, but not necessarily of the publisher.

## Editorial Advisory Board

# Table of Contents

**Section 2**
**Curriculum Design**

## Section 4
## Creating and Managing Learning Resources

# Detailed Table of Contents

Technological platforms are used in a range of different activities within interprofessional learning. These activities include accessing learning objects, conducting situative learning in order to establish jointly developed knowledge, providing an opportunity to develop an online professional identity and creating links within a community. In this discussion, one particular platform is particularly referred to throughout. That is the immersive virtual worlds (IVW), also referred to as multi-user virtual environments (MUVE). This is to provide more information for those unfamiliar with immersive virtual worlds and because these particular environments form the basis of many of the case studies described in the remainder of the book. Experience of virtual worlds also indicates that they may have special relevance to interprofessional education, particularly around notions of 'presence', 'embodiment' and 'identity', and these are concepts that have been given a fuller explanation for this reason. Secondly, a range of different theoretical frameworks for understanding the education activities and interactions that take place using these technological platforms are introduced and discussed.



The chapter begins with an overview of the current momentum toward interprofessional education and practice, citing specific trends in Canada as reflections of a global emerging consciousness. Initiatives undertaken at Dalhousie University are discussed in setting the context for this pilot study. Next, the pedagogy of critical diversity education is introduced and explained, with particular relevance for interprofessional education and practice. Comparison of face-to-face and online delivery of an interprofessional module based upon critical diversity education principles is then detailed, including research design and findings. The chapter concludes with a discussion of implications from this study.



Many in the online learning field now promote the need for a social presence online, in addition to cognitive and teaching presence, in order to fully realize benefits of online learning. In this regard, two important concepts arise when adopting e-learning: community and collaboration. Within healthcare there has been a recent push towards interprofessional education (IPE). IPE is an approach in which health and social care professionals come together to learn "with, from and about each other". In this chapter the authors discuss the background of healthcare IPE and online learning. They examine the potential benefits and limitations of both IPE and e-learning as well as issues related to combining these approaches. They will discuss the theory of 'communities of inquiry' and apply a modified version of it as a way to think about and create online IPE.. Lastly, the authors introduce an ongoing innovative healthcare e-learning project in Canada that was based on this theory and has focused on bringing together both IPE and online learning using the "build-a-case" method.



This chapter presents the development, implementation and evaluation of a module on interprofessional communication skills that incorporates principles of problem-based learning, delivered entirely online. Learners focus initially on foundational concepts of relationship and patient centered care, the importance of self awareness and understanding their own professional values and biases, progressing towards teamworking to develop common patient care goals. The module faculty facilitator is essential to role model and foster interprofessional collaboration. Qualitative content analyses of discussion board postings across 29 students, supplemented by small-scale in-depth interviews and a focus group, reveal they are able to learn interprofessional communication skills online. The 10 students who undertook both module components completed a project evaluation form: there was 85.6% agreement that the module taught them about interprofessional education and 92.9% agreement that their knowledge of other health professionals' perspectives increased. An online module can support the development of communication skills, but is recommended as one component of an overall interprofessional education curriculum.



A study investigating the effectiveness of a curriculum innovation involving students as online facilitators of interprofessional learning (IPL) provides a focus for this chapter. The research aim was to investigate whether Year 3 health and social care students were effective in facilitating online discussion forums contributing to the IPL of their counterparts in Years 1 and 2. Findings suggest that they were equally as effective as academic staff while offering some additional benefits. The account provides evidence of a successful online interprofessional initiative involving students promoting IPL.

It is not always transparent how e-learning can facilitate the interaction necessary for individuals to learn with, from and about each other as advocated by CAIPE (2002). This chapter emphasises the critical importance of facilitators acquiring new techniques to enable learners to make optimum use of the opportunities provided by Electronic Interprofessional Education (EIPE). To understand how barriers to online interprofessional education can be overcome, it is necessary to understand approaches to the development of positive e-learning environments. With this understanding, e-learning in supporting Interprofessional Education can be implemented more smoothly for both academic staff and students.



This chapter will consider the educational benefits and challenges of introducing e-learning objects within an interprofessional curriculum. It examines the tensions of curriculum development as content or process-driven in the context of facilitating interactive learning using blended learning strategies which combine online and face-to-face interactions. This chapter draws upon an evaluation of student and staff experiences of an interprofessional curriculum incorporating health and social care users and carers as case scenarios in a web-based simulated community, Wessex Bay, and highlights congruent and disruptive factors in negotiating transformative learning and cultural change. It draws conclusions and recommendations for informing practice in interprofessional education and suggests directions for future research to inform the substance (interprofessional case scenarios) and spaces (discussion boards, chat rooms, classroom) for collaborative learning in an interprofessional curriculum.



This chapter presents a case-study of a large-scale programme of interprofessional health education delivered through a virtual learning environment (VLE) and reusable interprofessional learning objects. The project commenced in 2005/2006, and brought together 1,800 students from 14 different health and social care disciplines in an InterProfessional Learning Pathway (IPLP). Interprofessional e-learning pedagogy and resources were developed to permit asynchronous interactions between large numbers of students from different disciplines, operating in virtual learning sets based around a series of patient scenarios which had been developed with service-user involvement. After the IPLP had run for the first three years, the experience of two of the 14 participating professional groups, dietetics students and physiotherapy students, was explored in 2008 through the use of focus groups and questionnaires. Both student groups felt the IPLP learning experience had a positive impact on their awareness of other roles

This chapter explores the early development stages of an interactive interprofessional online learning package that updates and supports health and social care professionals who mentor students in practice settings. The package aims to present content that is relevant and useful to fourteen different disciplines accessing it. A benefit of online content is that learning can be undertaken when convenient for the mentor, 24 hours a day-7 days a week, with the facility to stop and restart as needed. Additionally the package is constructed so both individuals and groups can use it; this both meets a regulatory body's requirement for having a face-to-face update every year, and provides support for interprofessional learning between mentors from different disciplines.



This chapter describes an innovative interprofessional education (IPE) module delivered with a blended approach, incorporating face-face teaching and online problem based e-learning (PBeL) activities. This has been evaluated using qualitative and quantitative methods. The qualitative work provided an overall perspective on how students experienced IPE. The quantitative work demonstrated significant differences in how different professions experienced the same curriculum; it also provided a useful insight into attitudinal changes over the period of learning. The evaluation findings are presented here with discussion of their implications for future delivery of IPE. Key issues include: maximising student engagement by optimising the timing of the module, the development of IT skills and managing the professional ethnocentrism and differences in professional epistemology and philosophy, which impact on students expectations of, engagement with, and subsequently the delivery of IPE.



The School of the Built Environment at the University of Salford redesigned its undergraduate programmes to include multidisciplinary project work at all 3 levels. This chapter provides a case study of the development and implementation of the interdisciplinary module at final level catering for students from five different disciplines. Overall, students responded positively to the module and academic tutors and visiting practitioners were also positive about student performance, but thought insufficient

# Foreword

The combined impact of e-learning and interprofessional learning is transforming professional education in health and social care. Such, however, is the pace of change that teachers' understanding all too easily lags behind. Some may be familiar with e-learning, others with interprofessional learning; few with both. This book comes to their rescue, written with authority and conviction by contributors from Australia, Canada, New Zealand, Norway, Sweden and the United Kingdom at the cutting edge of innovation in interprofessional e-learning.

Expectations of interprofessional learning are multiplying as it becomes more widespread and more responsive to the political agenda (WHO, 2010), stretching teachers to the limits of their ability as they explore uncharted territory, encounter unfamiliar student groups, reconfigure curricula and introduce new approaches to teaching and learning. More must be done to support them.

Generating state-of-the-art learning materials is one way, but how are production costs to be met, when educational resources are being cut in country after country recovering from economic recession? Enter e-learning: its materials are being manufactured expeditiously and economically, accessing and optimising the best available teaching expertise. Reusable, transferable, flexible and adaptable, they respond to the many and varied learning needs of students in professional and interprofessional groups in the university and the workplace.

E-learning, like interprofessional learning, knows no boundaries. It challenges teachers and students to go the second mile, urged on by the acumen, imagination, ingenuity and intellectual energy of its exponents. Resist the temptation to set premature limits, but resist too the temptation to see e-learning as the answer to every ill besetting professional and interprofessional education. It is no panacea. It is still being developed, refined and tested. It has yet to demonstrate its full potential. It may best be 'blended' with face-to-face learning. Well planned and delivered e-learning neither threatens nor usurps. Rather it introduces new insights, new perspectives and new approaches to enrich interprofessional learning as the successive chapters of this book testify.

In reflective mood at the end of an absorbing read, I recalled how my schoolteacher in a current affairs talk circa 1950 foreshadowed the advent of television in the classroom which, he opined, would reduce his profession to knob twiddling. How I long to reassure him.

*Hugh Barr*
*President, the UK Centre for the Advancement of Interprofessional Education*
*Emeritus Professor of Interprofessional Education, University of Westminster, UK*

## REFERENCE

World Health Organization (WHO) (2010). *Framework for Action on Interprofessional Education & Collaborative Practice*. Geneva: World Health Organization.

# Preface

## ABSTRACT

This book is intended to enhance thinking in interprofessional and interdisciplinary learning by considering how such education might be approached differently, by capitalizing on the opportunities offered by technological advances. It showcases developments as well as stimulating ongoing debate about the practical, theoretical and pedagogical implications of technology enhanced interprofessional (IPL) and interdisciplinary learning. As such the target audience are educators, course designers and managers in academia, although large scale employers, particularly in the field of health and social care could benefit from insights with regard to workforce development.

Although sceptics might argue that there is no substitute for first-hand interaction with colleagues and learners from different professions and disciplines, it is clear from the discussions within this book that technology potentially offers a multitude of opportunities to interact and engage in dialogue that can mimic real time interaction in a relatively safe environment. Discussants can experiment with ideas, explore values and beliefs and even play with tentative identities online in ways that would not otherwise be possible. The debate is therefore more nuanced and concerned, with questioning the extent to which interprofessional or interdisciplinary learning can be mediated through information and communications technologies.

A distinguishing feature of the book is its international flavour. While to an extent it showcases the work of the Centre for Interprofessional e-Learning (CIPeL), based in the UK, the work of the Centre is placed in the context of global developments in the field. This has led to an international volume with contributors from Australia, New Zealand, Canada, Sweden and Norway, as well as from several higher education institutions in the United Kingdom.

A major challenge for the book is that it seeks to bring together three key concepts and attendant practices by exploring interprofessional learning, interdisciplinary learning and the use of e-learning. All are in some ways contested concepts; technology can be accused of tending to lead pedagogy if careful thought is not given to intended outcomes; interprofessional learning can reinforce negative stereotypical behaviour if not prudently managed; and interdisciplinary learning, because it is not 'owned' by the discipline, can be perceived as peripheral.

Any one of these aspects could fill a book in itself; and in attempting to capture this breadth, this book will inevitably disappoint some readers. However, the aim is to provide a snapshot of developments in interprofessional and interdisciplinary e-learning at the end of the first decade of the 21st century. Notwithstanding evidence of the use of Web 2.0 technologies, the approaches explored within this book tend to favour Virtual Learning Environments (VLEs) and Social Constructivist pedagogies.

Aside from these commonalities there is wide diversity, and no single definitive answer is suggested as the 'best' approach to facilitate online interprofessional or interdisciplinary learning. In bringing together these concepts, chapter authors have had leeway to use their own voices and present their own local meanings, with an emphasis on description or depiction. This allows insight into the constraints and affordances of the context in which they work, so the reader can appreciate some of the issues with which the authors have been confronted and consider these in relation to their own development and implementation plans.

## BOOK STRUCTURE

The book is divided into five sections that broadly reflect the process of developing, implementing, storing and evaluating resources and approaches through a series of case studies. The first section is introductory, exploring the nature of interprofessional and interdisciplinary working and education, and setting it in the context of the recent history of computers in education and how people relate to virtual environments from the World Wide Web to 3-D worlds. These chapters challenge the reader to consider where their own practice might be located within this matrix. The also provide an overview of IPL and the technological base of education in the early 21st century, a lens through which to view the rest of the book.

The second section considers issues of curriculum design, providing insights drawn from empirical and theoretical or conceptual research and evaluation that might impact on curricula designed with the purpose of promoting IPL. It addresses a range of concerns, from the centrality of communication, the promotion of authenticity in learning resources, inextricable links between community and collaboration, student leadership and the relationship between interprofessionalism, attitudes and expertise. These chapters offer food for thought about overarching issues for academics who are developing an interprofessional curriculum.

Section three is concerned with implementing change and dealing with its consequences. It offers insight into challenges and opportunities in attempts to integrate e-learning and interprofessional or interdisciplinary learning. It presents a suite of honest accounts of issues encountered by academic staff at various stages of implementing curriculum change, and in some cases, research and evaluation into the impact of e-learning initiatives. Perspectives on cross sector collaboration, developing resources, institutional processes and structures, teaching approaches and interprofessional e-learning champions will be influential in helping others involved in implementing change reflect upon how they themselves might avoid pitfalls and take short cuts to optimizing success.

The fourth section focuses on the nature and processes of creating and managing the use of e-learning resources that promote IPL, beginning with the notion of interprofessionalism as a design philosophy that can facilitate the creation of reusable e-learning objects. It progresses to discuss how these resources can be used and re-used across different educational contexts, the educational theories that academics consider when building e-learning objects, and their impact on both learners and tutors. These chapters offer a detailed grounding in the issues that one is likely to encounter when developing and deploying such resources in IPL.

The final section looks forward to explore future possibilities that take interprofessional e-learning (IPeL) and interdisciplinary e-learning (IDeL) beyond the physical university, and beyond the personal computer. In considering technologies already in use, it explores practical considerations when using

ICT to support interprofessional learning in the workplace, on mobile devices and in virtual worlds, and discusses the value of such learning to students. These chapters offer insights into the possibilities that e-learning offers, both now and in the near future. Indeed, the final chapter speculates on where technology in education can lead, and how emerging technologies might influence interprofessional e-learning over the next decade.

## OUTLINE OF THE CHAPTERS

The introductory section of the book sets interprofessional learning and its mediation through e-learning into context. In Chapter 1, Lynn Clouder and Steve Smith analyse differences between a 'profession' and a discipline', drawing on arguments in the literature claiming that 'profession' has connotations of ideology and carries a practice element, whereas 'discipline', 'can be understood as 'a field of study'. They argue that the challenges facing the promotion of collaboration, a concept of interest to both disciplines and professions, are similar. They discuss these challenges in terms of the nature of knowledge; structural challenges; issues of identity and practical and logistical issues. They conclude that mutual incomprehension across specialists from different subject domains is no longer tenable and that higher education institutions need to prepare graduates for the contemporary world of work.

In Chapter 2, Adrian Bromage argues that expertise in interprofessional working can be defined as effective team working, placing the client's needs as paramount. He explores this in terms of literature about the nature of interprofessional education, interprofessional working and psychological literature on the development of expertise. He argues that interprofessional working can be regarded as both attitudinal in nature, and also as a particular model of problem-solving expertise. He suggests that viewing interprofessionalism as both an expertise and an attitude highlights and brings into focus the potential power of finding ways to link it to learners' pre-existing beliefs and values, in particular shared professional values, and as such can address some of the challenges faced in developing interprofessional collaborative skills among different groups of learners.

Samuel Edelbring, in Chapter 3, focuses on the use of technology in learning. His chapter examines the challenges for educational institutions to not only make the best use of technology but also to keep abreast of technological change and the seemingly increasingly rapid rate of development and innovation. He considers the place of technology in interprofessional education in particular, employing the familiar 'learn with, from and about' mantra that is our interprofessional byword, but changing the order slightly to discuss learning from technology, with technology and about technology. He calls for plurality in researching technology in education and points to its importance to the education and practice of interprofessional care.

In Chapter 4, Martin Oliver reviews pedagogical developments around e-learning in interprofessional education focusing primarily on the UK and providing a context for the remainder of this book. He explores research approaches employed concerning relationships between technology and the curriculum, beginning with an orienting discussion of curriculum as a field of study. He discusses how both e-learning and interprofessional learning suffer from a lack of evidence for their value, and explores the opportunities and difficulties around how e-learning might contribute to solving pedagogical challenges of interprofessional education. He concludes that all involved in this area share the challenge to build understandings of the complexities that we face when using e-learning to deliver interprofessional education.

Chapter 5, by Mark Childs, provides a discussion of the use of technological platforms to support a range of activities used in online approaches to interprofessional education. He argues that the various technological platforms available for education can be best understood by considering them as grouped into two main types, those emphasising learning from content, and those emphasising learning with and from others. He acknowledges that this distinction is in some aspects artificial, but is a useful explanatory tool for those new to these technological approaches. He concludes with a discussion, from the perspective of interprofessional learning, of how different platforms might support social engagement, community and the development of professional identity.

The second section of the book focuses on curriculum design issues. In Chapter 6, Marion Brown compares face-to-face and online delivery of critical diversity education for health and social care professionals in an interprofessional learning context in Canada. She concludes that diverse 'classrooms' challenge educators to utilise a range of approaches to teaching and learning, and suggests that rather than seeing online and face-to-face campus learning as dichotomous, mixing 'methods, the media and the messages' is advantageous.

In Chapter 7, Heather Mac Neill, Scott Reeves, Elizabeth Hanna and Steve Rankin focus on community and collaboration in an IPeL context, considering online learning through a social constructivist lens. They explore potential gains and limitations of delivering IPL online and theorise their thinking with respect to the 'community of inquiry framework'. They discuss the novel "build-a-case" approach, and recommend its transferability to other health professional programmes.

Chapter 8, by Patricia Solomon and Sue Baptiste, presents evaluation findings on the extent to which students can learn the fundamental principles of interprofessional communication online. They authors conclude that while 'ultimately students need to practice their communication skills face-to-face', they acknowledge the benefits of an online approach to developing interprofessional awareness and communication skills, and advocate it as a useful component of an overall IPE curriculum.

Chapter 9, authored by Lynn Clouder, Marie Krumins and Bernie Davies, focuses on a curriculum innovation that involved students taking on the role of student facilitators for a large scale IPL initiative. Evaluation findings suggest that students make very capable online IPL facilitators, and offer some additional benefits to academic staff facilitators.

In Chapter 10, Helen Bradbury, Melissa Highton and Rebecca O'Rourke offer an honest and reflective account, in case study format, exploring the introduction of collaborative e-learning activities into an interprofessional teacher education programme, a Master of Education in Clinical Education (MEd CE). The findings, drawn from the design and implementation of an educational research methodology module, provide valuable insight to others interested in fostering collaborative and interprofessional learning.

The final two chapters in this section are both concerned with involving service users and carers in interprofessional learning. Chapter 11, written by Frances Gordon, Karen Booth and Helen Bywater provides insight into the ways in which technology can overcome many of the barriers to the effective engagement of service users and carers in interprofessional curricula. The authors share practical and theoretical ideas underpinning involvement, and provide several examples of how principles have been put into practice.

Chapter 12, by Tarsem Singh Cooner focuses again on overcoming barriers to involvement of service users and carers. This chapter explores the properties of Web 2.0 tools and processes to enable mental health service user and carer involvement. However, it advocates the need for the technology to be underpinned by a framework for communication, and collaboration and again recommends 'the 'community of inquiry' model explored previously in Chapter 7.

Section three is concerned with implementing and dealing with the consequences of change. Pat Bluteau and Ann Jackson begin this section in Chapter 13, exploring the challenges of implementing interprofessional e-learning using a case study of the implementation of an interprofessional e-learning pathway (IPLP) shared across two universities. The IPLP was a CIPeL-driven curriculum development, commencing in 2005/2006, to bring together 1,800 students from 14 different health and social care disciplines. The authors' findings highlight the importance of strong leadership at key points within this process, in order to drive the change process through difficult periods.

In Chapter 14, Stephen Timmons, Heather Wharrad, Paraskevas Vezyridis, Jacqueline Randle, Joanne Lymn and Fiona Bath-Hextall report on a qualitative evaluation of the process of building and sustaining collaborative reusable e-learning object development across three educational sectors, Higher Education (HE), the UK National Health Service (NHS) and Further Education (FE) Colleges. The authors strongly recommend that development teams reflect upon their own interprofessional team working, ensuring that all team members are aware that Media Designers are not mere translators of ideas into a digital format, but have much expertise to offer, on an equal footing with other team members.

Dawn Foreman and Marion Jones emphasise in Chapter 14 the critical importance of learning facilitators' own understanding and knowledge of e-learning in the process of implementing interprofessional e-learning. When facilitators understand approaches to the development of positive e-learning environments, they can more readily understand how to overcome barriers to online interprofessional education. Thus the implementation of interprofessional e-learning can proceed more smoothly than would otherwise be the case.

In Chapter 16, Maggie Hutchings, Anne Quinney and Janet Scammell draw upon an evaluation of student and staff experiences of an interprofessional curriculum, incorporating health and social care users and carers as case scenarios in a web-based simulated community, Wessex Bay. The authors highlight congruent and disruptive factors in negotiating transformative learning and cultural change. They make recommendations for practice in interprofessional education, and suggest future research directions that can inform the substance (interprofessional case scenarios) and spaces (discussion boards, chat rooms, classroom) for collaborative interprofessional learning.

Chapter 17, by Karen Harrison and Lorraine McFarland, reports on a case study carried out in the context of a CIPeL-driven curriculum development that commenced in 2005/2006 to bring together 1,800 students from 14 different health and social care disciplines in an InterProfessional Learning Pathway (IPLP). The authors report the experience of two of the fourteen professional groups, dietetics and physiotherapy students. In general, both groups felt the IPLP learning experience had a positive impact on their awareness of other roles and interprofessional issues, and offered greater insight into the total clinical pathway of individual patients.

Section 4 of the book includes examples from Europe and further afield of the creation and management of e-learning resources that promote IPL, including reusable e-learning objects. Laurence Habib and Marit Fougner start the section in Chapter 18 with an analysis of the use of a video trigger as a pedagogical tool to facilitate interprofessional understanding in several professional courses in health sciences in Norway. Using actor-network theory as an analytical tool, the authors discuss the existence of a dichotomous relationship between technology and pedagogy, which is the source of major challenges in such work. They also recommend that senior management are involved in such projects to facilitate communication between project teams.

Chapter 19 moves back to the UK; Frances Gordon, Karen Booth and Helen Bywater of the Centre for Interprofessional e-Learning (CIPeL), provide new insights into the underpinning pedagogy of e-learning that supports the development of collaborative skills in health and social care. Drawing on a qualitative case study undertaken by the authors, they explore the working theories employed by academics when creating e-learning materials to support interprofessional education initiatives in the context of undergraduate, pre-qualifying courses for health and social care.

In Chapter 20, Richard Windle and Heather Wharrad focus on the definition, development and characteristics of reusable learning objects (RLOs) and outline examples of how such objects are meeting the challenges of IPL. They concludes that RLOs represent an e-learning format with a great potential for use within the interprofessional setting, providing a community of practice model to harness and empower a range of stakeholders.

Chapter 21, also written by Heather Wharrad and Richard Windle, provides further case studies involving interprofessional RLOs, arguing that it is important to consider a development and evaluation framework, including technical expertise and quality control, at critical stages. They stress that the key to high quality materials is to bring the interprofessional community together during the design process, and to have a firm focus on underlying pedagogical design principles.

Helen Lynch and Kerry Trabinger concentrate in Chapter 22 on the customisation and interprofessional application of toolbox objects. They describe a class of pedagogically rich and sophisticated e-learning objects created for the Australian vocational education and training system. Their chapter focuses on the tools and techniques of customization, and presents a model of reuse that can be implemented elsewhere with any pedagogically rich web based e-learning object in inter-vocational and interprofessional settings.

In Chapter 23, Karen Ousey and Stephen White present a case study exploring a multi-disciplinary collaborative initiative that aims to meet the learning needs of fourteen different health and social care professionals. They stress that one benefit of online content is that learning can be undertaken when convenient for the mentor, with the facility to stop and restart as needed.

The team of Chris Dearnley, Melissa Owens, Pete Greasley and Caroline Plews describes in Chapter 24 an innovative interprofessional education (IPE) module delivered with a blended approach of face to face teaching and online problem based e-learning (PBeL) activities. They found significant differences in how different professions experienced the same curriculum, identifying several key issues, in particular the influence of module timing on student engagement, the development of IT skills and managing differences in professional epistemology and philosophy.

Finally Gerard Wood and Song Wu provide in Chapter 25 a case study of the development and implementation of an interdisciplinary e-learning module for students from five different disciplines. Overall, students responded positively to the module and academic tutors and visiting practitioners were also positive about student performance; however, the study highlights the challenging nature of implementing interdisciplinary working within the built environment context. The greatest challenges concern managing large numbers of students in teams, and composing clear requirements with associated assessment criteria.

The final section of the book explores possibilities that can take interprofessional e-learning (IPeL) and interdisciplinary e-learning (IDeL) beyond the physical university, and beyond the desktop and laptop PC to hand-held mobile computing. Chapter 26, by Nina Godson and Adrian Bromage, reports a small-scale feasibility study into the use of mobile devices, personal digital assistants (PDAs) to support student nurses' learning on clinical practice placements. The study focuses on the practicalities, including selecting suitable devices, training and supporting learners in using the devices, and 'usability' of the devices in a clinical setting.

The use of such devices is also explored by Christine Dearnley, Stuart Walker and John Fairhall in Chapter 27. The authors explore the use of hand-held mobile ICT to increase the accessibility of learning for students with disabilities. In this case, they demonstrate how the features of mobile devices were used to support the organisational, memory and writing needs of students with dyslexia. The authors make recommendations for implementing mobile assessment for practice learning that are based on a theory of mobile learning designed for accessibility

In chapter 28, Elinor Clarke documents a small-scale pilot action research study of the use of a 3-D multi-user virtual world in interprofessional education. Such virtual worlds have in recent years been recognised as a space in which learners in diverse physical locations can potentially meet simultaneously to role-play in scenarios, which have an authentic look and feel to them, as if a 'virtual movie set'. The chapter brings home the practicalities, both in terms of the necessary physical infrastructure to support such learning, and the challenges it presents to both facilitators and learners, and makes suggestions for future research in this emerging area.

In Chapter 29, the final of the book, Andrew brooks, learning technologist at CIPeL, writes a more speculative chapter that outlines recent and developing innovations in information and communications technologies, with an emphasis on getting education out of the classroom and into a world that is itself increasingly becoming integrated with digital environments. In doing so, he challenges the reader to speculate on where the adoption of these innovations in education could lead, and to reflect on how this might further influence interprofessional e-learning over the next decade.

## FINAL WORDS

How can one summarise the themes emerging within this book as a whole? While acknowledging that e-learning is not a panacea, and must be approached critically, it is clear that relying on face-to-face learning activities only can be limiting, and in fact is no longer necessary, as information and communications technologies (ICT) are embraced increasingly in the academy, and indeed, the workplace. It can be argued that by encouraging collaboration online, we mirror the demands and working practices that are becoming typical of the contemporary workplace. Furthermore, as curricula are adapted to take into account interprofessional or interdisciplinary concerns, it becomes clear that logistical problems are a major barrier if the aspiration is to achieve learning on any reasonable scale. This is where interprofessional e-learning comes into its own; its ability to transcend boundaries of time, location and academic discipline are persuasive.

We encourage the reader to reflect upon the overlap between the chapters; these areas of commonality are indicative of common themes and concerns for those who would seek to develop interprofessional and interdisciplinary e-learning. These overlaps also serve to reinforce the fact that curriculum development is an holistic activity, and should be approached as such if lasting and embedded change is desired at an educational institution. Indeed, CIPeL's approach was to develop and pilot e-learning innovations at Coventry and Sheffield Hallam Universities, and then to disseminate what was learned to the wider community of interprofessional and interdisciplinary educators across the United Kingdom.

It is worth reiterating that the book does not seek to give a single definitive answer for finding the 'best' approach to facilitate online interprofessional or interdisciplinary learning. The book offers a broad and diverse picture of how e-learning is beginning to impact on interprofessional and interdisciplinary learning at the end of the first decade of the 21st Century. It is the editors' hope that this will help individuals and teams to progress their existing practices to the next meaningful developmental step, given their own needs and local context, and contemporary pressures on higher education.

# Section 1
# Introduction

# Chapter 1
# Interprofessional and Interdisciplinary Learning:
## An Exploration of Similarities and Differences

**Steve Smith**
*Coventry University, UK*

**Lynn Clouder**
*Coventry University, UK*

## ABSTRACT

*This chapter begins by considering the words used to discuss collaborative education. Although it can be argued that "practice" separates "a profession" from "a discipline", the merit in separating theory from practice is highly questionable. The literature suggests that the challenges to interprofessional and interdisciplinary learning are very similar, for example, the "silo" mentality causes problems within both. In addition, it is evident that the reasons behind advocacy of interprofessional and interdisciplinary learning are also similar. The chapter demonstrates that successful interprofessional and interdisciplinary learning requires fundamental changes to both the curriculum and the organisation delivering it. The authors conclude that while subtle differences might exist between interprofessional and interdisciplinary learning their promotion is based on a similar rationale, which is to ensure that students are prepared for the real world in which collaboration, boundary crossing, adopting multiple perspectives and working with others to achieve optimal outcomes, is paramount.*

## INTRODUCTION

Advocates of interprofessional learning frequently claim that it is distinct from interdisciplinary learning. D'Amour, Ferrada-Videla, San Martin Rodriguez and Beaulieu (2005) argue that the two concepts are rarely clearly defined and are used rather loosely, causing conceptual confusion, which is well recognized. Gilbert (2005) is resolute that interdisciplinary is not synonymous with interprofessional learning. Notwithstanding inherent concerns for those who wish to promote interprofessional education and claim that such ambiguity is not helpful, we critically examine the arguments in the context of the functioning and development

DOI: 10.4018/978-1-61520-889-0.ch001

of the Centre for Interprofessional e-learning (CIPeL). The CIPeL is a Centre for Excellence in Teaching and Learning (CETL) funded by the Higher Education Funding Council for England (HEFCE) in the United Kingdom. Although its remit has been to enhance interprofessional learning amongst health and social care professionals in the first instance, there is a sound rationale for dissemination across other disciplines. Extending the Centre's work into areas such as geography, environmental management, business management and engineering brings into sharp relief the contextual, conditional and contested nature of collaborative learning and teaching across boundaries.

An equally pressing concern stemming from inquiry into wider disciplinary areas as well as the interprofessional literature has been for the coherence of this book. We have worried about, "Should we provide a glossary?" "Should we attempt to standardize the labels used?" "Would we expose ourselves to the risk of widespread criticism if we allowed authors to use their own chosen terms and stayed true to the author's usage?". Clearly we needed to gain a greater depth of understanding of the issues and concerns before making an informed choice as to our strategy.

This chapter provides a firm rationale for variance in the use of terminology and therefore a major signpost for the chapters to follow. Readers will notice that authors use various labels, to describe initiatives that bring students together to 'learn with, from and about each other' (CAIPE, 1997). They will also recognize a great variation in the extent to which interprofessional or interdisciplinary learning is embedded in curricula and in the ways in which e-learning is used as a medium. To set the scene for chapters to follow, this chapter begins by discussing a variety of definitions in common usage. We conclude that in fact, interprofessional and interdisciplinary learning, both of which endeavour to promote collaborative and integrative learning and despite subtle differences, meet with comparable challenges with regard to implementation. By exploring both interprofessional and interdisciplinary learning across a wide range of contexts we conclude by providing insight into the antecedents of successful initiatives, elements of which will recur in the chapters to follow.

## Useful Definitions

Conceptual ambiguity begins with the words 'discipline' and 'profession', which 'are often used interchangeably', despite the claim that they have 'distinctly different meanings' (Mu & Royeen, 2004, p. 245). Rawson (1994, p. 40) suggests that whereas 'profession' has connotations of ideology, 'discipline', 'although a useful term, is limited by its association with background theoretical and methodological contributions'. Preferring to consider a discipline to be 'a field of study' and a profession 'a calling requiring specialized knowledge and often long and intensive preparation', Mu and Royeen (2004, p. 244) argue unequivocally that 'occupational therapy is a profession, not a discipline' because it has a practice element.

A discipline can be readily defined as 'a branch of learning or field of study' (Chambers Dictionary of English, 2003). Disciplines emerged as loci of authority during the Enlightenment; social sciences such as history, anthropology, psychology and sociology were born and developed alongside the natural sciences and became powerful determinants of the types of knowledge that we still value in contemporary society (Maxwell, 2003). Definitions of a profession incorporate the notion of occupation or career, vocation or calling, which involves advanced or specialized learning or training (Oxford Encyclopedic English Dictionary, 1991).

Professions are further defined with respect to underpinning values and beliefs:

*a disciplined group of individuals who adhere to high ethical standards and uphold themselves to, and are accepted by, the public as possessing special knowledge and skills in a widely recognised, organised body of learning derived from education and training at a high level, and who are prepared to exercise this knowledge and these skills in the interest of others (Australian Council of Professions, 2007).*

The notion of profession is also imbued with certain defining features such as formal qualifications, regulatory bodies with powers to admit and discipline members, and some degree of monopoly rights (Bullock & Trombley, 1999). An emphasis on moral probity, service orientation and codes of conduct mean that professions have historically been revered in wider society. Professional status, once reserved for the clergy, architects, lawyers, doctors and the armed forces, has been moderated the professionalization of many other occupational roles, such as teaching and the allied health professions (Eraut, 1994), previously considered to be 'semi-professions' (Etzioni, 1969). Allied health professions have fought hard to escape medical dominance and subordinate status in the medical hierarchy by becoming graduate professions (Bines, 1992). The imperative to professionalize that has groups vying for professional status is powered by rewards of social esteem and prestige, traditionally bestowed on professions (Hoyle & John, 1995).

However, there are "professional" mathematicians, physicists and historians (see, for instance, Mooney Melvin, 1987) working in what we might term disciplinary fields that distinguish between their academic discipline and their professions. For instance, Mu and Royeen (2004) make a distinction between occupational science and the profession of occupational therapy. Similarly, Gibson (2003) argues for the separation of the academic discipline of physics from the profession of physics. The merit in separating theory from practice when they are so fundamentally related seems highly questionable, although other definitions such as Wilde's (2007), which refers to, 'the study, or practice, of a subject using a specific set of methods, terms and approaches' is inclusive. Reflecting the esteem and prestige associated with certain professions, disciplines are equally hierarchical. Maxwell (2003) discusses the creation of modern science, which has been hugely influential in the development of the social sciences that have sought to identify with the scientific method. In considering the detrimental effects on social inquiry in disciplines such as sociology, economics and anthropology there is arguably an analogy with medicine and associated health and social care professions.

Expanding on these basic definitions we add multiprofessional learning, which is commonly used interchangeably with interprofessional learning (Finch, 2000). Cornish, Church, Callanan, Bethune and Curran (2004) suggest that the prefix 'multi' infers an additive and non-integrative system, in which labour is divided through traditional hierarchical power-over structure and interaction is limited. CAIPE (1997) reinforces distinctions from interprofessional learning by referring to multiprofessional learning as 'learning side by side', limiting the development of shared understanding and collaboration.

The term interprofessional learning has a long pedigree, having been consistently promoted around the world over several decades and has become associated almost exclusively with health and social care education. It is most commonly defined as, 'the process, through which two or more professions learn, with, from and about each other to improve collaboration and the quality of service (CAIPE, 1997). Freeth, Hammick, Reeves, Koppel and Barr (2005) develop this definition by suggesting that interprofessional learning will 'enhance learners' understanding of other professions' roles and responsibilities, whilst fostering mutual respect and understanding between members of the health care team'. It is also variously termed integrative, collaborative and 'non-disciplinary' (Munro, 2000).

Interdisciplinary learning has had periods of being in vogue and is currently experiencing a revival. It is considered as interaction:

*involving collaborations between students from differing subject areas in pooling their disciplinary knowledge in addressing complex and significant, real world problems [leading to] the ability to understand and be understood by a diverse group of specialists (Woods, 2007, p. 854).*

Franks, Dale, Hindmarsh, Fellows, Buckridge and Cybinski (2007) identify a number of typologies of interdisciplinary teaching and research, one of the most interesting of which is Lattuca's (2003) that moves up through 'informed', 'synthetic', 'transdisciplinarity' and finally, 'conceptual interdisciplinarity', which integrates disciplinary perspectives and critique of disciplinary understandings but addresses questions that do not have a disciplinary basis.

Definitions of interprofessional learning appear to acknowledge more overt connections with practice and the workplace than definitions of interdisciplinary learning; however, both explanations stress the centrality of collaboration and integration (Van den Besselaar, 2001) and both focus on the desired outcome of collaboration in addressing complex problems, whether such problems are those of whole communities or of individuals. Decreasing emphasis on the distinctions between professions and disciplines by focusing on the prefix of 'inter' further emphasizes similarities between the two modes of learning. Rawson (1994) suggests that 'inter' infers a relationship both between and among elements and implies some notion of reciprocal operations, whereas the term 'trans' also signifies a relationship across or beyond but does not carry with it any indication of mutuality'.

Rawson's thinking has recently been challenged from within the health professions where 'transdisciplinary learning' has been used to refer to practice that is characterized by a deliberate exchange of knowledge, skills and expertise that transcends traditional disciplinary boundaries (Stepans, Thompson & Buchanan, 2002). Such practice is characterized by a highly integrated team, organized according to comprehensive constructs and methods (Wall & Shankar, 2008) The benefits are presented as: the valuing of the knowledge and skills of each team member, blurring of disciplinary boundaries, increased levels of trust and the creation of broad networks of professionals able to analyze multi-faceted factors influencing human health and well-being. It is arguable that the notion of transdisciplinary learning captures the essence of interprofessional and interdisciplinary learning. It might in fact, have contemporary relevance, especially given the current emphasis on graduate capability and employability, which means that experience in the workplace is no longer restricted to vocational programmes. Clearly, terminology is fluid; the reader will recognize in the chapters to follow that differences exist between fields and within and between countries and even between individuals from the same fields. We will leave the reader, to weigh similarities and subtle differences in the hope that ideas about interprofessional and interdisciplinary learning might be considered to be mutually reinforcing.

## CHALLENGES TO OVERCOME IN IMPLEMENTING INTERDISCIPLINARY AND INTERPROFESSIONAL LEARNING

Many of the case studies that follow discuss the establishment of collaborative and integrative learning opportunities from scratch. Others frame interventions in terms of improving on past efforts, not least by moving from face-to-face to blended or e-learning approaches. As a means of setting the scene for the case studies we turn to highlighting some of the challenges that will have undoubtedly been faced by many of the interprofessional and interdisciplinary initiatives presented in subsequent chapters.

## The Nature of Knowledge

The importance of the nature of knowledge and how it is communicated should not be underestimated as a barrier to promoting integrative and collaborative learning. Knowledge of professionals in a specific field is usually necessarily specialized into a moderately narrow spectrum, which although increasingly necessary as the knowledge base increases, is an issue that impedes interdisciplinary learning (Tarvainen, 2004). Stew (2000) suggests that as knowledge becomes more complex disciplines are divided into sub-disciplines each wishing to demarcate the body of knowledge within boundaries relevant to their practice leading to 'occupational imperialism'. Again this militates against collaborative learning, especially when there are perceptions that it might devalue professional or disciplinary learning if it leads to the creation of generic workers and generic knowledge. Clearly these are very real issues. However, they highlight how crucial it is to attempt to integrate learning to avoid already tenuous understanding being lost.

Creating dialogue and finding a common language is essential, although not straightforward, especially in the light of different disciplinary pedagogies. Law is an extreme case in point because it has traditionally been dominated by the study of primary sources of 'black letter law' in the form of statutes and judicial decisions. This model is not easily transferable to other disciplines (Soetendorp & Byles, 2000) and creates challenges for law and associated disciplines to find a middle ground. Undoubtedly other professions and disciplines experience similar problems, although as Wood (1999) infers many of the issues would be relatively easily addressed should there be a will and a readiness to do so.

## Structural Challenges

Every culture has its norms and ways of working that are deeply embedded and resistant to change. 'All educational contexts represent and replicate, within their own internal processes, external power relations' (Vince, 1996, p. 47). Vince captures the sense in which the learning experienced by students in the health and social care professions, as well as students across other disciplines in higher education, is fundamentally shaped by dominant discourses in the workplace. Despite suggestions that functions within health, rather than being hierarchically stratified are increasingly differentiated (Adams, 2005), health is characterized by a strong sense of hierarchy and authority, which holds both within and between professions. For instance, medicine is still perceived to be an economically and socially desirable profession (Whitehead, 2007) and doctors are acknowledged as particularly influential and dominant players in the health hierarchy (Allen & Hughes, 2002).

Within any hierarchy, power and status conspire to elevate some and diminish others; if the status of one group is to increase, that of another must decrease (Whitehead, 2007). Status is therefore, a relative ranking, and a particularly valuable resource because its availability in any given context is limited (Milner, 1994). Whitehead (2007) acknowledges the challenge of interprofessional initiatives, with their collaborative working emphasis and makes no pretence that the success of such initiatives will depend to a great extent on buy-in from doctors. In recommending clarification of what is meant by collaboration and suggesting it might be conceptualised on a spectrum, she acknowledges that medics will be reluctant to see power differentials eroded. One assumes that such resistance to fundamental change will do little to change current hierarchical relations, which we believe thrive at least partially because the players are bound by the close proximity, the context and history in which they operate, namely the National Health Service.

Interprofessional initiatives challenge notions of power and hierarchy that socially construct health and social care education and this is why it is so problematic for many. Territorialism in the health care professions is predictable as in any situation where there are issues of power and control arising from a capitalist competitive system (Ericson, 1991). Doctors have remained firmly in ascendancy in the health sector, which has not experienced the same level of changed power relations as has occurred in some industries (Eraut, 1994). For instance, the construction industry is characterized by interprofessional relationships defined by the increasing financial power of developers and construction companies. Architects and engineers, once in charge of building projects and responsible for employing surveyors, service engineers, builders and subcontractors, have themselves become service providers without managerial control (Eraut, 1994). Inevitably this suggests a less hierarchical structure where power is more diffuse, possibly because it is not confined within the rules and norms of an institution such as the UK National health Service (NHS).

Gilbert (2005) acknowledges that despite major moves to promote interprofessional education, nationally and internationally, there is still a reluctance to make structural changes. De Witt and Baldwin (2007) identify enduring barriers, originally identified by Baldwin and Baldwin (1979), which include conceptual confusion, language, communication and culture but also involve faculty attitudes, cost of curriculum change and professional regulation. Neither the language and cultural issues nor the structural constraints to integrative and collaborative learning are confined to health. Fry (2001) refers to the 'sociology of academia' with specialized languages, dominance of some disciplinary jargon and insular disciplines. Conceptual confusion is identified in engineering where in one learning context the approach claiming to be interprofessional, in fact demonstrates very little integration (Tarvainen, 2004). Demonstrating the power of academics' attitudes, Levinson (2003) points to a lack of clarity in learning objectives and cites a lack of evidence of the effectiveness of interprofessional learning to justify giving it low priority. Such indolent attitude to change is possibly a reflection of management values. For instance, Wood (1999) identifies ambivalence of senior academics at one UK University to the need to promote interdisciplinary learning in built environment education. This hesitancy, despite earlier agreement on the desirability of greater collaboration in response to criticism of the UK construction industry, is attributed to hindrances created by faculty structures and resource pressures.

Interestingly, the Built Environment programme at Leeds Metropolitan University, where these challenges appear to have been overcome, provides a stark contrast, notwithstanding the acknowledgement of the need for adequate resources and high levels of coordination (Higher Education Academy, 2008). Perhaps the vision of a creating something better, even without hard evidence is enough to promote change in some of the more forward looking institutions, while for others structures provide tangible evidence of the immovable.

## Issues of Identity

Becher and Trowler (2001, p. 47) suggest that 'being a member of a disciplinary community involves a sense of identity and personal commitment, a way of being in the world' that reflects particular epistemological and cultural attributes that define the community to the outside world (Geertz, 1983). The strong sense of identity that develops promotes what William Sumner identified in 1906 as, an 'in group' and 'out-group' dynamic that leads to jostling for position in inter-group competition (Brewer, 1979). Whilst the dynamic makes for strong professional commitment and solidarity, it does not make for ease of collaboration across professions or disciplines. Traditionally each profession has socialized its newcomers within its own boundaries, resulting in limited knowledge about 'the practices, expertise, responsibilities, skills, values and theoretical perspectives of [other] professionals' (San Martin-Rodriguez, Beaulieu, D'Amour & Ferrada-Videla, 2005, p. 137).

Becher and Trowler (2001) go as far as referring to the professions as 'warring tribes', while Eraut (1994) observes that interprofessional relations are absent from accounts of the ideology of professionalism, except in asserting the supremacy of the true professions over newcomers (Eraut, 1994). DeWitt and Baldwin (2007, p. 32) are disheartened that even after years of promoting interprofessionalism, 'overwhelming barriers of disciplinary territoriality and systems inertia persist' so that 'each generation seems to have to repeat the experiences and frustrations of the past'. The silo mentality has prompted a call for a need to cultivate 'professional plurality' (Glen, 1999). However, even where interdisciplinarity has thrived for some time, Franks et al. (2007) note that there is always a tendency to default to disciplinary positions where disciplinary structures are perpetuated.

### Practical and Logistical Issues

Norman (2005) highlights several of the practical challenges in implementing interprofessional education at pre-registration level. He highlights issues such as cost sharing and resources and securing joint validation and accreditation of courses that cross disciplinary boundaries. However, one of the major logistical issues is the difficulty of getting large numbers of students from different geographical locations together, which puts pressure on timetabling as well as on estates. Norman (2005) also identifies the issue of coping with unequal numbers across groups in a way that avoids students from smaller professions being overwhelmed by those from the larger professions. These sorts of issues have more recently led to the advent of the use of e-learning approaches, many of which are presented as case studies in this volume.

## IF IT'S SO DIFFICULT WHY BOTHER DOING IT?

The primary rationale for interprofessional learning and its corollary interprofessional working in health is the promotion of improved patient-centred practice (CAIPE, 1997). Mapping its development in the USA, DeWitt and Baldwin (2007) trace the origins back to World War II. They attribute resurgence in interest during the past decade to inadequacies in the health system, a need to reappraise how care is delivered due to rising costs and growth of health maintenance organizations and growing interest in continuous quality improvement as a means of enhancing quality, efficiency and effectiveness. The situation in the UK is not dissimilar, although Price (2005, p. 76) suggests that greater complexity of the health care system, 'with increased reliance on clinical team working, and the necessity for communication and cooperation between both teams and team members' is a primary driver. However, he too highlights inadequacies in the system and points to recent high profile failures (Bristol Inquiry, 2001; Department of Health, 2003) that have had a heavy influence on recent government policy (Department of Health 2000, 2001a, 2001b, 2002). It seems that the post-war social and cultural discourse underpinning interprofessional learning has been replaced by a discourse of economics tangled with the rhetoric of quality and political expedience that unfortunately do not guarantee positive changes in practice.

Outside the health context, interest in learning that crosses subject boundaries is a growing theme in recent higher education literature (Woods, 2007) and arguments in support of interdisciplinary learning are persuasive. Woods (2007, p. 854) is an advocate of the 'educational benefits of engaging critically with one's own discipline by viewing its limitations from another perspective', although she too identifies the need for changed working patterns that require team work. In the context of law, there is acknowledgment that 'greater complexity and interdependency within the workplace has required the creation of different kinds of knowledge': the need for interdisciplinary dialogue has never been greater (Soetendorp & Byles, 2000). Similarly, in engineering, interdisciplinary studies are recognized as promoting 'skills that help future engineers to cope in a changing environment' (Tarvainen, 2004, p. 1). Woods (2007, p. 854) argues that future graduates need to be prepared to tackle problems troubling society; 'pandemics, water politics, global warming, famine, migration, [and] international crime'. Added to this list are issues of social inclusion (Wilson & Pirrie, 2000). The complexities of the health service might be seen as a microcosm of this larger picture. Like every sector, change is constant and graduates must become flexible and adaptable to cope with the pace of change.

Regardless of whether the focus of collaborative learning is an individual patient or client, or a building or management project there is an imperative for contemporary graduates to acquire capabilities that allow them to cross boundaries, adopt multiple perspectives and work with others to achieve optimal outcomes. What name is put to this is almost inconsequential; the rationale is of prime importance. Again the Built Environment programme at Leeds Metropolitan University illustrates this point. The programme claims to be responsive to the 'interdisciplinary nature of complex design development problems, which have caused an increased blurring of the boundaries between professions' and involves finding 'comprehensive solutions to complex problems' (Higher Education Academy, 2008). It uses the full gambit of terminology as student's progress through their programme: multidisciplinary in Year 1, interdisciplinary in Year 2 and interprofessional in Year 3. The positive outcomes associated with the programme are an increased awareness of professional roles and recognition of students' own responsibilities and those of others, recognition of the benefits of a team approach and understanding of the complexities of the construction industry and the interdependence of construction disciplines.

## FACTORS THAT FACILITATE COLLABORATIVE AND INTEGRATIVE LEARNING

The number of case studies presented in this volume highlights a wealth of successful interprofessional and interdisciplinary learning initiatives. However, when we talk of facilitating factors we set the scene in terms of instigating change, engaging staff, launching and developing new programmes or altering delivery to achieve improved learning outcomes associated with collaborative learning. Freeth, Hammick, Koppel, Reeves and Barr (2002) provide a comprehensive critical review of evaluations of interprofessional education that covers a more detailed breakdown of success in terms of specific interprofessional attributes that goes beyond the scope of this chapter. However, readers of the chapters to follow will gain insight into evaluations of more recent initiatives. We turn now to focus on factors that have been deemed helpful in aiding implementation across a diverse range of programmes.

The literature supports the need for a multi-faceted strategy. Although external structural and cultural issues will endure, structural challenges within academia are within a scope of influence. Successful interdisciplinary learning for the Built Environment programme at Leeds Metropolitan University required fundamental curriculum change and restructuring across the entire School, adequate resourcing and staff commitment to the programme. These findings are supported by the achievements of a long established Interdisciplinary Foundation at Griffith University, Australia, which are also attributed to fundamental curriculum change and structural reorganization (Frank et al., 2007). The Australian School of Environmental Studies employed the 'lifeboat model' of selecting staff, grouping them into interdisciplinary units and separating them from their traditional disciplines. A problem-based learning approach was adopted. Deliberate mechanisms such as team teaching, random allocation of office space, creating applied research teams and providing a common room, were employed to promote interaction amongst staff. This approach not only solves the issues of hierarchies and jostling for disciplinary superiority by removing staff from their disciplinary contexts, it throws them together and increases the potential for developing a shared language and culture. Perhaps most importantly, we are told that staff self-selected themselves for the venture on the basis of having a commitment to working across disciplines. Inevitably this was likely to create buy-in to making the project work and is important in that it indicates a high level of pre-existing commitment to the adopted philosophy.

The adoption of a problem based learning pedagogy at Griffiths University clearly addressed the issue of potential differences about the teaching and learning approach to be used as would any other approach if imposed across the School. Interestingly, given the focus for this book, Walsh's (2007) findings suggest that online learning approaches can successfully overcome barriers of different methods of learning adopted by different professional groups. In fact, there is firm evidence that online delivery of interprofessional learning does successfully overcome some of the issues associated with logistics and large cohorts (Scammel, Hurchings, Quinney, Hean, Dumbrell & Pulman, 2008; Walsh, 2007) notwithstanding the fact that learning to use technology is an added consideration with an associated time commitment (Scammel et al., 2008).

Wall and Shankar (2008) who conducted a study of a group of doctoral students engaged in a transdisciplinary learning programme found that aside from adequate resources, readiness for change was a fundamental factor in the success of the programme. They suggest that an open attitude to developing a broader perspective than is available within disciplinary boundaries is crucial. Interestingly, Wall and Shankar stress that this does not mean that one's own perspective should not be maintained; on the contrary, being open but maintaining one's perspective is vital. There are important messages here which suggest that staff who are allocated interprofessional or interdisciplinary teaching without having a commitment to it are jeopardizing the programme before they start. In addition, the notion of preserving a disciplinary perspective is considered positive, suggesting that the removal of staff from their disciplinary roots is far more nuanced than we might at first think. Wall and Shankar (2008) also stress the importance of relationality on the basis that learning is a social activity. Gilbert (2005) favours the need for a dedicated core faculty drawn from across the disciplines (presumably with a commitment to interprofessional learning), time to facilitate learning and structures that facilitate formal and informal student learning interaction outside of disciplinary boundaries, which would promote relationality and sense of community.

Lattuca (2002, p.711) suggests that implementation requires not only policies and programmes but also a process of collective learning. It seems that structural change is clearly a fundamental precursor to promoting learning that gives rise to collaboration and integration. However, this is not enough. Academic staff must be open to the notion of thinking across disciplines or professions to be able to translate these principles into their teaching. Changing heart and minds cannot be achieved by writing policies and changing room allocations. Staff commitment appears fundamental. Franks et al. (2007) argue that it must take the form of a philosophical and practical willingness to embrace 'conceptual interdisciplinarity' (Lattuca, 2003). Only if staff engage with the concept of interdisciplinarity and believe in what they are trying to achieve at a fundamental level will they be able to communicate their commitment to students.

## CONCLUSION

'Ways of thinking and orientations to learning that can ultimately lead to mutual incomprehension when specialists from different subject domains try to collaborate', (Woods, 2007, p. 854) are no longer tenable. It is incumbent on higher education institutions to ensure students are prepared for the challenges they will encounter as graduates by promoting collaborative and integrative learning. Initiatives that do so are not without their problems and solutions involve fundamental change both structural and attitudinal. As the chapters that follow illustrate, tinkering around the edges will not work; only wholehearted commitment will do.

## REFERENCES

Adams, A. (2005). Theorising Inter-professionalism. In The Higher Education Academy, Health Science and Practice Network, The Theory-Practice Relationship in Interprofessional Education. (pp. 31-37). London: Higher Education Academy.

Allen, D., & Hughes, D. (2002). *Nursing and the Division of Labour in Healthcare*. London: Palgrave MacMillan.

Australian Council of Professions. (2007). *About Professions Australia*. Retrieved April 15, 2008 from www.professions.com.au/define profession.html

Baldwin, D. C. Jr, & Baldwin, M. A. (1979). Interdisciplinary education and health team training: A model for learning and service. In Hunt, A. D., & Weeks, L. E. (Eds.), *Medical education since 1960: Marching to a different drummer* (pp. 190–221). East Lansing, MI: Michigan State University Foundation.

Becher, T., & Trowler, P. (2001). Academic Tribes and Territories, (2nd Ed.). Milton Keynes, UK: SRHE/Open University.

Bines, H. (1992). Interprofessionalism. In H. Bines & D. Watson (Eds.), Developing Professional Education. (pp. 136-142). Milton Keynes, UK: Open University.

Brewer, M. B. (1979). In-Group Bias in the Minimal Intergroup Situation: A Cognitive-Motivational Analysis. *Psychological Bulletin*, *86*(2), 307–324. doi:10.1037/0033-2909.86.2.307

Bristol Inquiry. (2001). *Final Report of the Bristol Royal Infirmary Inquiry*. Retrieved April 15, 2008 from www.bristol-inquiry.org.uk/final_report/report/index.htm

Bullock, A., & Trombley, S. (1999). *The New Fontana Dictionary of Modern Thought*. London: Harper-Collins.

Centre for the Advancement of Interprofessional Education (CAIPE). (1997). Interprofessional Education - A Definition. *CAIPE Bulletin*, (13).

(2003). *Chambers Dictionary of English*. Edinburgh, UK: Chambers.

Cornish, P., Church, E., Callanan, T., Bethune, C., & Curran, V. (2004). *From Multidisciplinary to Interdisciplinary: The Evolution of Shared Training, Research and Mental Health Care at Memorial University*. Paper presented at Shared Mental Health Care Conference, Vancouver, Canada.

D'Amour, D., Ferrada-Videla, M., San Martin Rodriguez, L., & Beaulieu, M. D. (2005). The conceptual basis for interprofessional collaboration: Core concepts and theoretical frameworks. *Journal of Interprofessional Care*, (Supplement 1), 116–131. doi:10.1080/13561820500082529

Department of Health. (2000). *The NHS Plan*. London: The Stationery Office.

Department of Health. (2001a). *Working Together-Learning Together: A Framework for Lifelong Learning in the NHS*. London: The Stationery Office.

Department of Health. (2001b). *A Service of all the Talents: developing the NHS workforce*. London: The Stationery Office.

Department of Health. (2002). *Liberating the Talents*. London: The Stationery Office.

Department of Health. (2003). *The Victoria Climbie Inquiry: the Report of an Inquiry by Lord Laming*. London: The Stationery Office.

DeWitt, C., & Baldwin, J. R. (2007). Some historical notes on interdisciplinary and interprofessional education and practice in health care in the USA. *Journal of Interprofessional Care*, *21*(S1), 23–37. doi:10.1080/13561820701594728

Eraut, M. (1994). *Developing Professional Knowledge and Competence*. London: Falmer.

Ericson, D. L. (1991). Guest Editorial: Territorialism in the health care professions. *The International Journal of Childbirth Education*, *6*(3), 30–31.

Etzioni, A. (1969). *The Semi-Professions and their Organization*. New York: Free Press.

Finch, J. (2000). Interprofessional education and teamworking: a view from the education providers. *British Medical Journal*, *321*, 1138–1140. doi:10.1136/bmj.321.7269.1138

Franks, D., Dale, P., Hindmarsh, R., Fellows, C., Buckridge, M., & Cybinski, P. (2007). Interdisciplinary Foundations: Reflecting on interdisciplinarity and three decades of teaching and research at Griffith University, Australia. *Studies in Higher Education*, *32*(2), 167–185. doi:10.1080/03075070701267228

Freeth, D., Hammick, M., Koppel, I., Reeves, S., & Barr, H. (2002). *A Critical Review of Evaluations of Interprofessional Education*. London: Centre for Health Science and Practice.

Freeth, D., Hammick, M., Reeves, S., Koppel, I., & Barr, H. (2005). *Effective Interprofessional Education: Development, Delivery and Evaluation*. Oxford, UK: Blackwell. doi:10.1002/9780470776438

Fry, G. L. A. (2001). Multifunctional landscapes – towards transdisciplinary research. *Landscape and Urban Planning*, *57*, 159–168. doi:10.1016/S0169-2046(01)00201-8

Geertz, C. (1983). *Local knowledge: Further essays in interpretative anthropology*. New York: Basic Books.

Gibson, J. M. (2003). Arrogance - A Dangerous Weapon of the Physics Trade? *Physics Today*, *56*(2), 54. doi:10.1063/1.1564350

Gilbert, J. H. V. (2005). Interprofessional learning and higher education structural barriers. *Journal of Interprofessional Care*, *19*(S1), 87–106. doi:10.1080/13561820500067132

Glen, S. (1999). Educating for interprofessional collaboration: Teaching about values. *Nursing Ethics*, *6*, 202–213.

Hawkins, J. M., & Allen, R. (1991). *Oxford Encyclopaedic English Dictionary*. Oxford: Clarendon.

Higher Education Academy. (2008). *Case Study: Professional development through interdisciplinary and interprofessional learning activities*. Retrieved April 15, 2008 from http://cebe.cf.ac.uk/BPBN/casestudy/leeds_cdcli.htm

Hoyle, E., & John, P. (1995). *Professional Knowledge and Professional Practice*. London: Cassell.

Lattuca, L. R. (2002). Learning interdisciplinarity: sociocultural perspectives on academic work. *The Journal of Higher Education*, *73*(6), 711–739. doi:10.1353/jhe.2002.0054

Lattuca, L. R. (2003). Creating Interdisciplinarity: grounded definitions from college and university faculty. *History of Intellectual Culture*, *3*(1), 1–20.

Levison, D. A. (2003). *CHMS Position Paper: Interprofessional Education*. Council of Heads of Medical Schools and Deans of UK Faculties of Medicine.

Maxwell, N. (2003). Two Great Problems of Learning. *Teaching in Higher Education*, *8*(1), 129–134. doi:10.1080/1356251032000052375

Milner, M. (1994). *Status and Sacredness: A General Theory of Status Relations and an Analysis of Indian Culture*. New York: Oxford University.

Mooney Melvin, P. (1987). In Quest of the Professional Historian: The Introduction to Public History Course. *The Public Historian*, *9*(3), 67–79.

Mu, K., & Royeen, C. B. (2004). Interprofessional vs. interdisciplinary services in school-based occupational therapy. *Occupational Therapy International*, *11*(4), 244–247. doi:10.1002/oti.214

Munro, L. (2000). Non-Disciplinary Power and the Network Society. *Organization*, *7*(4), 679–695. doi:10.1177/135050840074010

Norman, I. (2005). Inter-professional education for pre-registration students in the health professions: recent developments in the UK and emerging lessons. *International Journal of Nursing Studies*, *42*, 119–123. doi:10.1016/j.ijnurstu.2004.11.003

Price, J. (2005). Complexity and Interprofessional Education. In The Higher Education Academy, Health Science and Practice Network, The Theory-Practice Relationship in Interprofessional Education (pp. 76-87). London: Higher Education Academy.

Rawson, D. (1994). Models of Inter-professional work. Likley theories and possibilities. In Leathard, A. (Ed.), *Going Inter-professional: Working Together for Health and Welfare* (pp. 38–63). London: Routledge.

San Martin-Rodriguez, L., Beaulieu, M. D., D'Amour, D., & Ferrada-Videla, M. (2005). The determinants of successful collaboration: A review of theoretical and empirical studies. *Journal of Interprofessional Care*, (Supplement 1), 132–147. doi:10.1080/13561820500082677

Scammell, J. Hurchings, M. Quinney, A. Hean, S. Dumbrell, A. & Pulman, A. (2008). *A virtual practice community for student learning and staff development in helaht and social work Interprofessional education; changing practice through collaboration*. School of Health and Social Care, Bournemouth University/Higher Education Academy Health Science and Practice.

Soetendorp, R., & Byles, L. (2000, January). *Law for non-lawyers: facilitating inter-professional dialogue*. Paper presented at 2nd HEA Annual Conference on Learning in Law Initiative, Warwick University, UK. Retrieved April 15, 2008 from http://www.ukcle.ac.uk/interact/lili/2000/soetendorp.html

Stepans, M. B., Thompson, C. L., & Buchanan, M. L. (2002). The role of the nurse on a transdisciplinary early intervention assessment team. *Public Health Nursing (Boston, Mass.)*, *19*, 238–245. doi:10.1046/j.1525-1446.2002.19403.x

Stew, G. (2000). Interprofessional Learning – An Evaluation. In Bourner, T., Katz, T., & Watson, D. (Eds.), *New Directions in Professional Higher Education* (pp. 52–59). Buckingham, UK: SRHE and Open University.

Tarvainen, M. (2004). *Engineering Education and Interdisciplinary Studies*. Paper presented at 2004 SEF Annual Congress: The XXI Century: The Golden Opportunity for Engineering Education. Finland.

Van den Besselaar, P. & Heimeriks (2001). *Disciplinary, Multidisciplinary, Interdisciplinary: Concepts and Indicators*. Paper presented at 8th Cnference on Scientometrics and Infometrics, Syndney, Australia.

Vince, R. (1996). *Managing Change: Reflections on equality and management learning*. Bristol, UK: Polity.

Wall, S., & Shankar, I. (2008). Adventures in transdisciplinary learning. *Studies in Higher Education*, *33*(5), 551–565. doi:10.1080/03075070802373008

Walsh, K. (2007). Interprofessional Education Online: The BMJ experience. *Journal of Interprofessional Care*, *21*(6), 691–693. doi:10.1080/13561820701436912

Whitehead, C. (2007). The doctor dilemma in interprofessional education and care: how and why will physicians collaborate? *Medical Education*, *41*, 1010–1016. doi:10.1111/j.1365-2923.2007.02893.x

Wilde, R. (2007). Europeanhistory.about.com/library/glossaryhist/ bldefdiscipl.htm [Retrieved 2nd October 2007].

Wilson, V., & Pirie, A. (2000). *Multidisciplinary Teamworking: Indicators of Good Practice. Spotlights, Scottish Council for Research in Education*. Retrieved 15th April 2008 from http://www.scre.ac.uk/spotlight/spotlight77.html

Wood, G. (1999). Interdisciplinary working in built environment education. *Education + Training*, *41*(8), 373–380. doi:10.1108/00400919910298648

Woods, C. (2007). Researching and developing interdisciplinary teaching: towards a conceptual framework for classroom communication. *Higher Education*, *54*, 853–866. doi:10.1007/s10734-006-9027-3

# Chapter 2
# Understanding Interprofessional Expertise in Terms of Beliefs, Values and Attitudes

**Adrian Bromage**
*University of Birmingham, UK*

## ABSTRACT

*It is arguable that expertise in interprofessional working can be defined as effective team working to place clients' needs as paramount: a particular kind of problem solving. Psychological definitions of expertise focus upon problem solving in general, so it is arguable that this literature differs only in emphasis rather than the underlying nature of activities described. This opens possibilities to draw upon literature concerning the development of expertise, which tends to concern how high performers differ from non-experts. For example, Dreyfus & Dreyfus (1986) conclude this is in terms of how experts think and solve problems, rather than their personal knowledge. It seems that experienced problem-solvers discern regularly recurring environmental patterns, drawing upon a body of domain knowledge and experience to do so. In exploring these parallels, this chapter examines beliefs and attitude change, in terms of how they might relate to each other, to outline some possible implications for educators.*

## INTRODUCTION

This chapter will take a conceptual stance, examining recent writing on the nature of interprofessional education (IPE), interprofessional working (IPW) in terms of psychological literature on the development of expertise. In doing so, it also explores beliefs and attitude change, in terms of how they might relate to each other, outlining possible implications. The relationship between interprofessional working and domain expertise is arguably an important consideration for interprofessional educators. It has potential to impact profoundly upon the curriculum design process, given it is arguable that one educational aim is to enable learners to develop expertise.

Interprofessionalism refers to a particular approach to team working. It is characterised by professionals who work together across professional and role-defined boundaries, underpinned by an attitude of mutual understanding and collabora-

DOI: 10.4018/978-1-61520-889-0.ch002

tion (Finch et al., 2000). Similarly, Peile et al. (2001) proposed an aspirational model for what they named 'The Renaissance School of General Medicine', which will:

*produce broadly educated doctors who think in terms of patients rather than organs and are strong, multiprofessional team players (Peile et al., 2001, pp.1454-1455).*

The Combined Universities Interprofessional Learning Unit (CUILU) (2006) examined interprofessional learning strategies in the UK and the related interprofessional 'capabilities'. Their report includes material which that suggests interprofessional working tends to rest to some extent on values and attitudes, rather than domain knowledge. For example, page 7 of the report notes a key finding that:

*The culture of the practice context is one of sharing and collaborative team working in which the contributions of all the team are recognised, understood and valued in the interests of providing high quality, integrated care for patients/clients and carers (The Combined Universities Interprofessional Learning Unit, 2006, p. 7).*

The above quotation appears to emphasise a particular attitude to working with others, recognizing, understanding and valuing their contribution. Given this emphasis, the question arises of how this attitude might act to facilitate "*sharing and collaborative team working*" the activity that achieves the overall aim of interprofessional working: high quality integrated client care. Arguably, this question must be examined from the point of view of tutors who are involved in interprofessional education (IPE), who aspire to help their learners to move towards developing an understanding of this model of working. This chapter will explore this question, using insights from the psychological literature on expertise and beliefs, attitudes and attitude change. The first task is to explore the nature of values and attitudes in order to set the scene for attempting to understand how this might relate to literature on the nature of expertise.

## Beliefs and Values

The interprofessional notion of recognizing, understanding and valuing one's fellow professionals contributions, which it has been argues is suggestive of an attitude to interacting with others, invokes questions regarding the consequences of holding particular attitudes. In particular, how attitudes are manifested in behaviours. This section considers these questions, issues that have attracted an exploratory literature from a variety of perspectives ranging from cultural anthropology to behavioural psychology and cognitive science. A particularly suggestive literature can be found in accounts that psychologists and cognitive scientists offer on the relationship between thought, beliefs and actions.

At the heart of this particular body of literature is an intractable problem concerning the epistemology of 'beliefs'. Abelson (1979) notes that the term 'belief' is hard to define in a clear-cut way, and is used in different ways across the social sciences. While the lay-person might imagine that there is a qualitative distinction between belief and knowledge, many academic writers follow Abelson in viewing them as sharing many features. For example, both could appeal to personal experience and both are unbounded to some extent. Hamlyn (1970),

Rappaport (1986) and Abelson argue that given a plausible truth claim, belief can be promoted to knowledge, although what counts as knowledge tends to be defined by the truth tests prevailing within specific domains and circumstances, which may be the product of a particular culture or community. Nevertheless Abelson concludes that belief and knowledge differ sufficiently to require separate treatment, a stance that will be adopted in this chapter. Hamlyn distinguishes two types of belief. These are '*beliefs that*', commonly referred to as knowledge, and '*beliefs in*', roughly corresponding to the views a lay person might suppose would be held by members of a faith community of 'believers'.

It is interesting to consider where the value-base of interprofessional working might sit in these terms. The Combined Universities Interprofessional Learning Unit (2006) links the professional aim of

*providing high quality, integrated care for patients/clients and carers" with "sharing and collaborative team working" underpinned by "the contributions of all the team are recognized (The Combined Universities Interprofessional Learning Unit, 2006 p. 7).*

While it is theoretically possible to test this statement empirically according to the accepted 'truth tests' of the social sciences, it is equally interesting to explore the statement in terms of Hamlyn's argument that, unlike knowledge, beliefs require no warrant as to their truthfulness, aside from personal conviction or confidence. For example, Religious educators may be satisfied with the warranty of 'revelation', promoting their beliefs to 'religious knowledge' that they treat as having similar status to knowledge based upon truth tests (Hirst, 1974). It will be seen that such unwarranted beliefs are often held to be 'sacred' truths, and can exercise a very powerful force.

Abelson sees beliefs, despite their low epistemological status, as having strong evaluative or motivational overtones and relatively unbounded in their scope. Beliefs of a normative nature can be sub-divided usefully in a number of ways, for example, Rokeach (1968) is particularly interested in beliefs in the 'worthwhileness', 'desirableness' or 'goodness' of a thing, which he distinguishes as 'values'. These in turn divide into 'terminal' or 'instrumental' values, dependent on whether they are attributed to ends or means. Terminal values concern the desirability of certain identified goals, and are susceptible to hierarchical arrangement, as in Allport et al.'s (1951) 'scale of values'.

In Allport et al.'s account 'wisdom' can be considered a terminal goal and 'intellectualism' as an instrumental one. Viewed in this light, "*the contributions of all the team are recognized*" would be an instrumental value, and "*providing high quality, integrated care for patients/clients and carers*" can be considered a terminal value. In further exploring what this might mean for educators, the function of beliefs that is implicit in Allport et al.'s (1951) notions of 'instrumental' and 'terminal' values, will next be considered. This is of particular significance, given the possible contrast identified between "*the contributions of all the team are recognized*" as an instrumental value, and "*providing high quality, integrated care for patients/clients and careers*" as a terminal value.

## The Influence of Beliefs and Values on Thinking

Part of the literature on the function of beliefs concerns their role in generating or allowing interpretative cognitive structures. Rokeach (1960) argues that values provide us with an interpretative framework for understanding the universe. He developed this line of argument to view beliefs as acting as evaluative maps that guide our actions (Rokeach, 1968). Armstrong (1973) has developed a similar theoretical position, following Ramsey (1931), to conceptualise beliefs as 'maps' that have the individual as a central reference point, and which act to mediate interpretations of reality. These positions are arguably kindred to the landscape metaphors that developed by Trowler (1998) and, in particular, Jenkins (1972) as a means to describe educational curricula. This convergence suggests that it will be fruitful to explore the further literature examining the mechanisms by which beliefs are thought to exert their influence.

One strand of this literature explores how beliefs exert 'sense-making' and thereby 'predictive' influences. In Schank's (1999) theoretical orientation, the essence of understanding the present is that it articulates memories of the past. It is generally held that beliefs can 'delimit' an individual's thoughts at least to an extent. Osherson et al. (1986) found that individuals can generate beliefs about events and situations that they have not actually experienced directly. Such beliefs are extrapolated from their existing beliefs, using inferential links depending on the psychological plausibility of relevant arguments, either put or posed internally. Taken together, Schank and Osherson et al.'s work implies that beliefs will act to delimit thoughts about both directly experienced events, and situations that an individual has not personally experienced.

Schoenfeld (1983) found that beliefs associated with a problem act as an intellectual context that is capable of influencing the quest for solutions. It seems that beliefs, whether subject to conscious reflection or held unreflectively, can restrict the subset of cognitive resources available to solve a particular problem, rendering those easily accessible in one context, inaccessible in another. There exists research evidence that appears to support this conclusion. Oakhill et al. (1989) explored syllogistic reasoning, arguments that have several parts, typically two statements (or premises) that are assumed to be true and which lead to a conclusion. Oakhill et al. found that beliefs can hinder problem solving, as correct moves rendered implausible by the belief system tend to be discounted, an effect that is stronger with so-called 'open-ended' problems.

The implications for interprofessional educators are threefold. Firstly, that on first exposure to interprofessional learning, the learners' existing values and beliefs are likely to exert a powerful delimiting influence on how they will approach interprofessional team working. It would seem that this effect will be particularly marked in more open-ended or ambiguous educational activities. Secondly those beliefs may, in some cases, act as conceptual barriers to understanding the nature of interprofessional working, and it is to the challenges that this might raise that discussion now turns.

There exists a socio-psychological literature dealing with the internal organisation of beliefs. Rokeach (1960) hypothesises a binary belief-disbelief system, this asymmetry being necessary to allow for the conditions in which a given belief accepted as true implies that certain others must be rejected as false. Rokeach views such systems as 'psychological' rather than 'logical', as the contradictions within are not necessarily open to revision under the challenge of logical analysis, and they tend to locate in clusters, one often evoking many others. Indeed, such beliefs are frequently sustained, when put under pressure, by a strong emotional component. It is a truism of cognitive dissonance that sacred beliefs are the hardest to shift (Festinger, 1956, 1957). For example, institutional change is known to be easiest when the change is purely technical and hardest when sacred beliefs are at stake (Watson, 1987). Interprofessional working as a concept may thus be challenging for individuals who, for whatever reasons, are wedded to hierarchical notions such as the relative importance of different professional groups.

This suggests that challenging individuals' naïve beliefs about interprofessional working in educational contexts may in some cases require something more than rational argument. Furthermore, the process may in some cases be emotionally grueling for both learners and their educators. Indeed, the process may be more akin to helping to generate in learners '*leaps of insight*' as outlined in Meyer and Land's (2003) notion of 'Threshold Concepts'. A final practical consideration for is that it is arguably not helpful to work on an individual's beliefs alone, given the evidence that beliefs influence behaviour in complex ways. Discussion will turn next to this relationship in more detail.

## Attitudes and Behaviour

It has been seen above that beliefs exert a powerful influence on cognition, and that through this route they are likely to influence our actions. This dynamic seems to be captured in Kelly's (1955) Personal Construct Theory, which holds that individuals overlay naive theories onto the world, which is assumed to exist in a 'real' or objective sense, and to the extent they match, the individual is able to make sense of that world. Kelly's model is not solipsistic, a view that the world only exists in one's own mind, as the 'real' world delimits those theories, forcing revisions where mismatches occur. Within the model, emotions are tied to the adequacy of the construct system, for example, mismatches give rise to feelings of anxiety. However, Kelly's model cannot account for the irrational side of humanity, such as notions such as having a favourite colour, or the conclusions of Watson (1987) and in particular Festinger (1956, 1957) regarding the resilience of 'sacred' beliefs in the face of overwhelmingly inconsistent evidence.

The phenomenon of irrationality has arguably been captured in psychological studies of attitudes. Secord and Blackman (1964) reviewed the way 'attitudes' had been defined in many studies, concluding that most operational definitions suggest a threefold structure with cognitive, affective and behavioural elements. In summary, they arguably regard the consensus on attitudes as indicating an integration of knowledge, values, emotions and behaviour. In effect as a clustering not only of particular beliefs, but crucially, also of perceptual, behavioural and emotional tendencies around those beliefs.

Turning in detail now to how attitudes interact with behaviour, empirical research has found that the stronger or more specific a belief about an attitude object, the more likely it is that behaviour and action will be consistent with it (Davidson & Jaccard, 1979; Weigel et al., 1974). However an individual is less likely to behave consistently with their beliefs when these are inconsistent with how they feel about the attitude object (Norman, 1975), or when their values are general rather than specific to a particular attitude object (Davidson & Jaccard, 1979; Weigel et al., 1974).

The relationship between attitudes and behaviour is not just one-way. For example, there is empirical evidence that in certain circumstances acting in a counter attitudinal way can have a retroactive effect in modifying one's existing attitudes (Zanna & Cooper, 1974), thus it can be seen that beliefs and actions are dynamically interrelated. However, it appears change only occurs if the actor feels they have acted of their own free will in some way (Brehm & Cohen 1962), or made some kind of personal commitment to their actions (Zanna & Cooper 1974).

These research findings have interesting implications for interprofessional educators. The interrelation between affect and beliefs suggests that one task is to nurture positive feelings within students towards interprofessional working. Another implication is that some kind of practical experience of interprofessional working may help to generate positive 'interprofessional attitudes' or even reformulate negative ones, particularly if there is an element of learner choice involved at some level. Indeed, these themes will be implicit within some of the case-study chapters within the present volume, for example, the emotive intent behind representing 'authenticity' of service users' and carers' experiences within Reusable e-learning Objects, and participant activities within cyberspace environments such as discussion forums and 3-d virtual worlds.

One question that this raises in terms of educational evaluation is that of how relationships between particular beliefs and overt behaviours might be identified. Maclaren (1976) cautions that attempts to infer belief from action can degenerate into interpreting others' actions in a manner that distorts them so that they become consistent with one's own belief systems. Clearly a fairly sophisticated approach is needed, and Loar (1981) provides some insights into what is required. Loar views behaviour as responses to perceptions of the world that are shaped by our desires and aims, the nature of which can only be identified given our beliefs. For this reason, Loar recommends considering the functional relationship between belief, desire, perception and behaviour. He proposes that the specific task of interpreting attitudes requires a 'realist' approach, focussing on the functional roles of attitudes, arguing that this higher-level interpretation allows beliefs and desires a causal role in behaviour.

Loar's conclusion chimes with Rokeach's (1968) assertion that the cognitive and emotional function together. Norman (1980) made a similar point in his critical evaluation of research into human cognition. In terms of how this might play out in practice, the notion that beliefs and desires have a genuinely causal role in behaviour is suggestive of why, for example, Trowler (1998) found evidence in academics' responses to planned change to suggest that institutional culture should be understood as 'substantive' and genuinely causal, rather than merely 'ideational'.

This further suggests a possible framework for understanding how the 'instrumental value' of "*the contributions of all the team are recognized*" might exert a genuine influence on professional behaviour. It too can be viewed as substantive and causal, rather than simply ideational in nature. The latter point raises the possibility that interprofessionalism represents more than just a loose collection of attitudes and vaguely related behaviours; indeed, whether it can properly be regarded as form of professional expertise. If so, this would open the relationship to close scrutiny by educators. These questions will be explored in the next section.

## The Nature of Expertise and its Development

The literature on the development of expertise has a somewhat different focus to that on beliefs, attitudes and behaviours. It typically explores how high performers differ from non-experts, for example, Anderson (1985) and Dreyfus and Dreyfus (1986, 2005). Dreyfus and Dreyfus (1986) conclude that this is in terms of how experts and non-experts think and solve problems, rather than in terms of their personal knowledge alone. It seems that experienced problem-solvers discern regularly recurring patterns in their environment, drawing upon a body of domain knowledge and experience to do so. They do not solve problems by applying rules, but do so by intuitively recognizing recurring patterns (Dreyfus and Dreyfus, 2005).

There exists literature that may illuminate how this plays out in practice. The question of learning to solve problems rather than learning a body of knowledge is essentially a question of meta-learning. Bereiter and Scardamalia (1993) argue that an efficient, intuitive approach to problems develops in anyone with sufficient experience, however, not everyone is successful at what they do or are seen by others as 'experts'. Bereiter and Scardamalia (1993) argue that expertise is not a state; rather it is a continual process over time, which they characterise as a career approach. As one gains experience and is able to routinize some aspects of one's work, this frees time and energy for continually engaging in learning to meet new challenges. The latter is characterised by Bereiter and Scardamalia as 'progressive problem solving'; the act of tackling problems whose nature is such that the process of tackling them will increase one's expertise.

Bereiter and Scardamalia's notion of meta-learning may be echoed in some of the literature concerning how experts develop their expertise, for example, in the research findings from Michael Eraut's study of mid-career professionals' learning in the workplace (Eraut, 1999; Eraut et al.,

2000). However, Eraut essentially explores the learning context, rather than learning processes, asking three questions:

- What is being learned in the workplace?
- How is it being learned?
- What factors affect the level and direction of that learning?

Other researchers have explicitly focussed on the development of expertise. Dreyfus and Dreyfus (1986, 2005), for example, describe experienced problem-solvers as able to recognize recurring patterns and to develop learned procedures in the light of those patterns. Also important is the ability to consciously and unconsciously read novel situations, events and so on in terms of these learned patterns, a process that is sometimes referred to as 'generalising'.

Dreyfus and Dreyfus (1986) propose a model of adult skill acquisition based upon their view of the nature of expertise:

- **Novice:** Established objective facts form the basis of action, but in a relatively context-free way. Decisions are analytical and the learner is emotionally detached.
- **Advanced Beginner:** Adapts learned non-situational rules to the particular features of a given context. The learner is emotionally detached, and decisions are analytical.
- **Competence:** Goals and plans are based upon rules and context, but as the individual gains experience, they turn from rules to reasoning; albeit with difficulties in committing to a course of action. While the learner is emotionally involved in the outcomes, their understanding and decision-making is emotionally detached.
- **Proficiency:** Sufficient experience has been accumulated to enable rapid decisions and action based more upon situational judgment than on planning and problem solving based upon rules and principles. Understanding is characterised by emotional involvement, but decision-making is detached.
- **Expert:** Their wide experience and their ability to discern patterns from that experience enables them to make very subtle judgments. They are characterised by 'intuitive" and involved decision-making as well as involved understanding.

Thus far the nature of expertise, and the processes by which individuals are thought to develop it, have been briefly outlined. The next question is whether interprofessionalism can in itself be regarded as a form of expertise. For this to be true, it must be demonstrable that interprofessional working is in essence solving a professional problem by intuitively discerning regularly recurring patterns.

The starting point for answering this question is Dreyfus and Dreyfus' (1986) characterization of an expert. Translated to interprofessionalism, this would mean an individual with wide experience of interprofessional working, an ability to discern recurring patterns within interprofessional working, making subtle judgments with reference to those patterns, and to be emotionally involved in both decision-making and understanding.

Any claim as to whether skilled interprofessional workers are '*emotionally immersed*' in the problems they solve would seem to rest upon what has been identified in this chapter as an 'instrumental value'; "*the contributions of all the team are recognized*". This has emotional overtones, both in terms of being something that is valued and, as argues earlier in this chapter, in terms of potentially being emotionally challenging to learners with certain pre-existing beliefs. So in theory, at least, it would seem that emotionally-involved understanding and decision making is likely to characterize those skilled in interprofessional working.

As to whether there are any discernable recurring patterns in interprofessional working, this would seem to rest upon identifying such patterns in "*sharing and collaborative team working*", and it is likely that such patterns would tend to be social in nature. For example, the social flow of a team that has over time developed into skilled interprofessional workers, or in the social skills that might be required to involve a fellow professional who was unskilled in, or even hostile to, interprofessional working. It is arguably likely that any such patterns would be intuitively recognized by those with sufficient experience.

## DISCUSSION AND CONCLUSION

The arguments above indicate that it is theoretically possible for interprofessional working to be regarded as both attitudinal in nature, and also as a form of problem-solving expertise: Indeed, the latter could arguably be read in Allport et al.'s (1951) terms as the 'instrumental value' underlying interprofessional working. And it seems likely that this kind of expertise is essentially 'social' in its nature. This raises the question of what educationalists might do to help learners develop such expertise. It will be seen that the generic literature on expertise and its development are suggestive of the way forward.

Bereiter and Scardamalia (1993) make suggestions as to the nature of an environment that might encourage learning to solve problems, rather than learning a body of knowledge. It has already been seen that Bereiter and Scardamalia describe this process as 'progressive problem solving'; the act of tackling problems whose nature is such that the process of tackling them will increase one's expertise. The obvious implication is that interprofessional education should enable learners to develop experience of interprofessional working by tackling authentic professional problems in authentic contexts. This would seem to suggest some kind of work placement in a professional team, or an authentic simulation of such a team.

On the other hand, since the aim is to develop learners' expertise in interprofessional working, one can be more flexible. For example, it would be enough to bring learners who seek different professional destinations together in any task where they must blend their general knowledge and understanding in order to succeed. Arguably the overt task would be less important than engaging in the patterns of social interactions associated with that task (as might be identified in future empirical research), for it seems that learning to manage such interactions is the key to interprofessional expertise. There is a role in this for approaches such as case studies depicting various patterns of social interaction between interprofessional workers, or Reusable learning Objects that are structured to reflect such patterns.

One possible hurdle that has been highlighted in the discussion earlier in this chapter of how beliefs and attitudes interact with behaviour is the potentially very powerful influence of the pre-existing beliefs and attitudes that learners initially bring to interprofessional learning. These could manifest themselves in terms of resistance to interprofessional working, perhaps due to differences in the ontological and epistemological foundations of different professions. Several possible strategies to reduce the possible negative impact of such beliefs were discussed earlier in this chapter, however, viewing interprofessionalism as both an expertise and an attitude both highlights and brings into focus the potential value of seeking ways to link it to learners' pre-existing beliefs and values, and to 'scaffold' (Vygotsky, 1978) these beliefs towards interprofessional attitudes.

One possible way is briefly outlined here. Irvine et al. (2002) argue that all contemporary health, welfare and educational professionals share the ethical stance that they should place the needs and interests of their clients and service users as primary and above their own, as documented, for example, in published codes of professional ethics and the Hippocratic Oath. Irvine et al. argue that a clear connection can be made between these ethical stances and the need to establish and foster effective cooperation between professionals working in different fields and disciplines. This connection can be reinforced with evidence, such as for example, the Bristol Royal Infirmary Inquiry (2001), which demonstrates that no single profession can adequately deal with every health and welfare need of their clients and service users. And the latter, providing high quality integrated care, can be read in Allport et al.'s (1951) terms as the 'terminal value' of interprofessionalism.

## ACKNOWLEDGMENT

This chapter was conceived of and written during Dr. Bromage's term of employment as Senior Research Assistant at the Centre for Interprofessional eLearning (CIPeL), between November 2007 and November 2009.

## REFERENCES

Allport, G. W., Vernon, P. G., & Lindzey, G. (1951). *Study of Values*. Boston: Houghton-Mifflin.

Anderson, J. R. (1985). *Cognitive Psychology and its Implications*. Boston: Harvard University Press.

Bereiter, C., & Scardamalia, M. (1993). *Surpassing ourselves: An inquiry into the nature and implications of expertise*. Chicago: Open Court.

Bristol Royal Infirmary Inquiry. (2001). *Learning from Bristol: the report of the public inquiry into children's heart surgery at the Bristol Royal Infirmary 1984–1995*. Norwich, UK: The Stationery Office.

Dreyfus, H., & Dreyfus, S. (2005). Expertise in real world contexts. *Organization Studies*, *26*(5), 779–792. doi:10.1177/0170840605053102

Dreyfus, H. L., & Dreyfus, S. E. (1986). *Mind over machine: The power of human intuition and expertise in the era of the computer*. New York: The Free Press.

Eraut, M. (1999). *Theoretical and methodological perspectives on researching workplace learning*. Paper for AERA Annual Conference, Montreal, April 1999.

Eraut, M., Alderton, J., Cole, G., & Senker, P. (2000). Development of Knowledge and Skills at work. In Coffield, F. (Ed.), *Differing Visions of a Learning Society* (*Vol. 1*, pp. 231–262). Bristol, UK: The Policy Press.

Irvine, R., Kerridge, I., McPhee, J., & Freeman, S. (2002). Interprofessionalism and ethics: consensus or clash of cultures? *Journal of Interprofessional Care*, *16*(3), 2002. doi:10.1080/13561820220146649

Loar, B. (1981). *Mind and meaning*. London: Cambridge University Press.

Meyer, J. H. F., & Land, R. (2003). Threshold concepts and troublesome knowledge: linkages to ways of thinking and practising, In C. Rust (Ed.), Improving Student Learning: Theory and Practice Ten Years On (pp. 412-424). Oxford, UK: Oxford Centre for Staff and Learning Development (OCSLD).

The Combined Universities Interprofessional Learning Unit. (2006). *The Combined Universities Interprofessional Learning Unit final report*. Retrieved July 10, 2008, from http://www.sheffield.ac.uk/cuilu

Vygotsky, L. S. (1978). *Mind in Society*. Cambridge, MA: Harvard University Press.

# Chapter 3
# A Threefold Framework for Relating to Innovations and Technology in Education:
## Learning from, with and about Technology

**Samuel Edelbring**
*Karolinska Institutet, Sweden*

## ABSTRACT

*In recent years computer technology has developed quickly as have cultural practices in society. However, educational practices with technology have not yet reached the point where educators and learners benefit optimally from innovative technology. In interprofessional education (IPE) there are examples of technology use that forms an integral part of education. To improve IPE there is a need for reflection on how today and tomorrow's technology can contribute. As technology and its practices melt together this reflection is however, not easily done. A framework intended as a starting point for such reflection is presented in the following chapter which involves learning from, with and about technology. To assess the benefit of technology for learning we need research on technology integration, on outcomes from learning with technology as well as processes of learning in conjunction with technology. This research has to use different perspectives involving various scientific traditions.*

## INTRODUCTION

Technology and education go naturally hand in hand, if by technology we mean a variety of tools ranging from the use of whiteboards to hi-tech simulation technology. The concept of technology used in the chapter has a broad scope encompassing man made tools with a strong bias towards Information and Communication Technology (ICT). Formal education institutions have traditionally been reluctant to adopt new technology into practice (Cuban, 1988). In IPE there are, however, initiatives taken where technology is a crucial part of the learning process, some of them reported in this book. For interprofessional practice the functions of overcoming logistic barriers and providing links between health care areas has shown valuable (Hubbard & Temessl-Huber, 2005).

In many respects IPE shares educational conditions with other aspects of Higher Education. A core

DOI: 10.4018/978-1-61520-889-0.ch003

concern of IPE, however, is forming professional identities in a way that is open to collaboration with others in the interest of providing care (Barr, 2005). This implies collaborative practice and frequent meetings between learners and educators from different contexts. As a basis for using technology for learning, educators generally draw on sources from distance education, e-learning and health-professions education. An interesting area of technology research is also the Computer Supported Collaborative Learning area (CSCL) that raises awareness of collaborative aspects of learning (Koschmann, 1996).

Computer technology is under constant development in terms of software and hardware, as well as the way we use it. Anticipations of what the future has to offer around the corner are seldom correct, but recent trend-reports suggest that hardware, the physical devices, are getting smaller and melt more seamlessly into our everyday world. Devices are also getting smarter in the sense that they "know" where they are positioned in time and space. Information and social network services are becoming even more accessible and less dependent on single hardware servers, now residing on systems of hardware connected in "clouds" (for more technology trends see the Horizon report: Johnson, Levine & Smith, 2009). This continues the extension of human capacities through technology as tools that we use for communication and learning.

Current technology practices in IPE comprise e-learning modules on collaborative practice skills (Chambers, Conklin, Dalziel, MacDonald & Stodel, 2008), re-usable learning objects centred towards concepts and real life cases and interactive patient journeys (CIPEL, 2009; Jonsson et al., 2006), as well as technology that connects people in virtual communities (Demiris, 2006). Policy makers and educators need to keep an eye on current practices as well as looking ahead towards how technology can be used tomorrow as a lever advancing IPE. However, to be beneficial for practitioners and patients, introduction of technology in IPE needs to be driven less by entrepreneurial or policy interests and more by everyday challenges in education and practice. These challenges consist of answering how technology can contribute in core areas of IPE: values, communication and social processes (Mallow & Gilje, 1999).

Technology can play a central role in the change of attitudes and practices necessary for interprofessional endeavours. There are therefore good reasons to consider technology and innovations seriously. The objective of this chapter is to provide food for thought concerning technology for IPE and to stimulate further deliberate use of new technology. The author draws from experience of educational software development and its integration in health care education as well as teaching experiences in e-learning and informatics to educators and doctoral students in the health professions.

## BACKGROUND

### Now You See It – Now You Don't: The Transparency of Technology in Practice

Deliberate reflection on how technology supports learning social contexts is not easily done. When we use innovative technology it gets incorporated into our habits and a natural part of our resources. An illustration of how technology becomes one with ourselves is the famous blind-man-with-stick example that Merleau-Ponty (2002, p. 165) uses to illustrate how technology functions as an extension of our body. Where does the extension of bodily senses really end? –At the end of the blind man's hand? –At the end of the stick? Even quite sophisticated technology becomes transparent to us when incorporated into our habits. It takes deliberate reflection to actually see it because it becomes incorporated in the way we communicate and go about our daily business. Think of one day in your week, for example next Tuesday, without your mobile phone, without email, and why not leave your bank card at home while you're at it? Would it be possible to visualize even a couple of hours of that day without high-tech devices?

In my teaching of computer support for learning attended by teachers, I usually ask how many of them use computer support in their teaching already. Not many of them acknowledge that they do that. The follow up question is: –how many of you use PowerPoint in your education? Almost everyone now responds positively to this question. Computer support for learning can take many forms but we don't tend to see what we utilize today, instead we invest our thoughts into tomorrow's technology, hoping that it will solve some of the educational, logistical or organisational problems that we struggle with.

## Realistic Expectations on Technology

It is easy to be impressed with new dazzling technology. Arthur Clarke, the technology visionary, catches this in what became known as Clarke's third law: "Any sufficiently advanced technology is indistinguishable from magic" (Clarke, 1982, p. 36, footnote). As educators we need to reach beyond this uncritical amazement. The uncritical adoption of technology, just because it is there, leads to a "lemming-like rush into the technological sea" (Mallow & Gilje, 1999, p. 251). Paul Ramsden, who has a thorough experience of research in higher education dampens the belief in magic by stating: "no medium, however useful, can solve fundamental educational problems" (Ramsden, 1992, p. 161). The assumptions about what we can expect from technology must relate to knowledge about how people learn and work together.

One of the philosophers of the IPE community, Steen Wackerhausen, stated that in order to reach a full understanding of different phenomena in our professional life, we need to go outside our own habits and learn about the views and experiences of others. We need to reflect on our practices, but also be able to reflect beyond our roles and in a way become a stranger to ourselves in order to see others' perspectives (Wackerhausen, 2007). If we are not ready to reflect beyond our own habits, there is no technology in the world that will forward interprofessional aspects of education or practice. But if we are willing to reflect and critique our own habits and experiences then technology may be a great help in that process.

## Innovations and Innovative Use

Historically the world has seen a lot of innovations that change the way we communicate, how we transport ourselves and how we extend our capacities as humans. Technical innovations, things/products, are how we normally conceive of technology. Innovative use and the cultural practices that follow on from innovations need also to be addressed when assessing technology for learning.

When people's minds are set on a specific goal something extraordinary happens. History is full of examples of technical innovations that have changed human practices over a long period of time. Gutenberg's printing press is one of the more striking examples. His innovation transformed the way we conceptualise communication from voice to printed text. Another invention, the bicycle, had a tremendous impact on how we conceive of our bodily capacities, the wheel being "an extension of the foot" (McLuhan & Fiore, 1967, pp. 30-32). Both these inventions have in common that it took years before the use of them became socially and commonly accepted.

An example of innovative use of 'old technology' is how athletes use their bodies to jump higher and higher. Today it is possible to jump over a horizontal bar at a height of 2.45 m (Javier Sotomayor, 1993, 07-27, Salamanca). When the Scots started with the high jump in the 19th century they used the scissors technique and approached heights of 1.80 m. In 1865 Mr. Sweeney modernised this technique by extending his back and flattening out over the bar and set the world record of 1.97m. The technique reached its perfection many years later. By actually turning the chest away from the bar it became possible to lower the center of gravity even more during the jump by "the Fosbury Flop". The technique was improved by degree, but more importantly, by a qualitatively different approach, facing the bar the other way around.

Maybe the equivalent of such a different approach is to be found in health care education? When we as individuals set goals and use principles of deliberate practice (Ericsson, 2008) we can reach incredible heights. As collaborative participants with a shared goal there is a chance that we could reach even higher. History shows that innovative technology and practices are to be found in all walks of life. Two parallel processes are discernible here: the continuous development of innovative technology and innovative use of existing technology.

## A Critical Perspective

Although we (at a cost) can do hi-tech simulations of reality it is not always recommendable from an educational point of view. More is not always better in educational settings and high-end technology does not automatically lead to high-end learning. When using computer simulations for learning the intuitive way of thinking is that the more realistic the better. An example from hand anatomy learning shows however that some students actually were hindered by adding more sophistication into the simulation model (Garg, Norman, Eva, Spero & Sharan, 2002).

We need to go beyond our everyday conceptions of the learning tools we use in order to design optimal learning situations. Not only the actual content of media, but the form and the media itself affects the way we communicate and in the long run how we conceive of the world. The media researcher Marshall McLuhan in his work entitled "The media is the massage", makes us aware that we are being entertained and transformed by the media itself by stating: "All media work us over completely" (McLuhan & Fiore, 1967, p. 26). Furthermore, social cognitivist Albert Bandura raises the concern that the more people's lives are dominated by media, the less they learn from each other (Bandura, 1986).

Observations of today's Internet use leads some researchers to conclude that the new digital technology actually changes the way we think, not necessarily for the worse–but differently from before (Carr, 2008; Prensky, 2001). Privacy and security aspects are of course always on the agenda when patients are involved; when Internet and wireless networks are used those aspects are increased. Software and hardware failure is less of a problem today than in the past but still remains an issue, because when they do fail the people involved become frustrated (Hubbard & Themessl-Huber, 2005). Innovative technology is no cure-all for educational problems and there are sometimes unwanted consequences to its use, but with informed use we should be able to make it work for reaching our common goals.

## Three Different Ways of Relating to Technology for Interprofessional Education

To be able to see benefits of technology we need to conceptualise the different roles that technology has in the IPE context. To help in that process a three-fold conception of the role of technology is presented here. It will be far from comprehensive and there are certainly more facets than are outlined here, but it is a starting point for conscious reflection on technology for IPE. We need to be aware of how we look upon learning in IPE, what learning consists of and how it is brought about. Although theories of learning are not in focus here, I would like to highlight the importance of making explicit our assumptions about learning when reflecting on learning tools.

As a guide we can use a simplified framework for organising learning theories that is presented by the educational researcher Anna Sfard (1998). Sfard outlines two main paths that she suggests as metaphors for learning: the acquisition metaphor and the participation metaphor. Learning is seen either as putting more and more knowledge into the minds of people, or as students collaboratively creating knowledge situated in the shared culture. To some extent they can be seen as competing, incompatible views. Alternatively they can be considered as different facets of the complex phenomenon of human learning, here the inter-professional one. Interprofessional education as defined by the Centre for the Advancement of Interprofessional Education is:

*Interprofessional Education occurs when two or more professions learn with, from and about each other to improve collaboration and the quality of care (CAIPE, 2002).*

Let us borrow the structure from this definition to organise different conceptions of technology for this purpose: learning with, from and about technology to improve IPE. As an entry point to the discussion we start off with the "learning from" aspect.

## Learning from Technology

One way to relate to technology for learning is to view it as an object containing knowledge. A knowledgeable author prepares it and learners subsequently interact with the object but have no direct interaction with the originator. The technology has in this respect conceptual similarities with the printed book.

The way we look upon learning material and tools for learning is grounded in our assumptions about learning. The everyday conception is the quantitative information acquisition one where a learner accumulates more and more information later to be retrieved (Dahlgren, 2005; Sfard, 1998). This view of learning is straightforward: You have learners in need of knowledge, you have the knowledgeable educators who can produce and deliver knowledge and you have the information that learners acquire as knowledge. With this view it is technology-wise more or less a question of making the information available, update its content and to facilitate its accessibility.

The IT revolution brought with it accessible technology, created to deal with information. As a consequence "learning from" is to a large extent how we have used e-learning. And for good reason: there is no reason to believe that the importance of acquiring information and facts will decrease in the future. On the contrary; with the World Wide Web now celebrating 20 years, we now see information lying easily accessible at our fingertips. Access to specialised and up-to-date information has led to a cultural change and new patterns for how we as humans go about looking for information. Eight out of ten Internet users looked online for health information and many patients say the Internet has had a significant impact on the way they care for themselves or for others (www.pewinternet.org). Nowadays health care professionals know they will meet patients who are well updated with up-to-date specialist information something that affects the power balance between professional and client and the professional role to some extent.

Internet sites that give democratic access to encyclopaedic knowledge (e.g. www.wikipedia.com) is a valuable continuation of the function of libraries. To help us process knowledge there are freely available tools that provide possibilities to conceptually understand processes and areas of complexity in a visualized manner. Examples of this are modelled 3-dimension visualizations (e.g. www.3d-brain.ki.se) and movie clips (e.g. www.youtube.com). Professor Hans Rosling is a good example of a scholar who has utilized modern technology to reach new audiences by posting presentations and talks on the web. In his talk he uses previously known socio-economic and health data. He visualizes this previously abstract data in a manner that actually changes people's conceptions from myths about the third-world to a way that better reflects the actual conditions (http://www.ted.com/index.php/talks/hans_rosling_shows_the_best_stats_you_ve_ever_seen.html).

Film media as such is well known for its observational learning capacities with children (Bandura, 1986) but film is currently not in vogue in adult learning, perhaps because we still have not found the optimal way of integrating this medium into educational situations. One way of using this medium is making the important voice of the patient accessible as stories online (e.g. www.patient-voices.org.uk and www.healthtalkonline.org). A further potential of film is to present role modelling with examples of professionals acting in different situations and the exposure of case problems to learners. Presentations with interspersed interactivity are furthermore a way to keep participants engaged and enhance the learning experience in patient journeys for IPE (e.g. http://www.cipel.ac.uk/learning_objects/learning_objects.htm). A way forward for IPE is consequently to use new technology to produce learning materials that are reusable in different contexts over a long period of time. Reusable learning objects with text and visuals, movie clips and oral presentations are some examples of technology in the capacity of *learning from* technology.

## Learning with Technology

Learning *with* technology involves using technology in learning together with others and to use technology to improve collaboration. When interacting with others we form our own professional identity as well as learning about others' roles. We may also discover that people hold stereotypes about the professional roles that we represent, which may not be apparent in learning *from* technology alone.

An alternative view of learning to the one of acquiring of information is the participation view of learning (Sfard, 1998). Learning as enculturation, or the situated perspective is another way of describing this (Brown, Collins, & Duguid, 1989; Hutchins, 1995; Wenger, 1998). The situated perspective does not focus on quantitative amounts of information that travels from one source to an individual, but on how learning is construed through interaction between people, not only in formal educational settings. A key aspect brought forward in this view of learning is the forming of identity and this comes about when people interact with each other as well as with the environment (Wenger, 1998). Focus is shifting from learning outcomes from technology use towards what is happening with collaborative use of technology (Kolodner & Guzdial, 1996).

Interprofessional education is about meeting across disciplinary boundaries. In practice physical and logistical constraints often raise barriers to meetings and joint encounters (Hubbard & Themessl-Huber, 2005). Technology in terms of videoconferencing, discussion boards and virtual communities are examples when technology that is used to bridge barriers of space and time. Internet utilities like "The Advanced Meeting Planner" (www.timeanddate.com) for planning of international meetings and Doodle event scheduler (www.doodle.com) for collaborative scheduling facilitates the practical arrangements around meetings.

Roy Pea discusses the way technology can restrict conditions for interaction, often into a one-way *transmissional* communication technology (1996). In order for *transformative* communication to take place learner and educator interaction must not be hindered by limitations set by the technology which means that we have to be conscious of the tools we choose and how people actually use them. With the developments that we now see in communicative media, participants are now able to use more of their capacities for human communication.

Technology in this interconnective capacity functions best when people are so accustomed to it that the technology itself becomes transparent.

Some determination and patience when introducing technology in this respect may be needed for that reason. The character of the World Wide Web has undergone a shift from static web pages to collaboratively produced living documents and is now more an arena of social networking than information source. This shift is so remarkable that a new label has been coined, Web 2.0. Communication and social aspects of the Internet saw its early beginnings with discussion boards, blogs and communities and its counterparts in the formal learning sphere with virtual learning environments (VLEs). They were succeeded by contemporary social network media such as Facebook and Twitter. The use of social networks has seen a rapid increase of use with more than 400 million active users on Facebook alone (http://www.facebook.com/press/info.php?statistics, April 2010). The rationale for using these tools is to make friends, socialise and build networks. To add people to your list of friends is easily done. Research into actual networking patterns shows however that in reality it is a very small number of friends that users interact with in a qualitative way (Bernardo, Daniel & Fang, 2008; Golder, Wilkinson & Huberman, 2007). In educational settings there are consequently reasons not to be satisfied solely with surface signs of interactions.

The use and value of games in education have been debated for years but have recently become more and more accepted (Gee, 2003). The distinctions between educational technology and entertainment are not always easy to see by looking at the technology itself, it is to be found in the context of its use. Gaming for fun can lead to valuable learning experiences and team building outcomes with potential to outperform formal courses. An interesting category of gaming is the 3D virtual worlds where learners meet in designed contexts taking on different roles. The worlds of Second Life (www.secondlife.com) and Activeworlds (http://www.activeworlds.com) provide a multitude of possibilities for building arenas for IPE, however as with most new technology it requires some familiarity with the interface. Characteristically the participation metaphor of learning views learning as a natural consequence of people interacting with each other. The educational structure is less formal and hence less controllable but with a bearing on practice. *Learning with technology* is to a great extent relying on this learning capacity.

## Learning about Technology

Educators need a solid knowledge base about their topic and curiosity about learners, but also awareness about technical possibilities that can facilitate the learning process. How do we get awareness about innovative technology that might benefit IPE? A point of departure is having an attitude that considers the existence of more efficient technology to serve your needs than what you use today. There are many sources for learning about technology that serve a common interprofessional goal. This book provides several examples of technology and its integration into practice. Other sources are Internet sites (e.g. The New Media Consortium www.nmc.org/horizon) and conferences that have explicit focus on new technology for educational purposes. Not every technical novelty will endure, and sound judgement is needed in the integration of such development, but an interest into new technology is a necessary part of modern life. The evolution of technology goes hand in hand with new ways of communicating and different cultural views of the world. Just to stand aside watching is not a positive option, to participate in societal change through technology use certainly is. Awareness of new technology is part of the modern way of life.

A part of learning about technology is also learning about its use. This is not done overnight, either for an individual or for an organisation. Many enthusiasts have learnt the lesson that organisational integration of innovations is far from straightforward and effortless. It needs (and deserve) careful attention and involvement of key people. Rogers (2003) describes the organisational integration of innovations as a social process where early adopters, the enthusiasts, comprise only one of several categories of people who must get over the threshold of using the innovation before it becomes fully integrated. Early adopters and later adopters have different needs for taking up the innovation. The needs they have concern, among other things, the degree of uncertainty that they are willing to accept before approving the innovation. This social process also involves leadership decisions in favour of the innovation. If it is so troublesome to integrate innovations into educational use, is it really worth it?

Do you remember how you struggled with learning how to ride a bicycle? Probably very few of us remember just how difficult it was. I doubt however that anyone would have this struggle undone and I believe we can conclude that it was well worth the effort. In every introduction of new technology there are moments of struggling to learn new concepts and new procedures that replace habitual routines. The benefit may sometimes only become apparent in retrospect.

But there is cause to pause and examine reasons for technology integration from time to time. Does the attraction arise from the technology itself or do educational needs drive integration? Larry Cuban in his thorough historical analysis of IT integration in schools in the USA is somewhat sceptical as to the actual educational benefits of media and ICT driven innovations. His overview does not concern the kind of social network technology, or the user involvement that comes with today's web 2.0 technology, but if we can learn one thing from his analysis it is the importance of involving practitioners and teachers in the integration process of innovations (Cuban, 1988). There is a gap between producers, policy makers and actual users of educational technology and many applications are short-lived (Berman, Fall, Maloney, & Levine, 2008). In IPE one can also raise the question whether the innovation in itself is biased towards one professional tradition and not others. A single (professional) biased perspective in terminology or structure of a software interface is sometimes difficult to discover by professionals themselves because people tend to be blind to their own culture (Rogoff, 2003). This is however something that is obvious to learners from other traditions and consequently a hindrance to optimal use. Involvement of learners and teachers from different professions at several stages of the design process would help both the integration and the educational use of the software.

The design of the technology itself is consequently important, but so is the design of the context in which it is going to be used. In aviation training, where simulation training has been going on for decades, researchers now conclude that the design of the *educational situation* is just as important, if not more important as the design and sophistication of the soft- or hardware in use (Jentsch & Bowers, 1998; Salas, Bowers & Rhodenizer, 1998). It is therefore not only sparkling new innovations that we need awareness of, but a thoughtful investigation of available "old-school" technology that may prove that to be just as effective as the one we're dreaming of. ICT and different media are powerful tools already available. In parallel to actively following new technical developments we have to learn to see what we have at hand–right here right now; *learning about technology*.

## Interparadigmatic Research Approaches

In order to assess the value of technological practices in IPE, several research strategies need to be pursued. Barr (2005) acknowledges contributions from anthropology, social psychology and sociology. Sfard (1998) identifies a need for patchwork and plurality rather than a global theory of learning when assessing the complex phenomena of human learning. In line with this, researchers now question the traditional partition in quantitative and qualitative paradigms (Hammersley, 1996). When used with methodological awareness, studies from a variety of paradigmatic approaches can contribute to a rich picture of how people learn and interact together in conjunction with technology.

In the IPE community different perspectives about professional roles and different aspects of education and practice will continue to live side by side. Research on the benefits of technology for IPE will also be rooted in different perspectives because of researchers' different backgrounds in social science, health science and informatics among others. Different research perspectives and their respective methods allow for certain aspects of a phenomenon to be visible but hide others. As noted by Jerome Bruner: "research on anything will yield findings that mirror its procedures for observing or measuring" (Bruner, 1990, p. 104). Some of the traditional methods will not suffice in finding innovative ways of using technology because they make assumptions about practices that now undergo change. In IPE we cannot strive to unify research into one research paradigm that will reveal the full picture of learning from and with technology in IPE. We have to acknowledge the plurality of perspectives that exist in IPE and benefit from this diversity. The objects of analysis will vary, but among them will certainly be the integration of innovations and outcomes from learning from and with and about technology.

As educators, developers and technology users we can expand and enrich our view of technology for IPE when using different vantage points, not only the familiar ones. With this attitude, scientific results about learning and reports of technology practices serve not as recipes or rulebooks, but as a "light for the eyes and the lamp to the feet" of today's educators (Dewey, 1929, p. 15). Where educators actually put their feet should be based on sound informed judgment.

## CONCLUSION

There is no doubt that technology and its practices are crucial to the education and practice of interprofessional care. Not to be used for its own sake but in valuing and making possible communication and social processes. To take IPE forward we need further orientation towards current technology practices, some of which are reported in this book. We need also to be aware of and curious about future developments in the technical domains and integrate the tools we believe will benefit the interprofessional perspective on learning. This process needs to be pursued carefully by different traditions of science. We need an open attitude towards innovative technology but we must also remain aware of why we are doing it: to improve collaboration and quality of care.

## ACKNOWLEDGMENT

I would like to express my gratitude to Professor Lars Owe Dahlgren for encouragement and contributions from his IPE experience. Thanks also to Dr. Ester Mogensen, Dr. Hans Rystedt and reviewers for valuable comments and advice on the manuscript.

## REFERENCES

Bandura, A. (1986). *Social foundations of thought and action: a social cognitive theory*. Englewood Cliffs, NJ: Prentice-Hall.

Barr, H. (2005). *Interprofessional education: today, yesterday and tomorrow: a review*, (Rev. Ed.). Working Paper, Higher Education Academy, Health Sciences and Practice Network.

Berman, N., Fall, L., Maloney, C., & Levine, D. (2008). Computer-Assisted Instruction in Clinical Education: a Roadmap to Increasing CAI Implementation. *Advances in Health Sciences Education : Theory and Practice*, *13*(3), 373–383. doi:10.1007/s10459-006-9041-3

Brown, J., Collins, A., & Duguid, P. (1989). Situated Cognition and the Culture of Learning. *Educational Researcher*, *18*(1), 32–42.

Bruner, J. S. (1990). *Acts of meaning*. Cambridge, MA: Harvard Univ. Press.

Carr, N. (2008). Is Google Making Us Stupid? *The Atlantic* (July/August 2008). Retrieved August 10, 2009, from http://www.theatlantic.com/doc/200807/google

Center for Interprofessional e-learning (CIPEL). (2009). *Learning Objects*. Retrieved August 10, 32009, from http://www.cipel.ac.uk/learning_objects/learning_objects.htm.

Centre For The Advancement Of Interprofessional Education (CAIPE). (2002). *Defining IPE* [Electronic Version]. Retrieved August 10, 2009, from http://www.caipe.org.uk/about-us/defining-ipe/

Chambers, L. W., Conklin, J., Dalziel, W. B., MacDonald, C. J., & Stodel, E. J. (2008). E-learning education to promote interprofessional education with physicians, pharmacists, nurses and nurse practitioners in LTC facilities: promising potential solutions. *International Journal of Biomedical Engineering and Technology*, *1*, 233–249. doi:10.1504/IJBET.2008.016958

Clarke, A. C. (1982). *Profiles of the future: an inquiry into the limits of the possible* (2nd rev. ed.). London: Victor Gollancz.

Cuban, L. (1988). *Teachers & machines: The classroom use of technology since 1920*. New York: Teachers College Press.

Dahlgren, L. O. (2005). Learning Conceptions and Outcomes. In F. Marton, D. Hounsell & N. Entwistle (Eds.), The Experience of Learning: Implications for teaching and studying in higher education. (3rd (Internet ed.). Edinburgh, UK: University of Edinburgh, Centre for Teaching, Learning and Assessment.

Demiris, G. (2006). The diffusion of virtual communities in health care: concepts and challenges. *Patient Education and Counseling*, *62*(2), 178–188. doi:10.1016/j.pec.2005.10.003

Dewey, J. (1929). *The Sources of a Science of Education*. New York: Horace Liveright.

Ericsson, K. A. (2008). Deliberate practice and acquisition of expert performance: a general overview. *Academic Emergency Medicine*, *15*(11), 988–994. doi:10.1111/j.1553-2712.2008.00227.x

Garg, A. X., Norman, G. R., Eva, K. W., Spero, L., & Sharan, S. (2002). Is there any real virtue of virtual reality?: the minor role of multiple orientations in learning anatomy from computers. *Academic Medicine*, *77*(10Suppl), S97–S99.

Gee, J. P. (2003). *What video games have to teach us about learning and literacy*. New York: Palgrave Macmillan.

Golder, S. A., Wilkinson, D., & Huberman, B. A. (2007). *Rhythms of Social Interaction: Messaging within a Massive Online Network*. Paper presented at the 3rd International Conference on Communities and Technologies (CT2007).

Hammersley, M. (1996). The relationship between qualitative and quantitative research: Paradigm loyalty versus methodological eclecticism. In Richardson, J. T. E. (Ed.), *Handbook of qualitative research methods for psychology and the social sciences*. Leicester, UK: BPS Books.

Hubbard, G., & Themessl-Huber, M. (2005). Professional perceptions of joint working in primary care and social care services for older people in Scotland. *Journal of Interprofessional Care*, *19*(4), 371–385. doi:10.1080/13561820500165167

Huberman, B. A., Romero, D. M., & Fang, W. (2008). Social networks that matter: Twitter under the microscope. *First Monday, 14*(1), 1-5. Retreived October 13, 2009, from http://uk.arxiv.org/abs/0812.1045

Hutchins, E. (1995). *Cognition in the wild*. Cambridge, MA: MIT Press.

Jentsch, F., & Bowers, C. A. (1998). Evidence for the Validity of PC-based Simulations in Studying Aircrew Coordination. *The International Journal of Aviation Psychology*, *8*(3), 243–260. doi:10.1207/s15327108ijap0803_5

Johnson, L., Levine, A., & Smith, R. (2009). *The 2009 Horizon Report*. Austin, TX: The New Media Consortium.

Jonsson, P., Edelbring, S., Sennerö, A., Fors, U., Kumlien, S., Olt, H., et al. (2006). *MICS, Interprofessional case simulation*. Paper presented at the Karolinska Institutet's Educational Congress, Stockholm.

Kolodner, J., & Guzdial, M. (1996). Effects with and of CSCL: Tracking learning in a new paradigm. In Koschmann, T. (Ed.), *CSCL: Theory and Practice of an Emerging Paradigm* (p. 353). Hillsdale, NJ: Lawrence Erlbaum and Associates.

Koschmann, T. D. (1996). *CSCL: theory and practice of an emerging paradigm*. Hillsdale, NJ: Lawrence Erlbaum and Associates.

Mallow, G. E., & Gilje, F. (1999). Technology-based nursing education: overview and call for further dialogue. *The Journal of Nursing Education*, *38*(6), 248–251.

McLuhan, M., & Fiore, Q. (1967). *The medium is the massage*. Harmondsworth, UK: Penguin.

Merleau-Ponty, M. (2002). *Phenomenology of perception*. London: Routledge.

Pea, R. (1996). Seeing what we build together: Distributed multimedia learning environments for transformative communications. In Koschmann, T. (Ed.), *CSCL: Theory and Practice of an Emerging Paradigm* (p. 353). Hillsdale, NJ: Lawrence Erlbaum and Associates.

*Pew Internet & American Life Project*. (2009). Retreived August 10, 2009, from www.pewinternet.org Washington, DC: Pew Internet & American Life Project.

Prensky, M. (2001). Do They Really Think Differently? *Horizon*, *9*(6).

Ramsden, P. (1992). *Learning to teach in higher education*. London: Routledge. doi:10.4324/9780203413937

Rogers, E. M. (2003). *Diffusion of Innovations* (5th ed.). New York: Free Press.

Rogoff, B. (2003). *The cultural nature of human development*. New York: Oxford University Press.

Salas, E., Bowers, C. A., & Rhodenizer, L. (1998). It Is Not How Much You Have but How You Use It: Toward a Rational Use of Simulation to Support Aviation Training. *The International Journal of Aviation Psychology*, *8*(3), 197–208. doi:10.1207/s15327108ijap0803_2

Sfard, A. (1998). On Two Metaphors for Learning and the Dangers of Choosing Just One. *Educational Researcher*, *27*(2), 4–13.

Wackerhausen, S. (2007). *Reflection as transformation Collaboration and reflection across boundaries*. Key note speech, Proceedings of the First International Conference of the European Interprofessional Network (EIPEN), Krakow, Poland.

Wenger, E. (1998). *Communities of practice: learning, meaning, and identity*. Cambridge, UK: Cambridge University Press.

# Chapter 4
# Setting the Scene:
## E-Learning and the Evolution of Roles and Practices in Post-Compulsory Education

**Martin Oliver**
*University of London, UK*

## ABSTRACT

*This chapter explores developments in e-learning in colleges and universities, providing a context for the work that follows. Pedagogical developments are reviewed, focusing particularly on the UK. Research is then considered, beginning with an orienting discussion of the curriculum as a focus for study. Research concerning the relationship between technology and the curriculum is then discussed. Implications of this for teachers, students and others are then presented. The chapter concludes by emphasising the need to engage with, rather than ignore, the complexity that a social account of technology in education provides.*

## INTRODUCTION

The purpose of this chapter is to provide an account of the development of e-learning in post-compulsory education, focusing primarily on Higher Education and formal in-service learning, which is historically where the majority of interprofessional education has taken place (Freeth et al., 2002). This is considered in relation to pedagogy, and its relationship to interprofessional education, as introduced in the preceding chapters, is considered. To provide focus, the chapter then concentrates on research into curriculum design and practice, illustrating how these ideas fit with research into new technologies. This is then discussed in terms of implications for various roles, before conclusions are drawn about the kinds of issues that remain important within this area.

DOI: 10.4018/978-1-61520-889-0.ch004

### Technology and Post-Compulsory Education

Technology has always been implicated in educational development and change, whether it be the development of writing, the printing press or the Internet. Any review of this relationship, therefore, will necessarily be partial. For the purposes of this book, a review is provided that focuses on developments across the last few decades, concentrating on pedagogic developments and issues of professional role. This review concentrates on the UK, in order to provide a manageably brief case context.

## Technology and Pedagogy

Technology is frequently associated with progressive pedagogies. However, simply adopting technology is no guarantee of pedagogic development, and indeed can be associated with less sophisticated pedagogic approaches (Mishra & Koehler, 2006). Certainly, theories such as constructivism have had a powerful impact on practices such as instructional design (Duffy, Lowyck & Jonassen, 1993). However, almost any theory would have something to offer, since each provides its own distinctive insights into the process of learning (Mayes & de Freitas, 2004).

Rather than seeing technology as a driver for particular pedagogies, it may be fairer to view it as a weathervane, revealing contemporary educational preoccupations and developments. In the form of educational television, it was touted as an answer to the problem of mass training of military personnel during World War II; in the guise of teaching machines, it was advocated by Skinner (e.g. 1958) as a means of making the educational system more efficient; by Piaget (1929) and Papert (1980) as a way of supporting children's creative thinking and self-expression in the face of a heavily structured curriculum; and more recently by a variety of authors as a way of formalising the tacit learning that arises from social interaction, such as learning in the workplace, as a way of making this amenable to formal educational support and recognition (Dyke et al., 2006). Clearly, there have been pedagogic developments – Thorpe (2002), for example, provides a useful overview of the way in which pedagogic thinking has developed from a focus on materials and instruction to one on social competence, collaboration and situated performance. However, as she also points out, this is not a simple progression towards some utopian future: for example, the increasingly prevalent view of the learner as independent and autonomous can also be seen as viewing the learner as isolated, unsupported and responsible for their failings.

Whether or not pedagogic thinking progresses, research into the use of technology has, historically, suffered from a limited perspective. Whilst educational technology as a field of study may date back over 40 years, much research is located within specific 'generations' of technology, drawing only on a few years' worth of work. As far back as 1995, Mayes lamented the ongoing cycles of hype and disappointment – which he illustrated by analogy to the film 'Groundhog Day' – that have come to characterise research in this field:

> *People who have been involved over any length of time with educational technology will recognise this experience, which seems characterised by a cyclical failure to learn from the past. We are frequently excited by the promise of a revolution in education, through the implementation of technology. We have the technology today, and tomorrow we confidently expect to see the widespread effects of its implementation. Yet, curiously, tomorrow never comes. We can point to several previous cycles of high expectation about an emerging technology, followed by proportionate disappointment, with radio, film, television, teaching machines and artificial intelligence (Mayes, 1995, p. 1).*

This continues to have resonance, over ten years later. Clearly, no technology to date has managed to "solve" education, and it seems equally likely that no new technology will manage to, either. Instead, research into each new technology tends to follow a familiar pattern: technical or specialist research specifying the technology; initial 'proof of concept' papers establishing its potential within education; and further case studies elaborating uses and exploring applications within other educational disciplines or contexts. Theorisation – which ought to build from this (Oliver & Conole, 2003) – does not always follow.

## Technology and Interprofessional Learning

Interprofessional education has only recently been introduced into curricula, so that there are still calls for evidence of its value, and a need to determine how best to pursue it in a sustainable and effective way (Wright & Lindqvist, 2008). This is also true for e-learning, where new technologies arise whilst evidence about the previous technologies is still being generated, creating the opportunity for complementary results to arise from single studies. There are also similar values between the two areas: the client-centred focus of interprofessional learning mirrors the recent emphasis on students' experiences of e-learning (Benfield et al., 2009). Similarly, the need to bring together complementary professional roles in the design and deployment of educational processes – an issue returned to later in this chapter – mirrors interprofessional practices in healthcare (Juntunen & Heikkinen, 2004).

These similarities between the educational process and the goals of interprofessional learning allow a reflexive perspective to be developed. Courtney (2008), for example, describes the support provided for a group brought together to develop interprofessional e-learning materials; the different roles and responsibilities involved in such a process need to be brought together in support of a common aim (supporting learners), aligned by shared values. The goals – as well as the challenges – allow understanding and insight developed in one context to be brought to bear on the other.

However, the relatively brief history of both developments means that literature addressing both areas is scant. For example, a review undertaken in 2002 (Freeth et al.) made no mention of e-learning or the use of technology as a teaching method at all. Where work has been published, it sometimes focuses on the development of the system or resource (e.g. Gilchrist, 2008) rather than on what students learn with it. In other cases, e-learning is mentioned but is incidental to the issues that are focused on in the paper – for example, in Cooper and Spencer-Dawe's study (2006), which used e-learning as part of an intervention, but which focused on the role of service users rather than the teaching approaches used, or in Wright & Lindqvist's evaluation (2008), in which e-learning is judged to have the potential for supporting discussions, although here it was only one of a number of opportunities for this to happen. E-learning also makes inter-institutional programmes more feasible (Juntunen & Heikkinen, 2004); can make discussions more fluid and enables participation whilst on placement (Miers et al., 2007); and gives students flexibility in the way that they make use of course resources, so long as they are able to make sense of this rather than finding the opportunities confusing (Cooper et al., 2005) or being overwhelmed by the technology (Miers et al., 2007). More substantively, in addition to the logistical advantages of e-learning, the use of electronic case studies such as 'patient journeys' can support the development of interprofessional care plans, the promotion of interprofessional learning and attitudinal change (Clouder, 2008).

There are challenges as well as difficulties in interprofessional e-learning, however. Studies such as that by Juntunen & Heikkinen (2004), for example, argued that teaching using web-based systems could increase the need for teacher support and supervision; would require teachers to develop facilitation skills; and would shift the emphasis onto students' learning and away from a focus on the teacher, all of which may make it challenging to teaching staff (Clouder, 2008).

## Technology and the Curriculum

The focus upon materials development in research has consequences for practice, and for further research. It suggests that curricula should be understood either simply in terms of content or else as content and planned pedagogy. This provides a very reductive account of curricula and curriculum practice.

However, this is not the only way in which curricula can be understood. In this section, different conceptions of the curriculum will be presented, providing a context in which to discuss research that investigates the role of technology in supporting curriculum work.

## Conceptions of the Curriculum

Barnett & Coate (2005) lamented the lack of attention that the idea of the curriculum has received in Higher Education, although it must be noted that within the context of interprofessional education, questions of disciplinarity, 'transdisciplinarity' and interdisciplinary work have been considered (e.g. Barr, 2002).

Perhaps because this topic is so frequently neglected, however, the term "curriculum" can be understood in very different ways. In a study of the way academics designed curricula (Oliver, 2003), it was found that some did not use the term at all; in some cases, it was used to denote the syllabus; in others, the programme of study across individual courses; sometimes it was used to describe pedagogy as well as the topics to be covered; and in a small number of instances it was invoked in terms of what could be described as the 'hidden curriculum'–the context of values, politics and practices that influence students' experience, even if they remain tacit. In effect, this analysis revealed a nested set of conceptions of the curriculum, with each layer adding new conceptual and contextual detail to those that precede it.

Importantly, however, these conceptions were complemented by another use of the term that did not fit within this model. The common element across all of these conceptions was that they describe intentions; they concern plans or, at the least, abstractions of practice. By contrast, academics also talked of a lived curriculum – the performance undertaken by teachers as they engage with students. This was understood both in terms of the practices they undertake and also the spaces that needed to be created in which engagement could take place. In this sense, the curriculum was something that emerged from the performance of planned teaching with students.

Even in terms of the tangible elements of the curriculum – the resources, plans and so on – it is important to recognise that choices and actions are politically as well as rationally shaped. Millen (1997), for example, describes examples of practice that illustrate ways in which the creation of curricula is a political act, influenced by the academic's desire to present themselves as well informed and credible, their relationship with influential sponsors, their differentiation from peers and potential rivals and so on. Even something as apparently straightforward as a course reading list can, she argues, be understood as an action that reflects and perpetuates the ongoing politics of academic life. This has implications for technologies intended to support curriculum work, which will now be explored.

## Learning Design and Designing for Learning

Much research about the use of technology in post-compulsory education describes the effect of changed 'delivery' – in other words, the effects of introducing new tools in support of existing classroom practices. It has been claimed that such developments have little effect on learning outcomes (e.g. Lockee et al., 1999). However, it is unlikely that sweeping claims such as this could ever be answered in a truly convincing way, for a number of reasons. There is a lack of clarity about what 'counts' for inclusion in each condition (when does use of email make teaching count as 'e-learning', for example?); a majority of studies focus on expressed satisfaction rather than learning outcomes; serious questions have been asked about whether changes in pedagogy require new forms of assessment, making comparative studies problematic; questions have also been raised about the ecological validity of experimental comparisons, for example because of the important situational effects of motivation on learning; there are practical and ethical difficulties in designing comparative studies in actual educational settings, not least because of the risk of disadvantaging groups of learners in some way; and there is a general dearth of published accounts focusing on the failure of technology, suggesting that research reporting may be distinctly one-sided.

Since simple generalisations about curriculum delivery are unlikely to be possible, this section provides a narrative account of research that focuses on the design of technology to support curriculum work. In many ways, this tradition can be traced back to the work of the psychologist Skinner, whose behaviorist theory prompted the development of 'teaching machines' (1958). These early attempts to automate the curriculum led to the development of programmed learning and instructional design (e.g. Merrill, 2001), which explicitly treated knowledge reductively, seeking to develop hierarchies of concepts and skills that could be delivered to learners in a controlled way, matched to their abilities and progression. Increasingly sophisticated approaches can be seen in subsequent work on intelligent tutoring systems, in which materials were presented to students based on a more sophisticated model of the learner, which might contain information intended to represent their emotional state, preferences for formats and conceptual development as well as their performance within the system. However, although such systems might be versatile and powerful in theory, they have proved difficult to implement in practice, particularly in domains where knowledge is contested or ill-defined; and even when they can be implemented, the learning gains associated with them remain at best about half that of working with a professional tutor (Hampton, 2006).

Although this line of work has proved problematic, a very similar approach has been taken in the interests of efficiency, drawing on work on reusable learning objects (Oliver, 2004). The idea of reusable learning objects – RLOs – is that educational content is complemented by metadata (literally, data about data) that help other teachers find materials that they can then use. Since metadata can be searched automatically, it was suggested (e.g. Wiley, 2000) that systems could be developed that would create customised pathways for learners through pre-prepared learning materials on the basis of a user model. Like earlier research efforts, this approach has yet to demonstrate its viability in a scalable and credible way.

Nonetheless the idea that there is value in tagging and sharing educational content persists, albeit with the emphasis shifted to teachers (rather than systems) finding such resources and incorporating them into their own designs. Even this, however, has proved much more complicated than was initially expected. Early ideas that making academics aware of resources would enable them to use them (e.g. Downes, 2001) gave way to the idea that reuse would have to be understood in terms of appropriation and adaptation, not simple adoption (Oliver, 2004). Moreover, it has since been argued that more work is needed to understand how teachers can be helped to generate good quality metadata or to use learning object repositories once they have been created (Currier et al., 2004). Reuse remains a problem, although recent work exploring the socio-cultural contexts of repository use does suggest that designing with communities of users in mind may help to address this (Margaryan et al., 2006).

Another challenge has been that educational content –RLOs, for example – is only half the problem. The remainder of the process of curriculum work is the shaping of materials into curricula. This part of the process has also been addressed by the design of new technologies. The field of learning design, which built on the use of metadata and RLOs, has led to the development of systems to support the structuring and delivery of content, such as the London Pedagogy Planner (San Diego et al., 2008). These tools provide ways of formally specifying the curriculum, typically understood in terms of the alignment between content, activities and assessment. This level of formality brings with it the promise of 'runnable' pedagogy: designs that can be shared, loaded and executed through the software.

However, as before, the simple elegance of formality has been questioned in terms of its relevance to practice. For example, Beetham (2007) has questioned whether 'good' design can ever be captured by 'sound' principles, proposing instead that it involves tactic knowledge and expertise. The alternative – described as 'designing for learning' to show its relationship to but difference from learning design – focuses on representing pedagogy in ways that other practitioners can make sense of, so that 'sharing' and 'reuse' are not understood as universal, automatable processes, but as examples of appropriation, reinterpretation and redesign. Under this alternative formulation, the teacher's role is central, in marked contrast to the marginalisation that can be seen in some areas of instructional and learning design.

## Technology and Educational Roles

As seen in the previous section, the introduction of technology is not a simple, neutral process characterised by efficiency and improvement, but must at least partly be understood in terms of politics, professional identities and personal priorities. It is these elements that will be explored next.

## The Contested Role of the Teacher

Redefining the curriculum and automating processes that support teaching and learning will necessarily have consequences for teachers. There are widespread calls for teachers to change their practice. In some cases, this is a reflection of the development of new conceptions of pedagogy (Thorpe, 2002), which have implications for the relationship between teachers and learners. In others, such calls are argued to provide an excuse for the failure of technology to create innovation within educational systems, treating the teachers as scapegoats (Cuban, 2002). In yet others, an almost moralistic position is taken, suggesting that teachers should change their role in order to change power dynamics in the classroom as well as to reflect the opportunities provided by new technologies (Salmon, 2000). Across these accounts, there is an increasing emphasis on the social and communicative practices that contextualise formal instruction, foregrounding the role of facilitation, moderation, support and discussion (Price & Oliver, 2007).

The phrase, "from the sage on the stage to the guide on the side", is often invoked as a way of summarising many of these demands. Yet this simple slogan ignores evidence that good teaching has always involved more than a 'sage on the stage' (e.g. Trigwell et al., 1994). Much writing uses this simple caricature for rhetorical effect – positioning 'traditional' practice as being about 'the sage on the stage' associating it exclusively with impoverished conceptions of teaching, ignoring the more sophisticated understanding and practices of good teachers. By associating all traditional practice with 'bad' practice, it is easy to make the case that something new is needed. As Cuban (2002) suggests, the promise of a revolution tomorrow is an appealing one, with only the recalcitrance of teachers and academics standing in its way. We would be assured a wonderful future, this rhetoric suggests, if only we could stop academics being so stubborn and entrenched.

However, not all accounts of the relationship between technology and education are so utopian. Noble (1997), for example, provides a classic dystopian alternative. His analysis describes "digital diploma mills", in which teachers are marginalised, students are processed in an industrialised system and "the administration" holds power in academia. His account relies on the idea that curricula are treated as reducible to materials – which academics provide, and perhaps maintain, but which are owned by the administration on behalf of their institution. Support for these materials is provided by low-paid tutorial staff, deprofessionalised and removed from involvement in the processes of knowledge production. Such an account might seem extreme, but there is clear resonance with the proposals for the UK's e-University (PriceWaterhouseCoopers, 2000), and also with some writing in instructional design, such as Merrill's assertion (2001) that "too often instructional designers leave these important what-to-teach decisions to so-called subject-matter-experts (SMEs)".

The identity and role of academics is clearly contested, with the political interests of technology-related experts, managers and the academics themselves all being represented in research. Many teachers themselves, however, continue to struggle to resist change – not because they are opposed to change *per se*, but because their current practice has been designed in a way that demonstrates the values and principles that they believe to be important, whether these are about their commitment to their students or to the characteristics of their discipline (Price & Oliver, 2007). This does not stop them changing their practice – but it does affect how they engage with change. In other words, the apparent reluctance of academics to change can be explained in terms of their commitment to principles that continue to be important. However, so long as these strategic principles are allowed to remain intact, academics are often open to developing or replacing the operational practices through which these are realised. Much of the contestation that arises in this area does so because – explicitly or implicitly – the operational changes that are proposed (such as introducing a new technology) are used as a vanguard to a change in the principles that guide current academic practice. This may explain why the most effective interventions in academics' practice are those which are done with, rather than done on behalf of (or 'done to'), academics (Sharpe & Oliver, 2007).

## The Experiences of Learners

The policies and technological developments outlined above have implications for learners as well as teachers. The expectations of teachers and designers of technology will influence students' experiences, even if it cannot entirely determine them. These expectations, and students' reactions to them, will be outlined here.

Teachers are often criticised for being didactic and transmissive in their pedagogy as an excuse for advocating new forms of technology – yet as noted, good teachers do not take such a position (Trigwell et al., 1994). Policies and systems, however, can be just as variable in the sophistication of their accounts of learners. The Dearing report, for example, portrayed learners almost exclusively as people who made rational choices about what course to purchase, but were then passively 'done to' by their educational experiences (Smith & Oliver, 2002). However, more recent policies have taken a more constructive view, treating learners as people who are actively engaged in creating their own learning experiences; there is also a clear interest in the diversity of learners and their experiences, not least because of the importance of widening participation in post-compulsory education.

Researchers too have provided oversimplified accounts of learners. There is a tendency to treat all learners as homogeneous in their relationship to new technology – by labelling them all as "digital natives" (Prensky, 2001), for example. As has been argued, however, there is little empirical or theoretical support for this position – Bennett et al. (2008) liken it instead to an academic form of 'moral panic'.

While it may be convenient to treat all learners as if they fit neatly into the same category, where evidence has been gathered, it shows a much more complex picture. Benfield et al. (2009), for example, illustrate how students within one institution were generally consistent in terms of preferred site for studying (their homes), but varied considerably in terms of their ability to fit study around other commitments, the technologies they used to participate in learning and the different patterns of study across disciplines, for example. In spite of current hype around Web 2.0, only half of students in this particular study reported making any real use of such technologies. Differences across geographical settings – such as between developed and developing nations, for example, or in areas with different degrees of transport infrastructure – may also influence learners' choices about their studies. What this demonstrates is the importance of knowing the particular learners for whom a curriculum might be designed, rather than relying on sweeping generalisations as the basis for decisions.

## The Emergence of New Roles

The familiar roles of teachers and learners are not the only ones that have evolved. Other groups who are arguably less visible in the way that they support learning, such as librarians and administrators, have also been affected by changes related to the introduction of e-learning (e.g. Orick, 2000).

Other roles have also emerged. The Dearing report (NCIHE, 1997), for example, described the emergence of 'new professionals' in UK post-compulsory education, a group whose role was often marginal, typically being located outside of the conventional hierarchies of the institution, but who nevertheless played an important role in informing, supporting and shaping both policy and practice in the use of technology. Learning technologists – who typified many of the features of this group – have since been recognised more formally, for example through initiatives to develop professional recognition (e.g. Oliver et al., 2004). Their style of work has also been documented (Oliver, 2002). Learning technologists' accounts of their work show them as collaborating with academics around curriculum development, using this process as a kind of pedagogy that first solves short-term problems (many of which are technical), but which goes on to raise educational questions that may challenge and develop academics' practice more generally.

Although the term 'learning technologist' is primarily found within the UK, comparable roles are visible on a much wider basis. There are clear analogies with some traditions of instructional design in the US (Oliver, 2002); the roles are strongly reminiscent of the 'shepherds' described in Scandinavian Higher Education (Oliver et al., 2005); and there are strong similarities with the Educational Technologists working in contexts such as South Africa (Hodgkinson, Williams & Gray, 2008).

Some of this commonality may reflect the fact that these 'new professionals' are not so new. Almost identical roles can be found described in research from 1976 (Lawless & Kirkwood, 1976). These three decades or more of history are rarely recognised however; it is more common for learning technologists to 'drift' into their roles, and learn their practices on the job, often in isolation from others working in the same role (Oliver, 2002). This has obvious implications for the development of professional knowledge in this area, which will often be bound to people and their immediate contexts rather than shared and used to develop negotiated, evidence-informed professional practices at a national or international level.

## CONCLUSION

The introduction of new technology to post-compulsory education is complex, drawing together people with different commitments, responsibilities and values. Ideas about what Higher Education should be for, how people learn and what professional roles should be undertaken all vary, and the topic of technology use illustrates this vividly. It has been suggested for decades that technology is about to transform Higher Education, and indeed, much changes every year – yet as has been argued here, good teaching has always been about understanding students' learning, supposedly 'new' roles can be traced back decades, and innovative uses of technology to support curriculum design bear strong resemblance to earlier ideas even if the technologies involved differ. In no small part, this is because the problems that are faced – the complexities outlined here – are actually well established.

Understandably, there is a temptation to simplify accounts of learning, teaching and technology so as to make this situation seem less problematic. Many policies, and many research projects, start with simple assumptions. Focusing on resources, rather than on what people do with them, is one example of how this happens. However, as this overview has demonstrated, the social contexts of education are increasingly recognised and represented in work in this area. This is true, for example, in terms of the development of social accounts of pedagogy, of learning design and of the use of technology. It may be necessary to start with a simple focus, but it is also necessary to move beyond this, to take account of the complexities involved in designing, using and re-using curricula materials such as RLOs.

The challenge for teachers, then, lies in appreciating the complexity of the educational situations for which they are responsible. Developing an understanding of specific groups of students, their experiences and priorities should enable more appropriate and successful educational interventions to be designed. For researchers, the challenge lies in balancing awareness of the complexities of education with the need to create manageable research projects and draw conclusions that are comprehensible and relevant to others. For all involved in this area, the shared challenge is to build our understanding of the complexities that we face, so that even if 'new' developments turn out to be similar to those already past, they can at least reach further and do better that the previous attempts.

## REFERENCES


Barnett, R., & Coate, K. (2005). *Engaging the Curriculum in Higher Education*. Buckingham, UK: SRHE and Open University Press.

Barr, H. (2002). *Interprofessional education: today, yesterday and tomorrow. LTSN occasional paper no. 1*. London: LTSN Centre for Health Sciences and Practice, King's College.

Beetham, H. (2007). An approach to learning activity design. In Beetham, H., & Sharpe, R. (Eds.), *Rethinking pedagogy for a digital age: designing and delivering e-learning* (pp. 26–40). Oxford, UK: Routledge.

Benfield, G., Ramanau, R., & Sharpe, R. (2009). Student learning technology use: preferences for study and contact. *Brookes e-Journal of Learning and Teaching, 2*(4). Retrieved August 12, 2009, from http://bejlt.brookes.ac.uk/evaluating_learners_experiences_of_e-learning/

Bennett, S., Maton, K., & Kervin, L. (2008). The 'digital natives' debate: A critical review of the evidence. *British Journal of Educational Technology, 39*(5), 775–786. doi:10.1111/j.1467-8535.2007.00793.x

Clouder, D. (2008). Technology-enhanced learning: conquering barriers to interprofessional education. *The Clinical Teacher, 5*, 198–202. doi:10.1111/j.1743-498X.2008.00243.x

Cooper, H., & Spencer-Dawe, E. (2006). Involving service users in interprofessional education: narrowing the gap between theory and practice. *Journal of Interprofessional Care, 20*(6), 603–617. doi:10.1080/13561820601029767

Cooper, H., Spencer-Dawe, E., & Mclean, E. (2006). Beginning the process of teamwork: design, implementation and evaluation of an inter-professional education intervention for first year undergraduate students. *Journal of Interprofessional Care, 19*(5), 492–508. doi:10.1080/13561820500215160

Courtney, K. (2008). Exploring virtual opportunities to enhance and promote an emergent community of practice. *Electronic journal of e-Learning, 5*(4), 261-270. Retrieved October 10, 2009, from http://www.ejel.org/Volume-5/v5-i4/Courtney.pdf

Cuban, L. (2002). *Oversold and Underused: Computers in the Classroom*. Cambridge, MA: Harvard University Press.

Currier, S., Barton, J., O'Beirne, R., & Ryan, B. (2004). Quality assurance for digital learning object repositories: issues for the metadata creation process. *ALT-J: Research in Learning Technology, 12*(1), 5–20.

Downes, S. (2001). Learning Objects: resources for distance education worldwide. *The International Review of Research in Open and Distance Learning, 2*(1). Retrieved October 10, 2009, from http://www.irrodl.org/content/v2.1/downes.pdf

Duffy, T. M., Lowyck, J., & Jonassen, D. (1993). *The Design of Constructivistic Learning Environments: Implications for Instructional Design and The Use of Technology*. Heidelburg, Germany: Springer-Verlag.

Dyke, M., & Conole, G. Ravenscroft, A. & de Freitas, S. (2006.) Learning theory and its application to e-Learning. In G. Conole & M. Oliver (Eds.), Contemporary Perspectives in E-Learning Research (pp. 80-95). London: RoutledgeFalmer.

Freeth, D., Hammick, M., Koppel, I., Reeves, S., & Barr, H. (2002). *A critical review of evaluations of interprofessional education. LTSN occasional paper no. 1*. London: LTSN Centre for Health Sciences and Practice, King's College.

Gilchrist, M. (2008). Issues around creating a reusable learning object to support statistics teaching. *Electronic journal of e-Learning, 5*(4), 271-278. Retrieved October 10, 2009, from http://www.ejel.org/Volume-5/v5-i4/Gilchrist.pdf

Hodgkinson-Williams, C., & Gray, E. (2008). *Degrees of Openness: The emergence of Open Educational Resources at the University of Cape Town*. Paper presented at EMERGE2008 virtual conference. Retrieved October 10, 2009, from http://emerge2008.net/access/content/group/emerge2008/PresentationFiles/Hodgkinson-Williams/DegreesOfOpeness.pdf

Juntunen, A., & Heikkinen, E. (2004). Lessons from interprofessional e-learning: piloting a care of the elderly module. *Journal of Interprofessional Care, 18*(3), 269–278. doi:10.1080/13561820410001731313

Lane, H. C. (2006). *Intelligent Tutoring Systems: Prospects for Guided Practice and Efficient Learning*. Whitepaper for the Army's Science of Learning Workshop, Hampton, VA. Aug 1-3, 2006. Retrieved October 10, 2009, from http://people.ict.usc.edu/~lane/papers/ITSProspectsLane-Aug06.pdf

Lawless, C., & Kirkwood, A. (1976). Training the Educational Technologist. *British Journal of Educational Technology, 1*(7), 54–60. doi:10.1111/j.1467-8535.1976.tb00188.x

Lockee, B., Burton, J., & Cross, L. (1999). No comparison: Distance education finds a new use for 'No significant difference'. *Educational Technology Research and Development, 47*(3), 33–42. doi:10.1007/BF02299632

Margaryan, A., Currier, S., Littlejohn, A., & Nicol, D. (2006). *Report on Learning Communities and Repositories. Community Dimensions of Learning Object Repositories project Deliverable 1*. Retrieved October 10, 2009, from http://academy.gcal.ac.uk/cd-lor/learningcommunitiesreport.pdf

Mayes, J. T. (1995). Learning Technology and Groundhog Day. In Strang, W., Simpson, V. B., & Slater, D. (Eds.), *Hypermedia at Work: Practice and Theory in Higher Education*. Canterbury, UK: University of Kent Press.

Mayes, T., & de Freitas, S. (2004). *Review of e-learning theories, frameworks and models. JISC project report.* Retrieved October 10, 2009, from http://www.e-learning.ac.uk/resources/modelsdeskreview/view

Miers, M., Clarke, B., Pollard, K., Rickaby, C., Thomas, J., & Turtle, A. (2007). Online interprofessional learning: the student experience. *Journal of Interprofessional Care, 21*(5), 529–542. doi:10.1080/13561820701585296

Millen, J. (1997). Par for the Course: designing course outlines and feminist freedoms. *Curriculum Studies, 5*(1), 9–27.

Mishra, P., & Koehler, M. J. (2006). Technological Pedagogical Content Knowledge: A framework for integrating technology in teacher education. *Teachers College Record, 108*(6), 1017. doi:10.1111/j.1467-9620.2006.00684.x

NCIHE - National Committee of Inquiry into Higher Education. (1997). *Higher Education in the Learning Society.* London: HMSO.

Noble, D. (1997). Digital Diploma Mills – Part 1: The automation of Higher Education. *FirstMonday, 3*(1). Retrieved October 10, 2009, from http://www.firstmonday.dk/issues/issue 3_1/noble/

Oliver, M. (2002). What do Learning Technologists Do? *Innovations in Education and Training International, 39*(4), 245–252.

Oliver, M. (2003). *Curriculum Design as acquired social practice: a case study.* Paper presented at the 84th Annual Meeting of the American Educational Research Association, Chicago.

Oliver, M. (2004). Metadata vs. educational culture: roles, power and standardisation. In Land, R., & Bayne, S. (Eds.), *Education in Cyberspace* (pp. 112–138). London: RoutledgeFalmer.

Oliver, M., & Conole, G. (2003). Evidence-based practice and E-Learning in Higher Education: Can We and Should We? *Research Papers in Education, 18*(4), 385–397. doi:10.1080/0267152032000176873

Oliver, M., Price, S., Boycheva, S., Dugstad Wake, J., Jones, C., Mjelstad, S., Kemp, B., Nikolov, R., & van der Meij, H. (2005). *Empirical studies of the impact of technology-enhanced learning on roles and practices in Higher Education.* Kaleidoscope project deliverable 30-03-01-F.

Oliver, M., Sharpe, R., Duggleby, J., Jennings, D., & Kay, D. (2004). *Accrediting learning technologists: a review of the literature, schemes and programmes.* ALT accreditation project report No. 1. Retrieved October 10, 2009, from http://www.ucl.ac.uk/epd/alt-accreditation/Initial_review.doc

Orick, J. (2000). The Virtual Library: Changing Roles and Ethical Challenges for Librarians. *The International Information & Library Review, 32*(3-4), 313–324. doi:10.1006/iilr.2000.0138

Papert, S. (1980). *Mindstorms: Children, Computers and Powerful Ideas.* New York: Basic Books.

Piaget, J. (1929). *The Child's Conception of the World.* New York: Harcourt, Brace Jovanovich.

Prensky, M. (2001). Digital Natives, Digital Immigrants. *On the Horizon. NCB University Press, 9* (5). Retrieved October 10, 2009, from http://www.marcprensky.com/writing/

Price, S., & Oliver, M. (2007). A Framework for Conceptualising the Impact of Technology on Teaching and Learning. *Educational Technology & Society, 10*(1), 16–27.

PriceWaterhouseCoopers. (2000). *Annex 3: Learning products and services for the e-U.* Retrieved October 10, 2009, from http://www.hefce.ac.uk/pubs/hefce/2000/0044/00_44a3.pdf

Salmon, G. (2000). *E-moderating - the key to teaching and online learning*. London: Kogan Page.

San Diego, J., Laurillard, D., Boyle, T., Bradley, C., Ljubojevic, D., Neumann, T., & Pearce, D. (2008). Towards a user-oriented analytical approach to learning design. *ALT-J: Research in Learning Technology*, *16*(1), 15–29.

Sharpe, R., & Oliver, M. (2007). Supporting practitioners' design for learning: principles of effective resources and interventions. In Beetham, H., & Sharpe, R. (Eds.), *Rethinking Pedagogy for a Digital Age* (pp. 117–128). London: Routledge.

Skinner, J. (1958). Teaching Machines. [from http://www.bfskinner.org/f/teaching_machines.pdf]. *Science*, *128*(3330), 969–977. Retrieved October 10, 2009. doi:10.1126/science.128.3330.969

Smith, H., & Oliver, M. (2002). University teachers' attitudes to the impact of innovations in ICT on their practice. In C. Rust (Ed.), *Proceedings of the 9th International Improving Student Learning Symposium* (pp. 237-246). Oxford, UK: Oxford Centre for Staff and Learning Development.

Thorpe, M. (2002). From independent learning to collaborative learning: new communities of practice in open, distance and distributed learning. In Lea, M., & Nicoll, K. (Eds.), *Distributed Learning: Social and Cultural Approaches to Practice* (pp. 131–151). London: RoutledgeFalmer.

Trigwell, K., Prosser, M., & Taylor, P. (1994). Qualitative Differences in Approaches to Teaching 1st Year University Science. *Higher Education*, *27*(1), 75–84. doi:10.1007/BF01383761

Wiley, D. (2000). Connecting learning objects to instructional design theory: A definition, a metaphor, and a taxonomy. In D. Wiley (Ed.), *The Instructional Use of Learning Objects* (online version). Retrieved October 10, 2009, from http://www.reusability.org/read/chapters/wiley.doc

Wright, A., & Lindqvist, S. (2008). The development, outline and evaluation of the second level of an interprofessional learning programme – listening to the students. *Journal of Interprofessional Care*, *22*(5), 475–487. doi:10.1080/13561820802210968

# Chapter 5
# Analysis and Description of Education Employing Technological Platforms:
## Terminology, Features and Models

**Mark Childs**
*Coventry University, UK*

## ABSTRACT

*This chapter aims to provide a background to two aspects that figure prominently in later chapters of this book, by introducing many of the concepts relevant to them, and establishing a consistent terminology with which to describe them. The first of these aspects is that of the technological platforms employed. Technological platforms are used in a range of different activities within interprofessional learning. These activities include accessing learning objects, conducting situative learning in order to establish jointly developed knowledge, providing an opportunity to develop an online professional identity and creating links within a community. In this discussion, one particular platform is particularly referred to throughout. That is the immersive virtual worlds (IVW), also referred to as multi-user virtual environments (MUVE). This is to provide more information for those unfamiliar with immersive virtual worlds and because these particular environments form the basis of many of the case studies described in the remainder of the book. Experience of virtual worlds also indicates that they may have special relevance to interprofessional education, particularly around notions of 'presence', 'embodiment' and 'identity', and these are concepts that have been given a fuller explanation for this reason. Secondly, a range of different theoretical frameworks for understanding the education activities and interactions that take place using these technological platforms are introduced and discussed.*

DOI: 10.4018/978-1-61520-889-0.ch005

## INTRODUCTION

This chapter aims to explain some of the elements of technological platforms to support learning, and some of the theoretical frameworks with which to analyse the communication that takes place within them, however, some of the terms used in the literature are used with more than one meaning. This first section aims to disambiguate some of these terms and to explain the terminology employed within this book.

### Two Definitions of "Mediated Communication"

The educational scenarios described within this book have one thing in common, are all mediated through a technology and can take place at a distance. The term "mediated" is ambiguous in this context, however. In one usage of the word, all learning is mediated to some extent using a tool, since learning occurs "not only inside the person, but in his or her ability to use a particular set of tools in particular ways and for particular purposes" (Littleton, Toates & Braisby, 2007, p. 202). This act of mediation through the employment of an external artefact is a cornerstone of Activity Theory. Activity theory, as formulated by Vygotsky in the 1920s, demonstrates this in the form of a triangle (Center for Activity Theory and Developmental Work Research, 2004), (Figure 1).

*Figure 1. The modern reformulation of Vygotsky's Activity Theory (Adapted from Center for Activity Theory and Developmental Work Research, 2004)*

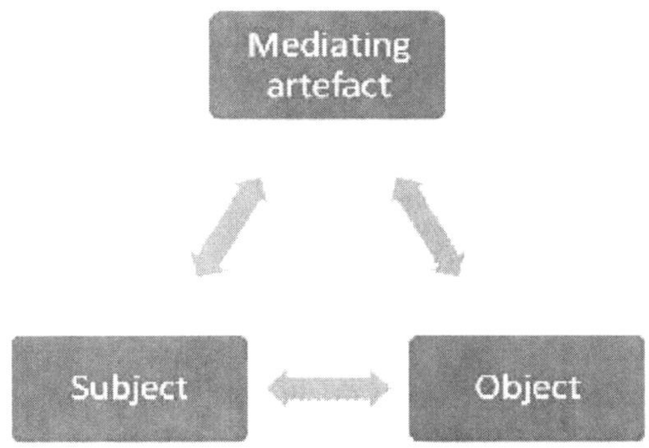


Activity theory is misnamed, in that it is not really a theory since it does not have (nor is intended to have) the explanatory or predictive properties of theories (Masterman, 2008, p. 224). However, what Activity Theory does is provide a means to describe an activity through a division of that activity into its constituent parts; considering actions to be both orientated towards some objective and mediated through an external tool or instrument. E-learning focuses on the use of digital technologies as tools, but the distinction between digital learning technologies and other tools is an arbitrary distinction; everything we use in learning and teaching, a chalkboard, a pen, even the words we use, is an information and communication technology in this broader sense.

However, the term "mediated communication" is also used with a different meaning, illustrated in reference to Figures 2 and 3. In Figure 2, Steuer (1995, p. 38) refers to a traditional view of communication occurring between two people A and B, either proximal or remote, where the communication is directly between them. The representation of mediated communication is shown in the next figure. In this scenario, the participants are always remote from each other; communication does not occur directly between sender and receiver, instead an intermediary technology platform (which Steuer calls a "mediated environment") is created. In this view, the participants then experience this platform, not any direct communication from the other (Figure 3) (Steuer, 1995, p. 37).

### Three Definitions of "Object"

The word "object" is another term that is used differently within different disciplines. Within Activity Theory as described above, the term object-orientated was used. This term (or more commonly the US English contraction "object-oriented") is used differently in different contexts and so this too needs to be defined. In Activity

*Figure 2. Traditional view of communication*

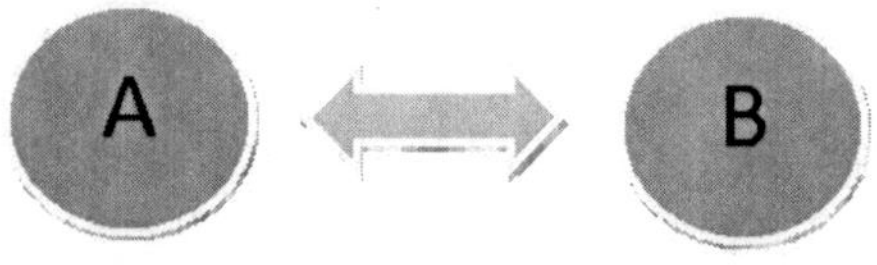


Theory, the activity is "object-orientated" in the sense that the activity is focused on the achievement of some objective.

"Object-oriented" also has a very precise meaning in computer science. Object-oriented analysis and design is a form of programming in which a system is divided into separate self-contained elements, which are then treated discretely. These elements are then organised into a structure with a formal set of relationships between them that determine how the separate elements act on each other (Eden, 2002, p. 380). This has an advantage over the creation of programs consisting of a single sequence of code in that individual objects can be treated individually, and reused in other programs.

Within these chapters the word "learning object" is also commonly used. Within this context, a learning object is a resource that can be used for learning and contains sufficient content to be understood on its own (though may be integrated with other resources). Learning objects should also be able to be reused, repurposed and aggregated into larger objects. Although learning objects are not necessarily digital in nature, in most instances they are, and in which case they are preferably also tagged with metadata so they are more easily discoverable and can be stored in a learning object repository.

## Technological Platforms for Learning

The technological platforms discussed within this book include one or more of the following functions:

They are the means through which communication occurs between learners and between learners and teachers, ("mediating" in Steuer's sense);

*Figure 3. Mediated environments view of communication*

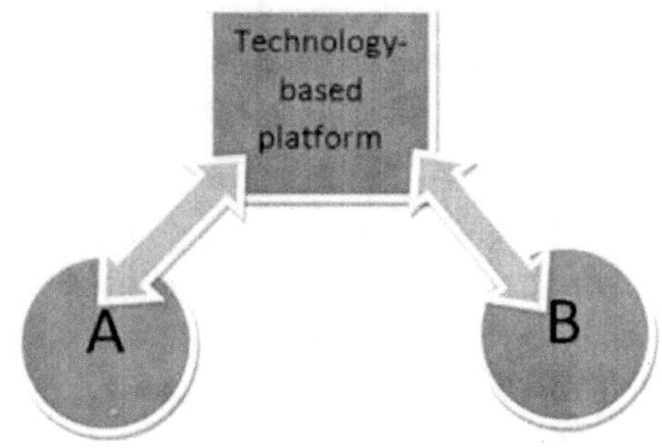


They are the location of the learning objects -They are the location of the tools used by the learners to support their learning. ("mediating" in Vygotsky's sense).

These functions respectively fulfil the following roles in learning and teaching (Kathy Courtney, personal communication):

- learning with and from others;
- learning about content;
- learning using tools.

Many different theories to explain how people learn exist, these are often held to be in competition with each other, and used to develop different teaching approaches. The range of different approaches will be covered here because many different teaching approaches are adopted in the case studies. A useful overview of the groupings of different forms of approaches to teaching is that given by the Learning Activities Reference Model or LARM (Falconer et al., 2006, pp. 16-21). Falconer et al. group these various approaches into associative, cognitive and situative.

## Associative Approaches

These approaches focus on the student acquiring knowledge and skills, the acquisition of these skills is then assessed and successful accumulation of knowledge rewarded with a grade. An example of an associative approach is behaviourism, in which learning outcomes are set which the student then attains (or not) (Falconer et al., 2006, p. 16).

## Cognitive Approaches

Cognitive approaches are those which view learning as a development of the learner's own internal thought processes, and in which students synthesise knowledge gained from the learning activities they undertake with their existing knowledge (Falconer et al., 2006, p. 17). Constructivism, experiential learning, problem-based learning and activity-led learning all adopt slightly differing versions of cognitive approaches but all see learning as a process by which the student acts on information and creates their own meaning.

## Situative Approaches

Situative approaches all facilitate learning through social participation (Falconer et al., 2006, p. 17). An example of a situative approach is social constructivism, in which knowledge is constructed jointly through discussion and collaboration (Bromage, 2004, p. 144). The "communities of practice" model (discussed later) is another situative approach to teaching. Possibly the earliest form of teaching, apprenticeship, is a situative approach to teaching.

Implicit within the design of the various technological platforms are different teaching techniques, and hence different approaches to teaching (and underlying these, different learning theories). Our two main definitions of learning platforms are described below, and the inherent approaches to teaching implied by their design are discussed.

## Virtual Learning Environments

Many institutions will have a single platform with which learners access course content, identify what modules or courses they are enrolled on, and catch up on announcements. These platforms include commercial systems and open source systems such as Moodle. These systems are referred to here as Virtual Learning Environments (or VLEs). VLEs tend to be largely used as repositories for course materials and as a means for passing on to students information about assignments and further reading. VLEs employed in this manner therefore lend themselves to assimilative teaching techniques. More sophisticated learning objects, however, require students to interact with them. These objects support experiential learning (normally associated with a cognitive approach to teaching). VLEs may contain within them potential for students to communicate with each other through text-based content such as wikis, blogs and discussion boards and so can provide elements of cognitive and situative learning approaches.

## Mediated Environments

Whereas Virtual Learning Environments provide information for learners to access and assimilate, mediated environments provide means for learners to communicate with each other. Examples of a mediated environment are videoconferencing (for example Skype) or webconferencing (which contains not only a video link but also a chat window and a shared whiteboard and other tools) or an immersive virtual world such as Second Life or Active Worlds (explained more below). The essential feature of mediated environments are that they enable users to communicate synchronously at a distance through a shared digital space, as opposed to a real physical space, using this spatial metaphor to create a sense of realness to the interaction (Zhao, 2003, p. 445; Towell & Towell, 1997, p. 593). Furthermore, the participants always have some sort of presence within these environments; this presence may be simply the text one writes (as in a chat room), one's voice (in audioconferencing), one's image (in videoconferencing) or a computer-generated version of oneself (an avatar) in immersive virtual worlds. As such, mediated environments lend themselves particularly to situative approaches, since they are particularly successful at enabling communication and collaboration.

## Presence in Mediated Environments

The creation of presence within these environments is a key feature of the means by which they support communication and hence learning:

- "presence" is used in the sense of being somewhere else, which can be either telepresence, when that other space is a real, but distant, location or virtual presence, if that other space is computer-generated.
- "copresence" is defined by Zhao (2003, p. 445) as either the sense of being together with other people in a remote physical environment, or the sense of being together with other people in a computer-generated environment.
- social presence is specifically the ability to project oneself socially and emotionally, a term used, for example, within the Community of Inquiry model (Arbaugh & Hwang, 2006, p. 10).
- A fourth form of presence is that of self-presence, or embodiment, described by Biocca (1997) as "*users' mental model of themselves inside the virtual world*". This is made possible within virtual environments by giving the participants a digital representation of themselves, this digital representation being referred to since the late 1980s as an avatar (Britt, 2008). Other important characteristics of mediated environments are outlined below.

### Identity

An online identity exists within mediated environments due to the capability of these environments to embody participants. However, since they do not necessarily have to reveal their offline identity, this means participants may choose to recreate their offline identity within the virtual environment, or create a new identity online. This online identity may be radically different from an offline identity, participants choosing to experiment with experiencing interaction as a person of different age, race or gender, or even as a member of another species or as an inanimate or abstract object (Cheng, Farnham & Stone, 2002, p. 98).

### Manipulability

Manipulability is the extent to which a participant can change the environment, both in terms of the fineness of the changes and the range of elements that can be changed. Dreyfus (2000, p. 57) states that "What gives us our sense of being in direct touch with reality is that we bring about changes in the world and get perceptual feedback concerning what we have done."

### Reciprocation

Reciprocation is the degree of feedback from the environment about the changes we have made, whether from objects, automated responses or other users within the environment. When this reciprocation is from other users, it can support the sense of social cohesion, without reciprocation we have no real sense that we are having an effect or that anyone or anything is taking note of our actions (Rourke, et al., 1999, p. 56).

### Responsiveness

Responsiveness is distinct from reciprocation, in that it refers to the speed of the feedback, rather than the degree to which it takes place. Within immersive virtual worlds where the technology will introduce delays in responses (called lag) this will act against sense of copresence with other participants (Wheeler, 2007, p. 111).

### Communication

The expressive aspects of vocal communication, such as tone and volume called paralinguistics (Farnell, 1999, p. 351) which are important for emotional exchange (Bailenson, Beall & Turk, 2004, p. 430) can be conveyed through paratextual elements, such as emoticons ☹ and typeface can stand in for these in text-based communication (Becker & Mark, 2002, p. 29). Within virtual worlds, expression can be conveyed through the movement and animation of the avatar, as well as through the use of text.

### Navigation

Navigation through the space is made up of various elements; wayfinding, motion and travel (Darken & Peterson, 2002, p. 494; Bowman, 2002, p. 281) and manoeuvring. Wayfinding is the cognitive element of navigation, an essential part of which "is the development and use of a cognitive map" (Darken and Peterson, 2002, p. 494). Motion is the act of movement through the space which can be either naturalistic (walking, vehicular etc.) or magical (flying, teleportation) (Bowman, 2002, p. 283). Manoeuvring is the smaller set of movements such as "changing the orientation or perspective, as in rotating the body or sidestepping" (Darken and Peterson, 2002, p. 494) and requires the environment to provide information enabling spatial awareness (Bowman, 2002, p. 282).

## The Role of Mediated Environments in Education

The difficulties and barriers that learners face when encountering these environments raises the question of why, then, would educators wish to use them in education. The authors of the following chapters will each raise their own particular rationales for engagement, but in general the value of these environments falls into one of the following categories (Childs, 2007).

## The Capacity to Create and Embed Learning Objects

These can be text, images, videos and audio files. Three-dimensional models can also be placed within any platform using flash or VRML models for example. Virtual worlds (as a specific example of mediated environments) also can contain any of these media objects, but since the entire environment is a three-dimensional environment, three dimensional learning objects are the norm. The scripting language within many virtual worlds enables learners to create their own content, enabling them to become fashion designers or artists through the creation of elements within the world.

## Communication at a Distance

Mediated environments enable classes to be run that bring students together and interact with each other and educators at a distance. As long as the individual has a broadband connection, and a sufficiently high specification computer to run the software (and other required peripherals depending on the type of environment, for example a camera for videoconferencing, or a headset for VoIP) then the students and lecturers can be located anywhere.

## Embodied Presence

The first two capabilities listed above are not, however, unique to mediated environments. Three-dimensional creations, and media objects, can be embedded in simple webpages. Distanced learning can be carried out simply through email, or through post and broadcast media. The essential difference, and the suggested advantage, that all mediated environments afford is the capacity to provide social, co- and embodied presence for those taking part. These forms of presence provide opportunities for participants to feel engaged and connected to others, and to experience the learning objects, to a far greater extent than a purely one-way, asynchronous or disembodied communication medium.

### Roleplay and Identity Tourism

The nature of these online identities is also that they can be modified to take on practically any form the participant wishes. This opens up the potential for the exploration of identity and for roleplay within these environments. Virtual worlds have the capacity to "rethink our fundamental assumptions about what personality is, what identity is, what it is to be a human being" (Mitchell, 2008). This can be conducted as a personal exploration of possible selves, an activity known as "identity tourism" (Taylor, 2002, p. 58) in which an avatar of different sex, or ethnicity or even species can be adopted by the participant.

### Learning in Different Technological Platforms

Identifying which of these various technologies is the most appropriate to use requires linking the affordances of any one technology with the nature of the learning one is attempting to enable. A methodology for matching the affordances of an environment to the most appropriate learning activities has been forwarded by Girvan, FitzGibbon and Savage (2009). This begins with identifying the affordances of an environment, in their case Second Life, which they state provides "a sense of self and presence resulting in immersion, socialising and collaborative learning, construction of objects and environments, flexibility" and persistence. From this they identified communal constructivism as an appropriate pedagogy and designed learning activities aligned with this, finding that the implementation of these was successful. All of the technologies listed above have their place, as do all of the teaching techniques. It is when educators fail to match the correct technology to the appropriate teaching technique for the learning being undertaken that e-learning can appear to be inadequate.

## MODELS OF COMMUNICATION AND LEARNING USING TECHNOLOGICAL PLATFORMS

There are a range of different forms of analysis that have been brought to bear on technological platforms for learning. The following are a selection of the leading ones, the intention being to provide a setting for the analyses conducted in the following chapters.

### Activity Theory

The original early 20th century theory propounded by Vygtosky has already been reviewed. Towards the end of the 20th century activity theory was expanded to include the contextual influences on the core triangle of subject, object and mediating artefact. These additional factors are:

- The rules and conventions governing or informing the activity;
- The community engaged in the activity and in which the activity is located;
- The division of labour between the participants in the activity (Engeström, 1999, p. 31).

The interconnection between the six elements of activity theory are shown in Figure 4.

For example, when applying activity theory to interprofessional education, each part of the process of IPE can be separated into these constituent parts. Interprofessional education is defined as occurring "when two or more professions learn with, from and about each other to improve collaboration and the quality of care" (CAIPE, 2002). Within this context, the subject is the individual learner, the object is to improve collaboration and the quality of care and the out-

*Figure 4. The essential elements of Activity Theory and their inter-relationships (Adapted from Engeström, 1999, p. 31)*

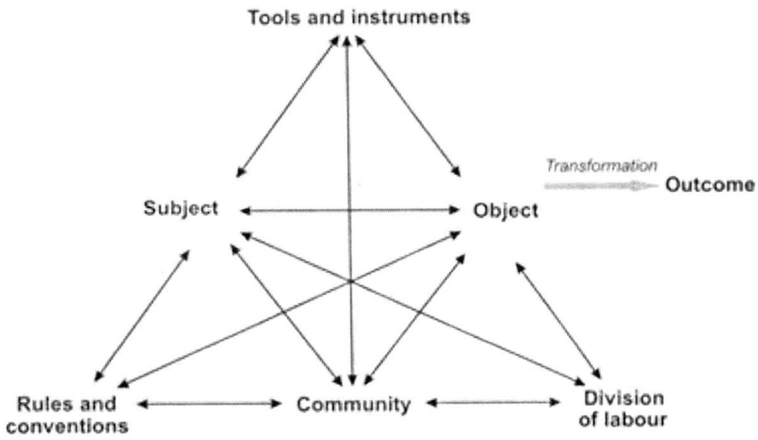


come is the actual improvement in collaboration or the quality of care. The tools and instruments may be a communication platform, or it may be an interprofessional learning object (IPLO) defined by Clarke (2008) as a "digital resource which supports professionals learning with, from and about each other".

The division of labour within this context are the different roles; for example, the role of learner, role of facilitator or shared roles and responsibilities. Cross-role interactions are at the core of much interprofessional communication, however, within a mediated environment, where it is common to create a new online identity and not reveal one's offline identity, the roles may not always be clear. This ambiguity "has the potential to unsettle or de-naturalise aspects of our roles (as teacher, learner or researcher)" (Carr, Oliver & Burn, 2008, p. 89) which educators may either see as a beneficial development (Carr, Oliver & Burn, 2008, p. 92) or hazardous, raising "significant issues of trust" (Robert, 2008, p. 28).

The community here is the community of professionals, perhaps even a community of practice (expanded in the relevant section below). In the specific act of working within the mediated environments, the community is all of the participants who are present within the activities taking place in that environment, although opinions differ as to whether this includes only those participants who are active or also those who only "lurk" (Kehrwald, 2007, p. 133).

The rules and conventions are those of the etiquette within the profession and the shared understanding of professional conduct. Also within the use of mediated environments, rules and conventions also develop that govern the ways in which interactions take place, and include the use of emoticons and abbreviations, etiquette about when to use public chat and when to instant message participants directly, where to position one's avatar in relation to other avatars and so on. Becker and Mark in a study that took in three mediated environments (two immersive virtual worlds and one text-only) (2002, p. 30) found that although the conventions differed in these environment, in all cases the "behaviours result from handling the available technology to navigate and communicate effectively" (Becker and Mark, 2002, p. 33) and thereby enable the greatest degree of copresence between the participants.

There have been some attempts to develop Activity Theory further to provide a structure for the description of the constituent parts of mediated environments. A mediated environments reference model, proposed by the author, adds the elements of presence and identity to the framework, forming a blend of Activity Theory and Communities of Practice (described below).

## Transactional Distance Theory

As with Activity Theory, the theory of transactional distance offers a structure for describing and analysing learning activities. It was first propounded by Michael Moore in 1972 (Moore, 1993, p. 22) and focuses on the psychological and communication distance between teacher and learner, a separation that occurs even in face-to-face situations. The position of transactional distance theory is that distance learning may introduce additional constraints to communication, but this only places it on a different point of a continuum of this transactional separation between participants, and is not essentially different. Many variables can affect this psychological separation; Moore groups these variables into three clusters; those of dialogue, structure and learner autonomy (Moore, 1993, p. 23).

Dialogue is essentially different from simple interaction in that it is "purposeful, constructive and valued by each party (Moore, 1993, p. 24). Achieving effective dialogue is dependent on the personalities and philosophies of the participants, their skill at communication and the subject matter of the course. Also important are environmental factors, of which the communication medium is just one element, although as Moore notes, this is often the main one focused upon in distance education.

The transactional distance that occurs in dialogue can be reduced through the use of effective structures for the learning and teaching. Moore (1993, pp. 28-30) offers a list of the elements of an effective structure as:

- Presentation of information through various media;
- Support of the learner's motivation;
- Stimulation of analysis and criticism;
- Giving of advice and counsel;
- Arranging for practice, application, testing and evaluation;
- Arranging for student creation of knowledge.

The third element, that of learner autonomy, assesses the degree to which the learning activities support the setting of a learner's own goals.

With the set of factors identified, the analysis of various communication media and how these support these different factors can then be tabulated and analysed. Transactional distance theory also places at the centre the various dialogues that a technology can permit and provides a means for assessing the effectiveness of a technology through the dialogues it supports. Transactional distance theory is still in use by many researchers particularly with the analysis of telematic systems (Wheeler, 2007).

## Media Richness Theory

Studies of the communication channels available within mediated environments have suffered from the preconception that successful media are those that mimic face-to-face communication the most closely; the greater the deficiency in modelling face-to-face communication, the less successful the medium (Jakobsson, 2002, p. 68). Those that contain the most information, and the most channels, the greater the "richness" (Dennis et al., 1998, p. 2) Many earlier studies assumed that communication would be for work-related activity, and participants took part in the studies with little previous exposure to the communication tools. An example of the work to come from these studies is the Media Richness Theory, by Daft and Lengel (Jakobsson, 2002, p. 68). However, as Jakobsson (2002, p. 69) points out the "whole point of a virtual world is that it is different". This basic flaw in Media Richness Theory has led to it being superseded by Media Synchronicity Theory.

## Media Synchronicity Theory

Media Synchronicity Theory (MST) (Dennis et al., 1998) examines various modes of communication according to what is most appropriate for different forms of communication in terms of conveyance and convergence. Conveyance is the MST phrase for transmission of information, for example, passing on information about the time and place of a meeting is conveyance. Convergence is when different opinions are exchanged and brought together to form a document or decision, for example, working out the best time for a meeting.

Media Synchronicity Theory further analyses different technologies by the extent to which they provide the following Dennis et al. (1998, p. 3):

- Immediacy of feedback, this is the speed of response to a message (and at its most immediate is a simultaneous, bi-directional communication). The ease with which a communication can be interrupted by the other parties is also contained within this category.
- Concurrency, that is, the number of simultaneous communications that can be conducted at any one time.
- Symbol variety, that is the range of different forms of communication that the medium can carry, this may be only text, or also contain images or video. An immersive virtual world contains possibly the highest form of symbol variety.
- Persistence, i.e. the length of time the information may be retained. A website, for example, will be available for a long time, oral communication does not persist (unless it is recorded).
- Rehearsability, this is the potential to check and re-check a communication before it is distributed. For examples emails can be reviewed (and indeed should be) before being sent. Instant messages also can be rehearsed before being sent (though with less opportunity for this). Within a dialogue it is not usually possible to plan out in advance what is to be communicated.

This approach gives a far more nuanced analysis than media richness theory, in that different media are more effective for different goals, no one medium is the "richest" in itself, not even face-to-face communication (Dennis et al., 1998, p. 4).

## Social Exchange Theory

Social exchange theory imports an economic model of exchange to make sense of the communication and knowledge sharing that occurs within a group of learners. The central idea of social exchange theory is that "individuals evaluate alternative courses of action so that they get the best value at the lowest cost from any transaction completed" (Hall, 2003, p. 288).

Thus within a computer-mediated communication platform, such as a discussion board, learners will communicate in increasing frequently with those peers who are perceived as possessing useful knowledge or who respond productively, and in decreasing frequency with those peers when those communications are unproductive. Social exchange theory describes the activity in a communication network in terms of:

- **Exchange actors:** participants within the group;
- **Exchange network:** sets of dyadic exchange relationships between actors;
- **Exchange resources:** can be money, but could be information, time spent supporting peers, etc.;
- **Exchange structures:** other relationships that support the exchange (for example, being members of the same learning set);
- **Exchange processes:** the actual tasks involved in making an exchange (uploading a post, making a comment on another's post etc) (Hall, 2003, p. 292).

This provides a framework for observing and analysing the communications that occur within a discussion board, for example, a set of blogs, or any platform which enables exchange between participants.

Social exchange theory also provides a process for encouraging communication within an online environment. As Hall (2003, p. 299) states "*To entice people to share their knowledge, in terms of a social exchange transaction, these actors need to be persuaded it is worth entering into a transaction in exchange for some kind of resource*". Viewing knowledge as a resource to be shared within an exchange network may provide additional methods for educators to encourage participation.

## Community of Inquiry

The Community of Inquiry model explores closely the concept of presence, and its importance in promoting critical thinking in online interactions (Rourke et al., 2004, p. 51). The model looks at three forms of presence:

- Social presence, the ability of learners to project themselves socially and emotionally, thereby representing themselves as "real people";
- Cognitive presence, the extent to which learners are able to construct and confirm meaning through sustained reflection and discourse; and;
- Teaching presence, the design, facilitation, and direction of cognitive social processes for the purpose of realizing personally meaningful and educationally worthwhile learning outcomes. (Arbaugh and Hwang, 2006, p. 10).

The imitation of the model from the perspective of a researcher investigating mediated environments is that it has only been developed in reference to asynchronous text-based interactions. This has led to it being very effective at incorporating and measuring those elements that can produce social presence, such as humour or self-disclosure (Rourke et al., 2004, p. 56) but less effective at distinguishing between those factors that lead to a sense of copresence as opposed to social presence; indeed it blurs the distinction between these two, counting both as forms of "immediacy" (Rourke et al., 2004, p. 52). The focus on teacher presence pre-supposes that online forms of learning are instructor-centred (Rourke et al., 2004, p. 54) and identifying objectively the occurrence of cognitive presence can be problematic. It may be possible, however to extend and update the model to include synchronous interactions.

## Communities of Practice

The communities of practice model is a framework for analysing and understanding learning by focusing on the relationship between the learner and their involvement within the group of other learners (Wenger, 1999, p. 4). The framework is made up of the following components (Wenger, 1999, p. 5):

- **Meaning (learning as experience):** How we as individuals and as a group describe our learning and create a narrative (either singly or shared) to make sense of it.
- **Practice (learning as doing):** The shared communications concerning the resources, actions and content of the joint activities.
- **Community (learning as belonging):** Communications about the general social network that comprises the community of practice and our place within it.
- **Identity (learning as becoming):** Discourse concerning our development as a consequence of our learning and our role within the community.

From an analytical point of view, the concept of practice is generic enough to be able to be applied to a range of cases, working within an organisation, functioning within a family, performing together in a play, and Wenger emphasises that learning occurs in all of these situations (Wenger, 1999; pp. 6-8). Within the field of online interaction, this has meant it has found applicability in not only looking at learners' interactions within a formal subject-orientated setting, but also to analyse and categorise the social, informal learning that occurs outside of the formal settings. Oliver and Carr (2009), for example, have applied the framework to describe the process by which gamers learn to be members of the community of World of Warcraft.

Central to the communities of practice model is the concept of reification, defined as "the process of giving form to our experience by producing objects that congeal this experience into 'thingness'" (Wenger, 1999, p. 58). Reification may produce terminology, tools, procedures, systems, etc., for example this chapter is an attempt at reification. Again Wenger uses the concept very generically. These reified experiences then inform the community identity, and the practice, of the community of practice.

Although, highly important for forming a community of practice, these acts of reification also form a boundary between participants of a community and non-participants. They are thus boundary objects. Other boundary objects can be skills and knowledge, personal narratives, shared experiences etc. In order to connect to a community those outside need to acquire understanding of these boundary objects, and can be helped to do so through a process of brokering i.e. by being introduced to these through a member of the community (Wenger, 1999, p. 105). A person at this boundary will not experience full participation, due to the length of time it can take to adopt these boundary objects, the communities of practice model refers to this stage as legitimate peripheral participation, one is on the edge, but that status has legitimacy within the community (Wenger, 1999, p. 100).

However, for the newcomer, distinguishing between peripherality (the first stage towards full participation) and marginality (exclusion from participation) is not always easy to do (Wenger, 1999, p. 164). For example, when acquiring the skills to function within an online learning environment such as an immersive virtual world, those learners with experience of other environments (for example MMORPGs) may not find those skills directly applicable. However, from their experience of other environments they can distinguish between peripherality and marginality, i.e. they know the inability to move their avatar, or know where to go, is simply part of the normal learning curve, whereas other students may be permanently dissuaded from further participation by the difficulties. Wenger also identifies a range of trajectories of participants through communities:

- **Peripheral trajectories:** The participant never becomes fully participative.
- **Inbound trajectories:** Newcomers become more and more full members of the community.
- **Insider trajectories:** Even when a full participant, the community continues to change, so one's status is constantly being renegotiated.
- **Boundary trajectories:** Some participants stay at the edge of a community and fulfil the role of brokering between various communities of practice.
- **Outbound trajectories:** Leaving a community also requires a negotiation of identity and relationships with respect to that community (Wenger, 1999, pp. 154-155).

Also, of course the community itself is changing, a process Wenger describes as a paradigmatic trajectory. The members renegotiate and reify meaning within the community.

## Interprofessionalism and E-Learning

The models presented above are alternative ways to deconstruct and analyse the content and processes of communication and learning in technological platforms. These models have different strengths, and some have particular weaknesses. From the perspective of interprofessional learning, the elements that have particular relevance are how these different platforms support social engagement, community and the development of professional identity.

Interprofessional Learning (IPL) is defined as "*a process in which different professionals learn from each other through interaction to develop collaborative practice*" (Howkins and Bray, 2008, p. XViii). These models can provide an understanding how interaction occurs, and how psychological distance can impede that interaction, or social exchange can encourage it. Collaborative practice requires a sense of the separate professionals sharing that collaboration, a cornerstone of fostering successful collaboration is an understanding how social presence and effective dialogue can be created. Developing collaborative practice also goes hand-in-hand with developing communities of practice, and this too can be understood through the descriptions provided by the theories here.

## CONCLUSION

Far from being abstract or solely academic concepts, all of the models presented here have practical value in providing inspiration and advice for the enhancement of effective collaboration at a distance, supported by communication, community and personal (or professional) development, examples of which are contained in the remainder of this book.

## ACKNOWLEDGMENT

The author would like to thank Adrian Bromage and Elinor Clarke of the Centre for Interprofessional e-learning (CIPeL) for their help in developing this chapter.

## REFERENCES

Arbaugh, J. B., & Hwang, A. (2006). Does "teaching presence" exist in online MBA courses? *The Internet and Higher Education*, *9*(1), 9–21. doi:10.1016/j.iheduc.2005.12.001

Becker, B., & Mark, G. (2002). Social Conventions in Computer-mediated Communication: A Comparison of Three Online Shared Virtual Environments. In Schroeder, R. (Ed.), *The Social Life of Avatars* (pp. 19–39). London: Springer-Verlag.

Biocca, F. (1997). The Cyborg's Dilemma: Progressive Embodiment in Virtual Environments. *Journal of Computer-Mediated Communication*, *3*(2).

Bowman, D. A. (2002). Principles for the Design of Performance-oriented Interaction Techniques. In Stanney, K. M. (Ed.), *Handbook of Virtual Environments; Design Implementation and Applications* (pp. 277–300). Mahwah, NJ: Lawrence Erlbaum Associates.

Britt, A. (2008). On Language: Avatar. *New York Times Magazine*, August 8th, 2008. Retrieved August 30, 2008, from http://www.nytimes.com/2008/08/10/magazine/10wwln-guest-t.html

Bromage, A. (2004). Atavistic Avatars: Ontology, Education and 'Virtual Worlds. In Seth-Preston, D. (Ed.), *Virtual Learning and Higher Education* (pp. 133–150). Amsterdam: Rodopi.

CAIPE. (2002). *Defining IPE*. Retrieved July 26, 2009, from http://www.caipe.org.uk/about-us/defining-ipe/

Carr, D., Oliver, M., & Burn, A. (2008). Learning, Teaching and Ambiguity in Virtual Worlds. In *Proceedings of the ReLIVE 08 conference*, November 20th and 21st, 2008, Open University, (pp. 83-93).

Center for Activity Theory and Developmental Work Research. (2004). *Cultural-Historical Activity Theory*. Retrieved February 6, 2009, from http://www.edu.helsinki.fi/activity/pages/chatanddwr/chat/

Cheng, L., Farnham, S., & Stone, L. (2002). Lessons Learned: Building and Deploying Shared Virtual Environments. In Schroeder, R. (Ed.), *The Social Life of Avatars* (pp. 90–112). London: Springer-Verlag.

Childs, M. (2007). Real Learning in Virtual Worlds. *Warwick Interactions Journal, 30*(2).

Clarke, E. (2008). *Use and reuse of Inter-professional Learning objects in Second Life*. Paper presented at the 3rd Plymouth e-Learning Conference, Plymouth, 4th April, 2008.

Darken, R. P., & Peterson, B. (2002). Spatial Orientation, Wayfinding and Representation. In Stanney, K. M. (Ed.), *Handbook of Virtual Environments; Design Implementation and Applications* (pp. 379–402). Mahwah, NJ: Lawrence Erlbaum Associates.

Dennis, A. R., Valacich, J. S., Speier, C., & Morris, M. G. (1998). *Beyond Media Richness: An Empirical Test of Media Synchronicity Theory*. Paper presented at the Thirty-First Annual Hawaii International Conference on System Sciences, Volume 1. Retrieved April 16, 2009, from csdl2.computer.org/comp/proceedings/hicss/1998/8233/01/82330048.pdf

Dreyfus, H. (2000). Telepistemology: Descartes last stand. In Goldberg, K. (Ed.), *The Robot in the Garden; Telerobotics and Telepistemology in the Age of the Internet* (pp. 48–63). Cambridge, MA: MIT Press.

Eden, A. H. (2002). A Theory of Object-Oriented Design. *Information Systems Frontiers, 4*(4), 379–391. doi:10.1023/A:1020835709566

Engeström, Y. (1999). Activity Theory and individual and social transformation. In Engeström, Y., Miettinen, R., & Punamäki, R. L. (Eds.), *Perspectives on Activity Theory* (pp. 19–38). Cambridge, UK: Cambridge University Press.

Falconer, I., Conole, G., Jeffrey, A., & Douglas, P. (2006). *Pedagogical Learning Activity Reference Guide*. Retrieved February 5, 2009, from http://athena.cs.man.ac.uk/apache2-default/r/LADIE/www.elframework.org/refmodels/ladie/guides/LARM_Pedagogy30-03-06.doc

Girvan, C., FitzGibbon, A., & Savage, T. (2009). *Identification and analysis of a pedagogy in action within Multi-User Virtual Environments*. Paper presented at the Virtual Worlds Best Practice in Education Conference, March 27-29, 2009, Second Life.

Hall, H. (2003). Borrowed theory: Applying exchange theories in information science research. *Library & Information Science Research, 25*, 287–306. doi:10.1016/S0740-8188(03)00031-8

Howkins, E., & Bray, J. (Eds.). (2008). *Preparing for Interprofessional Teaching: Theory and practice*. Oxford, UK: Radcliffe.

Jakobsson, M. (2002). Rest in Peace, Bill the Bot: Death and Life in Virtual Worlds. In Schroeder, R. (Ed.), *The Social Life of Avatars* (pp. 63–76). London: Springer-Verlag.

Kehrwald, B. (2007). *Social Presence and Learner Support: Understanding learners' experiences with mediated social processes in text-based online learning environments*. Unpublished doctoral thesis, University of Southern Queensland.

Littleton, K., Toates, F., & Braisby, N. (2007). Three Approaches to Learning. In Miell, D., Phoenix, A., & Thomas, K. (Eds.), *DSE212 Mapping Psychology Book 1. Milton Keynes*. UK: The Open University.

Masterman, E. (2008). Activity Theory and the Design of Pedagogic Planning Tools. In L. Lockyer, S. Bennett, S. Agostinho and B. Harper (Eds.), Handbook of Research on Learning Design and Learning Objects: Issues, Applications and Technologies (2 Vol., pp. 209-227). Hershey, PA: Information Science Reference.

Mitchell, B. (2008). Chaired debate, Anna Peachey, Roo Reynolds, Edward Castronova, Claudia L'Amoreaux, Bill Thompson, Ren Reynolds. *Re-LIVE08*, November 21st, 2008, Open University, Retrieved August 10, 2009, from http://stadium.open.ac.uk/stadia/preview.php?whichevent=1248&s=31&schedule=1468

Moore, M. G. (1993). Theory of Transactional Distance. In Keegan, D. (Ed.), *Theoretical Principles of Distance Education* (pp. 22–38). London: Routledge.

Oliver, M. & Carr, D. (2009). Learning in Virtual Worlds: Using Communities of Practice to Explain How People Learn From Play. *British Journal and Education and Technology, Special issue, 'Out of This World: 3-D MUVEs in Higher Education', 40*(3), April.

Roberts, G. (2008) Enhancing Second Life as an Educational Tool: Resolving Tensions Between Anonymity and Trust, poster presentation, Re-LIVE 08 conference, November, 20th and 21st 2008, Open University, http://www.open.ac.uk/relive08/documents/Detailed-final-programme.pdf

Rourke, L., Anderson, T., Garrison, D. R., & Archer, W. (1999). Assessing social presence in asynchronous text-based computer conferencing. *Journal of Distance Education, 14*(2), 50–71.

Steuer, J. (1995). Defining virtual reality: Dimensions determining telepresence. In Biocca, F., & Levy, M. R. (Eds.), *Communication in the Age of Virtual Reality* (pp. 33–56). Mahwah, NJ: Lawrence Erlbaum Associates.

Taylor, T. L. (2002). Living Digitally: Embodiment in Virtual Worlds. In Schroeder, R. (Ed.), *The Social Life of Avatars* (pp. 40–62). London: Springer-Verlag.

Towell, J., & Towell, E. (1997). Presence in Text-Based Networked Virtual Environments or "MUDS". *Presence (Cambridge, Mass.), 6*(5), 590–595.

Wenger, E. (1999). *Communities of Practice; Learning, Meaning and Identity*. Cambridge, MA: Cambridge University Press.

Wheeler, S. (2007). The influence of communication technologies and approaches to study on transactional distance in blended learning. *ALT-J, 15*(2), 103–117. doi:10.1080/09687760701470924

Zhao, S. (2003). Toward a taxonomy of copresence. *Presence (Cambridge, Mass.), 12*(5), 445–455. doi:10.1162/105474603322761261

# Section 2
# Curriculum Design

# Chapter 6
# Critical Diversity Education to Promote Interprofessional Understanding:
## A Comparison of Student Experiences Between Face-to-Face and Online Delivery

**Marion Brown**
*Dalhousie University, Canada*

## ABSTRACT

*The chapter begins with an overview of the current momentum toward interprofessional education and practice, citing specific trends in Canada as reflections of a global emerging consciousness. Initiatives undertaken at Dalhousie University are discussed in setting the context for this pilot study. Next, the pedagogy of critical diversity education is introduced and explained, with particular relevance for interprofessional education and practice. Comparison of face-to-face and online delivery of an interprofessional module based upon critical diversity education principles is then detailed, including research design and findings. The chapter concludes with a discussion of implications from this study.*

## INTRODUCTION

This chapter reports on results from a pilot study which explored student experiences of an introductory level interprofessional (IP) learning module on critical diversity, wherein module content and facilitation of IP group reflection were delivered both on campus and via a virtual learning environment (VLE). Data collected regarding student experience occurred through focus groups, both onsite and online, following participation in the module.

DOI: 10.4018/978-1-61520-889-0.ch006

### Interprofessional Education and Practice

There is growing evidence that IP education and practice are required to ensure the optimal health of citizens the world over (Kohn, Corrigan & Donaldson, 2005). The literature both supports the successes of IP collaboration and decries the actual and anticipated losses that result from inadequate communication, competing interests, partializing of patient need and territorial control over resources. Consequently, national health departments across the globe are committed to working with educational institutions to prioritize pre-licensure education as the ideal point of entry for learning "with, from and about" each other across the health care professions (CAIPE, 2002).

In 2003, Canadian federal and provincial health departments highlighted collaborative patient centred care as one of the cornerstones for optimal health care outcomes. Initiatives in Health Authorities across Canada have continued to prioritize opportunities for IP education to prepare for this collaboration once in professional practice. At Dalhousie University, the Faculties of Health Professions, Dentistry and Medicine formed the Tri-Faculty Interprofessional Academic Advising Committee (Tri-IPAAC) in 1997, with the responsibility to implement IP learning modules for students. For 12 years, over 3900 students annually, across 22 health professions, attended IP learning modules focussed on contemporary health care issues. While this programming served an important purpose, to date there has been no uniform means of providing IP experiences to students via an online forum. Given that one half of the professional schools within the Faculty of Health Professions at Dalhousie offer partial or whole degrees online, this gap was a key prompt in the development of the research project discussed below. The module '*Diversity Awareness within Interprofessional Teams*' (Baikie & Campbell, 2008b) became the first IP module to be delivered online.

## The Rise of Online Learning

The use of electronic media in Canadian health care education has been in place for twenty years, driven by the need to develop the capacity of health care providers in rural and remote areas (Roberts, O'Sullivan & Howard, 2005). In addition, patterns of out-migration have long been a threat to rural communities, with former residents, including student graduates, not returning to provide a range of helping services back at home. The required outcome – more educated health care professionals – forced greater creativity in delivery methods, given that onsite education housed the resources and relied on traditional pedagogical processes. Early electronic developments featured the use of teleconferencing, audio and videotapes, interactive television and video teleconferencing, often in clustered sites. But the Internet advanced distance learning technologies in unprecedented fashion, with unique opportunity, incredible speed and enormous capacity for the exchange of academic material (Allen, Bourhis, Burrell & Mabry, 2002).

Although Internet based technologies have significant contemporary appeal among students and many faculty, its welcome has not been unanimous. For some professions, such as social work, there is skepticism that online methods can meet the pedagogical demands, citing in particular the need for face to face instruction and evaluation of the person's attributes and attitudes alongside professional skills. On the other hand, online education meets the need for a broader geographic reach than available solely through onsite instruction and, importantly, can address educational access inequities that exist as a result of student experiences with disabilities, logistic and economic constraints, accommodation, and tuition fee strains, and the difficult balance of personal needs, study and work (Pigiapoco & Bogliolo, 2005). Moreover, as the results of the current research suggest, there are unique pedagogical benefits available through online methods.

Empirical evidence of the efficacy of online teaching and learning is growing. Chumley-Jones, Dobbie & Alford (2002) reviewed 206 articles detailing research on web-based learning in dentistry, medicine and nursing, with results suggesting that the outcomes are comparable to onsite educational methods. Another meta-analysis of 450 distance education articles concluded that replacement of traditional education with distance technology has little effect on student satisfaction regarding the quality of the educational process (Allen et al., 2002).

The need to develop online means for the IP learning modules delivered through the Faculty of Health Professions at Dalhousie was informed by the increasing challenges to coordinating and scheduling student participation from the large number of independent programs within the above-named three Faculties. Many programs include off-site practicum places (for example occupational therapy) while others deliver entire degree programs online (for example social work). A web-based platform through which to deliver IP learning modules would provide the means through which to make the material available to learners broadly and efficiently, supplementing and potentially replacing some existing face-to-face modules. A VLE supports the delivery of text, multi-media methods and surveys, as well as asynchronous (time and place independent) and synchronous (live "chat") communication. The nature of the module is described in the following section.

## Critical Diversity Education in Interprofessional Learning

In recent years, the concept of diversity has gained widespread attention among the health care professions, and it will be seen that it can play a significant role in IP education. Literature on the adaptation of diversity related curricula to an online environment is preliminary, however several studies report that students are less likely to be inhibited in their interactions than within the constraints of a typical classroom environment. Analysis suggests students are able to engage more openly with the often difficult and sensitive nature of the subject matter (Limburg & Clark, 2006; Littlefield & Roberson, 2005; Merryfield, 2001, 2003).

With Canada known for its multicultural mosaic rather than 'melting pot', many educational institutions and student-led initiatives from primary schools through to universities have adopted the slogan to 'embrace diversity', a movement that does indeed sound promising. Yet, at the School of Social Work at Dalhousie University, there has been a presiding concern that popular conceptualizations of diversity reinforce stereotypes through essentialist treatments of deeply meaningful cultures and customs, through performances of 'fun, food and frolic' or 'dress, diet and dance'. The curiosity and fascination aimed toward 'diverse peoples' has often led to experiences of exploitation and tokenism, a voyeuristic foray into embracing diversity. In so doing, faculty and students have often continued to marginalize and discriminate on the bases of race and ethnicity, class, sexual orientation, ability and gender, leaving the personally and professionally held assumptions that underlie these oppressive practices unexamined and well intact. As Canadian anti-racist educators Lopes & Thomas (2006) note, "Diversity, as a concept, does not provide a framework to examine power" (p. 266).

In contrast, the principles of critical diversity education provide a framework through which to examine assumptions, beliefs and values relative to oppressive practices of power and privilege. A critical diversity orientation demands disruption to one's ways of thinking, learning, being and knowing, because it questions how we have come to think, learn, be and know as we have. Figure 1 summarizes the differences between a critical diversity orientation and a non-critical diversity orientation.

Critical diversity education can contribute significantly to identity development within one's chosen profession, because it moves beyond the dominant discourse of professionalism, which typically prioritizes expert knowledge, objectivity

*Figure 1. Comparison of approaches to diversity (Adapted from Kumas-Tan, 2005).*

| **Non-critical Diversity Orientation:** | **Critical Diversity Orientation:** |
|---|---|
| -understands difference to be about individual attributes & characteristics. | -understands difference to be about social relationships, social structures, & processes of marginalization. |
| -focuses on the characteristics & attributes of others. | -focuses on our own values and beliefs. |
| -sees difference as neutral. | -sees difference as power laden & exercised in social relationships. |
| -understands oppression to be the result of individual acts. | -understands oppression as systemic & hegemonic. |
| -is outcome based. | -is process based: how we do it is as important as what we do. |

and outcome measurement. At Dalhousie University, one of the contributions of the School of Social Work to IP education has been the development of curriculum materials that explore the idea that how we define ourselves professionally hinges upon how we create meaning about ourselves, from within our unique experiences of race, class, gender, sexual orientation and ability, in particular. This ideological orientation challenges mainstream conceptualizations both in terms of diversity education and IP education, because it focuses on the 'self' rather than the 'other' in analyzing who we become as professionals and how we relate to members of our own profession, other professionals and service users.

The IP module *Diversity Awareness within Interprofessional Teams* was conceptualized, developed and implemented in accordance with critical diversity principles, by Gail Baikie and Carolyn Campbell (2008b), two faculty members at the School of Social Work at Dalhousie University, on behalf of Tri-IPAAC. It was delivered in two different modes, online and face-to-face on campus. Significant delivery distinctions necessarily existed: in the on campus delivery method, the module was delivered in a large room with approximately 300 students divided into groups of 10, at large tables. Students rotated the functions of timekeeper, note taker and chair. Agendas, module and case study materials were provided at each table and a primary facilitator led the delivery of the content on critical diversity through 15 minute lecture segments. Following each lecture segment, the small groups completed activities related to the lecture material. One roaming facilitator per 10 tables was available for clarifications or suggestions.

For the online delivery method, a small group of seven students took part in the module through the VLE over a two week period. Students did not take on the roles of timekeeper, note taker and chair, as these functions were not required in the online environment. The module agenda, module material and the case study were posted in the VLE. An online facilitator monitored the site to clarify procedures or material, as well as help move discussion along or remain focused. The components of the module include the following. Those of conceptual substance are discussed below in turn:

- Pre-module: Completion of an online 'Diversity Self Awareness' survey & reading of case study to be used in the module.
- Welcome, introductions and session overview (self-explanatory & logistical in nature).
- Defining our terms.
- Who we are as individuals and as a collective.
- Identifying assumptions from different perspectives.
- Exploring the significance of differing assumptions.
- Conclusion & completion of feedback sheets from students.
- Post-module: Completion of post-module reflection questions.

## Completion of an Online 'Diversity Self Awareness' Survey

The module begins with an online Diversity Self Awareness survey (Baikie & Campbell, 2008b), which was designed according to specific pedagogical principles of experiential education. Namely, the survey was designed to elicit the students' experiences regarding having to categorize oneself according to the Euro-western dominant hierarchy of identity categorizations. For example, the choice for sex was 'male' or 'other'; the choice for race was 'white' or 'other'; the choice for sexual orientation was 'heterosexual' or 'other'. The purpose of the survey was two-fold: first, to generate data on the demographics of the student group, and second, for participants to have the experience, in stark terms, of characterizing themselves in terms that replicate the Euro-western hierarchy of worth.

The survey design reflects a particular pedagogical orientation to diversity education, because, unlike the 'embrace diversity' approach of popular appeal, critical diversity education seeks to centre one's own assumptions, biases, values and identities as a means through which to learn about how one judges, categorizes, and possible marginalizes and condemns other people. Simply speaking, critical diversity education begins with looking inward and learning about one's inner conceptualizations and assumptions, well before looking outward, to assess others. Where so much of the popular notion of diversity is focused on learning about 'those people' and how 'they do their things' or how 'they' dress, worship and behave, critical diversity education is focused on learning about ourselves and how it is that we come to see and treat people in the ways we do.

## Defining our Terms

Drawing from the work of Comerford (2003), Baikie & Campbell (2008a) define the concept of critical diversity as a composite of the following principles:

- Commonalities and differences exist across individual and group identities and worldviews. Each dimension of difference exists along a continuum.
- Dimensions of difference are accorded ranked position within a hierarchy, with some identities and characteristics being granted higher status than others. Higher status on the hierarchy generally means that those dimensions are considered preferable, better, or worthy; lower status on the hierarchy generally means those dimensions are devalued, disadvantaged and considered 'less than'. While what constitutes commonalities and differences is understood as socially constructed, the personal and societal effects are experienced as real.
- Placement on this hierarchy is socially and historically constituted, meaning that

across time, place and culture, ordering at the higher or lower end can change and thus the relative valuing or marginalization of certain dimensions changes.

- A range of commonalities and differences is considered a strength within a group of people, a team, or an organization. It creates opportunity for innovation and possibility.
- Individuals differ in assumptions about, biases toward, exposure to and skills in responding to diverse identities and worldviews.

## Who We are as Individuals and as a Collective

During this segment of the module the results from the online Diversity Self-Awareness survey are presented to the students, to provide context for the discussion and once again encourage the surfacing of assumptions on hearing the statistics for each categorization.

## Identifying Assumptions from Different Perspectives

This portion begins with the students viewing a short video within which a service user is seeking health care services. The reaction of the attending health care provider is presented alongside the differing interpretations of her three colleagues, one of which is defined according to a critical diversity perspective. Students then discuss the three perspectives in terms of each one's positions and beliefs about difference, social relationships, personal identity, focus for change and appropriate action, as a means through which to identify and express the thoughts, values and beliefs they themselves hold. Small group discussion is facilitated so that each of the students has the opportunity to voice their own positions relative to the perspectives shared.

## Exploring the Significance of Differing Assumptions

In this section, the students review the case study that was available to them prior to the module and work through a series of questions. Purposefully, the focus of small group discussion is not to solve the problems raised or to make decisions, but rather to identify and discuss the assumptions of each of the IP team members in the case scenario while comparing them to the positions of the health care providers from the earlier video. Further, they are asked to consider how these assumptions may have originated and developed, and how these assumptions affect the IP team members' abilities to access resources in order to provide optimal care for the service user. Finally, students are asked to consider how the IP team members might address the issues in the case study from a critical diversity perspective, and what skills might be necessary to develop in order to have the team operate from a critical diversity perspective.

## Conclusion and Completion of Student feedback sheets

Brief concluding remarks bring the module to a close. Students are asked to complete the student feedback sheets immediately.

## Completion of Post Module Reflection Questions

The principle behind the post module reflection questions is a pedagogical one. It assumes that having some distance from an educational experience can lead to deeper or different reflection than that which is offered immediately.

The above outline constitutes the content of the IP module Diversity Awareness within Interprofessional Teams (Baikie & Campbell, 2008b), delivered in two different modes, online and face-to-face on campus. The findings of the comparison between the two delivery modes are detailed below.

## THE STUDY: COMPARISON OF ONSITE AND ONLINE IP MODULE ON CRITICAL DIVERSITY

### Research Design & Participant Sample

In this comparison of students' experiences between the on-campus and online delivery of an introductory IP module on diversity awareness, qualitative methodology was used. The aim was to gain understanding of student experiences, rather than an average response across all of the students in the modules, and for this reason a case study was conducted in which qualitative data was collected using focus groups and individual interviews, to gain an in-depth understanding of participants' experiences and interpretations with the module. The participant sample was one of convenience, with the sole criteria being attendance at either the onsite or online diversity awareness module.

The on-campus sample consisted of eight students: three were from medicine, two from nursing and three from social work. An on-campus focus group was held with this sample. The online sample consisted of seven students: two from health sciences administration, one from dentistry, two from social work and two from kinesiology. An online focus group was held for this sample, with questions listed for asynchronous response. The content of the module was identical across delivery methods, with the necessary formatting specifications detailed above. Data were inductively analysed using a method of constant comparison, with a focus on generating theoretical principles based on the lived experiences and perceptions of participants.

### Findings

The findings of the research are grouped by theme under the following: module structure and delivery; module content; and student interaction. These themes emerged from the analysis of the empirical data and were not pre-existing in the literature.

### Module Structure and Delivery

Onsite students discussed both positive and negative experiences with attending the session on the Dalhousie University campus. Positive experiences included having face to face contact with students in other professions, including meeting, talking and informal mingling; directly hearing other peoples' opinions; that it was well organized and ran on time; and having a common experience to discuss. Sometimes it was a surprise, this positive experience. One student said:

*It went much better than I had thought, due to my group. After all I'd heard about these things, it was okay.*

Another student said:

*The module facilitated an environment of interaction among those belonging to other professions. Plus I met some really great students from other professions. Our group had interesting discussions, mostly because we were open to moving from the exact question asked and opening up different topics.*

Other feedback indicated that students felt the module organisation could be improved. Reasons included poor timing relative to course examinations; hunger combined with frustration that refreshments were not offered; and that the session was too long, with too much lecture. Interestingly, feedback included both that there was not enough time for group discussion and that there was too much time for group discussion. Below are some comments that reflect the range of experiences among the students:

*Both IP [modules] I've attended seem to be too rushed and I leave thinking that I didn't take as much away from the experience as I feel I should have. The time constraints really take away from this experience which I know is something that is very difficult to do anything about.*

*The lecture dragged, the small group time went on too long and I resent being made to go right before an important exam.*

In contrast, the online students expressed satisfaction with the module structure and delivery available in the VLE, in particular the ability to access the VLE at times convenient to themselves, and the asynchronous nature of the online bulletin boards:

*I liked having time to construct responses - it allowed me to be more reflective and thoughtful in my postings.*

*Being online was very helpful because I was able to progress at my own pace. Instead of having to do everything in a specific time slot, I was able to think about the videos that were posted for as long as I wanted while doing other things before formulating my answers. This allowed me to process the content more effectively.... The effects of the asynchronous interaction were that we didn't feel forced into discussion, but it was allowed to happen naturally when people had things to say.*

## Module Content

While the focus group questions centred on the delivery *method* of the module, data was generated regarding the *content* of the module. Interestingly, this feedback was contributed only by the on-campus students. One concern raised was that the title for the module was misleading, and that students came to the module thinking they would learn about how to work in interprofessional teams with different ethnic cultures represented in them. Students reportedly had sought specific details about "other cultures" that would assist them in the health care professions with those "other cultures". There was great frustration expressed that the module "only" asked people to look into themselves and explore assumptions. Students typically reported:

*I think it would have been more effective to have had a seminar on cultural diversity so that we could have some solid information instead of just being "aware".*

*I did not learn anything about actual people's experiences with diversity. I think it would be more productive to have real people with their own experience to present to the group.*

In contrast, no concerns were raised regarding the content of the module by the students who participated online. Rather, typical comments included:

*It seems that society judges individuals based on their differences. Instead of embracing individual diversity, society tends to label and judge people that are not the norm and act like they deserve this treatment because they are different.*

*I learned in the Diversity Module is that some assumptions in certain situations are considered normal, where if they were only slightly adjusted in context, they would seem outrageous to anyone in our society.*

## Student Interaction

On-campus students expressed a range of opinions regarding the experience of student interaction. Several students expressed frustration with the lack of participants' attention to the tasks of the module, with interruptions and general disinterest evident. These were students who wanted to attend to the activities and felt others were disrespectful of efforts toward this end. Below is one representative comment:

*The discussions were unfocussed, and many of the participants did not take this IP [module] seriously. This is very problematic because these are issues that will be faced on a daily basis; I feel that this was not very beneficial and the issue of diversity was addressed in a very superficial manner.*

Other on-campus students experienced lively discussion with some if not all of their small group members. Face to face contact was particularly appreciated among the on-campus students, who frequently commented that it provides non-verbal cues and a freer flow of conversation. There was a range of feedback regarding the size of the discussion groups (approximately 10 students per table, which some students thought of as just right, while others thought too large) and meeting venue (similarly, just right and too large). Those with concern for size noted that this contributed to an overall setting that is not conducive to expressing personal ideas and feelings about diversity.

All of the online students had previously participated in the onsite IP learning modules, and consequently, some of their comments directly compare their experiences between the two delivery methods. Feedback regarding student interaction included:

*It was great discussing things via a web-based platform. It did not feel different from discussing it in person, just more efficient.*

*The online module ... has its benefits over the 2 hour module we do .... For me personally, I felt I could be much more honest over [the VLE] than I could be sitting in front of people. ... Discussing sensitive topics in this type of forum (where there is still a strong element of accountability) was great, because I was less afraid to say something that may be construed as controversial. Since I've done this module in BOTH formats, I'd say, to be honest, the discussion was more pointed, succinct, and focused than it was in the group setting in the [onsite venue].*

*I felt more open to write what first came into my mind, rather than sometimes altering it for the rest of the group.*

However, this positive review was not uniform. One person wrote:

*In terms of process, I found the online format challenging. I feel it limited back-and-forth discussion and participation was mainly individual question and answer with little or no follow-up/engagement.... Discussing critical diversity issues in this way felt somewhat impersonal which, on the one hand can free people up to express themselves differently than they would face-to-face, but for me, served as a barrier to feeling connected with others.*

## DISCUSSION

The literature reviewed in the opening of this chapter clearly addresses the question of why IP education is a valid inclusion in higher education, namely, that optimal patient outcomes are more likely to occur when professionals communicate and collaborate well together. While the 'why' of IP education is clearly established, it is the 'what' and 'how' of IP education that often continue be daunting, and toward which the analyses of this research may shed some light.

The results of this project offer insight regarding the questions of 'what' content gets taught and 'how' to engage in the teaching and learning processes. Indeed, review of these findings suggests that in critical diversity education, content and process merge. Considering and examining awareness of and perspectives on diversity were experienced in tandem with the delivery method: the medium and the message were inextricable. Thus, while the terms of this research were to evaluate the delivery method, these results serve as a reminder of the weaving of how we learn and what we learn; how we teach and what we teach.

This analysis was not found in the extant literature on IP education and practice; therefore it is suggested that the results of this small study can contribute to the IP dialogue in this way. Analysis of the merger of content and process led the author to the literature on pedagogy and the central position of educational design and process in higher education, for example, Boler & Zembylas (2003) and Lusted (1986), which again, is not readily found in the literature on IP education. Interestingly, such issues of pedagogy can get lost in curricular design as we struggle to fit more and more 'hard content' (so termed because of its reliance on its perceived external authority and positivist validity), for example, theories, objective knowledge and skills development, into curricula. But what of the 'soft content', the assumptions, values and beliefs that are borne of our interactions with complex social processes and frame our ideologies, ways of interacting in the world and evaluations of others? These are arguably of central concern to IP education, yet currently stand outside the IP scholarship. At the same time, this analysis does resonate with the research on critical success factors in the online learning environment, specifically those studies that centre social networking and interactivity as features required for students' satisfaction with the online community, for example Conrad (2005) and Soong, Chan, Chua & Loh (2001).

The results of this study make clear that inclusion of critical diversity content and process is valid within health care (inter)professional education. Dominant western thinking in the health care sector prioritizes examination, assessment and diagnosis of the 'other', with little reflective analysis of the 'self' and this pattern was indeed reflected in these findings. Analysis of the data from the students reveals this very way of thinking about diversity, evidenced in the frustration students expressed regarding not having learned from or about 'diverse people', meaning those *other than themselves*. Another example is the conflation of diversity as culture, a conceptualization that rests upon particular understandings of both terms that needs to be deconstructed. Indeed, many of the comments of concern regarding a critical diversity approach are indicative of the very issues and processes we are seeking to unpack.

Data regarding the nature and value of interprofessional contact (the tenet of learning 'with, from and about' each other) was evident primarily through the students voicing their appreciation for hearing the perspectives of their colleagues on the issues pursuant to the module. However, these perspectives were interpreted personally rather than via the particular professional lens. Thus, while students reflected on what they had learned from others, they embedded that learning within the personal rather than professional domain. Future study in the area of IP critical diversity education, then, would do well to refine the methodology so that these layers can be carefully deconstructed.

The findings of this study also reflect the range of expectations and experiences within the literature regarding the relevance and promise of online learning as well as ambivalence expressed by some. Student feedback mirrors existing findings regarding the overall student satisfaction that results from the accessibility, flexibility and portability of online technologies (Aminzade & Pescosolida, 1999); participants in this study voiced appreciation for the opportunities and choices available through online learning. At the same time, students also noted appreciation for face to face contact, particularly in the IP environment, citing an ease in learning about the professions through a personal encounter. Similarly, one online student expressed concern that deeper exchange of thoughts and feelings regarding critical diversity awareness was hampered by not being face to face. A sense of loss, felt through not having face to face contact, is an important consideration that needs to be taken into account by those who continue to believe that online learning, IP education and critical diversity education share conceptual and methodological terrain.

This shared terrain is evident in considering the conditions which enable reflection, processing and deeper consciousness. Here the research findings on the module delivery environment exert a strong influence. Feedback regarding time constraints, size of group, setting and ineffective facilitation in the on-campus group, compared with the online group, suggests that the conditions available in the virtual learning environment may be more conducive to a critical diversity learning experience. For example, time pressures that students face can be managed more easily in the virtual learning environment: if other course demands are imminent, students can choose an alternative schedule. Further, venue and large group distractions are not relevant with small online discussion groups. Comments regarding having time to ponder and compose responses are congruent with the literature regarding online learning's adaptability to a variety of learning styles and needs (Wolfson, Marsum & Magnuson, 2005; Oduaran, 2006).

Finally, these data highlight the challenges inherent in teaching and learning critical diversity perspectives, whether in IP settings or uniprofessional contexts. It could be interpreted that since there was some disparaging student feedback the module was not successful. However, a critical diversity approach begins on a different premise than one which assumes sound educational principles equate with student happiness or comfort throughout the learning experience. Rather, it is understood that confronting dominant assumptions, in both content and process and within oneself, will be discomfiting and reaction will ensue. An ideological commitment to the principles of critical diversity education is not easily shaken, because the possibility that lies ahead is the transformation of learning and professional practice available when we can "*suspend the ways of looking that orient us to our own world*" (McKee, 2003, p. 403) and enter respectfully into the lens of another.

## CONCLUSION

This chapter has reviewed the current context of IP education and practice, the promise of online learning, and the content of critical diversity education, three agendas to which academic institutions are accountable in meeting the many complex needs of students in the health care professions today. Based on these imperatives, a pilot study sought to explore the experiences of face to face on-campus and online students in an introductory level interprofessional learning module on diversity awareness. Findings and analysis were offered in the context of implications for future study and practice.

As we strive to meet the learning needs of a range of students, across expressions of 'classrooms', the results of this study suggest we need to incorporate a variety of teaching methods, means and media. Face to face on-campus learning and teaching is our heritage: we know it well and we continue to refine its possibilities. Online learning is with us to stay, with its many unique contributions that expand the potentials unmet in the onsite environment. Educational innovation needs to resist the dichotomous 'either/or' thinking that keeps us stuck in binary terms of engagement, and embrace the opportunities available through mixing up the methods, the media and the messages.

## REFERENCES


Aminzade, R., & Pescosolido, B. A. (1999). Introduction to the changing landscape of higher education. In Pescosolido, B. A., & Aminzade, R. (Eds.), *The social worlds of higher education: Handbook for teaching in a new century* (pp. 6–8). Thousand Oaks, CA: Pine Forge Press.

Baikie, G., & Campbell, C. (2008a). *Conceptualizing a Critical Diversity Model*. Unpublished manuscript.

Baikie, G., & Campbell, C. (2008b). *Diversity awareness in Interprofessional Teams: A module prepared for the Tri-Faculty Interprofessional Academic Advisory Council*. Halifax, Canada: Dalhousie University.

Baikie, G., & Campbell, C. (2009). *Critical Diversity within Interprofessional Education: Constructing Professional Cultures of Belonging*. Collaborating Across Borders II Conference, May 2009, Halifax, Nova Scotia.

Boler, M., & Zembylas, M. (2003). Discomforting truths: The emotional terrain of understanding difference. In Trifonas, P. P. (Ed.), *Pedagogies of difference* (pp. 110–134). New York: RoutledgeFalmer.

Centre for the Advancement of Interprofessional Education (CAIPE). (2002). *Interprofessional education – a definition*. Retrieved March 1, 2009, from http://www.caipe.org.uk.

Chumley-Jones, H. A., Dobbie, A., & Alford, C. L. (2002)... *Academic Medicine*, *77*(10), S86–S93.

Comerford, S. A. (2003). Enriching classroom learning about diversity: Supports and strategies from a qualitative study. *Journal of Teaching in Social Work*, *23*(3/4), 159–183.

Conrad, D. (2005). Building and maintaining community in cohort-based online learning. *Journal of Distance Education*, *20*(1), 1–20.

Kohn, L. T., Corrigan, J. M., & Donaldson, M. S. (2005). *To Err is Human: Building a Safer Health System*. Washington, DC: Institute for Health.

Kumaş-Tan, & Zofia Ö. (2005). *Beyond cultural competence. Taking difference into account in occupational therapy*. Unpublished Master's thesis, Dalhousie University, Halifax, Canada.

Limburg, F., & Clark, C. (2006). Diversity initiatives in higher education: Teaching multicultural education online. *Multicultural Education*, *13*(3), 49–55.

Littlefield, M. B., & Roberson, K. C. (2005). Computer technology for the feminist classroom. *Affilia*, *20*(2), 186–202. doi:10.1177/0886109905274676

Lopes, T., & Thomas, B. (2006). *Dancing on live embers: Challenging racism in organizations*. Toronto: Between the Lines.

Lusted, D. (1986). Why pedagogy? An introduction to this issue. *Screen*, *26*(5), 2–14.

McKee, M. (2003, July). Excavating our frames of mind: The key to dialogue and collaboration. *Social Work*, *48*(3), 401–408.

Merryfield, M. (2003). Like a veil: Cross-cultural experiential learning online. *Contemporary Issues in Technology & Teacher Education*, *3*(2), 146–171.

Merryfield, M. M. (2001). The paradoxes of teaching a multicultural education course online. *Journal of Teacher Education*, *52*(4), 283–299. doi:10.1177/0022487101052004003

Roberts, J., O'Sullivan, J., & Howard, J. (2005). The roles of emerging and conventional technologies in serving children and adolescents with special needs in rural and Northern communities. *Journal of Distance Education*, *20*(1), 84–103.

Soong, B. M. H., Chan, H. C., Chua, B. C., & Loh, K. F. (2001). Critical success factors for on-line course resources. *Computers & Education*, *36*(2).

Wolfson, G. K., Marsom, G., & Magnuson, C. W. (2005). Changing the nature of the discourse: Teaching field seminars online. *Journal of Social Work Education*, *41*(2), 335–361. doi:10.5175/JSWE.2005.200300346

# Chapter 7
# The Community of Inquiry Framework:
## A Pertinent Theory of Online Interprofessional Education?

**Heather Mac Neill**
*University of Toronto, Canada*

**Scott Reeves**
*University of Toronto, Canada*

**Elizabeth Hanna**
*Bridgepoint Hospital, Canada*

**Steve Rankin**
*eUnity, Canada*

## ABSTRACT

*Many in the online learning field now promote the need for a social presence online, in addition to cognitive and teaching presence, in order to fully realize benefits of online learning. In this regard, two important concepts arise when adopting e-learning: community and collaboration. Within healthcare there has been a recent push towards interprofessional education (IPE). IPE is an approach in which health and social care professionals come together to learn "with, from and about each other". In this chapter the authors discuss the background of healthcare IPE and online learning. They examine the potential benefits and limitations of both IPE and e-learning as well as issues related to combining these approaches. They will discuss the theory of 'communities of inquiry' and apply a modified version of it as a way to think about and create online IPE. Lastly, the authors introduce an ongoing innovative healthcare e-learning project in Canada that was based on this theory and has focused on bringing together both IPE and online learning using the "build-a-case" method.*

DOI: 10.4018/978-1-61520-889-0.ch007

## INTRODUCTION

*The exciting discontinuity, the exciting opportunity and threat, the exciting confusion now thrust upon us is an explosion of new ways of organizing, communicating, delivering, finding, modifying, and creating information. We have barely begun to see how to use these new ways for teaching and learning. It will take many decades to invent and wring out the very best uses of these new tools – even as newer tools continue to arrive, divert our attention, and offer ever greater possibilities (Gilbert, 2000).*

We are just beginning to scratch the surface of the advantages and shortfalls of online learning, and where it may fit in the 'toolbox' of resources for healthcare professionals' education. Most health professions' online learning to date has focused on providing information to learners in a text-like format. Some programs have progressed so learners can interact with the material and, at times, try "virtual" health care scenarios. Both are examples in which learners interact with the online material, but not with the other learners.

Many in the online learning field now promote the need for a social presence online, in addition to cognitive and teaching presence, in order to fully realize the benefits of this form of learning (Garrison et al., 2000; Palloff & Pratt, 2005). There is recognition of the importance of a learner's personal world to retain and reflect on knowledge, but equally important is the ability to socially construct this knowledge with others, to make practical sense of it and expand upon concepts from different perspectives and backgrounds. In this regard, two important concepts arise when considering adopting e-learning: community and collaboration. Palloff and Pratt (2005, p. xi) argue that "in the online environment, collaboration can be seen as the cornerstone of the educational experience…collaboration forms the foundation of a learning community online - it brings students together to support the learning of each member of the group while promoting creativity and critical thinking". Conversely, development of an online community helps foster collaboration.

Within healthcare there has been a recent push towards interprofessional education (IPE). IPE is an approach in which health and social care professionals come together to learn "with, from and about each other" (Freeth et al., 2005, pp. 4-6). Theoretically, IPE offers a very powerful educational approach; it allows learners to focus their learning towards a patient centered model of care and learn how to efficiently share limited health care resources and expertise. However, there are many roadblocks to delivering this form of resource-hungry, interaction-demanding form of education such as scheduling difficulties in bringing together several health professions and accommodating the diversity and different professional cultures of learner groups (Freeth & Reeves, 2004).

Online learning minimizes geographic and learner barriers that are difficult to accommodate in face-to-face IPE (Jairath & Mills, 2005; Barr et al., 2005). Potentially, it offers an "anytime, anywhere" learning, important to busy pre-licensure and post-licensure health professions learners alike. It is also able to respond, by multiple auditory, visual and tactile modalities, to various learning styles of adult learners, and allows them to skim material or explore in more depth as they wish (Jairath & Mills, 2005, pp. 4-6). It may also help accommodate different professional cultures, learning styles, and level of learners often found in an IPE forum (Barr, et al., 2005). In addition, online learning allows for sharing electronic resources to help support learners' opinions, professional backgrounds and promote evidence-based discussions. Lastly, an online environment may help "level the playing field" of healthcare hierarchy through mutual navigation of software and computer literacy, creating a more removed, safer learning environment to risk failure with colleagues. However, in online learning, individuals need to become part of a virtual learning community in order to promote connection and interactivity between each other and the material. Often this includes email, chat rooms and discussion boards, where non-verbal social cues that are so important in IPE may be lost.

In this chapter we discuss the background of IPE and online learning through a social constructivist lens. From an educational perspective, social constructivism examines how people create knowledge and meaning within a social context. Wenger (1998) argues that "learning is so fundamental to the social order we live by that theorizing about one is tantamount to theorizing about the other" (Wenger, 1998, p. 15). We examine the potential benefits and limitations of both IPE and e-learning as well as the issues related to combining these two approaches. We also introduce an innovative project in Canada that focusses on bringing together both IPE and online learning using the "build-a-case" method to illustrate how this can be applied to health professions' education. Lastly, we will discuss a modified theory of 'communities of inquiry' as a way of gaining a more comprehensive understanding of online IPE.

## BACKGROUND

Effective collaboration between various health and social care professions has long been regarded as essential for delivering high quality patient care (Evers, 1981; Gregson et al., 1991; Penthybridge, 2004). Often, however, collaboration is undermined by boundary infringements, a lack of understanding of one another's roles, limited communication and poorly coordinated teamwork (Skjørshammer, 2001; Pethybridge, 2004). In an effort to improve interprofessional practice, policy documents have repeatedly advocated the use of IPE (Department of Health, 2002; O'Neil 1993; World Health Organization, 1988). The collective view of these policies is that because professionals are trained separately, they are not equipped with the attitudes, skills or knowledge to work effectively together to deliver patient care. It has therefore been argued that IPE offers a way to resolve this problem. As a result, there has been a steady growth of IPE within the health and social care systems in the United Kingdom, Canada, the United States, continental Europe and Australia (Barr et al., 2005).

IPE is now a widespread and diverse activity. Initiatives have been reported in both pre-licensure and post-licensure education. It can be university-based or undertaken within clinical practice. It can also be employed as a part of a continuous quality improvement initiative. IPE can also be found both in primary health care settings and hospital/long-term care settings and can involve a range of different health and social care professions such as nurses, speech therapists, nutritionists, physicians, occupational therapists, physiotherapists and social workers. Encouragingly, findings from systematic reviews of the effects of IPE have indicated that this form of education can make a positive contribution to collaborative knowledge and skills as well as contribute to an improvement of patient/client care (Barr et al., 2005; Reeves et al., 2008).

Traditionally, IPE has been delivered as an interactive face-to-face (F2F) activity. Increasingly though, health professions' course planners are developing and delivering online IPE programs to help overcome limitations such as geographic distance between learners, scheduling difficulties due to heavy clinical workloads and overcoming the relatively high costs of offering IPE in a conventional manner. E-learning methods, such as electronic bulletin boards and video conferencing, are increasingly being used within IPE (Cornish et al., 2003; Farrell, 2005; Miers et al., 2007). Often, IPE initiatives employing e-learning methods also use traditional learning methods, such as seminar discussions, in a 'blended learning' approach (Richardson & Cooper, 2003).

## Online Learning and IPE-Marriage of Convenience or a Match Made in Heaven?

Arguably, both e-learning and IPE require rich financial and human resources (in the form of IT support for the former and IPE facilitation/coordination for the latter). However, if implemented effectively, both will, over time have possible long term financial benefits (e.g. recycling e-learning material, improved patient centred care). It may be argued that matching a fast paced, often distant medium of online learning with a socially rich and process driven method of IPE may be problematic. We feel, however, that they make a good match. Here we argue several advantages of online IPE learning over the traditional F2F IPE format and consider further theoretical synergistic effects of online IPE learning.

Probably the most obvious advantage is the ability to reduce scheduling and logistical problems of training vast numbers of multiple professional learners in one place at one time required of F2F IPE. However, both e-learning and IPE can be seen as representing a radical change from traditional educational methods (where lecturers impart one-way information to learners). Adopting both IPE and e-learning requires a radical shift in traditional views of content delivery, interactions in the classroom (learner-to-teacher, learner-to-material, and learner-to-learner), facilitation techniques specific to both IPE and online situations, technical know-how and support, and recognition and assessment of collaboration and group "knowledge making" for both learners and teachers alike.

However, as the world evolves, there is no denying the influence of technology and the importance of developing collaborative skills. Today's learners and professionals need to know how to manage multiple sources of information and to communicate and collaborate with others to achieve a high level of professional and interprofessional performance (Reeves & Freeth, 2003). Therefore, the enormous currency that both IPE and e-learning afford demands a realistic appraisal of the new skills and processes needed for this type of education to occur.

Arguably, learners and educators alike will need to adapt to the teacher's altered role, who is acting more as a coach and facilitator to an individual's own self-directed learning rather than a stand-alone resource. Educators will need to develop new rubrics of assessing learner/group participation. Both will need to become comfortable with the technology they use; however, this may allow a unique opportunity to promote IPE, by "levelling the playing field". Older learners may be more proficient in healthcare knowledge, but may feel inept when using technology. Conversely, younger learners may feel quite comfortable with technology, but lack the healthcare expertise. Learning to mutually navigate the software as a group may help establish the social presence needed to build the community and communication required for a prosperous IPE experience.

The unique anonymity that e-learning affords may help to increase participation in IPE learning experiences. IPE is best fostered in a safe environment, where learners feel they can contribute without negative judgement. Dark and Perrett (2007) note that "learners are more willing to risk failure and use that experience for learning when there is a degree of anonymity within the e-learning environment, compared with face-to-face teaching." Learners in a F2F setting may feel pressurised to portray a certain persona related to their profession. In healthcare, there is significant pressure to develop and maintain a respected identity. In fact, Dornan et al. (2007) found that "developing a professional identity is so fundamental to the process of becoming a doctor that having to be credible in the eyes of practitioners caused young medical students more stress than encountering illness and death". Portraying oneself in a somewhat remote and more anonymous medium may help break down some walls of identity protection.

The relative anonymity of e-learning may even encourage participants to "try out" different professional identities. Professional identity is often difficult to discern for pre-licensure learners, and the same can be true even for seasoned post-licensure professionals joining a new team. Payne (2006) argues that professional identity is not static, but fluid; shaped by social relationships within the healthcare environment. He states that while identity may be defined globally (by principles and theory) it is actually experienced locally (with different professions within their interprofessional community). For example, in a profession as multi-faceted as social work in the palliative care setting, role identity fluctuates depending on the needs of the health care team and the strengths of the health care worker. Working in an online IPE environment may allow learners to trial modified professional identities with less risk of failure than in a F2F format. So, working from the example above, a social worker in an online IPE environment may adopt a somewhat different professional persona than they assume in a F2F palliative care work setting, allowing them to explore different aspects of their role and increase their learning.

Wesch (2008) when speaking on his YouTube ethnography project also raises interesting points about the use of video and webcams on the internet that may pertain to video captured online IPE experiences. He speaks of YouTube video postings as a way to see ourselves through others eyes and ourselves through our own reflections, allowing a new self understanding. It is personal, yet still distant; anonymous yet everyone can watch. It allows us to "watch others without staring at them" and connect without "social constraint" or "responsibility". This medium may allow exploration of identity (both self and others) as needed to foster IPE without the social awkwardness or repercussions that may occur in a face to face setting.

However, the anonymity of e-learning will always be a double-edged sword. While it may promote safety, the mechanism by which it does so may reduce the "human connectivity" crucial to the IPE process and fidelity of the online experience may not as easily transfer to "real life" situations. So the anonymity that e-learning allows also presents the greatest challenge to online IPE.

The fidelity of online experiences to real life may be enhanced by creating them within a particular "community of practice" or situational environment of interprofessional learners (Wenger, 1998). This is where IPE may help foster the online experience (Howkins & Bray, 2008). In this environment, learners can come together to negotiate problems and solutions to local issues important to them all. The negotiation of a treatment plan, may for example, expand learning from, with and about each other. This may also allow the development of community and identity that may strengthen collaboration. This is why the process of learning in an online IPE format is equally important, or arguably more so, than the actual implicit IPE content delivered.

Lastly, one of the greatest challenges of IPE is to teach similar content to multiple learners, comprising multiple professional backgrounds and levels with multiple preconceived (often unconscious) biases towards each other, without overwhelming, under stimulating, alienating or focusing on any one learner (Howkins & Bray, 2008). E-learning, by nature, is highly individualized as it can be modulated by personal experience, profession, practice needs, preferred learning styles and interests (Jairath & Mills, 2005). E-learning can be self directed: learners can focus on their perceived needs, skip what is not relevant and revisit material as often as they wish. This allows learners to return to the material multiple times without feeling they are holding back other learners. It also provides background resource material for group discussions, decreasing the risk of insecurity a learner may feel without this resource.

While scanning the supporting documents, learners may also be exposed to different professional views and literature they would otherwise not have pursued directly. This may pique their interest to learn more about an area they were previously unaware of. Also, online resources available at the learners' fingertips (web, online journals, etc.) may support their respective professional points of view and promote evidence based medicine and knowledge translation. This type of multi-source learning allows for accommodation of different levels and backgrounds of learners without the need to simplify the material to the "lowest common denominator". It pushes learners to be aware of the depth and breadth of related resources in healthcare and be able to aptly select the most relevant and time efficient sources- a crucial skill in today's world of "information overload".

## Community of Inquiry Framework

Having discussed concepts of IPE, e-learning and their theoretical synergies, we now explore a potential theoretical model to use when approaching the development and implementation of online IPE. The Community of Inquiry (CoI) framework was first developed as a model of online learning by Garrison et al. (2000). Since then, it has gained significant momentum in the field of online learning and has been proven a valid theory in online course development (Swan et al., 2008). It consists of three elements: social, teaching and cognitive presences. It has been confirmed through factor analyses (Garrison et al., 2000; Garrison et al., 2007).

Social presence has been defined as "the ability of learners to project themselves socially and emotionally, thereby being conceived as "real people" in mediated communication" (Garrison et al., 2007). Garrison (2007) notes that research demonstrates a relationship between social presence and learning outcomes, learner satisfaction, and faster assimilation of technology. Its importance lies in collaboration required for knowledge making, but less so in pure information gathering. Its role is not just in the quantity of interaction, but rather the quality of interaction to achieve educational goals. In the case we will describe below, this could be likened to the process of learners synchronously coming together to "build-a-case".

Cognitive presence has been described as "the extent to which learners are able to construct and confirm meaning through sustained reflection and discourse" (Garrison et al., 2007). It is divided into 4 phases, moving from a collaborative process of probing and discussing ideas to one where individuals further reflect, create individual meaning and collaborative usage of created knowledge. These four parts include: (1) a triggering event, (2) exploration, (3) integration where meaning is constructed, and (4) resolution or application of the new knowledge. Arguably, the greatest challenge lies in moving from the exploration to integration phase. This may partly due to the role of teaching presence and the need for learners to acquire a collaborative solution to the triggering problem. In our "build-a-case" example described below, this is exemplified by students learning facts through the self-directed learning modules, then further integrating and exploring these facts with other learners in the build-a-case forum.

Teaching presence is defined as "the design, facilitation and direction of cognitive and social processes" in order to obtain optimal learning outcomes (Garrison et al., 2007). This is further subdivided into the following three components:

- Instructional design and organization: constructing framework, processes, evaluation, etc. of learners and associated support.
- Facilitating discourse: clarifying, encouraging, focusing, constructing and deconstructing group ideas, pointing out metacognitive process occurring during learning, etc.

- Direct instruction: guidance to resources, making connections, correcting wrong assumptions, etc. (Anderson et al., 2001).

It is noted that teaching presence is even more important in online learning than in face-to-face interactions due to the novelty of this approach and the loss of non-verbal social cues and interactions online. Teaching presence has been noted as a "significant determinant of student satisfaction, perceived learning and sense of community" (Garrison et al., 2007). In summary, social presence provides the creativity, construction and exploration of ideas, while teaching presence provides the organization, structure and leadership required to harness these ideas productively.

## Adaptation of the Community of Inquiry Framework: A Meaningful Theory for Online IPE?

Although online learning and IPE initially appear to be unlikely complimentary entities, on further inspection there are many areas of mutual benefit that this "marriage of opposites" imparts, as argued above. Both require the elements of community and collaboration in order to flourish, and both advance each other in doing so. Therefore, there is an inextricable link between community and collaboration which must occur to promote online IPE. It is acknowledged that "a great deal of interprofessional learning is implicit and informal" (Freeth et al., 2005, p. 4). Therefore, the ultimate goal of online IPE is to promote interprofessional knowledge, skills and behaviour through implicit and informal processes of IPE as well as the explicit content of the course objectives.

The community of inquiry framework is a natural fit for a structure on which to construct online IPE experiences. The framework uses the three interlinking concepts of social, cognitive and teaching presence. The overlapping intersections of these three presences have been termed:

Supporting discourse (at the junction of social and cognitive presence).
Setting climate (at the junction of social and teaching presence).
Selecting content (at the junction of cognitive and teaching presence).

We argue that these three intersections, when viewed with an IPE lens, translate into the following terms, respectively:

Collaboration
Community Building
IPE competencies

Together, these support the main elements required of both IPE and e-learning; namely collaboration, community and IPE competencies (see Figure 1).

Research is needed to substantiate this theory of a modified community of inquiry framework to support online IPE development and implementation. Also, this theory assumes an ideal situation of comprehensive IT support, highly skilled online facilitators, learners with some computer literacy and access to computer requirements and developed software free of technical glitches.

*Figure 1. Adaptation of the community of inquiry framework to online IPE*

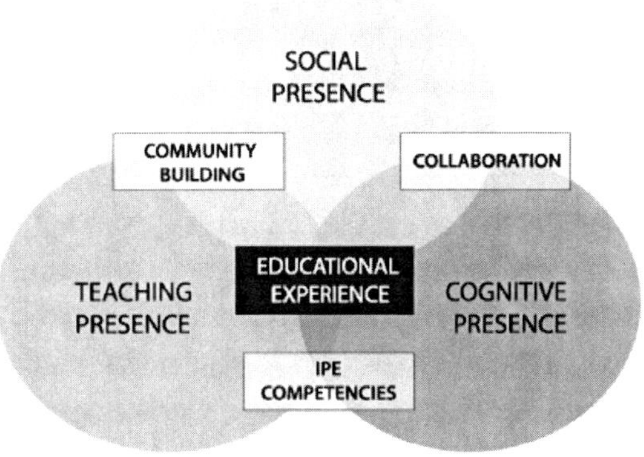

## Experience with the Online "Build-a-Case" Technique

Having discussed the relative merits and shortfalls of online IPE, and a theoretical approach on which to develop online IPE, we will now present a developmental account of a novel online IPE learning experience currently under development in Canada. Bridgepoint Health, is a publicly funded, University of Toronto affiliated, health care organization specializing in the field of rehabilitation and complex chronic disease prevention and management. Recently we were faced with of challenge of how to create online learning for practicing interprofessional healthcare workers in 3 areas:

- Chronic disease.
- The socially vulnerable population, who experience chronic disease with greater intensity and frequency (World Health Organization, 2009).
- Interprofessional delivery of education (Barr et al., 2005).

The course is an 8 week, blended methods approach in which teams of 6-8 interprofessional post-licensure learners work together online. The course is sandwiched by two face-to- face sessions at the beginning and end of the course. In between are 6 weeks of online learning as described below. Evaluation of this pilot study is currently in progress. Research methods include IPE attitudinal scales, pre and post tests on the content of the e-learning, qualitative focus groups and key informant interviews. The evaluation examines the impact of the online "build-a-case" learning method and interprofessional knowledge and attitudes.

This project utilized the theory surrounding the modified community of inquiry framework in its development. Cognitive presence was created by the development of online interactive modules. Teaching presence was created by use of a co-facillitation model of two healthcare professionals of different backgrounds. The greatest challenge was how to incorporate the social presence and the socially rich interactivity required of online interprofessional learning. Examples of online learning in healthcare in the current literature only incorporated text options of communication such as chat rooms and discussion boards.

There are potential advantages to online, asynchronous, text-based IPE (e.g. discussion boards). Hughes et al. (2004) argue that asynchronous (chat-based) online environments allows more fluidity and time to absorb, reflect upon and respond to the material, rather than a fixed time and place of a face-to-face session. This allows learners to move beyond simply sharing knowledge to deepen their understanding of the subject matter and explore different ideas. However, in real life situations at the pre-licensure level, courses using asynchronous text-based discussions do not seem to progress to critical debate or creation of new knowledge (Miers et al., 2007; Paulus & Phipps, 2008). This may be because these learners lack the clinical know-how to deeply discuss situations and instead focus more on knowledge acquisition.

Lobel et al. (2005) found more interactions between learners (as opposed to learner to teacher interaction) in synchronous chat than in F2F environments. This suggests that synchronous online interaction may promote more collaboration between students, possibly because the teacher acts to facilitate learning rather than provide answers. Also, Paulus and Phipps (2008) found more interaction between learners in synchronous than asynchronous computer-mediated communication.

However, there are also obvious limitations created by synchronous or asynchronous text based interaction. Most individuals have experienced a well intentioned email being misinterpreted. In a socially rich form of education like IPE, many non-verbal social nuances may be lost or misinterpreted, leading to further alienation rather than understanding of different professional cultures and backgrounds. Part of the success of IPE revolves around learners bonding and creating a shared repertoire and experiences, thus breaking down pre-conceived prejudices of other professions. This experience may be more removed and difficult to achieve in purely text-based environments.

It is possible that newer modalities, such as wikis and blogs, may promote more connectivity and online learning among multiple users; however, many synchronous, interactive social cues so important to interprofessional learning are still missing, causing potential for misinterpretation. We strove to find a synchronous form of computer-mediated communication that preserved the non-verbal social cues present in F2F sessions. We also looked for a method of collaborative learning that this form of communication would facilitate.

Many forms of collaborative learning activities have previously reported in online education literature, such as role playing, simulations, case studies, debates, fishbowls, webquests, small group activities, group blogs, as well as virtual teams (Palloff & Pratt, 2005). One of these collaborative tasks, the case study, is familiar to many health care professionals. Most will have experienced problem-based learning (Camp, 1996) during their formal training, and all have discussed difficult cases with colleagues. Case-based studies are also paramount to promoting healthcare interprofessional care, as they highlight the main benefit of this form of education - patient centered care.

Building cases for educational purposes requires considerable time and resources and is often based on knowledge and guidelines that continually change, or are unique from learner to learner. Furthermore, pre-made cases are less relevant to post-licensure/continuing education learners, who bring self-directed questions and goals of learning based on personal questions and area of practice. A recent study of case-based learning, for example, used the "build-a-case" format (Ryan & Marlow, 2004). Small groups of learners worked face-to-face with a facilitator to build a patient case by consensus, discuss opinions on treatment options, and lastly, to compare their case to established guidelines. The technique's authors argue that this process allows learners to explore and reflect on the typical pattern or "implicit scripts" of their respective practices, as compared to their fellow learners, revealing "practice surprises". These, in turn, lead to reflection on action and opportunities for behavioral change within their practice to "explicit models" such as practice guidelines if appropriate.

It has been documented that learners often "don't know what they don't know" and are therefore poor self assessors (Davis et al., 2006). They are frequently drawn to learn more about areas they are already proficient in, and fear exploring unknown areas of practice. This can limit their ability to learn new practices that may impact on patient care the most. This "build-a-case" technique allows learners not only to highlight personal questions relating to their area of practice, but also to understand issues they may not have considered before through collaboration with other learners in building the case and comparing their composition to guidelines. The final content (the fictitious patient case) is not as important as the process that learners use to get there. This process allows reflection on the "art" and practical aspects of healthcare, as well as evidence-based practice guidelines. Team members are exposed to "clinical pearls" from other learners and professions that might not yet be published or in the sphere of common knowledge. This process also allows participants to increase their interprofessional competencies. By working together to create the case, they learn about each other's areas of expertise, how to negotiate tasks and meaning and, in the end, have a shared repertoire of patient cases and experiences.

Lastly, in contrast to traditional case-based and problem-based learning, the build-a-case technique actually creates new cases and resources. These reusable learning objects can then be presented to early pre-licensure clinicians to introduce scientific and "clinical pearls" of practice along with the process of working interprofessionally.

## Methods/Description of the Courseware Product

Through our review of available courseware products, we found the majority of web-based education software solutions continued to chase technical possibilities and remain, at their foundation, educationally poor. Existing solutions delivered content with no opportunity for learners to challenge their comprehension, promote their understanding and develop their problem-solving skills. Simply put, most were one-way course content delivery systems lacking an interactive implementation of proven education and learning theories. This one-way education delivery model has been shown to fail in the classroom and over the web, primarily due to the lack of support for real-time interaction with learners and facilitators. When interaction between learners (and facilitator to learner interaction) is not the primary focus, questions can remain unanswered, information can be misinterpreted and possibly never corrected.

We searched for a product that supported an on-line IPE environment for the "build-a-case" model that would incorporate non-verbal social cues (using real-time communication for all learners through streaming video and audio) that are so important in healthcare and interprofessional learning. We also needed software that supported the ability to easily load and prepare clinical assets (such as medical images, MS PowerPoint files, blood work and videos), the ability to use medical mark-up, measurements and annotations necessary for quality collaboration and a learning environment where all learning objects and learner actions are fully shared. To support strong adoption, the authoring process had to be simple.

No such courseware existed; however, we worked with eUnity, a company developing a similar approach for sharing medical imaging online. One of the biggest challenges was to create a guided template for users that would respond to each of their internal professional templates they use when seeing and analyzing a patient case. In the end, the use of narrative examples was selected as the best option for this application, since it was truly the process, more than content, which we wished to focus on. Another challenge was deciding on the facilitation/teaching presence for the learners. We wanted to support learners in an interprofessional way that would not stifle their drive to freely express opinions and experiences. Two facilitators from different interprofessional backgrounds were incorporated into the build-a-case sessions. As well, "on-demand" facilitators in the areas of content, IT and IPE were accessible to the learners. A screenshot of the online build-a-case technique can be seen in Figures 2 and 3.

Resources were needed for the participants to facilitate their background knowledge in order to build cases. Typical online learning modules and webquests were created for each week of the build-a-case to help support this learning and were made as interactive as possible by utilizing visual, auditory and tactile learning techniques (see Figure 4).

Therefore, the online learning created for this project is partly asynchronous (self directed learning resources where learners interacts only with the material through multiprofessional and uniprofessional learning), and partly synchronous (socially rich interaction between learners online: through the use of webcams) to build a case together. Results on the outcome of this application are to follow.

## ONGOING AND FUTURE RESEARCH DIRECTIONS

The results of this ongoing pilot project are still to come, and include both qualitative and quantitative data. Participants included 13 practicing healthcare professionals divided into 2 teams from 2 facilities representing both acute care and rehabilitation.

*Figure 2. Screen shots of the Build-a-case technique: Building the case through the use of a narrative template.*

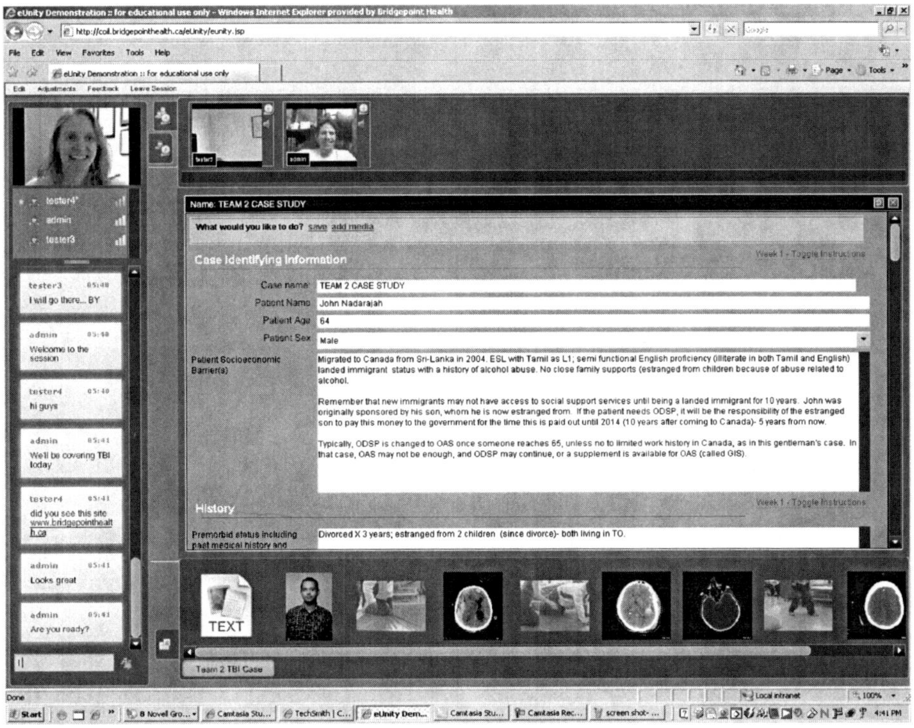


*Figure 3. Screen shots of the build-a-case technique: mutual navigation of images*

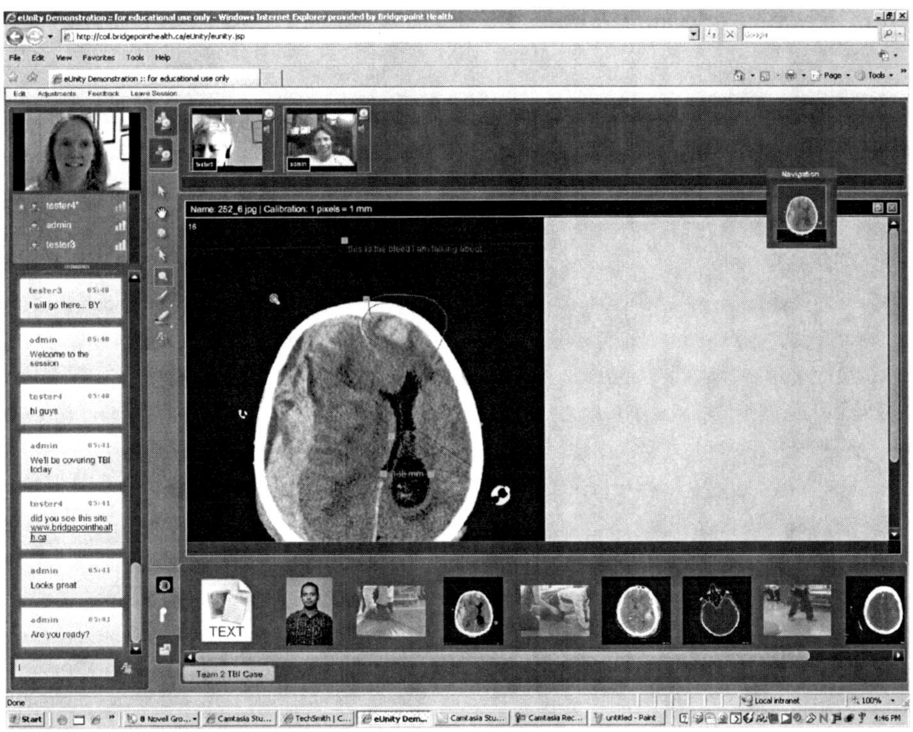

*Figure 4. Self, directed, learner focused interactive e-learning- example of a 'roll-over' to discover associated functional anatomy of the brain*

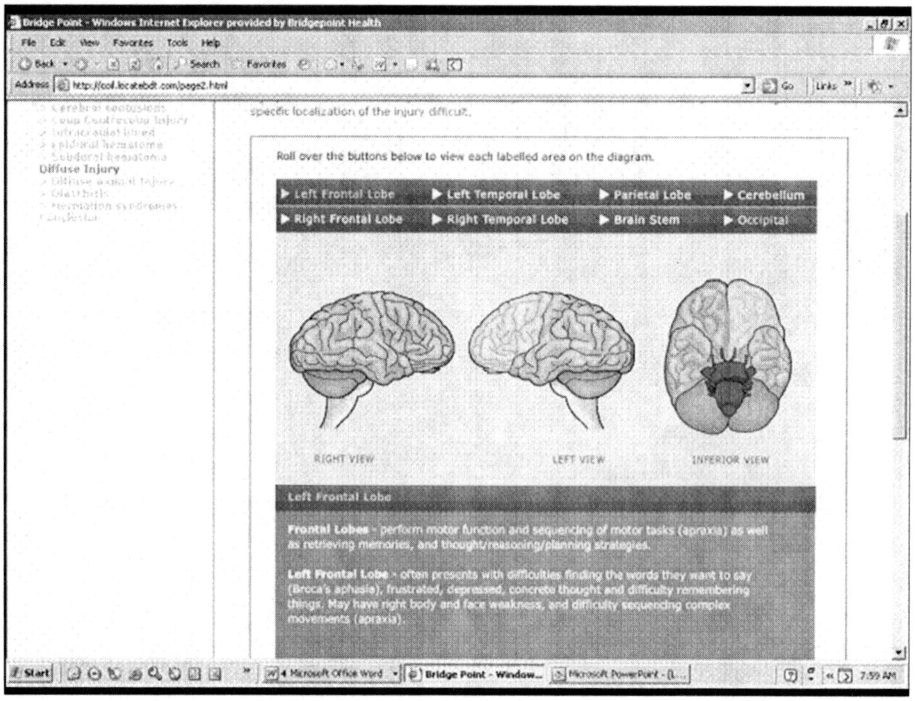


The teams included professionals from nursing, physiotherapy, occupational therapy, social work, speech language pathology, and pharmacy.

Many questions remain about how to effectively incorporate interprofessional education into an e-learning format, including the use of different forms of e-learning such as the build-a- case technique, and the use of theories such as the community of inquiry theory. What are the advantages and disadvantages of online IPE over a face to face format? What are the challenges and important aspects of online communication in promoting IPE? What is the appropriate balance of process and task in online IPE learning, and how do we facilitate each of these? Further research is needed to understand the importance, amount, and nature of techniques to be used in online synchronous IPE facilitation.

Our hope is that the results of the pilot will provide more clarity on the advantages and disadvantages of online IPE, inform positive changes to the COIL project and build-a-case technique, as well as foster the research planned for the final project roll out. In the final phase, learners will be randomized to either create the build-a-case individually or do so interprofessionally. Each compositon will be analysed in terms of the quality, accuracy, and richness of their learning and case creation, and the benefits and disadvantages to each technique.

Research undertaken to date on the "build-a-case" technique has been completed in a one time, face-to-face setting with uniprofessional learners, so the results are hard to generalize to interprofessional e-learning. However, in a study of 52 physicians, more than 75% felt the build-a-case technique better reflected their own practice than the practices of experts and that it prompted them to think about their own cases. Only 36% preferred to be given the details of a case to solve, rather than building it (Ryan and Marlow, 2004). Our research-in-progress will compare these results to a more intensive, prolonged, interprofessional and online usage of the build-a-case technique.

## CONCLUSION

As we have argued, IPE and e-learning may be mutually beneficial when combined. Indeed, they potentially offer significant advances for the development of community and collaboration. The community of inquiry framework may be a useful tool in helping guide the creation of online IPE experiences. At the intersection of its social, cognitive and teaching presences lay the core concepts of collaboration, community and IPE competencies. However, further research is needed to substantiate this in practice.

## ACKNOWLEDGMENT

We would like to acknowledge Dr. Ivan Silver and Dr. John Parboosingh for their mentorship and guidance, and Dale Min for his administrative assistance throughout the project. The project reported in this chapter was funded by an educational donation from HSBC bank and grants through University of Toronto (EDF, ITCDF and EIT summer studentship, 2008).

## REFERENCES

Anderson, T., Rourke, L., Garrison, D. R., & Archer, W. (2001). Assessing teaching presence in a computer conferencing context. *Journal of Asynchronous Learning Networks*, *5*(2).

Barr, H., Koppel, I., Reeves, S., Hammick, M., & Freeth, D. (2005). *Effective Interprofessional Education: Assumption, Argument and Evidence*. Oxford, UK: Blackwell Publishing. doi:10.1002/9780470776445

Camp, G. (1996). Problem-Based Learning: A paradigm shift or a passing fad? *Medical Education Online, 1*(2).

Cornish, P., Church, E., Callanan, T., Bethune, C., Robbins, C., & Miller, R. (2003). Rural interdisciplinary mental health team building via satellite: a demonstration project. *Telemedicine Journal and e-Health*, (9): 63–71. doi:10.1089/153056203763317666

Dark, G. G., & Perrett, R. (2007). Using e-learning as a tool for multidisciplinary education. *European Journal of Cancer Care*, (16): 90–93. doi:10.1111/j.1365-2354.2007.00777.x

Davis, D., Mazmanian, P. E., Fordis, M., Harrison, V., Thorpe, K. E., & Perrier, L. (2006). Accuracy of Physician Self-assessment Compared with Observed Measures of Competence. A Systematic Review. *Journal of the American Medical Association*, *296*(9), 1094–1102. doi:10.1001/jama.296.9.1094

Department of Health. (2002). *Learning from Bristol: The DH Response to the Report of the Public Inquiry into Children's Health Surgery at the Bristol Royal Infirmary 1984-1995*. London: Department of Health.

Dornan, T., Boshuizen, H., King, N., & Scherpbier, A. (2007). Experience-based learning: a model linking the processes and outcomes of medical students' workplace learning. *Medical Education*, *41*, 84–91. doi:10.1111/j.1365-2929.2006.02652.x

Evers, H. (1981). Multidisciplinary teams in geriatric wards: myth or reality? *Journal of Advanced Nursing*, (6): 205–214. doi:10.1111/j.1365-2648.1981.tb03214.x

Farrell, M. (2005). Enabling interprofessional education: the potential of e-learning. In Carlisle, C., Cooper, H., & Mercer, D. (Eds.), *Interprofessional Education: an Agenda for Healthcare Professionals*. Salisbury, UK: Quay Books.

Freeth, D., Hammick, M., Reeves, S., Koppel, I., & Barr, H. (2005). *Effective Interprofessional Education- Development, Delivery and Evaluation* (pp. 11–13). Oxford, UK: Blackwell Publishing. doi:10.1002/9780470776438.ch1

Freeth, D., & Reeves, S. (2004). Learning to work together: using the presage, process, product (3P) model to highlight decisions and possibilities. *Journal of Interprofessional Care*, *18*(1), 43–56. doi:10.1080/13561820310001608221

Garrison, D. R., Anderson, T., & Archer, W. (2000). Critical Inquiry in a Text-Based Environment: Computer Conferencing in Higher Education. *The Internet and Higher Education*, *2*(2-3), 87–105. doi:10.1016/S1096-7516(00)00016-6

Garrison, D. R., & Arbaugh, J. B. (2007). Researching the community of inquiry framework: review, issues, and future directions. *The Internet and Higher Education*, *10*(3), 157–172. doi:10.1016/j.iheduc.2007.04.001

Gilbert, S. W. (2000). *A New Vision Worth Working Toward – Connected Education and Collaborative Change*. Retrieved February 27, 2009, from http://www.tltgroup.org/gilbert/NewVwwt2000--2-14-00.htm

Gregson, B., Cartlidge, A., & Bond, J. (1991). *Interprofessional collaboration in primary health care organisations*. London: Royal College of General Practitioners.

Howkins, E., & Bray, J. (2008). *Preparing for Interprofessional Teaching: Theory and Practice*. Oxford, UK: Radcliffe Publishing.

Hughes, M., Ventura, S., & Dando, M. (2004). On-line interprofessional learning: introducing constructivism through enquiry-based learning and Peer Review. *Journal of Interprofessional Care*, *18*(3), 263–267. doi:10.1080/13561820410001731304

Jairath, N., & Mills, M. E. (2005). Online Health Science Education- Development and Implementation (pp. 4-6, 85, 106-114). Philadelphia, PA: Lippincott Williams & Wilkins.

Lobel, M., Neubauer, M., & Swedburg, R. (2005). Comparing how student collaborate to learn about the self and relationships in a real-time non-turn taking online and turn-taking face-face environment. *Journal of Computer-Mediated Communication*, *10*(4), 18.

Miers, M., Clarke, B. A., Pollard, K. C., Rickaby, C., Thomas, J., & Turtle, A. (2007). Online interprofessional learning: the student experience. *Journal of Interprofessional Care*, *21*(5), 529–542. doi:10.1080/13561820701585296

O'Neil, E.Pew Health Professions Commission. (1993). *Recreating Health Professional Practice for a New Century*. San Francisco, CA: Pew Health.

Palloff, R. M., & Pratt, K. (2005). Collaborating Online- Learning Together in Community (pp. 3-18, 57-97). San Francisco, CA: Jossey-Bass.

Paulus, T. M., & Phipps, G. (2008). Approaches to case analyses in synchronous and asynchronous environments. *Journal of Computer-Mediated Communication*, (13): 459–484. doi:10.1111/j.1083-6101.2008.00405.x

Payne, M. (2006). Intentity Politics in Multiprofessional teams: Palliative care social work. *Journal of Social Work*, *6*(2), 137–150. doi:10.1177/1468017306066741

Pethybridge, J. (2004). How team working influences discharge planning from hospital: a study of four multi-disciplinary teams in an acute hospital in England. *Journal of Interprofessional Care*, (18): 29–41. doi:10.1080/13561820410001639334

Reeves, S., & Freeth, D. (2003). New forms of information technology, new forms of collaboration? In Leathard, A. (Ed.), *Interprofessional Collaboration: From Policy to Practice in Health and Social Care*. London: Routledge.

Reeves, S., Zwarenstein, M., Goldman, J., Barr, H., Freeth, D., Koppel, I., & Hammick, M. (2008). Interprofessional education: effects on professional practice and health care outcomes. *Cochrane Database of Systematic Reviews*, (Issue 1), CD002213. doi:.doi:10.1002/14651858.CD002213.pub2

Richardson, B., & Cooper, N. (2003). Developing a virtual interdisciplinary research community in higher education. *Journal of Interprofessional Care*, *17*, 173–182. doi:10.1080/1356182031000081777

Ryan, D. P., & Marlow, B. (2004). Build-a-case: A brand new continuing medical education technique that is peculiarly familiar. *The Journal of Continuing Education in the Health Professions*, *24*(2), 112–118. doi:10.1002/chp.1340240208

Skjørshammer, M. (2001). Co-operation and conflict in a hospital: interprofessional differences in perception and management of conflicts. *Journal of Interprofessional Care*, (15): 7–18. doi:10.1080/13561820020022837

Swan, K. P., Richardson, J. C., Ice, P., Garrison, D. R., Cleveland-Innes, M. & Arbaugh, J. B. (2008). Validating a Measurement Tool of Presence in Online Communities of Inquiry. *E-mentor, 2*(24).

Wenger, E. (1998). *Communities of Practice: Learning, meaning and identity*. Cambridge, UK: Cambridge University Press.

Wesch, M. (2008). *An anthropological introduction to YouTube*. Presentation at the Library of Congress. Retrieved September 29, 2009, from http://www.youtube.com/watch?v=TPAO-lZ4_hU&feature=fvw

World Health Organization. (1988). *Learning together to Work Together*. Geneva, Switzerland: World Health Organization.

World Health Organization. (2009). *Chronic Diseases and Health Promotion*. Retrieved September 30, 2009, from http://www.who.int/chp/chronic_disease_report/part2_ch2/en/index1.html

# Chapter 8
# Fundamentals of Interprofessional Communication:
## A Case Study of an Online Facilitated Learning Experience

**Patricia Solomon**
*McMaster University, Canada*

**Sue Baptiste**
*McMaster University, Canada*

## ABSTRACT

*This chapter presents the development, implementation and evaluation of a module on interprofessional communication skills that incorporates principles of problem-based learning, delivered entirely online. Learners focus initially on foundational concepts of relationship and patient centered care, the importance of self awareness and understanding their own professional values and biases, progressing towards teamworking to develop common patient care goals. The module faculty facilitator is essential to role model and foster interprofessional collaboration. Qualitative content analyses of discussion board postings across 29 students, supplemented by small-scale in-depth interviews and a focus group, reveal they are able to learn interprofessional communication skills online. The 10 students who undertook both module components completed a project evaluation form: there was 85.6% agreement that the module taught them about interprofessional education and 92.9% agreement that their knowledge of other health professionals' perspectives increased. An online module can support the development of communication skills, but is recommended as one component of an overall interprofessional education curriculum.*

## INTRODUCTION

A number of authors have outlined the competencies required for successful interprofessional collaboration (Barr, 2001). Of central importance are those addressing competencies related to communication. Arguably the ability to communicate clearly and effectively with others is foundational to any collaborative interaction. Courses devoted

DOI: 10.4018/978-1-61520-889-0.ch008

to the development of communication skills in prelicensure learners have evolved in recognition of the foundational necessity of these skills (Priest, Sawyer, Roberts & Rhodes, 2005).

Communication skills are thought to be best taught experientially; in consequence most of the courses described in the literature are face to face (Priest et al., 2005). While online and virtual learning activities have been used to promote inter-physician communication (Sijstermans, Jaspers, Bloemendaal & Schoonderwaldt, 2007; Johnsen et al., 2005; Stevens et al., 2006) and the characteristics of virtual teams have been explored (Suchan & Hayjak, 2001), the ability of students to learn communication skills with, from and about each other in an online environment is largely uninvestigated. A recent review and meta-analysis of evidence-based practices online provided only tentative conclusions regarding online learning approaches with potential for further application (Center for Technology in Learning, 2009). This extensive document focused on studies in which the effects were based on objective learning measures and met minimum criteria of quality. One clear recommendation was the value of encouraging students to reflect on their own levels of understanding of the material and concepts under discussion. Although the inclusion of quizzes did not reap the positive rewards expected, the inclusion of simulations appeared to be value-added.

This chapter will present the development, implementation and evaluation of an asynchronous online course on interprofessional communication skills consisting of two related modules: Communication I: Establishing and Understanding Relationships, and Communication II: Making the Most of Groups and Teams. The modules were developed through the Institute of Interprofessional Health Sciences Education, a virtual learning center, which is a joint initiative of McMaster University, University of Western Ontario, Laurentian University, and University of Ottawa in the province of Ontario, Canada.

## BACKGROUND: CURRICULUM DEVELOPMENT

The importance of establishing interprofessional teams to develop curricula focused on collaborative skills has been noted (Priest et al., 2005). Our team included faculty from medicine, nursing, physiotherapy, occupational therapy and social work who had interest and experience in teaching communication skills within their specific professional program. From the outset the team determined an important collective belief; to work effectively in a team, learners need to understand their own personal beliefs and values. Prior to development of the curriculum, the team felt it important to reaffirm beliefs and core values related to communication and relationships. Several meetings were devoted to developing an Interprofessional Practice Model; this served to promote discussion among the team about the process and content of the curriculum, foster mutual understanding about others experiences and beliefs, and ultimately serve as a guiding model for our curriculum (see model in Figure 1).

The team also determined pedagogical principles that would inform curriculum development. Problem-based learning using a problem-stimulus to facilitate student learning and discussion would be the central focus. This approach was thoughtfully chosen. One of the key values of PBL is the opportunity it offers to work in small groups: wrestling with common scenarios, determining common and individual learning goals, searching for information to address these goals, then synthesizing and integrating the learning within the group (Gijbeis, Gochy, Van den Bossche & Segers, 2005; Taradi, Taradi, Radic & Pokrajac, 2005). Based on the team's extensive experience, problem based small groups are perfect crucibles for practicing how to work with diverse personalities and achieve predetermined objectives within a safe environment, applying values of mutual respect, honest communication, and thus develop-

*Figure 1. Interprofessional Practice (IPP) Model*

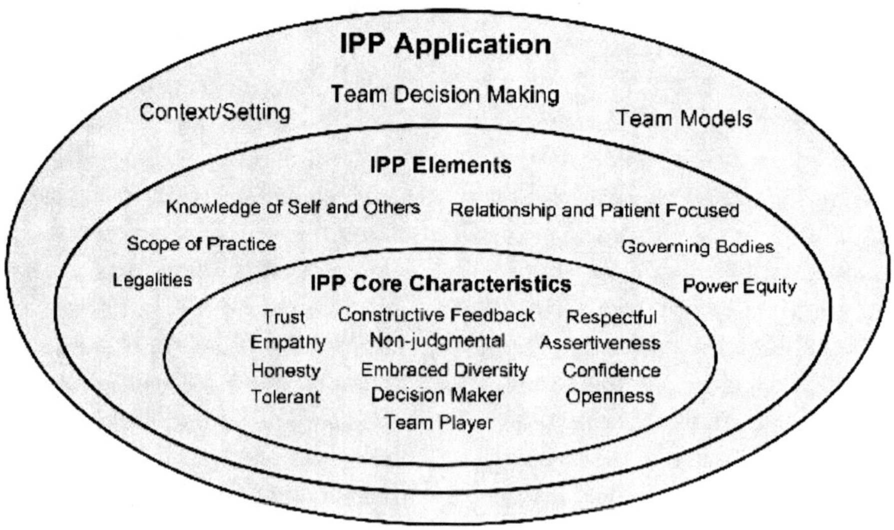


ing trust. These values and processes reflect the agendas of teams within the workplace struggling to realize individual professional missions while aiming to support and achieve the team goals and expectations.

The other principle that the development team agreed upon is the need for a faculty facilitator who would explicitly role model interprofessional collaboration to learners. Facilitators were selected for their mentoring experience and ability to provide support while guiding learners towards enhanced self awareness and integration of new skills in developing their emerging professional selves.

## THE CURRICULUM

This module is divided into two 6-Week parts, Communication 1 and Communication 2. Learners can participate in Communication 1 without continuing to Communication 2; however, Communication 1 is a prerequisite for Communication 2. Both parts are presented online through written narratives, videos, and case scenarios. Throughout the modules, students are guided by the Interprofessional Practice Model (Figure 1), learning the foundational concepts of relationship-centred care and patient-centred care. Each participant builds a learning portfolio that provides a means to follow individual learning processes. Learners participate in online discussions with their interprofessional colleagues and complete self-assessment assignments and personal reflections. There is one group-based project based on an interprofessional exploration of a case scenario chosen by each group from those offered. The weekly themes for each module are outlined in Figures 2 and 3, followed by further detail and examples.

### Communication I: Establishing and Understanding Relationships

Communication I gives students the opportunity to explore the importance of developing positive relationships with others and examines communication across professions. Learners also focus on exploring their values and beliefs and their decision-making process in selecting the career direction that led them to their educational program. The curriculum design is intended to progress

*Figure 2. Communication I: Establishing and understanding relationships*

| | |
|---|---|
| Week 1 | Self-awareness, relationship centered care, and patient centered practice |
| Week 2 | Values, beliefs & assumptions: self and others |
| Week 3 | Group environment for understanding values and beliefs: The group as a tool for self-awareness |
| Week 4 | Transitioning from group to team |
| Week 5 | The important elements and impact of forming a new team |
| Week 6 | Introductory understanding of challenges faced when working in inter-professional teams |

students through gaining better self understanding, understanding group formation, transitioning from group to team, to understanding the challenges of interprofessional teamwork.

Students are shown the importance of self-awareness as a foundation for relating to others in Week 1. Students reflect on the five "most remarkable and memorable experiences or "stepping stones" that brought them to where they are now." In Week three, students post on the online discussion board two stepping stones" which they feel comfortable sharing and think will help other group members of understand them a bit more. They then provide feedback on three stories that were shared online. Through these exercises students increase self-awareness of their own and their colleagues' personal values, start to form as a group through sharing of stories, are encouraged to reflect on judgments and assumptions they make, and have opportunities to practice giving feedback to other group members.

In Week 2 students explore their personal values and beliefs and how these could influence their professional practice. In one innovative exercise, students view a video of a patient's encounter with the health care team without sound. During the video, they reflect on what they see, who they think is involved, what is happening and their immediate response. They then share their interpretation online and discuss their viewpoints. They then view the video with sound and discuss how their personal beliefs and assumptions influenced their previous interpretation. Although on the surface the exercise appears simple, the discussion around assumptions based on dress, perceived role and

*Figure 3. Communication II: Making the most of groups and teams*

| | |
|---|---|
| Week 7 | Effective Team Communication |
| Week 8 | Team/Client/Patient Communication |
| Week 9 | Responding to Strong Emotions |
| Week 10 | Putting it all together: Self-awareness, relationship centered care, and patient centered practice as inter-professional practitioners. |
| Week 11 | Effective delivery of interprofessional patient centred team care in the practice setting. |
| Week 12 | Applying our learning to the Real World: Interprofessional team communication, relationship centered and patient centered care using real patient scenarios as the medium for learning |

*Figure 4. Problem-based scenario from Communication I, Week 6*

**Janet, having practiced for three years post graduation, has just started work on a new community health team as the first occupational therapist to be hired by this agency. She is bright, energetic and fully committed to practicing in an interprofessional manner. That is why she accepted this job – at the interview, she was assured that everyone believed in working in a respectful and collaborative way.**

**At her first team meeting, she introduced herself and started to talk about how she saw her role, asking people for their feedback and suggestions. Her input was received by a long silence, during which there was much shifting around on seats with people carefully avoiding eye contact with each other and Janet.**

**The physician, who is the meeting chair and the team leader, eventually stated that what Janet was hired to do was already spelled out in the job description; that they (the existing team) had spent time thinking about the gaps in service that they thought an OT could fill. What she was outlining was not what they needed. What was this client-centred approach anyway? No; she needed to get on with what was expected and not what she thought she would like to do. Did she think that she could just talk to patients – no, they needed splints made and raised toilet seats provided.**

**No one else at the table said anything; they also did not look at Janet. She felt very uncomfortable but stayed quiet, excusing herself from the room at the earliest possible opportunity.**

gender, for example, are rich, engaging and often startling to learners.

In Week 3 students develop a set of group norms for working together. As students become more comfortable with their learning group and in giving and receiving feedback, they reflect on their particular professional role within the team. An online discussion focuses on the different professional roles and responsibilities in a team, relating the role descriptions shared online to academic literature on team building and functioning. The final exercise is a problem-based scenario (Figure 4) in which students write a 500 word reflection on two specific team challenges emerging from the scenario.

## Communication II: Making the Most of Groups and Teams

In communication II, students build on the foundational knowledge and skills from Communication I. Short video recordings are used extensively to reinforce learning and promote online discussions. Students are introduced to differences between team communication tasks and skills, and use the Team Communication Observation Guide (deLottinville, 2006) to evaluate recorded scenarios. Students use an evidence based approach for learning communication skills (for a teaching module example, visit: http://www.skillscascade.com/), searching for and sharing research findings related to effective use of communication skills. The content becomes increasingly complex as students examine their own reactions and responses when confronted with a challenging clinical situation and learn strategies for dealing with strong emotions.

In Week 9 students share and respond to each others' "corrective scripts" for one scenario which includes three different skills constituting a more effective communication strategy. In Week 11 students write an interprofessional group care plan for a challenging patient/client who through homelessness and addictions raises issues related to stigma and stereotyping, response of the health and social care systems and collaborative patient centered care. The video clip is a powerful true life interview of a homeless man struggling with an addiction to 'Listerine' (alcohol-based mouthwash). In the final week students choose one of three scenarios, then develop an interprofessional care plan which is submitted for evaluation.

## Role of the Faculty Facilitator

We have found that successful online facilitation encapsulates a positive attitude to the module curriculum as well as modeling respect, regard for all participants, a desire for understanding and clarification of discussions and conversations, and a natural ability to embrace a student-centred approach to the learning experience. There is emerging agreement within published literature about the challenging role of 'online educator' and the need to be flexible, moving between elements including: skills in managing the learning environment through consideration of operational details and troubleshooting any arising technical difficulties, as well as exemplifying optimal abilities in overall facilitation (Kemshal-Bell, 2001; E-learning Facilitators, 2006; Gustafson & Gibbs, 2000). While facilitators often play a central role in developing the learning module content, this is not always so, as illustrated from our experience. We recruited faculty who were experienced in working in interprofessional team environments and in facilitating small group problem-based learning face to face. Facilitators also had experience as learners in an online environment.

While recruiting the facilitators, the module team recognized that expertise in facilitating small groups face to face did not necessarily translate into a comfort when applying those skills to the online environment. Our facilitators rapidly identified areas where they found some dissonance between their expectations of learners and what learners actually delivered. For example, with face-to-face groups, non-verbal communication and the general context are critical and importance sources of information for facilitators and tutors. When these cues are unavailable, how does one compensate? Our facilitators quicky appreciated that the style of learners' postings provided information about who they were and how they felt about the learning experience. These also provided opportunities to engage in "conversation" to probe assumptions, interpretations and statements about new knowledge and understanding.

## Evaluation

Evaluation of the module, which was approved by the Research Ethics Board of McMaster University, included both qualitative and quantitative measures. These methods are summarized below, and will be dealt with in turn:

1. Content analyses of the online discussions across the programme.
2. Focus groups or individual interviews of students and online facilitators at completion of the project. Students were asked to reflect upon their learning, provide examples of when and how they learned about the roles and scope of practice of others, the value of the collaborative learning activities and their perception of their most valuable learning moments. These were audio-recorded and transcribed verbatim and analyzed using a structured qualitative content analysis.
3. Interviews with academic facilitators to explore challenges and strategies that were unique to online facilitation of interprofessional communication skills.
4. An overall project evaluation form, completed after the modules and prior to focus group sessions to avoid contamination with the other group members. Students indicated the extent of their agreement with several statements on a 5 point scale ranging from strongly agree to strongly disagree. This questionnaire elicited feedback on the overall look and layout of the modules, the online discussion and the facilitation and interprofessional educational content, however for the purposes of this chapter, only statements related to the latter are presented.

## Participants

A total of 29 students volunteered to participate in Communication Skills I, which was a prerequisite for Communication Skills II. Not unexpectedly, as this extra-curricular, non-credit volunteer experience was held over the summer months, only 10 students completed all aspects of the course and evaluation components, although all 29 participated in the module to some degree. Four of these enrolled in Communication II and all completed the course components. Students were from medicine (n=1), nursing (n=4), occupational therapy (n=6) and physiotherapy (n=3) programs. This chapter focuses primarily on the qualitative content analysis of discussion postings across both parts of the programme. These data are supplemented by personal interviews with 4 students and a focus group of 7 students. Furthemore, 2 academic facilitators were interviewed. The 10 students who finished both course components also completed project evaluation forms.

## Findings

### Analyses of Online Discussions

The online discussions between all 29 students were downloaded and analyzed using an open-coding technique involving a line by line review of the online discussions to identify processes and content related to learning. All codes were reviewed, themes related to interprofessional learning were developed and representative quotes identified. This analysis identified and coded themes related to both process and content.

### Process Themes

Themes related to process included students' strategies to promote group work, how they addressed each other, the attitudes portrayed and how they proceeded through their learning activities. Seven specific themes were articulated and are offered here with relevant illustrative quotes:

*Asking another's perspective* emerged in situations in which dialogue went beyond posting relevant content. Students acknowledged the professional perspectives of peers and asked for other's viewpoints. The following posting illustrates a student giving positive feedback to a nursing student prior to asking others for their thoughts:

*I was curious about your nursing diagnosis as "unbalanced nutrition" and the evidence being BMI>30. The rest of the team labeled it "obesity". Interesting, because I think your approach is more sensitive (client-centered) and less objectifying perhaps. It made me think about labels and stigmas folks endure in society in general. What do others think about this?*

*Collaboration* was exemplified when students attempted to collaborate in their learning activities. This student portrays a willingness to partner and collaborate by volunteering to synthesize the information from a long discussion:

*I have combined the input from all members into one table. The further tasks as I see them are to edit the content to be concise and prioritize the problem areas (perhaps by numbering). Hope what I have done is useful. I'll let someone else take it from here.*

Not unexpectedly, rehearsal through collaborative communication soon bore fruit in the areas of *Honesty* and Openness. As students progressed through the module, they appeared increasingly comfortable with the process and were able to share personal experiences. Following a long discussion on whether those living in poverty have a "choice" one student remarked:

*Personally I've been on the poverty line twice in my life. Both times were connected to my 2 immigrations to the new country. I don't think it says a lot about me as a person, rather about the circumstances my life put me in. The best you can do in these situations is to set your priorities right, to have a goal and not to give up.*

Attitudes of *respect* were also apparent. This posting shows a positive sense of respect and partnership with other learners:

*It is amazing how we are all from different work places, may be different cultures we all were able to focus on the importance of communication and its importance of to the effectiveness of an outcome for a client.*

There were many instances in which students demonstrated *decision-making* by building on each other's expertise and knowledge to make team decisions. This entry illustrates a student agreeing with others, making suggestions and volunteering to take on organizational responsibilities:

*I agree that our approach should consider short and long term goals. Since Peggy has just presented to the clinic with these issues, we need to identify her priorities and help her with strategies to cope now with consideration of her long term goals i.e.: weight loss, improved coping with personal and work responsibilities and roles. It seems that we have consensus on column #1: collaborative problems being sleep apnea, painful knees, unbalanced nutrition/obesity and depression. How do we feel about this list? If we are in agreement on these, I will volunteer to streamline/ combine the concepts in the expected outcomes and submit.*

Students often demonstrated the ability to build on others' thoughts through acknowledging others' contributions, providing positive feedback, and agreeing with the direction of the discussion. Thus the skills of *relating and agreeing* were part of the students' tool kit:

*I agree with your identified problem areas regarding the sleep issues and painful knees. I like how you introduced the idea of family centered goals and education. I think the inclusion of the family is a crucial strategy in this case with respect to lifestyle changes (meal planning, diversion activity, increased physical activity, healthy choices, and support). The idea of education on reducing co-morbid conditions is also a great strategy to motivate clients.*

There was a willingness to explore *sharing resources* in order to enhance the learning of others:

*I found this article which I think could help to deal with the last scenario (PT, OT, client meeting). It describes several different models of decision making and highlights the pros of shared decision making (such as improved patients' outcomes and satisfaction).*

## Content Themes

Content themes are those relating to discussions on the substance of the module, communication and also on information provided about specific professional roles and responsibilities. The content analysis articulated five content themes.

Students commented on the dynamics of employing *communication strategies* and other specific strategies that could be utilized in team settings. In this posting a student comments on a videotaped team meeting:

*Physically people were not sitting at the same eye level which also doesn't give a flow to the meeting. In a meeting all the team members have to sit around the same table, be at the same level and discuss appropriate issues. That shows respect to the client and to each other at the team meetings.*

Although this module was not specifically focused on *ethical issues*, these became apparent as the discussions often involved sharing of personal values and beliefs:

*If a patient is non-compliant it becomes more of a moral issue than anything. Removing the patient from the practice can spare [the physician] some legal trouble but he has to deal with what he may have done to the patient.*

The module emphasized the importance of client centered practice. Hence there were frequent discussions about communicating with clients and their families as part of the team process and development of *client relationships*:

*I agree that in this team meeting the nurse should be able to express honestly the feelings of her colleagues. It's just that I felt it was framed so negatively. Perhaps the nurse could have reported the daughter's behavior to be of concern and list it as a problem to be discussed more fully. That way the focus is on the effort to understand more about the daughter's experience which is manifesting in a negative way, through her behavior. The team could establish a goal to include the daughter and her concerns about her mother on her own or in a family meeting, and therefore better support the client and family during the discharge planning phase.*

An important component of interprofessional education is learning about the *professional roles*, functions and scope of practice of others. This quote provides the perspective of an OT student who is trying to clarify roles:

*Other similarities are that both OT and PT promote wellness, prevention and optimum quality of life. OTs and PTs endeavor to practice in a client-centered and holistic approach to care. Because the therapeutic process can vary in length, the OT and PT, let's say in rehab or community setting, can have the time to develop the therapeutic relationship and better facilitate a positive, meaningful outcome for the client.*

*Teamwork or collaborative strategies* were recognized readily by participants as important to learning.

As students progressed through the module they offered strategies to improve their collaborative practice:

*I think that in a well functioning team, members might rotate the role of facilitator. It doesn't matter who, as long as there is some degree of process and structure that is followed to promote successful communication. I feel that someone needs to be responsible for this i.e. the rotating chairperson, and they need to have a process to follow, i.e. using a tool like the Kalamazoo.*

The online discussions represent 'real time' dialogue. Thus they differ in nature to responses during focus groups and interviews. We discuss the latter next, however, in order to reduce repetition, we present a brief summary below.

## Student Focus Group and Interviews

Overall, the content analyses of 4 student interviews and the 7-student focus group reinforced the students' perceptions of valuable learning in the content area of communication, and about interprofessional roles and scope of practice. Students became aware of overlapping roles, and the implications for practice:

*There was evidence of role blurring. It caused me to think more about what happens when that is occurring in the interprofessional team and that there's different ways of resolving that.*

Students were surprised at the lack of knowledge about their specific roles:

*It confirms that most people don't know about my role as a nurse practitioner.*

Students recognized that this might require them to advocate for their professional role in their future work:

*What I did learn is how others see my profession. So I would be working with an occupational therapist and seeing how [she] perceived what we did and how I was able to advocate for what we did.*

Students identified certain particularly helpful exercises or areas of reading and discussion. For example, one student discussed how important it was for her to realize the varying perspectives that each person brings to the team relationship:

*...the stepping stone really helped because it shows how each person kind of came to be, how they are, had their own unique way of looking at things...so that you can kind of understand where everybody's coming from but still kind of stay true to what you believe.*

For another student, recognizing the difference between groups and teams was important:

*I hadn't realized about the difference between teams and groups...I think it might be useful for me if I work in the future on the interprofessional team... it might be useful to help me to get adjusted and probably to improve the practices and function of the team. It was very helpful.*

Surprisingly given that this was an online experience, one student spoke of how in any communication "listening was a big component." Another student realized that communication also emcompasses non-verbal cues:

*[communication] is just everything from body language, facial expressions...*

The exercise of watching a video scenario both with and without sound reinforced this and helped them become aware of their own assumptions.

Some students indicated that occasionally the discussion wasn't as rich as it could have been.

*Everybody sort of came to the agreement that it's all a good idea and so we think everyone should do it but when you talk about some of the challenges people face and how to work around them...I didn't feel the discussion was...or went as in-depth as it could have.*

## Interviews with Facilitators

Making the transition to online facilitation, as opposed to teaching in a classroom setting or in small group PBL brings unique challenges, regardless of the content. We conducted in-depth interviews with 2 online facilitators following 3 instances of the module. We were interested in the challenges and strategies that were unique to online facilitation of interprofessional communication skills, and summarise our findings below.

It requires a skilled facilitator to promote in-depth dialogue in an open and non-defensive way. Both facilitators were very experienced as small group problem-based tutors. Despite their previous preparation, in comparing their online experience with the face to face experience, they felt it was harder to promote an in-depth level of discussion. Because of the comparative anonymity of the online environment:

*uncomfortable discussions can just be dropped – and it is very frustrating and really hard to keep them going.*

As the focus was on communication, difficult discussions about communication styles, relating and power were anticipated; however:

*we didn't get at the power relations quite as much.*

Conversely, the anonymity provided a degree of safety for some students who may have been reluctant to discuss communication issues face to face:

*[some students] were more forthcoming about their opinions. They had more motivation or determination or confidence to assert their opinions when they are typing in the computer than if they were facing someone who was giving them all kinds of power body language.*

The facilitators felt that the safety of the environment also provided an opportunity for students to:

*reflect on the experiences that they've had of good or bad functioning teams and to be able to work that out among other students rather than people who are supervising them.*

The facilitators worried about bringing in their own professional biases and their confidence in knowing about the specific roles of others' professions. Facilitator guides can provide insight into key professional roles in a given scenario, outline areas of controversy and provide strategies and questions to promote in-depth learning. For example, a physiotherapy student felt it was inappropriate to talk to an obese patient about her diet, and the facilitator, an occupational therapist, was unsure what the role of the physiotherapist should entail in this situation. Similarly, the facilitators worried that students' learning may be compromised if a specific profession was not represented in the group.

Facilitators found that it was important for them to reflect personally on their personal impressions of other professions, and how they presented their own. One facilitator felt it was important to "remain neutral" as facilitators are not there to "promote (their) own professional biases." One facilitator spoke of the importance of providing positive interprofessional experiences from her own clinical background to give a balanced perspective when students engaged in negative stereotyping.

Another benefit to the online environment was ease of participation for students in whom English was not their first language:

*"There was a woman from China who I did have a chance to meet and she was a very quiet person. She tended to post last in the beginning and wait for other people's responses and so I imagine if it had been a classroom situation she probably wouldn't have come out with some of the rich responses that she did come out with eventually..."*

## Project Evaluation Form

The project evaluation form was completed by 10 students who participated in both Communication I and Communication II. Figure 5 outlines the students' perceptions of their learning. For ease of reporting, statements which indicate students agree or strongly agree are combined into one category indicating the percentage of students agreeing with the statement. Similarly the percentage that disagree or strongly disagree with the statement have been combined. As can be seen, student feedback was very positive with 85.6% agreement that the module taught them about interprofessional education and 92.9% agreement that they had improved knowledge of other health professionals' perspectives.

## DISCUSSION AND FUTURE DIRECTIONS

Our course emphasized the importance of reflection, self-awareness and the principles of individual/team communications, as described to students in the module descriptor. In a 'traditional'

*Figure 5. Students' perceptions of learning following module completion (n=14)*

| | Agree (%) | Neutral (%) | Disagree (%) |
|---|---|---|---|
| **The module...** | | | |
| ...taught about IPE. | 85.7 | 14.3 | 0.0 |
| ...taught me about working collaboratively within a team for better patient outcomes. | 92.9 | 0.0 | 7.1 |
| ...provided me with opportunities to work with students from other health professions on finding solutions to joint problems. | 78.6 | 7.1 | 14.3 |
| ...helped me learn about other students' roles and responsibilities. | 92.9 | 7.1 | 0.0 |
| ...provided me with opportunities to work with students from other professions on identifying and defining problems. | 92.9 | 7.1 | 0.0 |
| ...helped me learn about other students' perspectives on the subject matter of the module. | 92.9 | 7.1 | 0.0 |
| **Compared to the beginning of this module...** | | | |
| ...I have a better knowledge of other students' perspectives on the content of this course. | 92.9 | 0.0 | 7.1 |
| ...I am better able to identify shared tasks that require joint solutions. | 85.7 | 14.3 | 0.0 |

communication course, learners would practice these skills face-to-face. Arguably these are foundational to all interprofessional communications, and the related knowledge could be delivered in any format. It is evident from the evaluation that students successfully learned with, from, and about each other in the context of a communication skills module delivered entirely online. Our findings offer evidence that asynchronous online learning can allow students more time for reflection. Learners have more time to monitor, compose and organize their thoughts and responses than in a traditional classroom (Salmon, 2002; Lai & Land, 2009).

Potentially online learning which specifically promotes reflection is more effective in developing these skills. This emerging theme merits further investigation (Center for Technology in Learning, 2009). The extent to which learning communication skills online impacts upon skill development compared to face to face interactions is, however, unknown. Ultimately, this question can only be answered by comparing the performance of learners from both modes of learning.

Given the costs of providing a faculty facilitator, a course in which learners are totally self-directed is appealing. However, in our view the faculty facilitator is essential. Chen et al. (2008) found that higher level questioning of learners promote higher levels of reflection in an online environment; this is a critical role of a facilitator. Facilitators also need to ensure students are engaged in meaningful interactions rather than monologues (Lai & Land, 2009).

However, facilitators also promote interprofessional education. Students did not always recognize when interprofessional learning occurred, highlighting the importance of a skilled facilitator to make this learning explicit. The facilitator asked questions to challenge the students' values and assumptions, promoted interaction among learners, and summarized and highlighted key learning issues emerging from online discussions on a weekly basis. Facilitators also played an important role in balancing objectives related to the communications content, with objectives specific to IPE. Some of the scenarios were challenging and warranted faculty guidance, particularly for students with limited experience in dealing with individuals having different values or life circumstances than themselves. Our faculty found that tutor guidelines helped them determine how to direct conversations and reinforce important learning. We recommend the use of such guidelines to direct both the content and process of learning.

We found that students are able to relate more meaningfully to online clinical scenarios if they have had fieldwork placements (Solomon & Geddes, in press). However, arguably the need for awareness of one's values, assumptions, and biases should come early in a student's professional program, being a foundation for all communication skills development. Undoubtedly, a student's maturity level and previous life experiences influence their self-awareness, and may have little to do with year of study. The question of optimal time for students to engage in this type of course is another area of future inquiry, as few existing studies explore this with sufficient clarity.

## CONCLUSION

This work was pilot in nature and the number of students completing the module was small, thus potentially representing a highly motivated group. However, the data presented in this chapter affirm that students can learn interprofessional communication skills in an online environment. Online discussions provide a direct way of assessing student interactions, and the qualitative analyses provided valuable insights not accessible in face-to-face learning.

The findings from this pilot work provide encouragement for those interested in overcoming timetabling barriers by utilizing asynchronous online learning activities. It is important to emphasize that the online problem-based learning module is one component of an overall IPE curriculum. This module provided the opportunity for students to learn foundational skills to support communication skills development throughout their careers.

Ultimately students need to practice communication skills face-to-face in the context of real patient care. Whether the socialization and communication strategies break down professional barriers and allow more open communication and dialogue amongst professionals following licensure is a question for further research.

## ACKNOWLEDGMENT

The authors would acknowledge the ongoing contributions and dedication of Gissele Damiani-Taraba, Suzanne King, Nicole Gervais and Caroline Phelan who ensured the successful delivery and evaluation of this module. Production of this chapter was made possible through a financial contribution from Health Canada.

## REFERENCES

Barr, H. (2001). *Interprofessional education: Today, yesterday and tomorrow*. London: Westminster University, Learning and Teaching Support Network Center of Health Sciences and Practice.

Center for Technology in Learning. (2009). *Evaluation of Evidence-Based Practices in Online Learning: A Meta-Analysis and Review of Online Learning Studies*. Washington, DC: U.S. Department of Education.

Chen, N., Wei, C., Uden, L., & Wu, K. (2008). *Effects of Reflective Teaching Strategies on Online Learners' Reflection Levels*. Eighth IEEE International Conference on Advanced Learning Technologies.

deLottinville, C. (2006). *Team Communication Observation Guide*. Adapted from Participants in the Bayer-Fetzer Conference on Physician-Patient Communication in Medical Education. (2001). Essential Elements of Communication in Medical Encounters: The Kalamazoo Consensus Statement. *Academic Medicine*, *76*, 390–393.

E-learning Facilitators (ELF). (2006). *The roles of E-learning facilitators: Skills, competences and training.* E-learning Facilitators (ELF). Retrieved July 11, 2009 from http://conseil-recherche-innovation.net/wp-content/download/ELF/ELF - facilitators roles report - Greta du Velay.pdf

Gijbeis, D., Gochy, F., Van den Bossche, P., & Segers, M. (2005). Effects of problem-based learning: A meta-analysis from the angle of assessment. *Review of Educational Research, 75*(1), 27–61. doi:10.3102/00346543075001027

Gustafson, P., & Gibbs, G. (2000). Guiding or Hiding? The Role of the Facilitator in Online Teaching and Learning. *Teaching Education, 11*(2), 195–210. doi:10.1080/713698967

Johnsen, K., Dickerson, R., Raij, A., Lok, B., Jackson, J., Min, S., et al. (2005). *Experiences in using immersive virtual characters to educate medical communication skills*. Proceedings of IEEE Virtual Reality 2005 (VR 2005), 179-186. doi: 10.1109/VR.2005.1492772

Kemshal-Bell, G. (2001). *The online teacher*. Report prepared for the Project Steering Committee of the VET Teachers and Online Learning Project, ITAM, ESD, TAFE NSW. Sydney: New South Wales Dept. of Education and Training. Retrieved March 29, 2009 from http://cyberteacher.onestop.net/final report.pdf

Lai, T., & Land, S. M. (2009). *Supporting Reflection in Online Learning Environments* (pp. 141–154). Educational Media and Technology Yearbook.

Priest, H., Sawyer, A., Roberts, P., & Rhodes, S. (2005). A survey of interprofessional education in communication skills in health care programmes in the UK. *Journal of Interprofessional Care, 19*(3), 236–250. doi:10.1080/13561820500053447

Salmon, G. (2002). Mirror, mirror, on my screen… Exploring online reflections. *British Journal of Educational Technology, 33*(4), 379–391. doi:10.1111/1467-8535.00275

Sijstermans, R., Jaspers, M., Bloemendaal, P., & Schoonderwaldt, E. (2007). Training inter-physician communication using the dynamic patient simulator. *International Journal of Medical Informatics, 76*, 336–343. doi:10.1016/j.ijmedinf.2007.01.007

Solomon, P., & Geddes, E. L. (in press). An interprofessional e-learning module on health care ethics. *Journal of Interprofessional Care*.

Stevens, A., Hernandez, J., Johnsen, K., Dickerson, R., Raij, A., & Harrison, C. (2006). The use of virtual patients to teach medical students history taking and communication skills. *American Journal of Surgery, 191*(6), 806–811. doi:10.1016/j.amjsurg.2006.03.002

Suchan, J. & Hayzak, G. (2001). The communication characteristics of virtual teams: a case study. *Professional Communication, IEEE transactions, 44*(3), 174-186.

Taradi, S., Taradi, M., Radic, K., & Pokrajac, N. (2005). Blending problem-based learning with Web technology positively impacts student learning outcomes in acid-based physiology. *Advances in Physiology Education, 29*, 35–39. doi:10.1152/advan.00026.2004

Warkentin, M., & Beranek, P. M. (1999). Training to improve virtual team communication. *Information Systems Journal, 9*(4), 271–289. doi:10.1046/j.1365-2575.1999.00065.x

# Chapter 9
# Leadership Online:
## Student Facilitated Interprofessional Learning

**Lynn Clouder**
*Coventry University, UK*

**Marie Krumins**
*Coventry University, UK*

**Bernie Davies**
*Coventry University, UK*

## ABSTRACT

*A study investigating the effectiveness of a curriculum innovation involving students as online facilitators of interprofessional learning (IPL) provides a focus for this chapter. The research aim was to investigate whether Year 3 health and social care students were effective in facilitating online discussion forums contributing to the IPL of their counterparts in Years 1 and 2. Findings suggest that they were equally as effective as academic staff while offering some additional benefits. The account provides evidence of a successful online interprofessional initiative involving students promoting IPL.*

## INTRODUCTION

This chapter considers the potential of students to provide leadership in promoting interprofessional learning (IPL) amongst their student peers. We draw on our findings from a study carried out at Coventry University in the United Kingdom (UK). Debate about the most appropriate models of IPL is ongoing and approaches are both diverse and numerous. However, the possibility of students being prime movers in inculcating interprofessional beliefs, values and attitudes in their peers has been largely unexploited. Our focus in this chapter is on the benefits that less well advanced students gain from online facilitation provided by a more advanced student. The chapter will be of interest to institutions developing either a blended or online approach to promoting IPL, which requires online facilitation and to those willing to consider that students can play a part in others' learning and are a valuable resource for advancing IPL as online facilitators.

DOI: 10.4018/978-1-61520-889-0.ch009

## STUDENT MENTORS IN HIGHER EDUCATION

Throughout the chapter we use the term 'student leadership', which invokes a sense of advocacy, guidance and direction. The general literature refers to 'peer leaders', 'peer tutors', 'peer facilitators' and 'peer mentors'. However, the term 'student proctor' has also been used in American contexts to describe students helping other students to learn (Saunders, 1992). Notwithstanding subtle differences and connotations to these concepts we adopt the terms 'leadership' and 'mentorship' and use them interchangeably throughout discussions. The model of student leadership adopted involves peer mentoring by senior students of less advanced students (Goodlad & Hirst, 1989). Leadership capability seems to be an increasingly important attribute sought by employers (Wagner, 2008) therefore evidence of having engaged in a formal programme of peer leadership might prove beneficial to students when composing their curricula vitae in preparation for employment.

The sound body of research evidence on peer-led learning highlights its academic benefits for all concerned (see for example, Gosser et al., 2001). Research by Goodlad and Hirst (1989) suggests that students benefit from peer mentors in that achievement is improved and is less variable. Students are generally positive about input from peer mentors, possibly because the learning feels less formal and the learning environment is less threatening without the presence of academic staff (Hayler, 1999). Indeed, Jamieson and Thomas (1974) had some time ago highlighted issues of power and conflict in the student-teacher relationship that has a profound effect on student satisfaction and learning. Micari, Streitwieser and Light (2006) suggest that students may be well placed to help other students understand material that they have recently learned which provides them with an innate ability to help students in a manner best suited to their developmental level. This point is supported by Saunders (1992, p. 216) who suggests that 'the value of peer tutoring is that recent student experiences are passed on to others, and that relatively more advanced students have valuable insights, which academic staff do not'. The criticism of academic staff suggests that students might be more up-to-date with current thinking and practice but may also relate to other issues such as staff being ill prepared or struggling to find time to offer support and guidance (McCall, 2007). Gallew's (2005) research findings on the benefits gained from senior occupational therapy students developing a lecture for first year students suggest that they perceived that the senior students were enthusiastic, confident, professional and possessed positive leadership skills.

Saunders (1992) raises several other important points in his critique of a peer tutoring initiative in engineering at what was Nottingham Polytechnic in the early 1990s, which still have contemporary relevance. He stresses the importance of thorough briefing about the organisation, aims and expected outcomes and need to attend to training requirements. Highlighting major decisions that must be made by course teams when setting up peer mentoring projects, Saunders encourages debate around issues such as whether schemes should be compulsory or voluntary, whether students should work on their own or with other tutors and whether tutoring should be assessed.

Despite recognised potential benefits of peer mentorship, its use in IPL has not been widely adopted. In fact, Micari et al. (2006, p. 285) point out that students 'rarely enjoy a formal opportunity to help other students advance in their intellectual development'. They attribute this to 'academia's reluctance to endow un-credentialed students with the authority to teach, or ...simply the absence of an infrastructure to support such an endeavour' (Micari et al., 2006, p. 285). Reflecting on the fact that students are generally positioned as learners and not as sources of learning, they suggest that this influences how they are viewed by Faculty and how they understand their own capabilities and purpose.

## STUDENT LEADERSHIP IN IPL

The literature on peer led IPL is scant. Gill et al. (2006) report on a peer assisted learning (PAL) initiative within a medical school in the UK that was expanded to include an interprofessional dimension. Medical students, who had completed an optional teaching module, engaged in small group teaching with post registration or graduate nursing students. The initiative was evaluated positively with feedback that the medical students were enthusiastic, showed interest and commitment and most importantly there was indication that interprofessional understanding and working had been improved. Gill et al. (2006) stress that the success of this type of initiative is dependent on preparation and ongoing support for peer facilitators and close working with academic staff.

A Canadian interprofessional initiative involving students in promoting IPL in selected Canadian health programmes provides insight into the benefits, challenges and implications of student leadership in IPL (Hoffman et al., 2008). Drawing on diverse evidence the authors stress the importance of student leadership in helping to enhance students' willingness to collaborate and in the long term vitality and sustainability of interprofessional efforts.

In discussing models for promoting student leadership, Hoffman et al. (2008) consider involvement in curriculum design and implementation through participation in working groups and steering committees, as well as student-initiated IPE. Students leading the design and delivery of IPL for their peers appears to be confined to relatively low risk discrete projects, such as running conferences or time-limited and stand-alone courses. The challenges identified in student initiated IPL include lack of funding, lack of institutional and/or administrative support, lack of faculty mentorship and/or guidance, lack of leadership opportunities and indeed interest in leading IPL activities as well as a lack of student interest in participating in IPL.

These factors are interesting in that they lead us to question whether in fact staff facilitated initiatives face similar challenges. Many of the structural issues are not insurmountable although, promoting interest in IPL might be a significant sticking point. Research suggests that passion and enthusiasm are essential qualities for teachers in general (Gibbs, 2003; Gibson, 2009), therefore one might assume that if this is lacking it may be even more of a challenge for IPL initiatives.

## LEADING LEARNING ONLINE: AN ADDED COMPLICATION?

To this long list of potential challenges a further complication can be added; that is the delivery of IPL in an online format, which characterises the programme at Coventry University. Online delivery of IPL has potential benefits in terms of overcoming some of the difficulties associated with face-to- face IPL. For instance, Scammell et al. (2008), reporting on a study of experiences of IPL using online discussion based on a virtual community, suggest the approach overcomes the difficulties of geographical distance and creates equity of experience for large numbers of students. Their findings suggest that students themselves particularly value the opportunity to explore contentious issues in what they perceive to be a safe environment. Findings from another study of the use of online modules for collaborative learning amongst health professionals, reinforces the suggestion that online delivery overcomes location barriers (Walsh, 2007). This study also illustrates the potential benefit of an online approach in helping to overcome the barriers of different methods of learning adopted by different professional groups (Walsh 2007).

The benefits offered by computer technology include 'new opportunities for learning collaborations and partnerships, strengthening and extending learning communities, promoting new ways of communicating and investigating, and providing better access to an increasingly wide range of discipline-specific educational and research-based pedagogical resources (Ellis et al., 2007, p. 83). Online discussion, in particular, is recognised for its 'reflective, collaborative and evaluative affordances' (ibid, 95). These benefits can be attractive to those wishing to promote an IPL strategy, which is scalable, manageable in terms of logistics and effective with respect to facilitating contact and exchange between students from the various professions, as well as across institutions. However, Ellis et al.'s (2007) research confirms that good discussions (whether they occur face-to-face or online) cannot be taken for granted; students need to understand how they should approach discussions and why they are discussing. This clearly requires careful attention to how resources are designed and discussions structured in order to optimize dialogue. It also has implications for how discussions are facilitated.

Online IPL clearly has some benefits although there are also challenges. The most obvious challenge is getting to grips with using technology. Scammell et al. (2008) found that both staff and students found familiarising themselves with the technology time consuming, although this was later valued as a transferable skill useful in other areas of learning. A major difference in online communication and a factor which characterises contemporary communication to an increasing degree is the lack of face-to-face contact that robs interaction of the subtleties of non-verbal communication, facial expression and intonation of voice. While this may be freeing for some students it is also experienced as limiting by others and possibly leads to the development of increased sensitivity about how words are used and can be interpreted by others. Certainly these factors point to a different sort of challenge for online facilitators attempting to create an atmosphere of trust and openness online.

The skills of the interprofessional facilitator have been identified as crucial to the success of face-to-face IPL initiatives (Oandasan & Reeves, 2005). Rees and Johnson's (2007, p. 550) research findings suggest that facilitation of IPL is 'anything but easy'; staff felt that they needed more support and found working outside their area of expertise challenging. Being able to 'think on your feet' was highlighted as vital for successful facilitation, which was perceived as an advanced skill. Online facilitation adds another layer of complexity to using facilitation skills effectively, not least in simply getting students to engage in posting basic messages. Due to its asynchronous nature it is arguably less demanding with respect to some challenges, such as having to think on one's feet, because it allows time for reflection. Nevertheless, Miers et al. (2007) highlight the importance of staff development in e-facilitation, especially if apparent poor analytical debate online is to be addressed.

Salmon (2004) uses the term 'e-moderation' to refer to the role of the online teacher or trainer and identifies a range of qualities and characteristics of successful e- moderators. She highlights a requirement to understand the online process, based on personal experience as an online learner, the need for reasonable technical skills, good online communication skills, knowledge, experience, motivation and determination to become a good online facilitator. Salmon suggests in recruiting e-moderators, it is important to find credible members of the learning community but not necessarily subject experts and quotes Knight's (2002) suggestion that ability to empathize and show consideration for learners are important factors for success. She is a firm advocate of staff development, suggesting:

*any significant initiative aimed at changing teaching methods or the introduction of technology into teaching and learning should include effective e-moderator training and support, otherwise its outcomes are likely to be meagre and unsuccessful (Salmon, 2004, 80).*

Much of the research in online facilitation of student discussion forums has involved academic staff rather than peer facilitators, which serves as a point of departure for this chapter, given that the initiative described presently involved student leadership in an online setting.

## RESEARCH CONTEXT AND APPROACH

The interprofessional learning pathway (IPLP) at Coventry University is underpinned by a social constructionist pedagogy, which focuses on 'acculturating students to the communities they wish to enter' (Thralls & Blyler, 1993, p. 251). The aim is to provide opportunity for collaborative discussion so that students come to understand how an interprofessional community uses discourse; in other words 'to come to consensus about knowledge affecting that community and adopt communal norms governing discourse practices' (Thralls & Blyler, 1993, p. 251). Within the IPLP, discussion focuses on a scenario and learning is facilitated through virtual interaction between students using asynchronous online discussion forums in the institution's virtual learning environment (Blackboard). With the exception of one introductory face-to-face meeting with their facilitator, students do not routinely meet one another during the four to five week period of interaction. Discussion forums have traditionally been facilitated by academic staff all of whom complete the in-house training course. By addressing both facilitation and IPL skills, this course discourages facilitators from acting in a uniprofessional manner, something that Scammell et al. (2008) identified as a challenge. The student facilitator project provided opportunity for students to be trained for and to take on the facilitator role; the asynchronous online nature of the IPLP discussion forums provided flexibility of time commitment, which meant that students could work around other demands, such as coursework deadlines.

The research aimed to explore the effectiveness of a small number of Year 3 students in facilitating online discussion forums contributing to IPL of their peers in Years 1 and 2. Ethical approval was sought and gained from the Coventry University Ethics Committees and the project was given full backing from Senior Management in the Faculty of Health and Life Sciences.

An open invitation for involvement was posted on the IPL site on the Blackboard VLE as well as being advertised on posters on student notice boards and through Course Directors. Student facilitators were then recruited in summer 2008 with a view to commencing facilitation of forums in October 2008 and March 2009. Student volunteers attended a face-to-face meeting at which they were briefed about the project, including details about payment on a pre-set hourly rate. Students were given a participant information sheet and completed a consent form for involvement in the research aspect of the project. Course Directors were asked to provide a short reference for each student as a means of ensuring that they were not experiencing any difficulties with their programmes that might interfere with the role, and students' own contributions to their discussions forums in previous years were checked to ensure that they had been active contributors with a commitment to IPL. In addition, students needed to feel able to commit the time to the project in the context of their own studies, to exhibit good interpersonal skills, to have access to a personal computer at home as well as at university and to be willing to complete a facilitator training course. Twelve Year 3 students were recruited to the project. Figure 1 shows their professional background.

*Figure 1. Student facilitators by profession*

| Course | Number of student facilitators |
|---|---|
| Adult Nursing | 5 |
| Medicine | 1 |
| Physiotherapy | 4 |
| Social Work | 2 |

Facilitator training commenced 3 weeks immediately prior to the first facilitation period, the final fourth week running into the facilitation period. The programme, which was a hybrid version of the staff training course that was running concurrently, was delivered totally online and supported by experienced academic staff IPL facilitators. Students had their own online forum within the training programme so they could share ideas and concerns with one another. The course was assessed through a written reflective account and all students received a certificate in online facilitation on completion. As a means of recognition of the students' input, agreement was gained to offset the Year 3 IPLP assessment with the training course assessment.

During the training and the facilitation period, a learning technologist provided face-to-face, telephone and email support for any technical issues as well as other queries raised by the student facilitators. Although additional academic staff were also on hand to provide support they did not routinely visit the online forums facilitated by students on the basis that we wished to convey a sense of trust in the students' commitment to their task. A total of 24 Year 1 forums and 22 Year 2 forums were facilitated by students over two four-week periods. Each forum contained approximately 15 students from a range of professions including: adult nurses, mental health nurses, learning disability nurses, children's and young people's nurses, midwives, medics, physiotherapists, occupational therapists, dieticians, paramedics, operating department practitioners, youth workers and social workers.

## DATA COLLECTION AND ANALYSIS

Several data collection methods were utilised to gain multifaceted insight into the project. In-depth interviews were conducted with student facilitators. On completion of each of the four-week IPLP periods the same online questionnaire was completed by all Year 1 and Year 2 students enabling comparisons across groups facilitated by both staff and by students. In Year 1, 163 from a possible 360 students (45%) from student facilitated groups responded to the questionnaire. In Year 2, 233 from a possible 308 students (71%) in student facilitated groups responded to the online questionnaire. Data from open questions in the Year 1 questionnaire were subject to qualitative data analysis and it is these data that form the basis of the findings presented. At the time of writing Year 2 data had not been fully analyzed, although some qualitative comments from Year 2 evaluations are used to support insights.

### Research Findings

A number of themes arose from the qualitative data that when combined seem to indicate the overall effectiveness of the student facilitators in enhancing the student learning experience as well as IPL. It appeared that student facilitators were equally capable as academic staff in promoting discussion that helped students understand their role in empowerment, (a key focus of the IPL module); helping students understand how their profession can collaborate to improve care; and recognising the value of interprofessional working. In fact they proved themselves able to cope with conversations about what are frequently sensitive and challenging topics with respect to both content and interprofessional context. In terms of skills, Year 1 students reported having developed ability to listen to others' points of view and confidence to discuss sensitive issues. There was some indication from student comments that student facilitators promoted greater insight into the role of other professions than academic staff facilitators and were better at encouraging students to reflect on their beliefs and values in relation to inequality and empowerment. For instance, one student suggested:

*I think it is good because they were able to tell us what to expect from IPL and how it helps in many ways in terms of learning about different professions and their jobs at the same time as keeping in mind the scenario and dealing with it.*

In a series of open questions, students were asked what they thought about having had a student facilitate their group. Notwithstanding the point made by several Year 1 students that they could not say whether student facilitators were a good, bad or indifferent idea, as they had no comparisons, the majority of student responses indicated a perceived benefit of having a student facilitator over an academic staff facilitator. Responses were limited with respect to why having a student facilitator might prove to be disadvantageous. However, several students did raise the issue of lack of substantial practice experience in comparison to academic staff as a potential drawback. In fact, this suggestion was in all cases speculative as none of the students appeared to have found it to be problematic and possibly reflects a lack of awareness of the facilitation role. This finding appears to support Salmon's (2004) suggestion that good facilitators do not necessarily need to be subject experts. It seems salutary that one student stated that s/he "didn't realise we were being facilitated by a student".

Year 2 students were in a position to be able to judge their student facilitators against their earlier experience of staff facilitators. The majority appeared satisfied; students appeared to have done at least as good a job as academic staff. However, there were comments that suggest that in some cases student facilitation had been better:

*Yes the facilitator did a very good job this year. Was very helpful and gave good sound help.*

*I think the facilitator was better than previously, as [s/he] spent more time on IPLP and emailed reminders.*

The overall impression from Year 1 students was that there had been several areas of added value from the student facilitator input.

## Empathy

The first benefit reaped from the student facilitators' was their ability to empathize with their peers. The students felt that the student facilitators acknowledged the external pressures that they were under, which they felt staff did not seem to appreciate. Whether or not this is an accurate point of view is uncertain, although it does possibly reflect McCall's (2007) observation that staff struggle to find time to fully appreciate non-academic concerns of students or are ill-prepared to do so. Alternatively staff may simply lack the degree of understanding that might exist between peers of a similar generation. Perhaps most significantly students perceived that the student facilitators had been through the same process and were aware of time pressures, workload and other sources of stress. For instance, one student stated:

*She knows exactly what we are going through regarding this and our other workloads having only done this last year.*

Student facilitators appeared to be well positioned to be able to empathise and connect with students as advocated by Knight (2002). The sense of shared experience provides a commonality that promoted improved understanding and a level of trust that that might not be so readily nurtured in student/academic relationships.

## Trust Based on Credibility

Another student expressed a strong sense of trust in her/his student facilitator and the task in hand:

*They have experienced IPLP for themselves and do know what it is like. Because of this they want us to pass and therefore help us as much as pos-*

*sible so we could be in the same position as them in the future.*

This confidence in being helped through the process was attributed by the students themselves to the fact that the student facilitators had been through the IPLP before, which gave them demonstrable insight and experience to enable them to successfully lead their peers. In other words, they were credible members of a learning community (Salmon, 2004). Comments such as, "they have recent experience in the IPLP and can guide us in the right direction" and "they understand the system" and are "able to give advice from their own experience" suggest that the facilitators were very much accepted in their role in leading learning. These findings support previous research (Micari et al., 2006; Saunders, 1999) that suggests students may bring innate abilities to help other students learn because they possess teaching skills as a result of having recently learned similar material. In addition, the students felt that as a consequence of their direct experience, the facilitators knew what practical tactics to adopt to engage them in discussions:

*They have experiences participating in IPLP and can share this experience with you and know best how to engage you with the subject.*

*It is reassuring to have someone who is in charge so to speak, to keep things flowing and in order.*

The data suggest that the Year 1 students recognised the skills exhibited by the student facilitators, developed through their online facilitator training, which contribute to them being seen as credible allies in the learning process. They also recognised the influence of the make-up of their groups and the extent to which attitudes, beliefs, levels of participation and commitment affected learning, highlighting the importance of good facilitation as a means of managing some of these issues. These findings lend support to the importance of the prerequisites for an online facilitator identified by Salmon (2004) such as the ability to understand the online process based on personal experience as an online learner, the need for reasonable technical skills, good online communication skills, knowledge and experience to share, as well as the motivation and determination to become a good online facilitator. This final point was illustrated by the level of enthusiasm with which the student facilitators embraced their role.

## Enthusiasm

Enthusiasm was highlighted as a quality needed to lead discussions. If the facilitator felt passionate about it, it was likely the students would too. One student reflected:

*[S/he] made me consider topics I wouldn't have before and will make me a better healthcare professional.*

Another student suggested that his/her facilitator's enthusiasm was infectious, stating "it gives you more motivation". These findings reflect Gallew's (2005) suggestion that, amongst other qualities, teachers need to possess enthusiasm and passion (Gibbs, 2003; Gibson, 2009). Research by Kunter et al. (2008) suggests that it is important to distinguish between enthusiasm for teaching and enthusiasm for the subject matter. The data did not distinguish whether students were enthused by facilitators who were perceived to be good facilitators, or by those who gave the impression of having a passion for interprofessional learning. The first student comment suggests that the enthusiasm might have been for IPL although clearly the two are intertwined (Gill et al., 2006). Notwithstanding our inability to discriminate between these two foci, perhaps most importantly our findings stress the significance of self selection for the role of the online facilitator based on motivation and genuine interest in teaching and/or IPL.

In-depth interviews conducted with the student facilitators confirmed that intrinsic motivation was more influential than the extrinsic motivation of payment, which was not substantial and as such was considered to be a 'perk'.

## Diffusion of Status

Several students' comments suggested that the student facilitated forums were less hierarchical than in the staff facilitated groups, simply because students were more approachable and used similar language. For instance, one student reflected that "they are on the same level as you" and another suggested:

*We were able to communicate effectively with them better on a similar level. I feel there wasn't any pressure to the work being assessed, and were able to give personal answers to activities.*

A further student commented:

*It helps that [the student facilitators] are from a similar generation as yourself so have a sound understanding about thought processes in that generation.*

These findings possibly resonate with Haylers (1999) suggestion that peer mentors can be beneficial because the learning feels less formal and the learning environment is less threatening without the presence of academic staff. In this context student facilitators were perceived to be similar 'beings' even though interestingly the age range amongst students meant that some were probably not of the same generation. The comment regarding assessment suggests that students are constantly aware of being judged by academic staff even in less formal settings. Acknowledging the deference with which a student facilitator operated, in stating, "he was respectful towards us and encouraged us to participate without being condescending", another student adds weight to the suggestion that diffusing power differentials by addressing the power imbalance between teachers and students (Jamieson & Thomas, 1974), was in this situation, beneficial for learning.

## Support and Challenge

Respect from the student facilitators was clearly influential in nurturing engagement and making students feel that their comments were worthwhile and of value. However, being able to help students explore further and develop the discussion at a higher level is a definite skill that some student facilitators clearly possessed. Comments such as:

*our facilitator tried to draw discussions out as some people were just posting the same things" and "she praised good points and asked additional questions we maybe hadn't thought about*

illustrate that student facilitators recognised their responsibility to enhance discussion. They also highlight the challenge that facilitators face in the light of evidence that analytical debate online can be lacking (Miers et al., 2007).

Effective facilitation involved achieving a balance between support, challenge, recognition of the need for praise and encouragement. However, perhaps most importantly, the Year 1 students appeared to greatly appreciate having some reassurance that they were doing things in the right way:

*She was able to reassure us that accessing and taking part in IPLP was straightforward and not something to worry about. This was believable as she herself had taken part.*

This comment illustrates how for some students simply getting onto the forums and finding courage to post even basic messages was a new and potentially anxiety-promoting experience made significantly easier by facilitators with whom they felt they could identify. A final comment, "IPLP is a very clever tool and has brought me out of my online shell" seemed to capture one student's sense of achievement in this respect.

## Promoting IPL

The student facilitators themselves identified that the experience had enhanced their own IPL, encouraging them to explore interprofessional aspects of the scenarios and related discussions. It appears that the students they facilitated also recognised the benefits of this. For example;

*[the student facilitator] had a good understanding of students' concerns and fears about becoming a professional.*

*[the student facilitator] opens students eyes to opportunities that arise from doing well in IPLP.*

Many students acknowledged that their facilitator kept discussions "on track" and encouraged them to explore "other perspectives" reflecting attention to the interprofessional nature of the discussion triggers within the learning materials. Indeed, during the facilitation process, students could be seen to be modelling several interprofessional capabilities identified within the Interprofessional Capability Framework (Combined Universities Interprofessional Learning Unit, 2004).

# DISCUSSION

The findings from the research explored within this chapter clearly demonstrate that student facilitators were not only as effective as academics in their abilities to promote IPL in an online learning environment, but may have also offered added value to student learning. The benefit of student facilitators was felt through a diffusion of status; they were enthusiastic and able to empathise; they created an atmosphere of trust and support, and challenged thinking. However, the preparation for such initiatives should not be undervalued and lead us to some practical recommendations. The following points that are grounded in the research findings may not only relate to student-led initiatives but could also be beneficial for academic staff facilitation.

## Self Selection

Self selection for the role of online facilitator appears to be an important factor influencing commitment and enthusiasm for facilitating well. For student facilitators, being given the responsibility of leading an online group to develop their ideas about IPL may enhance their own experiences, skills and understanding of interprofessional working to take into practice. The process of being a facilitator should be of mutual benefit to the student group being led, and the facilitator in their own personal and professional development.

## Train Them

Salmon (2004) identifies the characteristics of a good online facilitator and the need for relevant training. Our experience suggests that providing training for online facilitators is crucial, not least because of the differences between online and face to face, teaching and facilitation. It is essential that facilitators are familiar with the technology adopted, the structure of the learning environment, materials, content and crucially, the interprofessional nature of the learning. Timeliness of the training is also an important factor, and wherever possible should take place immediately before the period of facilitation to ensure that the learning is fresh in the facilitators' minds.

## Technical, Academic and Community Support

It is vital that facilitators can access appropriate support mechanisms both prior to and during the period of facilitation as their needs change. In any online platform, technical support provided by a technologist or equivalent is imperative to running a module or course in an online environment. Timely responses to facilitators about any technical challenges are likely to result in a less disruptive and more efficient experience from the student's viewpoint.

In the study featured in this chapter, student facilitators benefited from sharing experiences, tips and guidance in a central, online community area where they could talk to one another. Academic support for student facilitators was also important. This type of support could take the form of allocating 'buddy' academics, although in this study we used a central point of contact where questions could be asked or concerns raised to avoid students feeling like students; checked up on and closely monitored.

## Trust Them

Perhaps one of the most important messages in this chapter is the importance of trusting the student facilitators in the programme. Giving students' responsibility for their forums without close monitoring might seem to be a high risk strategy, however, in this instance it resulted in a high degree of commitment; certainly the trust was not abused and there was evidence that being trusted to do the job made students feel more responsible. One student facilitator admitted to having enjoyed being given a level of responsibility above what was usually given to students, while another reported feeling that she had undertaken an obligation and to ensure she could fulfil this, accessed the internet via her mobile phone as well as computer so that she could always respond tostudents needing support.

A key attribute of leadership and autonomy is the ability to take responsibility for a task, and to be trusted by senior colleagues to do a good job. Should peer review mechanisms be used in the online environment for staff, student facilitators could be included; we do not advocate that such a system should be implemented solely for student facilitators.

## Offset Assessments Where Possible

As recommended above, students that put themselves forward for leading online discussion forums should be trained to prepare them for the role. In this pilot study the training programme was assessed through a one-thousand word reflection on the IPL facilitator role, which was counted for students' Year 3 IPL assignment. It was attractive to the students to offset this training against the other assessment they were required to complete. Arranging offsetting assessment can be complex or straightforward depending on the content of modules and overlap of learning outcomes, which in this case was high. If feasible, it provides an alternative means of recognition of student effort.

## Give Recognition

It is important to provide recognition for student efforts in these types of initiatives to avoid accusation of exploitation. Recognition could be given in the form of certificates for their employment portfolio, or the opportunity to present at conferences and write for publication with academic colleagues. All of these approaches have been adopted in this project and taken up with enthusiasm by the students who recognize the importance of developing their curriculum vitae. Difficulties in the lack of student interest in IPL could potentially be overcome by ensuring that students are recognised for their commitment to leading online discussion forums to promote IPL. This is likely to have a snowball effect. If students are led in online discussions by fellow students who are motivated and excited about interprofessional working issues and debates, this could lead other students to realize its importance in other circumstances such as on placement.

## Gain Institutional Support

Hoffman et al. (2008) identify lack of support as a major challenge to this type of initiative. We were fortunate enough to have full support from the Faculty of Health and Life Sciences as well as some funding to carry out this project. Without full support and commitment this type of initiative would be very difficult.

## CONCLUSION

Implementation of a student led IPL initiative in this context provides evidence of clear benefits to student learning. For student facilitators, the ability to lead student groups is a fundamental achievement and one that could enhance employability. Having been given a status that students are not usually given, their commitment and enthusiasm for the role may promote interest not only in IPL, but also in leadership and accountability skills, which are arguably valuable interprofessional attributes in themselves. We hope that their counterparts may see them as role models and put themselves forward for similar opportunities later in their programmes, because we believe that both students and their facilitators learned more about IPL because students were involved in implementing the programme.

## REFERENCES


Combined Universities Interprofessional Learning Unit. (2004). *Interprofessional Capability Framework*. Retrieved August 10, 2009, from http://www.sheffield.ac.uk/cuilu

Ellis, R. A., Goodyear, P., O'Hara, A., & Prosser, M. (2007). The university student experience of face-to-face and online discussions: coherence, reflection and meaning. *ALT-J. Research in Learning Technology*, *15*(1), 83–97.

Gallew, H. A. (2005). Students Teaching Students: Learning Through Doing, Being, and Becoming. *Occupational Therapy in Health Care*, *19*(3), 105–117. doi:10.1300/J003v19n03_08

Gibbs, C. (2003). Explaining effective teaching: self-efficacy and thought control of action. *Journal of Educational Enquiry*, *4*(2), 1–14.

Gibson, J. (2009). The five 'Es'of an excellent teacher. *The Clinical Teacher*, *6*(1), 3–5. doi:10.1111/j.1743-498X.2008.00239.x

Gill, D., Parker, C., Spooner, M., Thomas, M., Ambrose, K., & Richardson, J. (2006). Tomorrow's Doctors and Nurses: Peer Assisted Learning. *The Clinical Teacher*, *3*(1), 13–18. doi:10.1111/j.1743-498X.2006.00087.x

Goodlad, S., & Hirst, B. (1989). *Peer tutoring: A Guide to Learning by Teaching*. London: Kogan Page.

Gosser, D. K., Cracolice, M. S., Kampmeier, J. A., Roth, V., Strozak, V. S., & Varma-Nelson, P. (2001). *Peer-led Team Learning: A guidebook*. Upper Saddle River, New Jersey: Pretice Hall.

Hayler, R. W. (1999). Assessing Proctors: solving problems in grading student performance in a peer support system. *Mentoring & Tutoring*, *7*(1), 35–39. doi:10.1080/0968465990070103

Hoffman, S. J., Rosenfield, D., Gilbert, J. H. V., & Oandasan, I. (2008). Student leadership in interprofessional education: benefits, challenges, and implications for educators, researchers and policy makers. *Medical Education*, *42*, 654–661. doi:10.1111/j.1365-2923.2008.03042.x

Jamieson, D. W., & Thomas, K. W. (1974). Power and Conflict in the Student-Teacher Relationship. *The Journal of Applied Behavioral Science*, *10*, 321–336. doi:10.1177/002188637401000304

Knight, P. T. (2002). *Being a Teacher in Higher Education*. Buckingham, UK: SRHE and Open University.

Kunter, M., Tsai, Y., Klusmann, U., Brunner, M., Krauss, S., & Baumert, J. (2008). Students' and mathematics teachers' perceptions of teacher enthusiasm and instruction. *Learning and Instruction, 18*, 468–482. doi:10.1016/j.learninstruc.2008.06.008

McCall, B. (2007). Cry for help from tutor 'counsellors'. *Times Higher Education Supplement, 1785*, 4–5.

Micari, M., Streitwieser, B., & Light, G. (2006). Undergraduates Leading Undergraduates: Peer Facilitation in a Science Workshop Program. *Innovative Higher Education, 30*(4), 269–288. doi:10.1007/s10755-005-8348-y

Miers, M. E., Clarke, B. A., Pollard, K. C., Rickaby, C. E., Thomas, J., & Turtle, A. (2007). Online interprofessional learning: the student experience. *Journal of Interprofessional Care, 21*(5), 529–542. doi:10.1080/13561820701585296

Oandasan, I., & Reeves, S. (2005). Key elements for interprofessional education: Part 1: The learner, the educator and the learning context. *Journal of Interprofessional Care, 19*(Suppl. 1), 21–38. doi:10.1080/13561820500083550

Rees, D., & Johnson, R. (2007). All together now? Staff views and experiences of a prequalifying interprofessional curriculum. *Journal of Interprofessional Care, 21*(5), 543–555. doi:10.1080/13561820701507878

Salmon, G. (2004). e-Moderating: The Key to Teaching and Learning Online. Abingdon, UK: RoutledgeFalmer.

Saunders, D. (1992). Peer Tutoring in Higher Education. *Studies in Higher Education, 17*(2), 211–218. doi:10.1080/03075079212331382677

Scammell, J., Hutchings, M., & Quinney, A. (2008). *A virtual practice community for student learning and staff development in health and social work inter-professional education; changing practice through collaboration.* Bournemouth, UK: Bournemouth University, School of Health and Social Care. Retrieved August 15, 2009 from http://www.health.heacademy.ac.uk/publications/miniproject/scammell08.pdf

Thralls, C., & Blyler, N. R. (1993). The Social Perspective and Pedagogy in Technical Communication. *Technical Communication Quarterly, 2*(3), 249–270.

Wagner, T. (2008). RIGOR redefined. *Educational Leadership, 66*(2), 20–24.

Walsh, K. (2007). Interprofessional Education Online. The BMJ experience. *Journal of Interprofessional Care, 21*(6), 691–693. doi:10.1080/13561820701436912

# Chapter 10

# Learning from Each Other? Using Technology to Develop Collaborative Learning in Clinical Education

**Helen Bradbury**
*University of Leeds, UK*

**Melissa Highton**
*Oxford University, UK*

**Rebecca O'Rourke**
*University of Leeds, UK*

## ABSTRACT

*This chapter discusses and evaluates the introduction of collaborative e-learning activities into an interprofessional teacher education programme, a Master of Education in Clinical Education (MEd CE). The activity involved designing an educational research methodology module. The authors chose educational research methods as the task focus because, although the students had some familiarity with research methods, researching clinical education was new to them all. This task enabled participants to learn about educational research methods, course design and e-learning tools, resources and pedagogy. The process enhanced collaborative and interprofessional learning.*

## INTRODUCTION

This chapter offers a case study in interprofessional education that promoted collaborative learning through e-learning. It discusses and evaluates the introduction of collaborative e-learning activities into an interprofessional teacher education programme, a Master of Education in Clinical Education (MEd CE). Now in its 5th year, the MEd developed from, and continues to run alongside, a successful Post-Graduate Certificate. The activity involved designing an educational research methodology module and enabled participants to learn about educational research methods, course design and e-learning tools, resources and pedagogy. We chose educational research methods as the task focus because, although the students had some familiarity with research methods, researching clinical education was new to them all.

Evidence of interprofessional learning is notoriously hard to capture (Barr 2007) and our experience was no different. It has been easier

DOI: 10.4018/978-1-61520-889-0.ch010

to identify and evaluate the engagement with e-learning than to identify interprofessional learning. Nevertheless, collaborative learning both within and across professional groupings clearly takes place, especially in relation to learning how to use the technology to complete the online task. This learning scaffolds the process of learning about educational research methods and course design for the participants. Learning about technology and educational research methods creates trust across the group which makes it possible for them to challenge and debate their individual research studies on an equal but different professional standing.

We explore the introduction of this activity from the standpoint of the students and the course team, taking Salmon's Five Stage model of e-moderation (Salmon, 2000) as a guide. We draw on two sources of information for this chapter. First, a content analysis of material generated by the students on the programme, which is explored in relation to the experience of e-learning and interprofessional collaborative learning. These materials, used with permission from the health care professionals involved, include individual reflective accounts (which formed part of the module assessment) and records of the on-line interactions from the collaborative task. Second, we used appreciative inquiry methodology (Cooperrider & Srivastva, 1987) to chart the experience of the course director as she worked with Gilly Salmon's principles (Salmon, 2000, 2002) to introduce, review and refine elements of e-learning into what has now become a 'blended learning' programme (Littlejohn, 2007).

The chapter begins by outlining the educational and professional contexts within which the course is situated. We briefly describe the course and the online activity which forms the basis of this chapter before discussing the main themes and issues which emerge from our evaluation. Finally, we reflect on the implications of our experience for others.

## Educational and Professional Context

Integrating technology into practice requires a considered, reflective approach to course and curriculum design to ensure learning needs and teaching aims are met. Blended learning means more than having some face-to-face and some online elements. It describes the extent to which these elements work together to give a cohesive learner experience and the choice of the best tools for the task. Litttlejohn (2007) suggests that the design and subsequent success of a blend is subject to three contexts: the learner's context, the teacher's context and the technology context. Each of these also exists in relation to interprofessional learning, which is itself a far from straightforward practice.

There is a small body of work exploring the practice of introducing e-learning in clinical education. The majority of this work (Connor, 2003; Morgan, 2006) is concerned with the experience of under-graduates, as are studies of interprofessional education, which tend to centre on clinical practice rather than clinical practitioners as clinical educators (Barr, 2007; Peacock & Hooper, 2006). We explore the extent to which findings from this work are replicated or challenged in the specific context of our case study, which focuses on the introduction of collaborative e-learning to an interprofessional group of experienced clinical practitioners undertaking a clinical education programme.

The themes and issues we address include: the experience and preconceptions of participants with regard to their knowledge of e-learning; the need for clinical educators to develop a critical stance towards the dominant forms of e-learning available in their professional clinical practice; the need to design an authentic online activity which promotes critical thinking; and the development of collaborative learning across interprofessional groups. In their study of a postgraduate course delivered entirely online, Nicol and Sinclair (2003) identified three themes. These were: text-based versus face to face communication; asynchronous discussion and contributions via written rather than spoken contributions. These themes were also major concerns in the reflective accounts we analysed. While our findings were broadly in line with those of Nicol and Sinclair, we expand on their findings in relation to the absence of verbal cues and the concept of lurking.

## The Clinical Education Programme

The clinical education programme discussed here is part time and provides a strongly work-related, academically robust route for all healthcare professionals engaged in clinical teaching and learning. In the UK, this is approximately 16 professions. Although doctors are numerically dominant on both the PGCE and MEd programmes we actively recruit across the professional range. Each cohort normally includes health care practitioners from between 5 and 9 distinct professions, including dentists, nurses, radiographers, occupational therapists, orthoptists, midwives, speech and language therapists and pharmacists. The multi-disciplinary backgrounds of staff on the course team contribute to the interprofessional ethos of the programme. Their experience of e-learning also varies greatly and includes sceptics and 'technophobes' as well as enthusiasts.

The programme seeks to embody Senge's imperative – 'Be the change you wish to see' (Senge, 1990, 156) through its philosophy of interprofessional education and commitment to develop participants as change agents within practice in promoting learning across, as well as within, professions. This framework of experiential learning informs the design and delivery of the whole programme, drawing on the work of Knowles (1980), Schon (1983) and Kolb (1984) and the later extensions of their work into the field of lifelong and professional learning pedagogy (Boud, Cohen & Walker, 1993; Bradbury et al., 2009). The programme is delivered through work-based assessment and study with intensive 3-day residential units, all supported by use of the VLE.

## The Learners' Context

The introduction of e-learning activity into the 'blended' (Littlejohn, 2007) design, delivery and assessment of the MEd CE took place alongside the introduction of e-learning into clinical practice and clinical education in the UK. Its introduction was linked to a major government initiative (the National Health Service (NHS) Improvement Plan) which targeted interprofessional working as a means to improve patient care and maximise resources (DoH, 2004, pp. 60-61). E-learning was highlighted as central to achieving interprofessional working and the focus on interprofessional education was central to achieving this. This initiative consolidated the 2000 NHS plan which had emphasised multi-disciplinary learning (DoH, 2000).

The nature of e-learning in clinical practice and clinical education means that students arrive on the programme with differing conceptions, and experiences, of e-learning. A common perception is of e-learning, or VLE use, as a repository of resources to support face to face and independent learning. This is unsurprising, since the ability of the web to offer access to learning materials 'any time, any place, anywhere' has been widely promoted as a benefit of e-learning approaches. This 'availability' model however, leads to use of materials in an individual 'on demand' way. The use of virtual learning environments for group work and collaboration challenges this flexibility, as participants are required to work together at an agreed pace to an agreed timetable in an agreed online 'place'. This creates tension for students whose prior experience of e-learning, if they have any at all, had been interaction with a computer (e.g. MCQ testing, training package for health and safety, information repository for clinical skills etc.) rather than a dialogic interaction between people where knowledge is created and learning occurs.

In relation to interprofessional learning, most students have had some experience of training alongside other health care professionals, e.g. Advanced Life Support Training, but for the majority their only experience of structured interprofessional learning was their previous year with us as students on the PGCE. Only a minority have experienced online working with other health care professionals.

## Developing a Critical Stance

Clinical educator roles often involve wearing multiple 'hats' requiring staff to possess a range of skills and abilities, and the arrival of new technologies for e-learning, administration and research offers a training challenge for all those involved in professional development. New and emergent technologies mean that colleagues need to constantly maintain and improve their skills to remain up-to-date and make the best possible use of technological innovations in their field. Clinical educators must develop a critical stance towards the dominant forms of e-learning available in their professional clinical practice; a process similar to the challenge for the programme as a whole, where experts in clinical practice confront and work through their positioning as novices in pedagogy. It is not unusual for clinicians to have difficulty identifying which technologies are of genuine use; access to some of these technologies as learners helps ensure that they can make informed decisions in these areas.

In the same way that students are challenged to question and critique e-learning technologies and practices, their experience on the programme involves exploring the distinctions between uniprofessional, multiprofessional and interprofessional learning (Barr, 2007, p. 50). The emphasis throughout on collaborative learning means that they are constantly learning with, from and about each other's professional practices as clinicians and educators.

## Designing an Authentic Online Activity

The teaching on our programme models a range of teaching strategies, enabling students to experience a variety of approaches to learning, some of which they may already use in their professional lives. We broaden their learning by giving them the experience of designing the educational research methodology module through an online discussion room, a tool which facilitates conversations between learners and teachers, and which supports our philosophy of engendering collaborative approaches to learning.

Salmon (2000) offers a model to support online learning novices become confident and effective in this method of learning. Each stage involves a key role for the moderator.

**Stage 1--Access and motivation:** At this stage, students need to be able to set up and get into the conferencing system. They also need to be motivated to use it, and to be clear about its role within their programme of study.

**Stage 2--Online socialization:** At stage two, students gain confidence in using the online forum and begin to learn how to engage with it as a social environment.

**Stage 3--Information giving and receiving:** Now the user starts to appreciate the potential of the online forum for the exchange of information, feedback and support.

**Stage 4--Knowledge construction:** This is a key stage: moderators and students work together to generate and make new meanings through their collaboration.

**Stage 5--Development:** At this stage the student is beginning to be independent online.

Salmon's 5 stage model was used to introduce students to the task shown in Figure 1. The task

*Figure 1. Online Interprofessional group task*

**We would like you as part of your case study to use a discussion room on the VLE to undertake the following group activity. You will be placed in an interprofessional group of 4/5 students.**

**The Activity:**

-You are the course management team for a clinical education programme. The next module you need to develop is about educational research methodology. You have the details of the module objectives, syllabus, teaching hours, assessment, and reading list already – they are in your Programme Handbook.

-You can decide how the course is delivered; traditionally the taught element of the course has been delivered over two units of three days. Drawing on this unit's sessions and your reading about curriculum development and course design you need to design the teaching for this module.

-The learning outcome for this activity is to plan and develop a programme of study, drawing on theoretical approaches to teaching and learning and theories of education management as appropriate.

-We would like you to agree a three-week period in your group where you all will be available to communicate using the VLE. At the end of the three weeks please upload your agreed design into the discussion room. This task needs to be completed before the next unit of teaching. At this unit we will critique your design.

teaches students about e-learning, by allowing them to experience it, and about group processes. It adopts Barr's principles of interprofessional education:

*The application of principles of adult learning to interactive, group-based learning that relates collaborative learning to collaborative practice within a coherent rationale informed by understanding of interpersonal, group, inter-group, organisational and inter-organisational relations and processes of professionalism (Barr, 2002, p. 23).*

We maintain the interprofesional ethos when allocating students to working group. As the ratio of doctors to other professional groups is always higher, a typical group consists of two doctors, a nurse, a pharmacist and an allied health professional e.g. radiographer.

The access and motivation stage started with a face to face discussion orientating students to the activity's purpose and objectives. They were encouraged to devise their own ground rules for communicating in the discussion room. This was followed by a practical session in the computer cluster at the university to introduce the VLE. The students had all met face to face on a number of occasions; however the on-line socialisation stage was still important as the students had to adjust to communicating through this new medium. A set task had been given earlier in the course which involved designing and then running an objective structured clinical examination (OSCE), which made use of an on-line discussion room in the initial design stages. We intended the experience of the first two stages to enable participants to achieve the next stage; that of interacting with course content and other participants. This is a precondition for the knowledge construction stage, where participants work together to generate knowledge and make new meanings through their collaboration.

The development stage, where the learner becomes more self-directed, leads into the assessment, where participants write a reflective analysis. They are directed to consider some of the following in this account:

- A reflection on their group's development and of the activity itself.
- Why use on line learning?
- What are the pros and cons?
- How does this differ from other ways of learning?
- What theoretical approaches to teaching and learning are being used in this context?
- Is discussion important and why?
- What is the best way to structure on line discussion?
- Compare the use of synchronous and asynchronous discussion in collaborative learning.
- Does the technology support learning?
- What is the potential for this method in healthcare education?

So, in summary, we wanted our clinical educators to collaboratively design the teaching for a module they would be taking themselves. We hoped we had provided a useful activity which they could see the benefits of doing, firstly by becoming familiar with the requirements, syllabus and reading for the next module, secondly, developing their knowledge about programme design, thirdly, reflecting on e-learning and its potential application for their practice as clinical educators by writing a reflective essay about their experience.

## Technology Context

The technology used in this case was the virtual learning environment offered by the University of Leeds. For the early cohorts this was an open source, locally developed VLE. In 2008 the University migrated to Blackboard LMS. Both VLEs offered asynchronous discussion rooms and a file upload facility, Blackboard offers an additional synchronous 'chat' facility and the use of wikis and blogs. Although access to on-line technology, in workplaces and homes, is much more widespread today than when the on-line activity was first introduced, it is by no means straightforward. For example, firewalls in place on NHS Trust systems restrict social software. Even those using the VLE in more traditional ways experience problems as NHS firewalls block article downloads from the VLE.

The careful interventions of the course team in facilitating participants at the 'access and motivation' stage were undermined later by changes in accessibility beyond the tutors' control. Students are introduced to the VLE on campus, where the technology invariably works, but they then use it off-campus, frequently via dial-up rather than broadband. This can lead to timing out of connections and the frustration of losing contributions which are composed directly into the VLE rather than uploaded as a word document. There were additional problems when the university, as part of its security policy, required students to change their passwords. One cohort found their access to the VLE denied for the final week of their task because participants were notified of this change via university email accounts which, as part-time professional route students, they never used. With the most recent cohorts, more sophisticated and experienced users of technology were frustrated by the limitations of the technology available via the university as the synchronous chat facility was very unpredictable.

## Evaluation: Course Director

The Clinical Education programme Course Director first explored e-learning in 2000, when the university started to promote its in-house VLE. She saw its potential to develop communication with students on a clinical programme who came in for blocks of teaching but who were otherwise scattered around the country. Her first experience of establishing a discussion room for students followed a pattern familiar to many early advocates of e-learning. After introducing the discussion room in a face-to-face session a few initial postings followed and the room quickly fell into disuse. An opportune conference presentation from Gilly Salmon introduced the 5-stage model just at the point, in 2004, where e-learning was introduced in the Clinical Education programme because of its introduction into clinical practice. Salmon stresses the need for online activities to include motivating factors such as assessment, clear deliverables, designated roles and timescales (Salmon 2002). It was decided to assess the new activity in the programme, recognising that assessment drives engagement; but it was decided to assess the students' reflection on the activity, not the online activity itself.

## Learning the Technology

In pursuit of these objectives, the students' experience of learning is often stressful as it pushes them outside their comfort zones. Getting to grips with the technology is a major preoccupation – and a source of frustration:

*Initially they feel totally at sea. They only fully understand the process as a result of going through it. Some are quite motivated to make it work and they 'drag along' the ones who are less motivated. It is only when they start to reflect that they realise how much they learnt from the task. Even the ones who don't enjoy the experience learn from it. Some do enjoy the experience, but a lot don't – or they don't to begin with – but it is very unusual for them to feel they did not learn anything (Course Director).*

Over the years there have been attempts to prepare students for this experience but nothing does quite prepare them for it. The focus on how they are working together as a group and how they understand the e-learning resource are as much a part of the task as the course design which is its ostensible focus. These are complementary not competing demands, although students do not always experience them in this way.

The students' earlier experience of the objective structured clinical examination (OSCE) seeds the use of the on-line discussion room (Salmon's stages 2 and 3) but does not prevent the majority of students from hitting problems with the technology during the 3-week period in which they use it to collaboratively design the research methods module. For the first few cohorts it seemed that their capacity to learn from the experience was bound up with the frustration and uncertainty of that experience; it needed to be uncomfortable in order to prompt reflection and learning. The different levels of interest in and experience of e-learning that participants brought to the programme strengthens the collaborative impulse and the dynamic between confident and less confident is what pulls the group as a whole through the activity. Interestingly, confidence does not correlate with professional groupings. The tutor deliberately keeps her distance from the educational research methods task group activity

*I do some motivating – well done for finding your way here, this discussion is going fine sort of posts – and keep them on track. It takes them a week to get in and get started. If that wasn't happening I would need to do more but I have never had to (Course Director).*

## Interprofessional Collaborative Learning

Groups are not self-selecting but chosen to promote interprofessional working and to balance gender, within the constraint of doctors almost always being the majority professional group. When viewed from the standpoint of the programme as a whole, its interprofessional dimension is distinctive but it is harder to pinpoint its impact and significance within discrete tasks. Students' course evaluations and reflective accounts repeatedly cite the benefits of collaborative working and peer support but there are no explicit references to the benefits of interprofessional working or learning. Group working, which is consistently evaluated as the best aspect of the programme, clearly - if often implicitly - draws on the strengths of the interprofessional group. A more explicit engagement with the interprofessional context of clinical education can be seen in the micro-teaching activities, which are peer-assessed, and in the various forums where students present and discuss individual research projects. Here, it is not unusual for students to identify differences in approach to both clinical practice and clinical teaching associated with discrete professional groupings or interprofessional teaching. A comparison with earlier cohorts, which were uniprofessional, suggests that individuals are more prepared to be collaborative when they are conscious of representing a profession amongst other professions.

## LEARNER PERSPECTIVES

### Perceptions of E-Learning

Our evaluation of learners' perspectives draws on four iterations of the task, beginning with the 2005-06 cohort. Throughout all cohorts the predominant attitude towards the task was uncertainty and anxiety in anticipation and frustration during its completion. These emotional responses centred partly on the task itself and partly on the technology required to complete it. There were two significant changes in attitude between earlier and more recent cohorts. First, whereas earlier cohorts complained that the technology was difficult to access, use and understand, some members of the more recent cohort were frustrated because it was not sophisticated enough. Second, earlier cohorts made few references to the wider context of e-learning, and where they did, it was to express scepticism about the supposed advantages of e-learning for busy professionals. They predicted that it would become just another chore and made no specific references to e-learning in relation to their clinical practice or clinical education. In contrast, recent cohorts talked about being scared by the prospect of e-learning but motivated by its potential usefulness in the work context, citing undergraduates' expectations that computer mediated communication should form part of their teaching and learning.

Successive cohorts display increasingly positive engagements with e-learning and recognise it is becoming an established feature of clinical practice. The earlier cohorts comment almost exclusively on negative aspects of e-learning, including concerns that by front-loading the tutor's work load it actually undermines collaboration and interactivity. With each cohort, these negative comments diminish and examples from personal and professional practice support the view that e-learning is beneficial. Many of these examples relate to information and the repository dimension of e-learning, but even these identify opportunities it would be hard to replicate other than as e-learning. For example, an online subscription to a textbook which is regularly and frequently updated; building community knowledge through a professional association's annual online Q&A session between trainees and registered members; the capacity to accommodate irregular work and family commitments or dispersed colleagues and networks.

In the most recent cohort negative preconceptions surrounding e-learning – that it is inefficient, a trend, promotes didactic learning and teaching, renders the teacher redundant – were introduced only in the context of how to challenge these attitudes. There were imaginative suggestions for incorporating e-learning into clinical education. For example, one participant talks about the pressure of organising clinical placements within the constraints of the lecture and seminar programme in an area of clinical practice where opportunities for placement are limited. The activity alerted her to the possibility of using computer-based simulation to counter placement shortages and podcasts to create flexibility in placement timings.

Although students often struggle with the online activity, they increasingly see its relevance once the course has finished. Students more frequently report that their university is including interactive sessions for undergraduates and asking for facilitators. When this happens they realise that the experience of the online activity can help them. When it was first introduced, students could not relate the e-learning activity to their own practice because it was so rare. In 2004 the programme was ahead of what was happening in clinical education but now clinical education makes greater use of e-learning and students appreciate its inclusion. The 2009 cohort clearly understood the value to them in the future of the online activity.

## Participation in E-Learning

Although more positive approaches to e-learning emerged with each successive cohort, students continued to be frustrated by asynchronous methods of communication. Conversation using the asynchronous message board was slow and stilted – taking days rather than minutes to receive a response or reach decisions. Several participants commented that this stifled debate, creating passivity rather than the engaged, active learning intended:

*We successfully completed the task. However, we did this using our own experiences to date rather than questioning them and looking at other models (Student C3H).*

*Waiting for someone to propose something and then everyone else agreeing (or failing to disagree) that this was the best way (Student C2Y).*

In the most recent cohort, where one group used synchronous chat, there was appreciably more evidence of debate and discussion.

The use of text-based contributions and the absence of social or verbal cues led to the uneven participation noted by Nicol and Sinclair (2003) but it was reflected upon at great length by participants. Several commented that the absence of non-verbal and visual cues was de-motivating and blocked communication. They noted how challenging this was for professional groups whose entire professional learning is by observation within a social relationship between the teacher, other learners and patients in which verbal communication is led by the skilled reading of non-verbal clues. Working within an asynchronous, text-based environment poses additional challenges to this group of interprofessional learners, especially when, as several note, there is heightened anxiety about making mistakes in front of peers. Nevertheless, a minority of contributors saw advantages in asynchronous communication, comparing it favourably to group work because, for example, ideas could be mulled over or time was available to consider a response.

Participation links to the theme of leadership, and the setting and working to ground rules. Groups were encouraged to set ground rules during the initial face-to-face meeting, including who would act as chair and what their role would involve. However, once the online discussions started, the ground rules disappeared from view. Groups did not record them online or return to them, as they might have in a face-to-face group. There were frequent references to the absence of conventions and protocols for volunteering or being volunteered in the online environment but no attempt to establish new ones. Equally frequent comments, especially in the earlier cohorts, about the limited input from the tutor / moderator suggest that participants needed a clearer framework within which to participate. Leadership was problematic for almost all the groups and it was discussed in almost all commentaries. Leaders emerge by default; usually those most confident with technology. This confidence was not specific to any professional group; leadership roles were taken by doctors, dentists, midwives, pharmacists and nurses. This emphasises that interprofessional working involves differences within, as well as between, professions.

## Rethinking Lurking

Nicol and Sinclair (2003, p. 274) accept the conventional negative view of lurking – a term used to describe members of online forums who observe but rarely actively participate. Lurking is understood as a form of disengagement but our analysis of the reflective accounts led us to reframe lurking. There is a continuum of online presence in which being online and being 'in charge' or actively and frequently posting is only one end of the spectrum. Participation needs to be redefined to include reading, working out what you are going to say, preparing off-line for on-line exchanges and encounters and following up on-line encounters with off-line activities. They are all part of the participation continuum but they are not ' noisy' or visible forms.

One student made the point that it was not always possible to tell how much time people were spending on the task - somebody might not post much but be doing a lot of work offline. Silence and lurking triggered assumptions from others which were not necessarily accurate. Students may be absent or not posting for legitimate reasons; and unaware of the impact this has on group morale. For student C2Z the realisation that his assumption "of indolence until proved otherwise" had impacted negatively on working relationships within the group led to a reappraisal of this personality trait and its role in his wider professional relationships.

## Learning Continuity

There is a contrast between the continuity students experience in the online community compared to the course units. Online communities potentially offer part-time learners greater opportunities to engage with the course content and with their peers. This might be expected to make it easier for learners to maintain an active connection to the course between the residential units. Surprisingly, there are no references to this aspect of their e-learning activities in any reflective accounts or online discussions and several students actually found it harder in the online community than the course as a whole:

*Usually I find myself thinking, reflecting and developing my ideas about the course whilst driving or engaged in some completely unrelated activity but I found that once I logged off, I switched off (Student C3G).*

*The group did not seem to be there if I did not switch the computer on (Student C2W).*

This highlights the need to include scaffolding activities which will enable students to make explicit the connections between their independent, online and group learning.

## Lessons Learnt and Implications for Others

It has been an interesting experience to evaluate the educational research methodology task from the standpoint of collaborative learning through technology. Overall, we feel that it is a well-designed activity which successfully introduces e-learning into our programme. It certainly encourages the students to think about, and participate in, e-learning in new ways. Although it models the use of technology to enable collaborative working, it is less clear that students understand the implications of this for their own practice as clinicians and educators. Reviewing each cohort since the task was introduced in 2004 we see that opportunities for collaborative learning are heightened when students are positioned as novices. This holds true whether they are novices in relation to knowledge, for example educational research methods, or practices, for example the use of e-learning tools and resources.

The e-learning context, however, is dynamic and what worked well for the first student cohort became less relevant to each successive cohort. Where the students' e-learning experience was minimal, the requirements of the substantive task were less important as the main focus was to provide an experiential perspective on e-learning and material for their reflection on group dynamics and processes. As students enter the course with greater experience of e-learning in clinical practice and education the task needs to reflect these changed expectations.

We will change the focus of the on-line activity. Instead of using e-learning as the method to collaborate on a course design task, students will in future collaborate by researching examples of how e-learning is, or could be, used in their own field of clinical practice for presentation to and discussion by the group as a whole. This will enhance the interprofessional dimension of the programme, as well as integrating the content of the e-learning task with the development of e-learning practices. This should counter the tension between the task of course design and engagement in e-learning which students often express and which sometimes impacted negatively on their motivation.

The evaluation highlighted the contrast between face-to-face sessions, where students frequently connect discussion themes to their own professional context and invite interprofessional comparison, and the online discussions where comments of this kind were almost completely absent. This led us to reflect on where and how we explicitly emphasised the programme's interprofessional dimensions. For example, students are not directed to interprofessional issues in the list of topics to consider in their reflective account of the e-learning task. When the task was first introduced this was not the case. The change happened because earlier cohorts were so daunted by the technology challenges that their attention was unhelpfully split between this aspect and the interprofessional issues, which resulted in them doing justice to neither. It was decided to simplify the task by foregrounding learning the technology, a strategy which, as we have seen, does implicitly build collaborative learning across what are always richly interprofessional groups. However, there is greater scope for making this aspect of the task explicit by directing students to consider, for example, the extent to which professional cultures influence group dynamics.

## CONCLUSION

In summary, we have three lessons for ourselves which might also be useful to others. First, to be mindful of complacency. In evaluating what we – and our students and external examiner – considered to be a successful activity, we identified several ways in which it falls short of our expectations, in relation to both technology and interprofessional learning. Second, to be aware that e-learning is in constant and rapid change, and that this impacts on students and course teams in uneven ways. Making technological change the focus, as well as the medium, of learning activities provides scope for greater engagement with this important aspect of professional life and learning. Finally, we have reflected on the challenge of interprofessional learning itself – necessary, challenging but elusive (Barr 2007). We recognise the extent to which we inadvertently make the interprofessional dimensions of the programme invisible. For the course team, educating and facilitating learning with a range of different professionals has become so much the norm that we do not always acknowledge what a novelty this is for our students; nor the extent to which they need direction in understanding the impact of this on their development as clinical educators.

## REFERENCES

Barr, H. (2002). *Interprofessional Education Today, Yesterday and Tomorrow. A review*. London: Learning and Teaching Support Network, Centre for Health Sciences and Practice.

Barr, H. (2007). *Interprofessional Education in the United Kingdom 1966-1997*. York, UK: Higher Education Academy.

Boud, D., Cohen, R., & Walker, D. (1993). *Using Experience for Learning*. Buckingham, UK: Society for Research into Higher Education and Open University Press.

Bradbury, H., Frost, N., Kilminster, S., & Zukas, M. (2009). *Beyond Reflective Practice: New Approaches to Professional Learning*. London: Routledge.

Connor, C. (2003). *Interprofessional Computer Assisted Learning*. Retrieved July 6, 2009, from http://www.health.heacademy.ac.uk/projects/miniprojects/connor.pdf

Cooperrider, D. L., & Srivastva, S. (1987). Appreciative Inquiry in Organizational Life. *Organizational Change and Development, 1*, 29–169.

Department of Health (DOH). (2000). *The NHS Plan*. London: The Stationery Office.

Department of Health (DOH). (2004). *NHS Improvement Plan: Putting people at the heart of public services*. London: The Stationery Office.

Knowles, M. S. (1980). *The Modern Practice of Adult Education: from Pedagogy to Andragogy* (2nd ed.). Upper Saddle River, NJ: Prentice Hall.

Kolb, D. A. (1984). *Experiential Learning: Experience as a source of learning and development*. Upper Saddle River, NJ: Prentice Hall.

Littlejohn, A., & Pegler, C. (2007). *Preparing for Blended e-Learning*. London: Routledge.

Morgan, J., Rawlinson, M., & Weaver, M. (2006). Facilitating online reflective learning for health and social care professionals. *Open Learning, 21*(2), 167–176. doi:10.1080/02680510600715594

Nicol, D., & Sinclair, C. (2003). The Social Dimensions of Online Learning. *Innovations in Education and Teaching International, 40*(3), 270–280. doi:10.1080/1470329032000103807

Peacock, S., & Hooper, J. (2006). *How Do Diverse Groups of Learners in the Health Sciences Respond To A New Virtual Learning Experience?* Retrieved July 6, 2009, from http://www.health.heacademy.ac.uk/projects/miniprojects/completeproj/peacock.pdf

Salmon, G. (2000). *E-moderating: the key to teaching and learning on-line*. London: Kogan Page.

Salmon, G. (2002). *E-tivities: the key to active online learning*. London: Routledge Falmer.

Schon, D. (1983). *The Reflective Practitioner*. London: Temple Smith.

Senge, P. (1990). *The Fifth Discipline: The Art and Practice of the Learning Organisation*. New York: Doubleday.

# Chapter 11
# Involving Service Users Through Digital Means to Enhance Interprofessional Learning

**Frances Gordon**
*Sheffield Hallam University, UK*

**Karen Booth**
*Sheffield Hallam University, UK*

**Helen Bywater**
*Sheffield Hallam University, UK*

## ABSTRACT

*This chapter will provide guidance for educational practice founded on theory and on the experience of involving service users and carers in student education. Whilst this is an accepted philosophy and practice it is not necessarily easy to achieve. There are numerous ways of including service users in education but the era of digital media has added a means of bringing the service user into the learning environment and of overcoming many of the barriers to their effective engagement. The Centre for Interprofessional e-learning (CIPeL) has been engaged in developing e-learning materials which address some of the barriers to interprofessional education and issues related to user involvement in education. This experience is outlined and some examples from practice are given.*

## INTRODUCTION

Service user involvement in the education of health and social care professionals is part of a general drive towards user involvement in health and social services. The involvement movement has emerged powerfully from the perceived necessity for the public to be engaged in determining their own health and wellbeing agendas. The aim of involvement is to improve health and social services by focussing on the needs of the service user (DoH, 1994, 1998, 2001, 2005; Waldegrave, 1996; Sainsbury Centre for Mental Health, 1997). This chapter will look at the issues arising from engaging service users meaningfully in learning experiences and will offer some solutions, through the use of digital formats, to challenges to this engagement.

DOI: 10.4018/978-1-61520-889-0.ch011

CIPeL is a collaborative Centre for Excellence in Teaching and Learning (CETL) funded by the Higher Education Funding Council for England (HEFCE). The main purpose is to produce e-learning materials which address some of the barriers associated with interprofessional learning, including service user involvement in the production of e-learning. In the light of our experiences, this chapter will critically examine the challenges of collaborative involvement and of giving the service user a real voice in the development of learning materials. It will identify the main issues in the production of learning materials to ensure that they accommodate the multiple voices in interprofessional education. The intention is to provide guidance for effectively involving service users in educational practice through the use of digital formats.

## BACKGROUND

The service user involvement movement has gained momentum over the last two decades (DoH, 2001). Increased access to knowledge and an increasingly customer focussed value base within society have both influenced the growing concept of consumerism as a basis for health and social care. The Patients' Charter (DoH, 1991) attempted to transform traditional conceptualisations of the NHS patient as a passive recipient of what the health service had to offer to that of a discerning customer of the service. Although the Charter was discontinued as part of the changes to health services under the NHS Plan (DoH, 2000a) the underlying ethos of encouraging empowerment of citizens specifically in health and social care services was reiterated.

A natural extension of this movement is within the education of students for health and social care professions and several policy documents have extended service user involvement into education and training (DoH, 2000b; Sainsbury Centre for Mental Health, 1997; Levin, 2004). The policy of user involvement in student learning represents a move away from the notion of professional knowledge as supreme towards the notion of the service user/carer as holding their own expert knowledge about their condition or situation and that those who use health and social services are well placed to illustrate to students concepts such as human rights, anti-oppressive practice and the promotion of equality. As Taylor (1997, p. 117) points out, "*learning from service users challenges the whole notion of expertise, who holds it and on what basis*".

The contribution of service users to education is not new. Involvement of service users in the undergraduate education of nurses, midwives and health visitors (UKCC, 2000) and of social work students has been a requirement for some years (DoH, 2000a). There have been a variety of modes of involvement over many years from using the 'interesting' patient as teaching material to inviting the service user into the classroom. However, what has recently emerged is the recognition of the impact that service users can have on individual professional practice and perceptions (Levin, 2004) and that they can develop effective partnerships in service delivery and education (Barnes, 2000).

A further step is in interprofessional learning where user involvement in presenting 'real' life experiences helps to show students the importance of working interprofessionally with other disciplines and agencies to provide care that is truly focussed on people and their families. Essentially, the service user provides the link between theory and practice and enables students to look at their practice from a perspective beyond professional knowledge. This is especially important where collaboration between services and professionals is essential to the overall wellbeing of the service user.

It could be argued that during their work experience, students are continually learning from service users. In many respects this may be the best way to learn interprofessional collaborative practice. However, the reality of practice may make meaningful experiences difficult because of barriers such as: increasing clinical demands on staff time (Morison, Boohan, Jenkins & Mountray, 2003); crowded clinical placement curricular with an emphasis on professional competencies (Nisbet, Hendry, Rolls & Field, 2008; Lloyd Jones, Hutchings & Hobson, 2007); lack of exposure to interprofessional experiences (Nisbet et al., 2008) and differences in professional language and values (Mellor, Heyer & Howe, 2002).

Additionally, the placement is an uncontrolled learning environment in the sense that learning opportunities may not be predictable or consistently available and where there may be limited opportunity to reflect on the experiences and explore different approaches. However, it would seem that there is a place for a variety of learning experiences to provide a holistic approach to interprofessional learning with an emphasis on placing the service user at the heart of those experiences.

## SERVICE USER INVOLVEMENT IN INTERPROFESSIONAL EDUCATION

### Issues, Controversies and Problems

Whilst accepting the philosophy of the involvement agenda in student education, our experience reveals that achieving the aims of inclusive, empowering forms of involvement is no easy task. This section of the chapter will discuss some of the problems associated with promoting effective service user involvement in the educational context and describe solutions we have implemented through digital means.

Service user involvement in student education refers to a variety of processes and approaches. However, criticism is frequently aimed at developments that are perceived as merely 'tokenistic' (Felton & Stickley, 2004). Tokenism can be described as involvement with no real meaning; situations where people are involved but have little or no choice about how they participate. Masters et al. (2002, p. 314) assert that a sure way to ensure tokenism is to "ask people to be involved in tasks where their role is unclear and without adequate time and resources to develop these skills". Moving away from tokenism is a complex task and arguments around what can be portrayed as tokenistic approaches seem to centre on the issue of power. Felton & Stickley (2004) suggest that power inequalities inherent in our systems serve to marginalise service users and this challenges the implementation of meaningful change.

However, we argue that change can be facilitated, albeit slowly and a movement can be made towards more collaborative forms of involvement. Whitaker and Taylor (2004) use work by Hickey and Kipping and by Consumers in NHS Research to introduce various ways of categorising involvement. They describe a continuum of approach as beginning with a consumerist perspective that aims to provide service users with information and allow them to make informed choices. This, they claim, results in the type of participation and involvement where information is provided and service users are consulted in order to inform decision making. This approach, however, does not necessarily allow service user decision making nor are their views necessarily utilised. Hickey and Kipping (1998) refer to an approach at the other end of a participation continuum that embraces democratisation in which decision making power is incrementally shifted from the service provider (educationalist) to the service user. At the extreme end of this continuum lies "user control" (Hickey and Kipping, 1998, p. 85) where power is redistributed to the user who makes the decisions or chooses to involve others in the decision making process.

In the educational context, there are inspiring examples of this user control perspective in the creation of digital materials (see www.ceimh.bham.ac.uk). This is discussed further in the chapter by

Tarsem Singh-Cooner within this book. Barnes et al. (2000) assert that in order to promote the user's perspective in the learning process the value base that locates the service user at the centre of care and interventions should be mirrored. Clark (2002) indicates that the underlying premise of using service user case studies is about teaching critical analysis and promotes both the development of learning the 'facts' of scientific practice and the 'art' of being able to deal with ethical conflicts.

An additional consideration is the multi-professional perspective of interprofessional curriculum design. In a uniprofessional context addressing the professional relevance and perspective is relatively unproblematic. In interprofessional learning however, multiple professional voices must be represented. The values held by different professions affect how problems and solutions are framed (Clark, 2002) and pose difficulties in ensuring relevance for a diverse student audience. This increases the need for collaborative approaches to address the complexity of such curriculum design in that not just two voices, that of service user and professional, have to be accommodated, but multiple voices need to be heard.

This discussion so far has centred on a form of partnership with service users that implies there is a need to promote equality from a power perspective and to value the experiential or 'local' knowledge held by service users or their carers about their past, present and reaction to illness (Turner, Sheldon, Coles, Mountford, Hiller, Radway & Wee, 2000). However, there are individuals whose 'local' knowledge is inaccessible directly to students through difficulties for marginalised groups to engage and for those whose involvement is not feasible due to a variety of factors.

So, whilst we have argued that involving service users in student education is undeniably a positive contribution, we also recognise that facilitating participation for a number of groups or individuals, who could offer significant learning to students, can be difficult. The aim of student education in health and social care is to learn to care for people from diverse populations. In some ways it may be relatively easy to support the participation of certain groups of service users, for example those who are already active in support groups or patient advocacy. However, there are groups who may be defined as 'hard to engage' and are thus excluded. Being hard to engage results from various factors such as: being part of a marginalised group that is difficult to locate; being too ill, too disabled, too young, too vulnerable (Atkinson & Flint, 2001; DoH, 2002; Le Var, 2002).

Even where groups or individuals would not be defined as hard to engage, there are identified barriers to involvement. These include service users' lack of experience or concerns about speaking to the large groups of students who undertake interprofessional studies; discomfort in telling personal stories to groups of strangers; having to repeatedly tell the story and having the potential of distressing feelings being reactivated and lack of control over what students may ask them in the classroom setting. There are also difficulties for students who have reported being anxious about saying the wrong thing, or making a mistake or tiring the service user (Gordon et al., 2005).

## SOLUTIONS AND RECOMMENDATIONS

Despite the acknowledged benefits of promoting service user involvement in education (Wood and Wilson-Barnett, 1999; Spencer, Blackmore, Heard, McCorie, McHaffie, Scherpbier, Gupta, Singh & Southgare, 2000; Le Var, 2002) it is not a straightforward enterprise. We have identified three main issues for interprofessional education. These are: the problematic issue of ensuring partnership approaches with service users; of promoting the inclusivity of all groups despite their condition or situation; and in ensuring that the multiple voices inherent in interprofessional education can be accommodated and heard. The following section will describe work undertaken by CIPeL, in producing interprofessional e-learning materials that help to resolve the issues identified.

## Partnership Approaches in Service User Involvement

In terms of involving service users in the design of these materials, our work moves along the participation continuum previously described by Hickey and Kipping (1998), where involvement at one end is described as consumerist and involves information giving and explanation, through consultation and partnership, to democratisation at the other end, with user control as the main feature. The location on this continuum of the production of learning materials is determined by the pedagogical purposes of the individual learning encounters.

In our experience, effective learning situations do not always call for an equal balance between the science and art of practice so the pedagogical design of the learning materials will reflect the intended purpose of the learning experience. For example, in situations where students are expected to meet an outcome which requires them to understand theoretical constructs, the learning object may be designed to provide the student with information of a more factual nature. However, if the learning outcome is about appreciating service users' experiences of accessing services, learning materials that present the service users' personal perspectives about the experience would be appropriate. In such a situation where an authentic representation of the service user perspective is central to the pedagogical intent, and where the learning design involves "a story behind the facts" (Clark, 2002, p. 298) we focus our approach to service user engagement at the "partnership" stage of the continuum (Hickey & Kipping, 1998, p. 85) where power is shared and decisions are made jointly.

This partnership approach ensures a blending of the expert knowledge of the service user, in terms of their own experience, and the expert knowledge of the professional. We take heed of Clark's (2002, p. 303) warning that as professionals we need to be "aware of the power we wield in choosing what we include and what we do not, and for what purposes we design and use them". Our particular focus is how challenges to inclusive service user engagement in student education can be accomplished through digital means. We aim to locate service user perspectives centrally and to do this we frequently use service user case studies.

To ensure engagement is at a partnership level, we use guidelines for service user involvement in education developed in previous work by the authors (Gordon, Walsh, et al, 2004). The guidelines were developed as part of a research study that involved service users, students and lecturers in exploring how service users can be supported to engage effectively in student learning. These guidelines have been disseminated widely (Gordon et al., 2004; Gordon, Walsh, Marshall, Wilson & Hunt, 2005; Gordon, Walsh, Marshall, Wilson & Hunt, 2006). The study was contextualised to service users being involved in classroom teaching, however, we have found that they are applicable when promoting our partnership approaches in developing digital materials. The guidelines include drawing attention to the areas of:

- Collaboration
- Preparation
- Communication
- Support
- Debrief

**Collaboration:** Provides the groundwork to building good working relationships and is fundamental to the development of a partnership. It relies on ensuring time and resources are available to support this relationship. The activities involve building trust to enable participants to feel safe to contribute, and joint consideration of the aims and processes of the teaching.

*Vignette:* 'Bob' is a vulnerable adult who has been cared for by his parents for the forty years of his life. His parents, James and Geraldine are now in their eighties and are becoming increasingly anxious about his future. They have a powerful story to tell but they too are becoming frail. We needed to work collaboratively with this family to help them create learning materials that allowed them to tell their story and to communicate key messages to students about their needs. To do this, we have developed close relationships in order to build trust and partnership working. This has involved home visits, email and letters and supported trips to the university to make recordings. We have spent time helping them to select photographs and other items they wished to share with students.

**Preparation:** Is essential for all involved in the partnership. Practical and supportive aspects need to be addressed, such as: supporting training needs, confidentiality and consent; accessibility issues including the use of appropriate language, risk assessment, health and safety, remuneration and hospitality.

Recording an individual's experience using technology can be particularly daunting if s/he is not adequately prepared for what is required and how the technology is to be used. Using video, for example, may seem to be a simple way of recording an individual's story but this may be challenging for an individual presenting their own story for the first time on camera.

*Vignette:* Tom was invited to our recording studio to see the environment in which the recordings would take place. Time was allowed to set up the camera and make a simple recording to show Tom what was involved, what he was required to do and then to see an example of a recording. This allowed Tom to give his informed consent for participation in a supportive, non-threatening environment. During this preparation time we were able to establish the difficulty Tom experienced talking directly into the camera. In order to put Tom more at ease, we arranged for a member of the team to be present and Tom was filmed speaking to this individual as if he were having a conversation.

**Communication:** Is about making sure that all involved are fully informed about the processes of the learning initiative. The service users need to be familiar with the purpose of the learning materials and preferably be involved in setting the learning outcomes. Care should be taken to ensure that the service user is able to communicate effectively and this may involve the use of communication aids or in some cases, interpreters.

*Vignette:* Owen was very willing to share his experiences with us and was keen to ensure what he was contributing would be beneficial to the students who would be viewing his video recording. A member of the team discussed with Owen the content of the module the students were studying. Owen asked that we structured questions for him so that he could focus his answers towards relevant and appropriate examples of his experiences. He felt this would give him more confidence and a more structured approach on the day of the video recording session if he felt nervous.

**Support:** Ongoing support and training to participate in the production of learning materials facilitates a partnership approach. Being mindful of the service user perspective and needs throughout the process promotes positive participation. Advice and guidance may be required about how to present the story and about the strategies of sharing aspects of their story that are potentially negative in a positive way. The rehearsal of question and answer scenarios might also be a useful strategy and help in dealing with elements that might cause discomfort. It is also necessary to consider how to support individual disengagement at the end of the initiative.

*Vignette:* Bob, a vulnerable adult, required assistance to take part in recording his story to enable him to have a positive experience. By meeting with Bob and his family in the familiar environment of his own home, we were able to establish how we could provide support for Bob in order for him to convey his story effectively. Bob's disability meant he was unable to read easily from a prepared script. On discussion with his parents a plan was devised for his father to write Bob's story in large print in order for him to practice and read clearly on the day of the audio recording with his father present for support. This audio recording allowed Bob to share his story using a method with which he was comfortable and familiar.

*Vignette:* John had the opportunity to visit the recording studio and make a short video before he made his final recording. On the day of recording the video, John arrived at the studio nervous but excited about this new experience. As filming commenced and John began to recall a particular incident he became emotional. We were able to stop the recording and support John until he regained composure. However, with his permission we were able to capture this emotion allowing the students to experience the impact the incident had on him. This may not have been possible if John had felt intimidated and vulnerable in a face-to-face situation.

**Debrief:** concerns gaining feedback from service users following their participation. This provides another element of support and enables a response to issues and closure of the session. Recognising the emotional impact involved when service users or carers share real life experiences is important and time should be allowed to sufficiently prepare and support individuals before, during and after their involvement in teaching and learning.

*Vignette:* The importance of debriefing should never be underestimated when service users or carers have been asked to participate in teaching and learning. This is a carer's personal account: "When I was asked to share my experiences as a carer I had not anticipated the emotions my account would unearth. I was more concerned about what I wanted to contribute and if it would be beneficial. As a carer I was living in the "here and now", doing what was necessary to care for my husband. Dealing with each part of his illness as it unfolded and the necessary care he required. I recalled my story from the day he was diagnosed with a brain tumour to his current condition and the care he required in great detail. After being asked to tell my story the reality of the world in which we were now living hit me hard. Although I was given a brief about why I was there and given the opportunity to share what I thought was important to me, the potential impact on me personally had not been discussed. However, the staff involved had anticipated the effect of this experience and had prepared a debrief session for me if required. This allowed me time and opportunity to talk about my feelings in a safe and supportive environment."

## Promoting Inclusivity of Service User Involvement

We have discussed the fact of certain groups being 'hard to engage' in student education. However, health and social care students will inevitably care for people from these groups in their practice. This fact emphasises the need to involve marginalised, disadvantaged and vulnerable groups' perspectives into student learning. The issue of promoting the inclusivity of all groups despite their condition or situation is one that we contend is usefully addressed via digital means. Our philosophy is to have the service user at the heart of their learning experience through exposure to authentic service user and carer experience. However, where it is difficult or impossible to have service users involved in person, our solution is to bring the service user's story and experience to the students through digital means.

One example of this is Ted's Story, which is represented as a scenario incorporating video and audio clips tracing Ted's journey who, as an older adult, experiences the loss of his wife, loss of his health and loss of independence and ultimately becomes very depressed and potentially suicidal. The scenario features Ted, represented by an actor, speaking to the students directly about his situation. The perspectives of various professionals are also introduced, as is the story of Ted's daughter.

What this digital solution achieved, was the ability to present students with a 'lived experience' from the service user's and carer's point of view. Ted's story was based on a number of actual cases with which the creators of the scenario had been involved. An actual person in Ted's situation, being frail and vulnerable would not be available for students in a classroom situation, People like Ted, however, would be a familiar feature of the students' practice experience and the facility to discuss his situation as though he were real but in the safer more controlled environment of the classroom is a useful one. Stories like Ted's enable the students to explore their professional knowledge in the light of the individual's needs and experiences. An added dimension is how the contributions of students from other professions, who are also part of the learning experience, shape an interprofessional understanding of the care situation presented.

In our institution we deliver a core interprofessional curriculum to approximately 3000 health and social care undergraduate students each year. The use of technology in this way, enables the delivery of such service user focussed learning to be delivered consistently to the whole cohort.

## Interprofessional Student Learning

At the beginning of this chapter we identified three issues we see as pertinent areas for consideration in involving services users in interprofessional education: the avoidance of tokenism; ensuring students are exposed to rich learning derived from the involvement of all those who use services, particularly marginalised groups, and the problem of representing the multiple professional voices inherent in interprofessional education. This section explores the contribution that digital solutions can make to interprofessional student learning. The resources we have developed help students to explore a situation from their own perspective, and collaboratively within the context of a multidisciplinary team. This raises issues of ensuring relevance and meaningfulness from a professional practice perspective across a very diverse student population. The idea is to produce learning situations for students that are authentic to their practice world. Such approaches have been a feature in education for some time and can be traced to Dewey's (1916) early work in 'learning by doing' in which the relationship between learning and life was explored. These issues are pertinent to the development of students to operate in the real world of practice, and rely on interacting with representations of the 'real world'.

It has been argued that the development of professional knowledge, in our case interprofessional knowledge, is enhanced by active learning processes such as collaboration (Posey & Pintz, 2006). When planning our interprofessional curriculum, we developed two digital scenarios: Ted's story and Jane's story. The design of these encompassed theoretical perspectives around transformative learning through collaborative processes. Our intention was that through engaging in a transformative learning experience, learners would examine their own beliefs, assumptions and expectations alongside those of others (Mezirow, 2000). For these processes to take place we believed that the Ted and Jane scenarios needed to be authentic through representing the 'real world' of practice. We assumed that all students accessing these materials could engage with one or the other. Indeed, student feedback in evaluation processes indicated a positive response to the e-learning in that they considered it reflected issues they dealt with in practice:

*I love Ted, and I could see myself working with people like him in practice (third year physiotherapy student).*

This student's quote revealed that she had encountered 'Ted' as a real person and that the practice issues raised were perceived as pertinent to her learning. This had provided the motivation to engage with the collaborative processes in the learning situation. However, for some students this was not as positive or as straightforward. Even though the scenarios were clearly authentic to the real world, presented an authentic service user voice, and provided plausible situations that all students would understand and relate to at some level, it became clear from student feedback that if their own profession did not feature, a barrier was presented:

*Nor did they (the scenarios) reflect every profession. The most annoyance that people got was that everything seemed to be torn between midwifery and nursing. We know that's the majority of people who do the module but it's hard to tailor it to your area if it's just a nursing case study (focus group quote from student).*

It seemed to us when considering the students' feedback, that there was more to authenticity than just being true to the real world of practice or representing 'realistic' people. Barton et al. (2005) contend that the factors surrounding authenticity of the learning task, such as context, learner motivation, social interaction and social presence need to be carefully managed in any curriculum utilising e-learning.

*It became clear to us that the management of our curriculum needed to recognise that authenticity had to have a component of being directly associated with the individual student's profession and practice and that authenticity of the practice arena was not enough. The aim of our curriculum was to enable interprofessional collaboration in practice. The original designers of Ted and Jane, considered that these two scenarios would help all students achieve learning outcomes around collaboration:*

*There were lots of triggers within the scenarios that got them to discuss a range of professional issues that required them to collaborate interprofessionally (interview quote from lecturer).*

However, student feedback strongly indicated the need to have stories that mirror professional practice in a plausible way so that they felt able to collaborate in making decisions about the service user's care. Where students felt that their voice would not be heard in what was purported to be a 'multiple voice' situation, motivation was lost and the social interactions were challenged.

Our response was to accept the need for this aspect of authenticity and the solution was to create two more stories: 'Bob's story' and 'Sarah's story'. The resulting stories were successful in providing an opportunity for all students' voices to be heard in the social interactions of the learning situation. This resulted in relevant learning across nine different professional groups, and subsequent feedback showed that students had been able to engage effectively with the scenarios.

## FUTURE RESEARCH DIRECTIONS

The use of digital technology is a growing trend in all aspects of education. A key principle of the HEFCE strategy for e-learning (2009) is the enhancement to learning and teaching through the use of technology. There is likely to be an expansion in its use particularly for the large cohorts involved in interprofessional education. To date there is a paucity of reported studies in interprofessional e-learning. Our work indicates that e-learning can provide solutions to the large scale implementation of interprofessional education and the effective engagement of service users. In our work we have investigated how theories of learning and curriculum development translate into the use of digital resources. It is important to understand the pedagogical principles underpinning the use of e-learning and to use technology in a way which enhances learning. This is an area that requires further examination and in particular there is a need to articulate more specific theoretical understandings of how service users can be involved at a partnership level in producing e-learning experiences so that their authentic voice is heard.

## CONCLUSION

Service user involvement is a valued concept in the education of health and social care students. This chapter has identified some of the challenges to meaningful involvement in interprofessional learning and has suggested that part of the solution is adopting a partnership approach with service users in the development of e-learning materials. We have also discussed how digital means can offer solutions to other problems in this context, specifically how 'difficult to engage' groups can be represented in student learning and how the multiple voices inherent in interprofessional education can be accommodated. We believe that advances in digital technology present an exciting future for interprofessional education, providing that consideration is given to its role in enhancing learning.

These are the issues we have raised that may be resolved through a partnership approach with service users and the appropriate use of digital technology:

- Logistical problems of providing a consistent experience for very large cohorts of students.
- Achieving involvement of services users who cannot come into the classroom and/or are from 'hard to engage' groups.
- Promoting service user control of what is represented and being comfortable with the end product.
- Reducing barriers of speaking to others, enabling the service user to tell their story once and not having to repeat it in order to accommodate different groups of students.
- Creating a safe learning environment for students (free to make mistakes, question issues, challenge value bases).
- Making materials available for use at any time, any place and enabling students to revisit the materials.

We believe that advances in digital technology present an exciting future for interprofessional education, providing that consideration is given to its role in enhancing learning. This will require a focused research and evaluation strategy.

## REFERENCES

Atkinson, R., & Flint, J. (2001). Accessing hidden and hard to reach populations: snowball research strategies. *Social Research update, 33*. Retrieved September 7, 2009, from http://www.soc.surrey.ac.uk/sru/SRU.HTML

Barnes, D., Carpenter, J., & Bailey, D. (2000). Partnerships with service users in interprofessional education for community mental health: a case study. *Journal of Interprofessional Care, 14*(2), 189–200. doi:10.1080/713678554

Barton, K. McKellar, P. & Maharg, P. (2005). *Authenticity in learning: transactional learning in virtual communities*. Draft paper submitted to the sixth International Clinical Conference. Retrieved August 7, 2009, from http://www.uwex.edu/disted/conference/Resource_library/handouts/03Info_H03.pdf

Clark, P. G. (2002). Values and voices in teaching gerontology and geriatrics: Case studies as stories. *The Gerontologist*, *43*(3), 297–303.

Department of Health (DOH). (1991). *The patients charter*. London: HMSO.

Department of Health (DOH). (1994). *Working in partnership*. London: HMSO.

Department of Health (DOH). (1998). *Modernising mental health services*. London: HMSO.

Department of Health (DOH). (2000a). *NHS plan*. London: HMSO.

Department of Health (DOH). (2000b). *Expert patient*. London: HMSO.

Department of Health (DOH). (2001). *Involving patients and the public*. London: HMSO.

Department of Health (DOH). (2002). *Requirements for social work training. Department of Health*. London: HMSO.

Department of Health/Government Strategy Unit. (2005). Independence, Well-being and Choice: Our vision for the Future of Adult Social Care in England. London: DoH.

Dewey, J. (1916). *Democracy and education: an introduction to the philosophy of education*. New York: Simon and Schuster.

Felton, A., & Stickley, T. (2004). Pedagogy, power and service user involvement. *Journal of Psychiatric and Mental Health Nursing*, *11*, 89–98. doi:10.1111/j.1365-2850.2004.00693.x

Gordon, F., Walsh, C., Marshall, M., Wilson, F., & Hunt, T. (2004). Involving patients and service users in student learning: developing practice and principles. *Journal of Integrated Care*, *12*(6), 28–35.

Gordon, F., Walsh, C., Marshall, M., Wilson, F., & Hunt, T. (2005). Good practice guidelines for patient and service user participation in teaching and assessing. Sheffield, UK: The Combined Universities Interprofessional Learning Unit: Sheffield Hallam University and University of Sheffield.

Gordon, F., Walsh, C., Marshall, M., Wilson, F., & Hunt, T. (2006). *The Combined Universities Interprofessional Learning Unit: Final report*. Sheffield, UK: Sheffield Hallam University and University of Sheffield.

Hickey, G., & Kipping, C. (1998). Exploring the concept of user involvement in mental health through a participation continuum. *Journal of Clinical Nursing*, *7*, 83–88. doi:10.1046/j.1365-2702.1998.00122.x

Higher Education Funding Council for England (HEFCE). (2009). *Enhancing learning and teaching through the use of technology: a revised approach to HEFCE's strategy for e-learning*. London: HEFCE.

Le Var, R. M. H. (2002). Patient involvement in education for enhanced quality of care. *International Nursing Review*, *49*(4), 219–225. doi:10.1046/j.1466-7657.2002.00128.x

Levin, E. (2004). *Involving service users and carers in social work education. Social Care Institute for Excellence*. London: Policy Press.

Lloyd-Jones, N., Hutchings, S., & Hobson, S. (2007). Interprofessional learning in practice for pre-registration health care: Interprofessional learning occurs in practice – is it articulated or celebrated? *Nurse Education in Practice*, *7*, 11–17. doi:10.1016/j.nepr.2006.03.002

Masters, H., Forrest, S., Harley, A., Hunter, M., Brown, N., & Risk, I. (2002). Involving mental health service users and carers in curriculum development: moving beyond 'classroom' involvement. *Journal of Psychiatric and Mental Health Nursing*, *9*, 309–316. doi:10.1046/j.1365-2850.2002.00493.x

Mellor, M. J., Hyer, K., & Howe, J. (2002). The geriatric interdisciplinary team approach: challenges and opportunities in educating trainees together from a variety of disciplines. *Educational Gerontology*, *28*, 867–880. doi:10.1080/03601270290099868

Mezirow, J. (2000). *Learning to think like an adult: core concepts of transformation theory. Learning as transformation: critical perspectives on a theory in progress*. San Francisco: Jossey Bass.

Morison, S., Boohan, M., Jenkins, J., & Mountray, M. (2003). Facilitating undergraduate interprofessional learning in healthcare: Comparing classroom and clinical learning for nursing and medical students. *Learning in Health and Social Care*, *2*(2), 93–104. doi:10.1046/j.1473-6861.2003.00043.x

Nisbet, G., Hendry, G. D., Rolls, G., & Field, M. J. (2008). Interprofessional learning for prequalification health care students: An outcomes-based evaluation. *Journal of Interprofessional Care*, *22*(1), 57–68. doi:10.1080/13561820701722386

Posey, L., & Pintz, C. (2006). Online teaching strategies to improve collaboration among nursing students. *Nurse Education in Practice*, *6*(2), 372–379. doi:10.1016/j.nepr.2006.07.018

Sainsbury Centre for Mental Health. (1997). *Pulling together*. London: Sainsbury Centre Publications.

Spencer, J., Blackmore, D., Heard, S., McCorie, P., McHaffie, D., & Scherpbier, A. (2000). Patient-orientated learning: a review of the role of the patient in the education of medical students. *Medical Education*, *34*, 851–857. doi:10.1046/j.1365-2923.2000.00779.x

Taylor, I. (1997). *Developing learning in interprofessional education: Partnerships for practice*. Buckingham, UK: Open University Press.

Turner, P., Sheldon, F., Coles, C., Mountford, B., Hiller, R., Radway, P., & Wee, B. (2000). Listening to and learning from the family carer's story: an innovative approach in interprofessional education. *Journal of Interprofessional Education*, *15*(4), 387–395.

UKCC. (2000). *Registrar's Letter 17/2000 Requirements for Pre-Registration Nursing Programmes*. London: United Kingdom Central Council for Nursing, Midwifery and Health Visiting.

Waldegrave, W. (1996). *Chief Secretary William Waldegrave speech at the Chartered Institute of Public Finance and Accountancy's conference*: Building Partnerships - Working Together. Held in Brighton 11-13 June 1996.

Whittaker, K. A., & Taylor, J. (2004). Learning from the experience of working with consumers in educational development. *Nurse Education Today*, *24*, 530–537. doi:10.1016/j.nedt.2004.06.005

Wood, J., & Wilson Barnett, J. (1999). The influence of user involvement on the learning of mental health nursing students. *Nursing Times Research*, *4*(4), 257–270. doi:10.1177/136140969900400403

# Chapter 12
# Designing for Enquiry:
## Using Web 2.0 to Enable Mental Health Service User and Carer Involvement

**Tarsem Singh Cooner**
*The University of Birmingham, UK*

## ABSTRACT

*This chapter sets the imperative for service user and carer involvement in the processes of educating mental health professionals. It begins by outlining some of the traditional barriers higher education institutions have faced in encouraging service user and carer involvement in teaching and learning. It then outlines the properties that Web 2.0 tools and processes can offer to overcome some of these obstacles. In developing effective interdisciplinary blended learning opportunities it is argued that the use of Web 2.0 alone will not ensure effective learning outcomes. The Community of Inquiry model is introduced to explore how the processes of enquiry, collaboration and communication can be embedded into the heart of interdisciplinary blended learning designs.*

## INTRODUCTION

The Centre of Excellence in Interdisciplinary Mental Health (CEIMH) is a UK Centre of Excellence in Teaching and Learning (CETL) funded until 2010 by the Higher Education Funding Council for England. One major aim of CEIMH's work has been to promote interdisciplinary mental health teaching and learning across several different disciplines within the University of Birmingham, UK. This chapter is based on the lessons we have learned through working on over seventy funded projects to promote this aim. The term interdisciplinary was adopted as a means of formally acknowledging that in all areas of our work the contributions of service users and their carers would be respected and valued (on an equal basis) to all other disciplines involved in the training of mental health professionals.

DOI: 10.4018/978-1-61520-889-0.ch012

Our project work has illustrated that successful interdisciplinary mental health teaching and learning occurs when academics and students from different disciplines working with service users and carers come together to access, contribute and share in the processes of creating social, collaborative and meaningful constructions of knowledge aimed at developing effective practice. One major outcome of this work is the recognition that for successful interdisciplinary teaching and learning to take place, communication, collaboration and enquiry must be at the heart of the process. The practicalities of organising such learning opportunities between students, service users and carers has proved challenging in higher education settings in the past due to issues of time, place and access. This chapter provides an overview of some of the lessons we have learned in the process of addressing these challenges through creating enquiry-based interdisciplinary blended learning designs. In sharing the processes we undertook, this chapter has three major objectives.

The first is to contextualise the imperative for mental health service user and carer contributions to interdisciplinary education while briefly outlining some of the traditional barriers higher education institutions have presented to their involvement. The second is to examine how the particular properties of Web 2.0 tools and processes can potentially help overcome some of these barriers. The third and final part of the chapter illustrates how a pedagogic framework such as the Community of Inquiry model (Garrison & Anderson, 2003) can be used to guide the social, cognitive and teaching processes required to blend Web 2.0 and service user and carer involvement into an effective enquiry-based interdisciplinary blended learning design. Due to the ongoing development of this work, the issues raised and discussed here are offered as a contribution to thinking and practice based on our experiences to date, rather than as a finished article.

## INTERDISCIPLINARY EDUCATION

### Definition and Ethos

It is important to contextualise the work of CEIMH by clarifying our definition of, and ethos towards the development of interdisciplinary education and practice. Building on the UK Centre for the Advancement of Interprofessional Education definition (CAIPE, 1997 & 2006), we broadly define interdisciplinary education as occasions where two or more disciplines (the term includes mental health service users and their carers) learn with, from and about one another to improve collaboration and the quality of mental health services. Because the service user and carer community may not be recognised as a "distinct profession" we use the term interdisciplinary as opposed to interprofessional to formally acknowledge their contributions to the learning process. In all other respects there is no difference between the principles of interprofessional and interdisciplinary education and practice.

Project applications to CEIMH have to clearly outline how mental health service users and carers will be involved in every aspect of the planning, design, delivery, assessment and evaluation of projects. Our development process has evolved over time and we now actively work alongside applicants throughout the duration of projects to help them create enquiry-based blended learning designs that have interactive learning elements aimed at maximising collaborative working and cross disciplinary communication. Our initial project evaluation work has indicated that a successful learning design will contain situations where students from two or more disciplines do more than simply listen to the same lectures. As Hammick, Freeth, Koppel, Reeves and Barr (2007) explain, enquiry-based interactivity in the learning process is key if it is to help develop the circumstances that will lead to effective collaborative practice and promote better care.

## Involving Service Users and Carers in Teaching and Learning

Mental health service users and carers tend to bring a vast pool of authentic experiences and knowledge to the task of teaching professionals. By virtue of their past involvement with services they can provide insights that can enable students from different disciplines to understand the holistic nature of mental distress and its impact on the service user, their carer and those around them. Utilising this experience in the education and training of mental health professionals can help to produce practitioners capable of working in partnership to provide improved and appropriate provision for service users and their carers (Tew, Gell & Foster, 2004). Service user and carer involvement is not a new phenomenon, in the late 1990s consumers of a broad range of health services and their carers sought increased power, control and influence in the way services were delivered and professionals trained (Rose, 2001). Service users played a major role (Hickey & Kipping, 1998) in bringing about a situation in which the development of service delivery began to be characterised by an ethos of partnership working where the expertise of the consumer began to be recognised and valued. In the United Kingdom, government policy began to support the involvement of service users and carers in the development and delivery of health service provision and education (DOH, 2000).

The principles underpinning these directives aimed to transfer power or control of aspects of service delivery, planning and education from professionals to service users and their carers (Wright & Rowe, 2005). Part of the reason for involving consumers in this process was to challenge the causes of bad service delivery based mainly on entrenched attitudes and negative perceptions held by health professionals. To begin to break down the routines that supported paternalistic and hierarchical service provision, the ultimate aim of this partnership process was to improve interdisciplinary service delivery (Happell & Roper, 2002). In the field of workforce training in mental health, the principle of collaborative working plays a prominent role, for example the first of the Ten Essential Shared Capabilities set out by the National Institute for Mental Health in England actively promotes Working in Partnership (Hope, 2004).

The Centre's approach to interdisciplinary education aims to remain faithful to the culture and value base of mental health education and practice. Through its funding of projects it actively seeks to encourage the concept and practice of partnership working (NIMHE, 2003) between students, practitioners and service users and their carers. We have found that involving service users and carers in the education process can play an important role in developing communication, clinical reasoning and reflective practice skills as well as enabling an exploration of professional attitudes and behaviour. These experiences are mirrored in the work of Spencer, et al. (2000). The Centre through curriculum development aims to reduce a culture that creates a division of "us" and "them" in which the assumption is that people with direct experiences of mental distress (such as service users and carers) are somehow different from practitioners, educators and students (Tew, et al., 2004).

The aim is to "chip away" at the remnants of a professional education value base that assumes professionals need to be the "experts", and who as a consequence of their knowledge can force their understanding and interventions upon service users and carers who are assumed to lack any insight, resources or ability to promote their own recovery. We have found that where mental health service users and carers are actively involved in the teaching role, learners can benefit from the expertise their experiences bring into the learning encounter. This approach is recognised and research suggests that being taught by trained mental health service users and carers can generate new insights, reduce anxiety and increase student confidence (Babu, Law-Min, & Adlam, 2008; Gell, 2003; Wykurz & Kelly, 2002). The Centre's interdisciplinary approach seeks to promote a more holistic understanding of mental distress, its impact and possible alternative roads to recovery. A major aim of our project work is to prepare trainee health and affiliated professionals such as nurses, social workers, psychologists, medics and teachers for partnership working so that they see collaborative working with service users and carers "as the norm" in their post-qualified practice (Ikkos, 2003).

## Barriers to User/Carer Involvement and Shared Learning

Whilst the arguments and principles supporting service user and carer involvement are clearly stated, putting them into practice in higher education has proved quite challenging. Tyler (2006) highlights how the structural and internal logistics of a University can create a number of practical barriers that can prevent service users and carers from getting involved in the process of educating students. These barriers can include practical issues of time, place and access where involvement can be prevented because of problems such as organising payments for substitute carers, providing or funding payment for travel to and from the University, paying child care costs or offering a form of payment that will not interfere with eligibility for benefits. Basset, Campbell and Anderson (2006) also outline how the culture and processes of a University can exclude full participation. They give examples such as the notion of hierarchies of knowledge where the lecturer may feel hesitant about letting go of the "expert" role, a "wheel em on" approach where input is limited to a marginal role precluding any involvement in the development of teaching objectives or learning designs, a lack of support for new service user and carer trainers/educators, a closed shop where access to participation can be limited by who you know, rather than what you know and can offer.

The situation in relation to bringing students together for shared learning must not be overlooked. Even when there is a will to develop collaborative teaching and learning opportunities between disciplines, our earlier experiences mirrored some of the issues raised by Ross and Southgate (2000) who point out that practical difficulties in finding rooms to accommodate large cohorts, timetabling restrictions, different professional requirements and staffing issues all accounted in many cases for the lack of collaborative interdisciplinary learning opportunities in many universities. Research also suggests that in most pre-registration professional training, a "silo" approach develops hindering the professionals' ability to work in an interdisciplinary manner once qualified (Hall & Weaver, 2001; Horsburgh, Lamdin & Williamson, 2001; Stumpf & Clark, 1999; Tunstall-Pedoe, Rink & Hilton, 2003). Our experiences suggest that some of the underlying concepts of Web 2.0 tools and processes when creatively used to develop enquiry-based interdisciplinary blended learning designs can help overcome a number of these practical barriers to service user and carer involvement.

## WEB 2.0

## Collaboration and Communication

The term Web 2.0 was first coined by Tim O'Reilly (2005) and has no agreed definition. Our work in this area suggests that Web 2.0 should be thought of more as a concept than as a technology. Terms such as the "read/write web" and the "social web" have been used to describe some of its main properties. Other than its ability to connect people synchronously and asynchronously to overcome some of the barriers of time, place and access, there are a number of collaborative learning concepts behind Web 2.0 tools and processes that make it an attractive proposition for academics looking to develop interdisciplinary curricula. For example, it can enable a wide range of people to engage in the processes of enquiry by freely accessing and sharing knowledge through collaborative processes with others to consume, remix, repurpose and collectively generate new knowledge and ideas. These processes are essential for developing enquiry in learning, however before looking at these learning concepts in more detail, it is worth briefly exploring the mechanics of engaging with these sites.

Access to most Web 2.0 resources is generally achieved through a web browser such as Internet Explorer, Firefox, and Safari. Because these sites enable people to write to them, they normally require users to create an identity and log in before they can comment or upload content. An interesting point about this process is that users do not have to use their real name and thus can post anonymously. Some of the most commonly recognised Web 2.0 activities are blogging, media sharing, social networking, and collaborative working using Wikis or GoogleDocs. Unfortunately a full description and analysis of the educational properties of each of these tools is beyond the scope of this chapter (For a more detailed account, see Mason & Rennie, 2008; Richardson, 2009).

## The Learning Concepts: Web 2.0

Whilst there are many advocates for using Web 2.0 in education (CICLE, 2009; Mason & Rennie, 2008; Richardson, 2009; TLRP, 2008) there are also voices that warn against the uncritical adoption of new technologies and processes (Brabazon, 2008, Keen, 2008). They argue that a clear understanding of the strengths and weaknesses of the methods and tools used must be achieved before any wholesale adoption. Our own critical appraisal of Web 2.0 tools and processes (in developing enquiry-based blended learning designs) suggests there are four related properties that can aid the design of effective interdisciplinary teaching and learning. Although these properties are divided into four segments and explored here with illustrative examples, they should not in any sense be considered independent. Whichever way they are divided up it will always be necessary to see the influence of one property in relation to the others. The first property is the ability of Web 2.0 tools and services to support a high degree of communication and collaboration. Using these freely available facilities, service users, carers, students, practitioners and academics can plan teaching, share information, coordinate research/problem-based activities, engage in debate and build learning communities within a flexible environment that can be free from the traditional constraints of time and place.

We found for example that using GoogleDocs online word processor as the main medium for planning and designing teaching events, reduced the need for face-to-face meetings and enabled service users and carers to participate fully in these processes by working remotely. Functions such as the share option on the online word processor allowed group members to collectively access, create, view, add, amend, save and delete content in planning and design documents. All edits were displayed in real time ensuring that the most up to date versions were always visible along with a revision history option allowing members to easily access previous documents to review progress, or amend mistakes. These functions enabled service users and carers to contribute to the planning and design process at a time, place and pace convenient for them. (For more on GoogleDocs see http://tinyurl.com/6j7ds7.)

The second property Web 2.0 offers is the ability to easily create content such as text, video or audio files and then freely publish these to a select group or potentially worldwide audience. We found that the "write, publish and discuss" simplicity of Blogs (such as Google's Blogger.com) provided an excellent medium for mental health service users and carers to develop a supportive community of practice by creating an online space to share, discuss and archive their experiences of engaging in and seeking to improve interdisciplinary teaching and learning processes. (For more on Blogs see http://tinyurl.com/2kobxs.)

The third property, developing the skills of enquiry can be achieved by combining problem-based case studies with the simple collaborative functions of a Wiki (such as PBWorks.com). Utilising the Wiki link function, students were guided to additional content on the Internet. Working alongside mental health service users, carers and practitioners they learned to develop the skills required to retrieve, interrogate and assess the validity of sources they intended to use. Working as a team they used the write, edit and save functions of a Wiki to collaborate and collectively produce potential solutions to complex life like mental health situations. Exposure to these learning opportunities covers the fourth property, that of developing literacy (through experience) in using these tools and reflecting on how they could be used to aid effective interdisciplinary practice in the future. (To find out more about Wikis see http://tinyurl.com/23y3wp.) The collective impact of engaging with these properties also provided students with opportunities to develop the skills and confidence required to become independent learners.

In terms of designing for enquiry and enabling communication and collaboration, these properties are in sharp contrast to our experiences of using institutional virtual learning environments (VLEs). Although there are a number of attributes that VLEs possess over Web 2.0, such as a consistent user interface, the same tools for all learners, permanence in that they are not likely to go out of business without warning, and access to technical support, there are also a number of elements that can reinforce the old barriers. One example is that service users and carers have to be registered with the VLE before they can take part in any teaching and learning encounters, registration can be quite a complicated and time-consuming process and prevents immediate participation. Also, Attwell (2007) points out that higher education institutions have used VLEs to "manage learning" rather than encourage open networking and creativity, the material on these systems are invariably created by tutors with little room or facility for the collaborative creation of content available in open networks. The design behind VLEs tend to mirror the old processes of teaching and learning in higher education by recreating the walls of the classroom and institution in the virtual world. These types of restrictions perpetuate the isolation of the university from the wider outside community. The philosophies behind the designs of these systems (Beetham, 2007) tend to struggle to compete with the collaborative enquiry-based teaching and learning opportunities offered by Web 2.0 tools and processes.

## THE COMMUNITY OF INQUIRY MODEL

### Designing for Learning

Although Web 2.0 presents viable opportunities for greater service user and carer participation, the development of effective enquiry-based interdisciplinary blended learning designs will not automatically result as a consequence of bringing these two elements together. Our experience has highlighted that the use of any new technology and educational process has to be guided by a sound pedagogic model to ensure that overall aims and objectives are met. After considerable experimentation, most of the recent interdisciplinary teaching and learning designs developed by CEIMH projects have been guided by Garrison and Anderson's (2003) Community of Inquiry (CoI) model.

We believe the CoI model has produced higher levels of learner satisfaction because it embeds blended design and enquiry into the heart of the teaching and learning design. The model has a constructivist underpinning and promotes the idea that the individual brings his or her personal history, knowledge and experiences into the learning encounter. It supports the view that learning is "emergent" rather than "given" or "discoverable" (Vygotsky & Cole, 1978). The model encourages the perspective that learning arises from social practices and contrasts vividly with didactic forms of teaching that may primarily view knowledge as something that exists "out there" external to the student and therefore can be "given" through the processes of "transmission" (Goodman, Lillis, Maybin, & Mercer, 2003; Scollon, 2001).

This model compliments the collaborative and social learning concepts underpinning Web 2.0 tools and processes and supports the incorporation of authentic learning experiences that service users and carers can bring into the interdisciplinary mental health teaching and learning encounter. For deep and meaningful learning to occur, Garrison and Anderson (2003) argue three elements must be present when creating effective learning designs. Within the CoI framework these three elements are identified as social, cognitive and teaching presence. It is important to note that each of these three elements are interdependent and must not be seen in isolation from each other. The following account provides a brief introduction to each of these three elements. It also demonstrates how in the process of creating an enquiry-based interdisciplinary blended learning design; this model can help to structure thinking around development to overcome a number of potential organisational and cultural barriers to service user and carer involvement.

## Social Presence

Social presence:

*the ability of participants in a community of inquiry to project themselves socially and emotionally, as 'real' people (i.e., their full personality), through the medium of communication being used (Garrison, Anderson & Archer, 2000, p. 94)*

Creating a climate that encourages open communication is important for service user and carer involvement in a community of enquiry. In creating a teaching and learning design, thought must be given to the processes required to create a situation in which participants can feel free to express themselves in a risk free manner. In a community of enquiry this is one of the first steps required to ensure that collaboration can be encouraged to enable the sharing of experiences and practices within groups of learners. Feeling emotionally secure can enable individuals to engage in open and purposeful discourse as group cohesion and camaraderie build. In respect to service user and carer participation, developing this environment is essential on two levels.

The first relates to the wider issues of involvement in the learning design process. Our experiences show that for service users and carers to project their real identities into the learning encounter, they must feel that the learning activities they engage in are authentic. To ensure authenticity means that they must be involved in every aspect of the learning and teaching design process. In an environment where planning through joint working takes place, academics must ensure that the contributions of service users and carers are valued and treated on par with other professionals. The second level concerns engagement with students. All of our recent project designs have been based on creating interdisciplinary learning groups where students from different disciplines and service users and carers from day one work as a team. In these teams they engage in exercises that involve online and face-to-face learning activities aimed at encouraging collaboration and communication. Embedding service user and carer participation in this manner right from the start means that they have the same opportunities to develop relationships within the group and ensure that their participation is present throughout the learning process.

To ensure that a successful level of sharing, debate, discussion and learning takes place, the design has to ensure that everyone involved is orientated to this method of learning so they are clear about their role in the process and the expectations of the tutor. A supportive learning environment will also ensure that the well being of all participants within the learning process is monitored throughout the duration of the activities and if needed, access to support services is made available. From a practical point of view funding for service user and carer participation has to be in place along with appropriate training and support. Although very rare, where required the Centre has provided service users and carers with access to equipment and finances to ensure there are no barriers to taking part in online activities. All of these elements can play a major role in creating a socially proactive environment in which service users and carers can actively engage in the processes of teaching and learning.

## Cognitive Presence

Cognitive presence:

*an environment that enables learners to construct and confirm meaning through sustained reflection and discourse in a critical community of enquiry (Garrison, Anderson & Archer, 2001, p. 11).*

In the evaluation of our recent projects most students have stated that they find mental health service user and carer involvement has led to personally meaningful and educationally worthwhile learning outcomes (http://tinyurl.com/pfgu4a). To achieve these outcomes has required the careful design of the learning activities. To encourage the processes of discussion and enquiry where students have the ability to challenge existing knowledge, stereotypes, beliefs, opinions and values requires learning activities that trigger a sense of puzzlement and debate. Interdisciplinary groups presented with authentic problem-based case studies or digital Photo Story triggers (created by service users and carers http://tinyurl.com/qnglom) can be enabled to develop their learning by exploring a wide range of issues related to mental health practice.

*Figure 1. Image of "My Experience of Night Medication Photo Story Trigger". (© 2009, www.ceimh.bham.ac.uk, used with permission).*

The strengths of communication and collaboration inherent in Web 2.0 technologies can be exploited to encourage groups to explore the issues presented through the exchange of information. Exploiting the time and place flexibility provided by online tools such as Wikis, mental health service users and carers can engage in sustained discussion with students to help them reflect on their learning and connect new ideas. One of the benefits of using problem-based case studies or digital triggers is that they can enable students in a safe environment to apply new ideas. Where the resolution to a particular issue needs to be reconsidered, discussion and debate in a community of enquiry can lead to the development of reflective practices and new skills and knowledge. This process can enable group participants to share personal meaning and ask questions to confirm understanding within a wider mental health community. Creating a climate that enables open communication and risk free expression is essential for this part of the learning to be successful since it can impact on the ability of students and service users and carers to engage in the processes of supporting meaningful discourse. In selecting the content for discussion and debate, teaching presence plays an important part.

### Teaching Presence

Teaching presence:

*the design, facilitation and direction of cognitive and social processes for the purpose of realizing personally meaningful and educationally worthwhile learning outcomes (Anderson, Rourke, Garrison & Archer, 2001, p. 5).*

Linking the possibilities offered by Web 2.0 to enable service user and carer involvement in the teaching and learning process required a radical rethink in how we approached our projects learning designs. The CoI model was particularly helpful in guiding the development of blended learning. It encouraged a fundamental change to our teaching and learning approaches and led to a number of blended redesigns enabling a restructuring of class hours so that interdisciplinary engagement could be extended through web-based learning opportunities. The CoI model helped to support the development of learning designs that allowed us to thoughtfully examine and optimally integrate the strengths of face-to-face oral and online media communications. The aim for us was to use the strengths of each to produce learning experiences consistent with the aims and objectives of the different projects.

Project partners were encouraged to think beyond the classroom walls and explore the possibilities that Web 2.0 offered to connect all the actors involved in the learning process. They were asked to consider how these facilities could change the role of students from passive recipients of knowledge, to active agents in the co-construction of knowledge and meaning. In training service user and carer colleagues to take part in this process, they were encouraged to think about guiding discussion and the implications of sharing personal meaning in an online and face-to-face environment. Being part of each process, particularly setting up the curriculum and design meant that they had a greater say and understanding of the teaching and learning processes, aims and objectives.

With the CoI model guiding the development of our learning designs, we have been able to utilise the collaborative and communication facilities built into Web 2.0 tools and processes and involve mental health service users and carers in a way that would not have been possible before. The combination of these elements has enabled us to create effective learning encounters for students because enquiry, collaboration and cross discipline communication have been at the heart of the learning design.

## FUTURE RESEARCH DIRECTIONS

With the popularity of social networking and other Web 2.0 based activities there are a number of areas that need further exploration to help build our understanding of the impact these tools and processes can have on higher education. Embedding enquiry-based blended learning practice across higher education institutions using Web 2.0 is one important area for study. At Birmingham, the Centre has worked with the Central E-Learning Team and the equivalent of the Staff Development Unit (APOD) to collectively create and deliver a two-day Designing for Enquiry-based Blended Learning (DiBL) training event to a cross section of academics.

The DiBL event had the support of APOD and the E-Learning Team because the CoI model was consistent with one of the key goals of the institution's teaching and learning strategy, namely that of promoting the processes of enquiry-based teaching and learning. The Centre used its experiences to create a set of resources to make explicit, implicit knowledge about the CoI model and then trained APOD and E-Learning Team partners to use these tools. The DiBL event was designed to adhere to the principles of enquiry-based learning and members of the E-Learning Team and APOD took on the role of coaches to help the participants (working in teams) to learn how to create enquiry-based blended learning designs (for more information about the DiBL event see http://tinyurl.com/r3hzap).

The motivation for holding this event was created through a series of presentations by the CEIMH team to senior managers. The aim was to raise their awareness of how the institution could potentially employ Web2.0 technologies to widen participation, meet the needs of a diversifying student population and address some of the challenges of funding and competing in a global education market (Bradwell, 2009). CEIMH is currently in the process of collating and disseminating a number of pertinent lessons gleaned around developing communities of practice focused on creating enquiry-based blended learning designs. Suggested future research directions related to this topic include exploring organisational cultures, processes and attitudes that can either help or hinder the development of these communities.

Another area requiring further investigation is the role and methods of assessment in enquiry-based blended learning designs. Currently all our projects were restricted to the assessment methods outlined in original module specifications by project partners, these consisted primarily of assignments, examinations and presentations. If we are to use technology enhanced processes, particularly where students are expected to produce work collaboratively online, then assessment methods need to adapt and recognise the potential learning student's achieve in the production of this work. Research into student motivation, engagement with tasks and whether deeper learning occurs through these processes could help inform changes in future assessment methods. In relation to wider issues raised in this chapter, one area of particular importance is an examination of how Web 2.0 tools and processes can be used to engage service users and carers in developing and improving the delivery of mental health services out in the field. With the strengths this medium provides to enable communication and collaboration, issues such as the impact it can have on influencing future policy and practice decisions is a key area for further research.

In education there is still a need for good practice guidelines in using Web 2.0 tools and processes. Research into the impact of accessibility, security, permanence, privacy, copyright and design from an educational institution's point of view, is still in its infancy and needs to be further developed to inform good practice. Finally, from a practical point of view, answers to questions such as: "does exposure to collaborative working with mental health service users and carers during the education process result in partnership working in post-qualified practice?" need exploration along with an examination of the factors that can encourage or impede this process.

## CONCLUSION

With the current level of freely available access to Web 2.0 technologies there has never been a better time to create opportunities for service user and carer participation in interdisciplinary education. The possibilities Web 2.0 tools and processes offer in enabling students from different disciplines to share ideas, opinions and experiences with service users and carers has never been greater. The very nature of the tools and processes provided by this medium can encourage learning through the processes of collaboration, communication and enquiry. It has been argued that the existence of these properties within Web 2.0 tools and processes alone cannot guarantee that effective and meaningful learning opportunities will occur. It's argued that the development of blended learning designs has to be guided by sound pedagogic models. Working with mental health service users and carers requires careful planning at a number of different levels so students can benefit from their involvement. Developing these interdisciplinary experiences for students in the early part of their educational processes should at the very least positively impact on two areas. The first is to make them aware of the powerful opportunities Web 2.0 can offer for future consultation in the development of services once they qualify. The second is to normalise the principle of partnership working so once in practice they can work collaboratively to deliver effective mental health services.

## REFERENCES


Anderson, T., Rourke, L., Garrison, D. R., & Archer, W. (2001). Assessing teaching presence in a computer conferencing context. *Journal of Asynchronous Learning Networks*, *5*(2).

Attwell, G. (2007). *Web 2.0 and the changing ways we are using computers for learning: what are the implications for pedagogy and curriculum?* Retrieved May 23, 2009, from http://www.elearningeuropa.info/files/media/media13018.pdf

Babu, K. S., Law-Min, R., & Adlam, T., V. B. (2008). Involving service users and carers in psychiatric education: what do trainees think? *Psychiatric Bulletin*, *32*(1), 28–31. doi:10.1192/pb.bp.107.015354

Basset, T., Campbell, P., & Anderson, J. (2006). Service User/Survivor Involvement in Mental Health Training and Education: *Overcoming the Barriers. Social Work Education*, *25*(4), 393–402. doi:10.1080/02615470600593675

Beetham, H. (2007). An approach to learning activity design. In Beetham, H., & Sharpe, R. (Eds.), *Rethinking Pedagogy for a Digital Age: designing and delivering e-learning*. Abingdon, UK: Routledge.

Brabazon, T. (2008). *The University of Google: Education in the (Post) Information Age*. Aldershot, UK: Ashgate Publishing Limited.

Bradwell, P. (2009). *The Edgeless University Why Higher Education Must Embrace Technology*. London: DEMOS.

CAIPE. (1997 & 2006). Interprofessional Education - a definition. London: CAIPE.

CICLE. (2009). *Higher Education in a Web 2.0 World. Committee of Inquiry into the Changing Learner Experience*. Retrieved May 18, 2009, from http://www.jisc.ac.uk/media/documents/publications/heweb20rptv1.pdf

Department of Health (DOH). (2000). *The NHS Plan*. London: HMSO.

Garrison, D. R., & Anderson, T. (2003). *E-learning in the 21st century: a framework for research and practice*. New York: RoutledgeFalmer. doi:10.4324/9780203166093

Garrison, D. R., Anderson, T., & Archer, W. (2000). Critical inquiry in a text-based environment: Computer conferencing in higher education. *The Internet and Higher Education*, *2*(2-3), 87–105. doi:10.1016/S1096-7516(00)00016-6

Garrison, D. R., Anderson, T., & Archer, W. (2001). Critical thinking, cognitive presence and computer conferencing in distance education. *American Journal of Distance Education*, *15*(1), 7–23. doi:10.1080/08923640109527071

Gell, C. (2003). *Involving service users in the training of graduate primary care mental health workers*. Nottingham, UK: Trent Workforce Development Confederation.

Goodman, S., Lillis, T., Maybin, J., & Mercer, N. (2003). *Language, Literacy and Education: A Reader. Stoke on Trent*. UK: Trentham Books.

Hall, P., & Weaver, L. (2001). Interdisciplinary education and teamwork: a long and winding road. *Medical Education*, *35*(9), 867–875. doi:10.1046/j.1365-2923.2001.00919.x

Hammick, M., Freeth, D., Koppel, I., Reeves, S., & Barr, H. (2007). A best evidence systematic review of interprofessional education: BEME Guide no. 9. *Medical Teacher, 29*(8), 735–751. doi:10.1080/01421590701682576

Happell, B., & Roper, C. (2002). Promoting consumer participation through the implementation of a consumer academic position. *Nurse Education in Practice, 2*(2), 73–79. doi:10.1054/nepr.2002.0068

Hickey, G., & Kipping, C. (1998). Exploring thc concept of user involvement in mental health through a participation continuum. *Journal of Clinical Nursing, 7*(1), 83–88. doi:10.1046/j.1365-2702.1998.00122.x

Hope, R. (2004). *The ten essential shared capabilities: a framework for the whole of the mental health workforce*. London: Department of Health.

Horsburgh, M., Lamdin, R., & Williamson, E. (2001). Multiprofessional learning: the attitudes of medical, nursing and pharmacy students to shared learning. *Medical Education, 35*(9), 876–883. doi:10.1046/j.1365-2923.2001.00959.x

Ikkos, G. (2003). Engaging patients as teachers of clinical interview skills. *Psychiatric Bulletin, 27*(8), 312–315. doi:10.1192/pb.27.8.312

Keen, A. (2008). *The Cult of the Amateur: How blogs, MySpace, YouTube and the rest of today's user-generated media are killing our culture and economy*. London: Nicholas Brealey Publishing Ltd.

Mason, R., & Rennie, F. (2008). *E-learning and social networking handbook: resources for higher education*. London: Routledge.

NIMHE. (2003). *Cases for Change: User involvement*. Leeds, UK: NIMHE.

O'Reilly, T. (2005). *What Is Web 2.0?* Retrieved May 14, 2009, from http://www.oreillynet.com/pub/a/oreilly/tim/news/2005/09/30/what-is-web-20.html?page=1

Richardson, W. (2009). *Blogs, wikis, podcasts, and other powerful web tools for classrooms* (2nd ed.). Thousand Oaks, CA: Corwin Press.

Rose, D. (2001). *Users' Voices: the Perspectives of Mental Health Service Users on Community and Hospital Care*. London: Sainsbury Centre for Mental Health.

Ross, F., & Southgate, L. (2000). Learning together in medical and nursing training: aspirations and activity. *Medical Education, 34*(9), 739–743. doi:10.1046/j.1365-2923.2000.00626.x

Scollon, R. (2001). *Mediated Discourse: The Nexus of Practice*. London: Routledge.

Spencer, J., Blackmore, D., Heard, S., McCrorie, P., McHaffie, D., & Scherpbier, A. (2000). Patient-oriented learning: a review of the role of the patient in the education of medical students. *Medical Education, 34*(10), 851–857. doi:10.1046/j.1365-2923.2000.00779.x

Stumpf, S. H., & Clark, J. Z. (1999). The Promise And Pragmatism Of Inter-Disciplinary Education. *Journal of Allied Health, 28*, 30–32.

Tew, J., Gell, C., & Foster, S. (2004). *Learning from Experience: Involving Service Users and Carers in Mental Health Education and Training: A Good Practice Guide*. Higher Education Academy/NIMHE/Trent Workforce Development Confederation.

TLRP. (2008). *Education 2.0? Designing the web for teaching and learning. Teaching and Learning Research Programme*. Retrieved May 18, 2009, from http://www.tlrp.org/tel/files/2008/11/tel_comm_final.pdf

Tunstall-Pedoe, S., Rink, E., & Hilton, S. (2003). Student Attitudes To Undergraduate Interprofessional Education. *Journal of Interprofessional Care*, *17*(2), 161–172. doi:10.1080/1356182031000081768

Tyler, G. (2006). Addressing Barriers to Participation: Service User Involvement in Social Work Training. *Social Work Education*, *25*(4), 385–392. doi:10.1080/02615470600593394

Vygotsky, L. S., & Cole, M. (1978). *Mind in Society: The Development of Higher Psychological Processes*. Cambridge, MA: Harvard University Press.

Wright, C., & Rowe, N. (2005). Protecting Professional Identities: Service User Involvement and Occupational Therapy. *British Journal of Occupational Therapy*, *68*(1), 45–47.

Wykurz, G., & Kelly, D. (2002). Developing the role of patients as teachers: literature review. *BMJ (Clinical Research Ed.)*, *325*(7368), 818–821. doi:10.1136/bmj.325.7368.818

# Section 3
# Implementing Change and its Consequences

# Chapter 13
# Developing an Interprofessional E-Learning Pathway:
## Leading Academics Through the Change Challenge

**P. A. S. Bluteau**
*Coventry University, UK*

**J. A. Jackson**
*The University of Warwick, UK*

## ABSTRACT

*Implementing interprofessional education is fraught with challenges, developing an e-learning approach whilst overcoming some challenges, is still not smooth. This chapter using a case study approach offers an insight into the successful development and implementation of an interprofessional e-learning pathway. The chapter draws on the authors' experiences of the challenges and constraints in setting up this model. By unpacking these challenges the reader will be offered the opportunity to explore the impact on two Higher Education Institutes and the professional groups working within these organisations. It highlights the difficulties associated with managing change and the importance of strong leadership at key points within this process.*

## INTRODUCTION

Interprofessional education (IPE) has become a key focus in the Higher Education institutions (HEI) that provide education to health and social care professional groups. In the last 20 years evidence has increased which supports the inclusion of IPE in health and social care curricula (Barr, 2009). It is also apparent within the literature surrounding IPE, that acceptance of interprofessional initiatives have grown, and with this a growing hope and belief that IPE might be part of the solution to improve the experience of patients, service users, clients, carers and their families (Department of Health [DH], 2000a; General Medical Council [GMC], 2003, 2009; Bluteau & Jackson, 2009).

IPE, without doubt, has caused tension during its initiation, gradual introduction and development within curricula in Higher Education Institutions (HEIs) (Barker et al., 2005; Bluteau

DOI: 10.4018/978-1-61520-889-0.ch013

& Jackson, 2009). This chapter explores a case study of the implementation of IPE within an e-framework across two universities, beginning with the thinking behind the choice of an e-learning experience in contrast to the more prevalent face to face encounters within IPE. Whilst this is only representative of two institutions, neither the challenges or experiences are unique (Freeth et al., 2005; Gilbert, 2005). This account provides the reader with an insight into this journey highlighting potential pitfalls associated with the implementation of IPE. Next it explores the challenges experienced in developing IPE and offers a model of the challenges within the context of an Interprofessional e-learning pathway (IPeLP). Finally, by using a deconstructive approach, the chapter illustrates the maturation process of this model in the context of IPE.

## BACKGROUND

Interprofessional working is dependent on individual professionals having a sound understanding of the roles, skills, knowledge and guiding philosophies of other professions (Gordon & Walsh, 2005). In the Faculty of Health and Life Sciences at Coventry University in collaboration with Warwick Medical School at the University of Warwick, there are fourteen separate courses leading to registration as a qualified health and social care professionals, and hence a great opportunity to learn with, from and about students from other professions. The process of the different groups discussing their own roles and insights with others helps to celebrate the unique skills and knowledge that each profession brings to an integrated model of patient, client, care or service user care.

This case study draws on the authors' experiences of the first year of developing and implementing an Interprofessional e-learning Pathway (IpeLP) in which eleven of these different professional groups participated, with Social Work, Youth Work ans Operating Department Practitioners joining later. (Figure 1).

The process of developing a model in relation to the challenges that have surrounded the inception and development of IPE demands the inclusion of some context in which to frame the important components of IPE. Collaboration is a major component of IPE (Barr, 2009; Barr, 2005; Yan et al., 2007; Jackson & Bluteau, 2009). The underlying concept of IPE is to teach future professionals to work collaboratively within health and social care teams so that they are able to deliver quality care to each and every patient/client/service user (Barr et al., 2005). Whilst there is good evidence (Reeves et al., 2008, Hermsen et al., 2005, The Audit Commission 1999, Wilcock et al., 2000) to support the belief that collaboration through teamwork does indeed improve the quality and continuity of patient centred care, there are many challenges that hinder this process.

In terms of the educational process these challenges could be viewed in terms of a two strand continuum. At one extremity, within the HEI, there are issues such as lack of faculty support (Gilbert, 2005), timetabling difficulties (Cooper et al., 2001; Oandasan et al., 2005) and cost (Pecukonis, 2008) which are frustrating, time consuming and generally irritating, but on the whole practical

*Figure 1. Participating students in the level 1 IPLP module – September 2005 to January 2006 Cohort.*

| Professional group | No. of participants |
|---|---|
| Adult Nursing | 261 |
| Medical | 191 |
| Midwifery | 24 |
| Dietetics | 42 |
| Physiotherapy | 162 |
| Occupational Therapy | 156 |
| Learning Disability Nursing | 32 |
| Clinical Psychology | 15 |
| Paramedic | 27 |
| Mental Health Nursing | 66 |
| Child Nursing | 27 |
| **Total number of students** | **1002** |
| Facilitators | 34 |

and eventually solvable, although at some human cost. At the other end, in the uniprofessional departments, the problems are more intractable and unmovable. Issues such as tribalism (Barker et al., 2005; Hall, 2005), refusal of ownership (Jackson & Bluteau, 2007) and 'uniprofessional silo mind set' (Rafter et al., 2006) threaten the sustainability of IPE initiatives and can lead to 'burn out' of the champions who support the developments.

In the practice arena, initiatives aimed at addressing IPE tend to be focused on the need for streamlining, improving the cost effectiveness, efficiency and quality of care within the service setting, Jackson & Bluteau's (2005) experience of developing a pilot interprofessional student experience in practice found the opportunity for collaboration was problematic. It is in practice that the theoretical teaching concerning IPE, which students have been exposed to, tests reality. It is important to consider that although students may experience a wide range of IPE encounters, it will be at times of crisis that they will meet actual interprofessional working, where tribalism and silo working disappear and collaborative practice takes precedence as humanitarian needs become the driving force.

For example, Tulloch (2006) describes his experience of seamless intervention within health-care following the terrorist attacks in the London underground on 7th July 2005. In everyday working, collaborative practice is more likely to be a 'happy coincidence' as opposed to a structured and planned intervention. Within higher education the pressures are not life threatening but create stressful situations and hinder collaborative working not only between students and their professional groups, but also between HEI institutions.

In the context of this case study the potential for face to face delivery was considered in the early stages of the development cycle of IPE, however the problems that were faced in considering this option were many. To introduce a face to face programme for in excess of 1,000 students at three points during each programme created several immediate logistical issues. Firstly, there was no space on campus large enough for an introductory session. The group would have to be broken down into four groups, and while this was possible, this raised anxiety and concern from the teaching staff who would be involved.

Further, the large introductory group would then be broken down into approximately seventy smaller groups, requiring space to work together, ideally seventy rooms. Student groups could work in the same room if there was enough space to work and enough distance to avoid distraction, however this was impossible with the constraints of other students within the university using the campus: there were simply not enough rooms.

In addition to the number of rooms needed each week for the IP learning, over a period of weeks the process would have had an impact on uniprofessional timetables in that students would need to be released from both theory and practice sessions for both a lead lecture and related small group work. That students should be released from theory caused great concern and in some cases outrage, but the greater issue was that many students would be attending practice placements across the UK. This became an insurmountable barrier.

Given these practical constraints, the decision was made to implement an e-based IPE programme rather than the more traditional face to face delivery. E-learning offered an alternative method of delivery which could overcome many of the issues. However, e-learning presented its own challenges, in particular in relation to the large numbers of students and facilitators across different locations with contrasting needs and abilities (Conole, & Oliver, 2006; Conole et al., 2006; Hiltz et al., 2007; Littlejohn & Peglar, 2007).

It quickly became apparent that although e-learning offered a solution to multiple locations and large numbers of participants, the individuals who would be participating had very varied expertise in using an e-approach. Some were comfortable and at ease with having discussions online, whilst others were anxious of the technology. In addition, many practice placements did not routinely use computers in the work settings, and mentors were concerned about the intrusion if students were required to take time out to use computers in practice placement time. Nevertheless e-learning appeared to offer a possible solution to the development and implementation of an Interprofessional e-learning Pathway (IPeLP), although it became increasingly obvious that there was much preparatory work to do with those who would be directly or indirectly affected by this development.

The asynchronous nature and access to the learning environment that e-learning offers, as well as the pedagogy of e-learning, were decisive factors in the adoption of this mode of delivery. The possibility of developing an IPeLP that could be built on a constructivist model was the first step in building the pathway. It was also important to establish a learning environment that could be accessed from a variety of places, at the students' and facilitators' convenience. Our perception was that the asynchronous nature of e-learning did pose a problem in that the development of IPeL was seen by many lecturers and course leaders as an addition to the existing courses; the concern was that there was a hidden agenda for students and facilitators to do this as an extra piece of work. This was a problem that took over three years to resolve and a curriculum rewrite before the embedding process started to develop.

## Paving the Way for Interprofessional E-Learning

IPE was formally introduced to Coventry University in 2002 following the achievement of a successful bid to become one of the thirteen First Wave sites to address the Modernisation of Allied Health Professions (DH, 2000b). Coventry chose to address the modernisation agenda by having two separate strands to the project. One strand concentrated on the development of individuals prior to employment and following employment, first by engaging education establishments in learning more about the potential roles on offer in health and social care, and second by developing the potential of the "skills escalator" (DH, 2001) with the notion that individuals have the potential to climb the escalator (DH, 2000a).

The second strand concentrated on developing IPE initiatives for undergraduates in practice, using electronic patient journeys as the focus for discussion (First Wave Team Final Report, 2005). The patient journeys created during the First Wave project were used successfully by students in practice, delivered in face to face workshops in practice settings. Following on from this sucess the use of the patient journeys was seen as the answer to the introduction of IPE into the undergraduate curricula; however this was not a straightforward process. For example, the patient journeys had been created for use in practice. Some success had been achieved in encouraging academic staff to use the journeys within more theoretical modules but this was only achieved in a uniprofessional context. It was apparent that resistance was growing within the academic community with regard to the concern that the inclusion of interprofessional content would require removal of uniprofessional content from each professional curriculum.

Strong leadership at senior management level was needed in order to make the necessary changes. Although resistance was evident at faculty level, management were clear that the process would be led by a top down approach. This model of change could be described as 'step-directive', that is, management-driven radical change (Ruddle & Feeny, 1997). Given the evident resistance, the style with which such change was managed was crucial, ideally this approach should not be necessary if the inclusion of interprofessional content had been embraced, but given the circumstances and time constraints it was essential if IPeLP was to be implemented.

A clear element of 'force-and-support' (Dutfield & Eling, 1994) was involved; it was decided that support would be given to those leading the initiative, but that equally the initiative would be implemented despite the resistance. However, this was to an extent tempered by the adoption of a visionary leadership style which was clearly in evidence throughout this period, with the recruitment of an IPeLP lead person to manage the development and implementation. This approach reflects the management style noted by Bromage (2006), who draws on the work of Simpson (1996) and Kotter (1990), in concentrating on the validity of the top down model:

*senior managers should act as champions of change, developing and maintaining an innovation-supporting culture (Simpson, 1996), albeit acting as 'leaders' rather than 'managers' (Kotter, 1990) (Bromage, 2006, p.3).*

This style of management can sometimes be restrictive and result in a delay in the process of inclusion and embedding of new initiatives into the faculty. This management decision led to the development of an e-based interprofessional curriculum which was accredited within the curriculum in May 2005, based around the patient journeys, with a view to embedding it in within the existing university virtual learning environment.

Further work took place following validation to incorporate medical students from the neighbouring medical school at the University of Warwick. Close collaboration between the two institutions was led by key academic tutors, firstly in face to face workshops, which had enabled the development of a climate of respect and trust allowing this collaboration to flourish. The authors, who led this initiative, were able to work together towards a common goal in a climate of support, open communication and a trusting relationship which was essential to achieve collaborative working.

As the first wave project drew to a close, a successful collaborative bid between Coventry University and Sheffield Hallam University led to the birth of the Centre for Interprofessional e Learning (CIPeL), the creative hub where the focus would be to develop both interprofessional e-learning objects and a culture in which creativity within the HEIs was encouraged and rewarded. The Interprofessional Learning (IPL) curriculum was introduced and ran in parallel to the CIPeL initiative. Emphasis was placed on separation of the IPL curriculum from CIPeL, thereby allowing both research and teaching development activities to become the main focus.

## INTRODUCING THE INTERPROFESSIONAL E-LEARNING PATHWAY (IPELP)

The Interprofessional e-learning Pathway has been running since September 2005 and has a presence in Years 1, 2 and 3 or 4 (the latter for medical and dietetics students only) of the curriculum (Bluteau & Jackson, 2009). At each of these points, students have opportunities to engage with other students and qualified professionals to discuss issues and scenarios directly related to patients, service users, clients and carers. These discussions helped to develop terms of reference and a cognitive framework that guides students when observing, exploring and reflecting on the multitude of models of both interprofessional and multi-disciplinary working, which they encounter in practice.

The interprofessional e-learning pathway is not a module in its own right, though it is linked to a designated module at each level of each course, through its learning outcomes and assessment. Students are invited to attend a face to face session, in which an overview of the IPeLP is given in the context of learning outcomes, assessment, navigating the online environment and the importance to future practice. The thematic approach means that for most courses, the IPeLP is launched and group work activities commence before, during or after the module which carries the learning outcomes and assessment.

There is a virtual learning environment for the Interprofessional e learning pathway for each year of the course, managed by an IPeLP learning technologist and a 'year lead' person. Within this web, each virtual learning set has its own dedicated discussion space where student deliberations are visible only to other members of the virtual learning set and facilitator. Students and facilitators are able to access their group from their University's open-access computers, from home or other locations where they have access to the Internet. If they are on placement while an online scenario is running, they are expected to continue to login regularly from campus, placement or home. The online discussions are particularly enriched by insights from group members on placement who have demonstrated that they are able to relate the discussion to live, current issues that they are experiencing.

Each year students are allocated to an interprofessional student group which forms their 'Virtual Learning Set'. Each learning set has a facilitator (a member of staff or professional practitioner) and in years 1 and 2, the students meet for one face to face meeting which provides an opportunity to meet the facilitator and members of the group, to get to know each other, to set the ground rules relating to how to work together and socialise before beginning any online group work. In year 3 the launch event provides details of the IPeLP, but students are expected to meet purely online, as they will have become familiar with each other during years 1 and 2.

Each of virtual learning sets includes a mixture of students drawn from different professional groups. There are no more than 15 students in each group and each group will be matched to the scenario that they will be discussing according to particular criteria. For example, in years one and two, each group reflects on a scenario in which their profession is evident; in year 3 short illustrative vignettes are used to demonstrate issues relating to team working drawing out general issues rather than specific role related issues.

The scenarios are revealed on a week-by-week basis over 4 weeks and students are able to contribute their insights to their group's discussion area, by initiating or responding to comments from other students or to questions from their facilitator. In each year of the pathway, assessment is based on a reflective assignment related to a selection of the student's relevant discussion contributions. Students are encouraged to log into the IPeLP web regularly to explore the scenarios and to read, reflect on and contribute to discussion postings.

## Exploring the Challenges

In describing how the IPeLP works, the process appears logical and straightforward, but this step by step account disguises the underlying challenges that simmered under the surface. Figure 2 illustrates the pre IPE era, where tension was evident. The two stakeholders of firstly, the HEI and secondly, the uniprofessional practice groups shared common goals but were governed by contrasting agendas. The academic modular focus of the institution had the power to impose a rigid framework on professional courses. Professional courses were dictated to by professional bodies that imposed another set of rigid guidelines that were often not entirely compatible with the institutions' demands.

*Figure 2. Pre IPE era - tension between stakeholders*

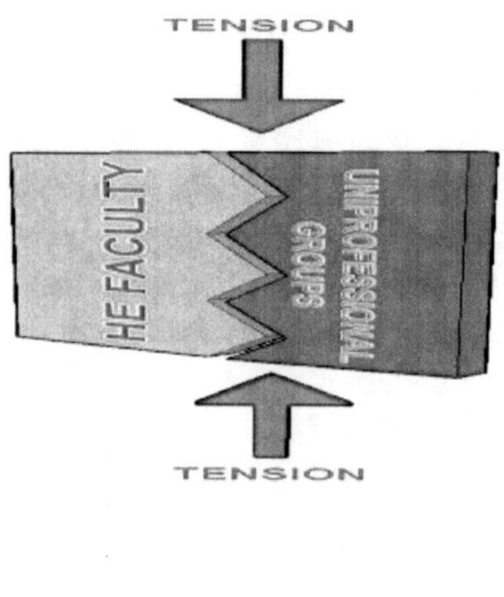

*Figure 3. Development of tensions between stakeholders*

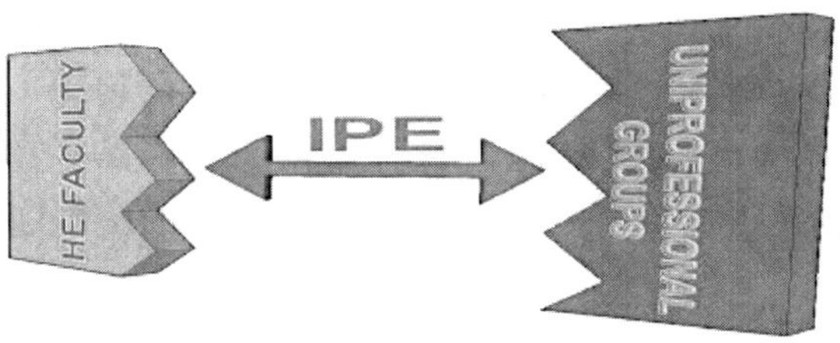


The tensions that were apparent pre IPE intensified as IPE arrived. The faculty, responding to government directives were aware that this must become a key focus in professional courses and were keen to introduce the concept. Figure 3 demonstrates the development of tension between the two stakeholders. The uniprofessional professional practice groups had concerns about introducing another component into their already full timetables. Course leaders reported that the amount of material that had to be taught, already exceeded the hours that students were allocated.

There was concern that to introduce IPE would mean that important uniprofessional material would be lost. IPE was seen as an irritating nuisance, and resistance was forming despite the initiative being led by a top down approach by the senior management team. Individuals were invited to consult with proposals, and a series of meeting were held at regular intervals with key personnel invited to attend. Poor attendance and time constraints in relation to curriculum validation meant that senior management forced the model through without full agreement at all levels. Tension increased and some parties were concerned that not enough time had been made available to thoroughly discuss the development of this initiative.

Figure 4 demonstrates the next stage where the tension had pushed the faculty and the uniprofessional courses 'over', and tensions can be seen running across a continuum covering the two stakeholders. The faculty produced logistical problems that led to difficulties in organisation of resources, use of technology and timetabling. In contrast the uniprofessional constraints were much more deeply seated. The constraints ranged from overt tribalism to silo mentality. Departments were concerned that the introduction of IPE would have an impact on their programmes and the time available for uniprofessional teaching. Overt tension was clearly apparent and at times, was difficult for those involved even though the antipathy wasn't aimed at any one individual, or team of champions, but was part of a process of change which would impact on all parties.

Figure 5 moves on to a calmer place, where IPE is beginning to become embedded and shared ownership has started to form. Memories can be very short and IPE is now starting to be embraced and be celebrated. Interestingly the

*Figure 4. Acceptance by stakeholders highlights challenges and constraints*

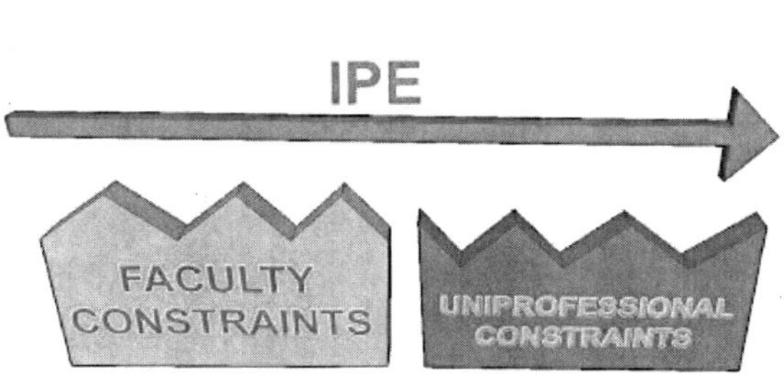


*Figure 5. IPE starts to embed in the faculty and uni-professional groups*

tension between the faculty and the uniprofessional departments regarding IPE has lessened. Faculty and the uniprofessional departments are moving together, each feeling more comfortable, shared ownership is developing with greater trust and IPE is starting to embed in both faculty and uniprofessional elements. As IPE starts to embed and settle, it is apparent that some elements are yet to be fully incorporated into the faculty and departments. These elements include fully embracing the pathway into the modules. The clear understanding that the teaching and learning materials have to be developed and maintained by the module team in which IPeL is embedded is yet to be fully achieved. The creative CIPeL team remains a focus of the development. These elements relate to the future progress of IPE between the two universities.

## DISCUSSION

The process of the development of the IPeLP focuses on the nature of the challenges that surrounded the development of IPE. Interestingly the focus of e-learning was not the primary challenge. It was sometimes used as a scapegoat and referred to as a problem. But e-learning has worked well, fitted the requirements of IPE, and created virtual environments where students across professional groups can be together, and experience the diverse worlds of other professional groups.

The authors believe that the top down approach to curriculum change has both advantages and disadvantages. The support that this brings is invaluable when working across such a wide variety of departments in addition to institutions. Forcing change upon academics is not ideal, however in this case there seemed to be no other alternative due to the pressure of time. Whilst concerns were raised initially regarding the introduction of IPE, over time academics at both of the universities involved have embraced the introduction, contributed to its development and taken ownership of the model. The importance of a small core team who are able to steer the process, support the individuals involved in delivering the curriculum and most importantly support each other cannot be underestimated.

Ongoing research and evaluation is currently underway examining both student and facilitator perceptions of an interprofessional e-pathway and its efficacy in terms of the use of e-learning. A longitudinal study is a key part of this process in which students complete quantitative evaluation tools across a five year period to establish their readiness for interprofessional learning (RIPLS) (Mattick & Bligh, 2009), and their attitudes towards other professionals (Luecht et al., 1990). The issue of gender is currently under discussion regarding the influence of gender and IPE, and how this may impact on the teaching of interprofessional education in the future.The aim is to follow up a small group of students by interview to examine the students' perceptions of working both in virtual teams and in face to face teams in practice. A group of researchers with an interest in discourse analysis are also reviewing the discussions within the virtual learning sets In addition, each offering is evaluated to ascertain a range of student comments regarding their experience of the IPeLP and will be disseminated in the future.

## CONCLUSION

IPE raises many challenges for a range of groups and individuals. It has the power to develop tension and unease. In the case study reported in this chapter, this was compounded by the introduction of e-learning. It was found that strong leadership is required to champion the change, to enable the development of new initiatives. However the top-down approach can be restrictive and may delay the process of inclusion and embedding into the curriculum. It is clear that strong leadership may conflict with new developments when stakeholders are unsure of the leader's vision. However, as

IPE develops and becomes securely embedded within curricula, problems resolve and the initial concerns and anxieties fade, as the change gains more champions, and initial offerings of IPE through metamorphosis become more accepted and owned. Change can be painful and difficult, in particular when the cornerstones of a professional group are challenged. But the introduction of new material should always be challenged: through challenge and questioning the importance and place of a new endeavour will emerge.

## REFERENCES

Barker, K. K., Bosco, C., & Oandasan, I. F. (2005). Factors in implementing interprofessional education and collaborative practice initiatives: Findings from key informant interviews. *Journal of Interprofessional Care, 19*(S1), 166–176. doi:10.1080/13561820500082974

Barr, H. (2005). Learning together. In Meads, G., & Ashcroft, J. (Eds.), *The Case for Interprofessional Education in Health and Social Care*. Oxford, UK: Blackwell. doi:10.1002/9780470776445

Barr, H. (2009). Interprofessional education as an emerging concept. In Bluteau, P. A. S., & Jackson, J. A. (Eds.), *Interprofessional Education; making it happen*. London: Palgrave.

Barr, H., Koppel, I., & Reeves, S. (2005). *Effective interprofessional education: argument, assumption and evidence*. Oxford, UK: Blackwell. doi:10.1002/9780470776445

Bluteau, P. A. S., & Jackson, J. A. (Eds.). (2009). *Interprofessional Education; making it happen*. London: Palgrave.

Bluteau, P. A. S., & Jackson, J. A. (2009). Interprofessional education unpacking the early challenges. In Bluteau, P. A. S., & Jackson, J. A. (Eds.), *Interprofessional Education; making it happen*. London: Palgrave.

Bluteau, P. A. S., & Jackson, J. A. (2009). An e-learning model of interprofessional education. In Bluteau, P. A. S., & Jackson, J. A. (Eds.), *Interprofessional Education; making it happen*. London: Palgrave.

Bromage, A. (2006). The management of planned change: an interdisciplinary perspective. In Hunt, L., Bromage, A., & Tomkinson, B. (Eds.), *The Realities of Change in Higher Education: Interventions to Promote Learning and Teaching* (pp. 1–14). London: Routledge.

Conole, G. White, S. & Oliver, M. (2006). The impact of e-learning on organisational roles and structures. In G. Conole & M. Oliver, (Eds.), Contemporary perspectives in e-learning research: themes, methods and impact on practice. (Open and distance learning series). Oxford, UK: Routledge Falmer.

Conole, G., & Oliver, M. (2006). Introduction. In Conole, G., & Oliver, M. (Eds.), *Contemporary perspectives in e-learning research: themes, methods and impact on practice. (Open and distance learning series)*. Oxford, UK: Routledge Falmer.

Cooper, H., Carlisle, C., Gibbs, T., & Watkins, C. (2001). Developing an evidence base for interdisciplinary learning: a systematic review. *Journal of Advanced Nursing, 35*(2), 228–237. doi:10.1046/j.1365-2648.2001.01840.x

Department of Health (DH). (2000a). *The NHS plan: a plan for investment, a plan for reform*. London: HMSO.

Department of Health (DH). (2000b). *Meeting the challenge: a strategy for allied health professionals*. London: HMSO.

Department of Health (DH). (2001). *Working together, learning together: a framework for lifelong learning for the NHS*. London: HMSO.

Dutfield, R., & Eling, C. (1994). Managing the individual's resistance to change. In Lovell, R. (Ed.), *Managing Change in the Public Sector*. Harlow, UK: Longman.

First Wave Team. (2005). *Meeting the Challenge Project Final Report*. Coventry, UK: Coventry University.

Freeth, D., Hammick, M., Reeves, S., Koppel, I., & Barr, H. (2005). *Effective interprofessional Interprofessional Education: development, delivery and evaluation*. Oxford, UK: Blackwell.

Gilbert, J. H. V. (2005). Interprofessional learning and higher education structural barriers. *Journal of Interprofessional Care, 19*(S1), 87–106. doi:10.1080/13561820500067132

GMC. (2003). *Tomorrow's doctors. Recommendations on undergraduate medical education*. London: GMC. Retrieved 03/04/09 from www.gmc-uk.org/education/undergraduate/tomdoc.pdf

GMC. (2009). *Tomorrow's doctors. Outcomes and Standards for undergraduate medical education*. London: GMC. Retrieved October 22, 2009, from http://www.gmc-uk.org/education/undergraduate/tomdoc.pdf

Gordon, F., & Walsh, C. (2005). A Framework for Interprofessional Capability: developing students of health and social care as collaborative workers. *Journal of Integrated Care, 13*(3), 26–33.

Hall, P. (2005). Interprofessional teamwork: professional cultures as barriers. *Journal of Interprofessional Care, 19*(S1), 188–196. doi:10.1080/13561820500081745

Hermsen, M. A., & Ten Have, A. M. J. (2005). Palliative care teams: Effective through moral reflection. *Journal of Interprofessional Care, 19*(6), 561–568. doi:10.1080/13561820500404871

Hiltz, S. R., Turoff, M., & Harasim, L. (2007). Development and philosophy of the field of asynchronous learning networks. In Andrews, R., & Haythornthwaite, C. (Eds.), *The Sage Handbook of E-Learning Research*. London: Sage.

Jackson, J. A., & Bluteau, P. A. S. (2007). At first it's like shifting sands: Setting up interprofessional learning within a secondary care setting. *Journal of Interprofessional Care, 21*(3), 351–353. doi:10.1080/13561820701233905

Jackson, J. A., & Bluteau, P. A. S. (2009). Creating a model: Overcoming the challenges of implementing Interprofessional Education. In Bluteau, P. A. S., & Jackson, J. A. (Eds.), *Interprofessional Education; making it happen*. London: Palgrave.

Kotter, J. P. (1990). *A force for change: How leadership differs from management*. New York: Free Press.

Littlejohn, A., & Peglar, C. (2007). *Preparing for blended e-learning*. Abingdon, UK: Routledge.

Lord Laming. (2003). *The Victoria Climbié Inquiry, report of an inquiry.* Retrieved April 3, 2009, from www.victoria-climbie-inquiry.org.uk

Lord Laming. (2009). *The protection of Children in England: a progress report*. London: The Stationery Office.

Luecht, R., Madsen, M., Taugher, M., & Petterson, B. (1990). Assessing professional perceptions: Design and validation of an interdisciplinary education scale. *Journal of Allied Health, 19*(2), 181–191.

Mattick, K., & Bligh, J. (2009). *Readiness for Interprofessional learning Scale*. In P. A. S.

Miller, C., Woolfe, C., & Mackintosh, N. (2006). Evaluation of Common Learning Pilots and Allied Health Professions First Wave Sites FINAL REPORT. Retrieved April 3, 2009, from http://www.smd-edu.qmul.ac.uk/student_info/undergrad/ipe/docs/CLAM_Final_Report_Revised_Sept_06.pdf

Oandasan, I., & Reeves, S. (2005). factors, processes and outcomes in IPE. *Journal of Interprofessional Care*, *19*(1), 39–48. doi:10.1080/13561820500081703

Pecukonis, E., Doyle, O., & Bliss, D. L. (2008). Reducing barriers to interprofessional training: Promoting interprofessional cultural competence. *Journal of Interprofessional Care*, *22*(4), 417–428. doi:10.1080/13561820802190442

Rafter, M. E., Pesun, I. J., Herren, M., Linfante, J. C., Mina, M., & Wu, C. D. (2006). A preliminary survey of interprofessional education. *Journal of Dental Education*, *70*(4), 417–427.

Reeves, S., Zwarenstein, M., Goldman, J., Barr, H., Freeth, D., Hammick, M. & Koppel, I. (2008). Interprofessional education: effects on professional practice and health care outcomes. *Cochrane Database of Systematic Reviews*, Issue 1.

Ruddle, K., & Feeny, D. (1997). *Transforming the Organisation: New Approaches to Management Measurement and Leadership. Oxford Executive Research Briefing*. Oxford, UK: Templeton College.

Simpson, V. (1996). The Hypertext Campus project. In TLTSN Case Studies 2: Managing the Adoption of Technology for Learning. Teaching and Learning Technology Support Network.

Smith, J. (2005). *The Shipman Enquiry: The Sixth Report. Shipman: the final report*. London: The Shipman Enquiry.

The Audit Commission. (1999). *Critical to success*. London: The Stationery Office.

The Bristol Royal Infirmary Inquiry. (2001). *The report of the public inquiry into children's heart surgery at the Bristol Royal Infirmary 1984-1995*. Retrieved April 3, 2009, from www.bristol-inquiry.org.uk/finalreport/the_report.pdf

Tulloch, J. (2006). *One day in July: experiencing 7/7*. London: Little, Brown.

Wilcock, P. M., & Headrick, L. (2000). Interprofessional learning for the improvement of healthcare: why bother? *Journal of Interprofessional Care*, *14*, 111–117. doi:10.1080/713678552

www.gmc-uk.org/static/documents/content/TomorrowsDoctors_2009.pdf

Yan, J., Gilbert, J., & Hoffman, S. (2007). World Health Organization study group on interprofessional education and collaborative practice. *Journal of Interprofessional Care*, *2*(6), 587.

# Chapter 14
# Building and Sustaining Collaboration in Cross Sector E-Learning Development

**Stephen Timmons**
*University of Nottingham, UK*

**Heather Wharrad**
*University of Nottingham, UK*

**Paraskevas Vezyridis**
*University of Nottingham, UK*

**Jacqueline Randle**
*University of Nottingham, UK*

**Joanne Lymn**
*University of Nottingham, UK*

**Fiona Bath-Hextall**
*University of Nottingham, UK*

## ABSTRACT

*This chapter will focus on the process of building and sustaining collaborative reusable e-learning object development across three educational sectors, Higher Education (HE), the UK National Health Service (NHS) and Further Education (FE) Colleges, using the LOLA project as a case study. A qualitative evaluation of 'process' ran alongside the entirety of the LOLA project. This chapter reports the findings of this qualitative research, and analyses how collaboration was achieved between the diverse institutions who were project partners. The strengths of this approach included the commitment of the team members to collaboration, while practical challenges included the location of the team members, but also wider issues in the institutions involved, and in particular, the role of the Media Developer and the perception of it by other team members.*

DOI: 10.4018/978-1-61520-889-0.ch014

## INTRODUCTION

The LOLA project explored learning objects for life-long learning, and was a cross-sector collaboration across three educational sectors, Higher Education (HE), the UK National Health Service (NHS) and Further Education (FE) Colleges. These sectors were represented by the University of Nottingham School of Nursing, Nottingham University Hospitals NHS Trust, and Castle College Nottingham respectively. The overall aim of the project was to produce a set of Reusable Learning Objects (RLOs) covering a range of health care topics. The intention was that these RLOs would be usable across the three educational sectors involved in the project.

While there is general agreement that cross-sector collaboration is a desirable aim, it has not always been achieved. As the Department for Education and Skills (2003) said:

*There is too little cross-sector collaboration in supporting learners as they move through the education system – we need a greater focus on linking our public sector systems to provide unified support for learners throughout life (Department for Education and Skills, 2003).*

For students there are clear benefits. A student on an Access to HE course at an FE college accessing a repository of learning objects to support their studies moves to a university nursing course and has access to the same repository and a similar set of learning objects (albeit at a different academic level). The same individual would be able to access the repository as a qualified nurse continuing their education in the work place. A common repository of learning objects to support learning helps to create a seamless transition as students move across learning sectors (Universities Collaboration in E-Learning (UCEL), 2009).

Successful communities of practice have taken this collaborative approach, including the Universities Collaboration in e-learning (UCeL) and the National Learning Network (NLN) in the FE sector. Though cross sector initiatives exist, they tend to be small scale. The universities of Bournemouth and Plymouth have partnered local colleges to produce resources for courses in Tourism and Arts. Cross-sector collaboration has also been achieved in language learning (Scottish Centre for Information on Language Teaching and Research, 2008). By widely disseminating the items in the repository and exchanging materials with other communities, time and funding constraints as well as project commitments were significantly reduced. This is because the production of new RLOs will often involve just reorganising assets into new ones or creating only those RLOs that are not yet available. It is worth noting that the development and maintenance of open repositories can be greatly beneficial for sectors like the FE where current conditions do not permit the development of them in-house (Fleming, Tammone & Wahl, 2002).

The LOLA project, the RLOs produced, their use and evaluation are discussed in more detail in chapters in this volume by our colleagues Richard Windle and Heather Wharrad. The LOLA project was funded by a grant from the Eduserv Foundation (2009). Instead of an external or top-down managerial introduction of a repository, the aim was to recruit members from various sectors who could be equally involved in the development and were willing to advertise and incorporate these materials into their daily teaching practice. This chapter will discuss the 'how' of building and sustaining collaboration, often a neglected topic in the e-learning literature.

A collaborative, cross-sector approach is thought to be essential for lifelong learning (Jameson & DeFreitas 2006), and they are well developed in nursing and health studies (Brady, 1997; Trim, 2001), though not specifically for the development of e-learning. Jameson et al. (2006, 964) report on two cross-sector collaborations (eLISA and CAMEL) to recommend a "controlled form of collaborative distributed-coordinated leadership within team-based e-learning projects". McConville (2007) reports an initiative where FE students were given access to HE e-learning resources, though this was designed principally as an aid to recruitment. One of the few papers to consider cross-sector collaboration in detail is by Connolly, Jones and Jones (2007), however, the study reported in this chapter is wider in scope in so far as the LOLA project encompassed Higher Education (HE), Further Education (FE) and UK National Health Service (NHS) partners.

We can therefore see that while cross-sector collaboration has much to recommend it as a model for e-learning development, it is not a straightforward process, and in particular, it is unclear the extent to which the development teams can work interprofessionally: as task-focussed team working between members of different professions who communicate and learn from each other in order to achieve shared objectives. While institutional and cross-sector collaboration are often cited as examples of good practice (Hanna, 1998; Reid, 1999; Mason & Lefrere, 2003), and have been positively supported at a policy level, (Harvey and Beards 2004) what is unusual about the LOLA project is that this research study, explicitly designed to investigate these issues, was designed into the project from its inception.

The University of Nottingham School of Nursing already has some experience in collaboration with other HE institutions through the UCEL collaboration (UCEL, 2009). The University of Nottingham Centre for International ePortfolio Development has successfully developed technology which supports lifelong learning (Hartnell-Young et al., 2006), though their paper concentrates largely on the technical, rather than the organisational issues. This evaluation of process was built in to the original grant application for the funding of the LOLA project and we are therefore able to present a distinctive perspective on the building and sustain of cross-sector collaboration in e-learning development.

## Methods

The research methods used to investigate collaboration within the project were qualitative. These included two rounds of interviews, one about half way through the project, and another round when the project was completed. The interviews were conducted by a researcher independent of the project, in order to give a degree of critical distance from the project and its participants. The interviewees were all the collaborators in the project (n=9) including Content Developers, the Project Leads and the Media Developers, drawn from all collaborating institutions. The same interviewees were used at each stage of the research, and their names were anonymised. Interviews were lasted about 30-45 minutes, were digitally recorded and then transcribed verbatim. The data were then analysed, also by an independent researcher. Analysis was conducted using QSR Nvivo 8, creating codes to describe sections of data, and then grouping those codes into the themes reported below.

## Findings

### Positive Interprofessional Experience

Participation in cross-sector collaborative teams for the development of RLOs was generally conceived as a positive experience by the members of the project. For uninitiated members it offered the opportunity to work collaboratively and exchange perspectives and teaching materials towards the development of innovative educational tools. For experienced members it reinforced and strengthened this attitude. In particular, we have indentified two main aspects which contribute to the perceived positive experience. These were the exchange of perspectives, and an increase in knowledge.

Working collaboratively offered a great opportunity for staff from different sectors to meet and exchange perceptions, ideas, thoughts and teaching materials about common areas of expertise. It also offers an opportunity for sectors (particularly FE) that do not have sufficient resources to proceed with such developments by taking advantage of those of other sectors which are more self-sufficient, especially in the area of media development:

*I think it has been really positive! You know, especially with working with [name] and I just wish it would continue (Interviewee 1).*

Furthermore, it offers an opportunity for colleagues to understand each other's working and teaching areas, as well as the requirements that need to be considered during the development of the RLO, and in relation to the different user groups and their educational needs. Although there is an unsurprising tendency for all representative of a particular sector to think first about its own students-users and then about cross-sector re-usability, the discussions during the development process were of great benefit to all the members and, in effect, the audiences, since the end products crystallise their various perspectives:

*I hope especially if positive results come from this, it will pave the way for other FE colleges and Higher Education Institutions to collaborate and not just for financial gain but for being able to get students to learn the value of education better by using better resources, I think it is a really good opportunity (Interviewee 3).*

Moreover, this opportunity was of great benefit to all members because the exchange of perspectives led also to the exchange of knowledge. In fact, the development of an RLO became a learning process for all the members involved, including the Media Developer, whose role was to produce the RLOs, principally in Macromedia Flash, based on designs and content produced by other team members, through sharing and challenging ideas and practices. They have indeed achieved interprofessional task-focussed team working, as members of different professions who communicate and learn from each other to achieve shared objectives.

One major positive outcome from this collaboration was the opportunity for all members to evaluate and rework their teaching style. The development of a RLO necessitates the content to be able to stand alone. Any assumptions that the content author has prior to this development, and in relation to the teaching materials in the classroom, need to be re-examined and refined for inclusion, since each RLO needs to be self-explanatory, with appropriate use of language and wording and coherent structure and meaning.

The team members' initial excitement and enthusiasm for working collaboratively with colleagues from other sectors was maintained throughout the development of RLOs. In many cases this was reinforced as the project came to fruition and the end products started being used by students. The wider dissemination of the produced materials promoted their work and enhanced the reputation of the institutions in which they are located. In addition, the wider dissemination of the developed RLOs contributed to the exchange of materials, ideas, thoughts and teaching styles not only within but also outside the collaborative team.

## Approach

Contrary to the dominant top-down approach in the development of RLOs, a bottom-up approach offered an alternative development process concentrated on creating a feeling of ownership and a community of practice for team members closer to the student-user. The bottom-up approach was welcomed by all members of this project team, who thought that the materials produced were more likely to be used as a result. It was observed that as soon as people from the sectors involved, who were not initially involved in the project, began to understand and appreciate the value of the RLOs produced by their colleagues' involvement, they expressed considerable enthusiasm for participation:

*So our approach was you know getting a bottom up approach, getting communities of people together, running workshops and deciding on you know what sort of resources as a group would benefit our different groups of students (Interviewee 2).*

This fact had a two-fold benefit for the project team. Firstly, they continued the advertising and incorporation of the RLOs into their teaching. Secondly, they provided the team with constructive feedback about the appropriateness of the RLOs for their students' learning requirements. For example, even though the project team included FE members, it was their colleagues who pointed to that fact that the language used for specific RLOs was above the standard that many of their students would be able to understand.

While the overall experience of collaboration was positive, a number of issues were raised in the interviews which elucidate some of the barriers to collaboration. These were 'physical location', 'managers and colleagues', 'time constraints', 'funding', 'RLO concepts', and importantly, the 'Media Developer's role'.

## Issues: Physical Location

Physical location had a major impact on communication between members. Sectors that were in close proximity (especially within the same building) were more likely to develop informal communication channels. In many cases, the location of the involved sectors often seems to offer additional informal channels of communication and may very well support informal and face to face communication especially when team members, even if they come from different sectors, are located within the same premises. As a result, the University and NHS Trust project members worked more closely together on the RLO production than did the FE college lecturers:

*sometimes getting hold of them is even more difficult because their schedules don't have that sort of inherent flexibility (Interviewee 4).*

## Issues: Managers & Colleagues

The development of RLOs can often be very time consuming. Therefore immediate educational benefits for the institutions involved might only arrive at a later stage. Managers and colleagues (particularly those who have had experience in collaborative teams) needed to be involved from the early stages of the development, in order for the members of the team to be able to negotiate effectively their contribution with regards to their pre-existing work commitments. Managers' involvement was also important where future reorganisations or unexpected increases in work affected and even jeopardised members' involvement. Team members thought that it is important for the people around them to understand and appreciate the value of the RLOs as early as possible, not only for the support of existing members, but also as an effective way of attracting new contributing members:

*I think if it was going to be done again I think it would need to be structured at the beginning so that the staff knew exactly what was expected of them, [and] their managers in order for the managers to appreciate the length of time it takes to help to support the members of staff (Interviewee 6).*

## Issues: Time Constraints

The most challenging aspect in forming cross-sector and cross institutional collaborative networks developing RLOs was undoubtedly time and the synchronisation of team members' availability. In general, academics tend to be presented with more inherent flexibility in terms of work commitments and schedules when compared to other sectors like healthcare organisations and FE. However, this flexibility is often underestimated or not appropriately utilised to support the project, especially in terms of academics being willing to visit other sectors for formal meetings when the members of these sectors are unable to travel to universities, in which the Media Developer is usually located.

There was also the need for all members to update their group regularly about their own whereabouts, and particularly about their availability for meetings and development work (such as photo shooting, filming, narration), in order to synchronise activities and tasks. For example, there were instances where the group made an assumption that a team member is under a heavy schedule at a particular time when in fact that was not (or was no longer) the case, wasting valuable time towards the development of an RLO.

## Issues: Funding

As always funding is needed for this kind of project. The major portion went on the media development of the RLOs to hire an experienced media technologist for the actual technical creation. However, all team members thought that it would have been helpful to buy their time out, particularly in sectors where heavy schedules and increased work commitments might detract from them from being active members of the development team:

*the funding, everything, whatever you would be compensated for, whether it's money for creation of the project or money to buy your time out, it needs to be sorted out right from the very beginning (Interviewee 6).*

## Issues: RLO Concepts

The conceptual development of a RLO for common use among educational sectors with various levels of learning was a great challenge. A satisfactory level of explanation and communication of concepts (especially between FE and HE) was sometimes hard to achieve. The appropriate choice of the RLO concepts determined the formation of sub-groups within the team. There was a tendency for members to elaborate on already-existing relationships with the aim of developing RLOs intended to benefit the students of their own sectors:

*I think I geared mine very much towards FE students (Interviewee 7).*

However, even within the same educational sector there was a danger that the RLO might be too specific and, thus inapplicable to other institutions of the same sector. For example, especially when developing RLOs with very specific practical instructions and guidelines this might prohibit other similar institutions from using them since they might not adhere to those institutions' internal policies:

*but the supporting material that is cited within the actual RLO itself which I think was a bit of a mistake doing it that way, that will change, that may become outdated (Interviewee 4).*

It was, therefore, important for the team to set clear targets about the kinds of RLOs expected to be produced. At this point, the team had to consider an educational level-based approach in the development of concepts into RLOs. In this way, the concept could progress incrementally in complexity and language, at various levels under a lifelong learning approach.

Therefore, the resulting RLOs concentrated on the explanation and communication of basic principles, theories and basic practical steps that are widely accepted, and then elaborate on this concept by building-in exceptions and specificities, perhaps via another accompanying RLO. The RLO repository developed as part of the LOLA project has a valuable role, in the sense that it can provide interested parties with assets to configure the RLO according to their institutional environment.

## Issue: Media Developer's Role

In developing RLOs for digital dissemination, the Media Developer has a central position in the collaborative team. This is because they usually have the experience and knowledge regarding the capabilities of current software packages for the development of RLOs which other team members lack.

However, their centrality is not always acknowledged: teams tend to assume that as long as the RLO is put on paper and the content has been created, the Media Developer will then be able to proceed with the digital construction of the RLO. This one-way, non-interprofessional collaboration may have a significant impact on both the time needed for the development and on the quality of the final product. This then, is the potential trap for teams producing RLOs; a 'blindness' towards inadequacies in their own interprofessional team working:

*I personally would have liked to have spent more time here actually with the media developers (Interviewee 6).*

In particular, members of the network who are unfamiliar with the relevant technology found it difficult to communicate their vision of a specific RLO to the Media Developer, and at the same time, it was for the latter to interpret it appropriately. Particularly for RLOs of extended size and scale, interpreting words and thoughts into suitable images, audio and interaction can be a very involved process. As a result the development of such RLOs became very time consuming, since there was a need for greater communication between the Content Author and the Media Developer.

Furthermore, there was the possibility that tensions would develop between these members of the team, particularly when the Media Developer's interpretation began to affect the accuracy of the information included in the RLO. It therefore needed to be constantly reworked by the Media Developer. Lastly, as more time was spent on the RLO by the Content Author, although they began to realise what was achievable and practicable, there was a danger that they might also begin to forget their initial vision of the RLO.

# DISCUSSION AND IMPLICATIONS FOR PRACTICE

This detailed study of collaborative processes in a cross-sector e-learning project has shown that collaboration between institutions in different educational sectors is indeed practical, and generates many of the benefits that the literature predicts (Hanna, 1998; Reid, 1999; Mason & Lefrere, 2003).

During this project, two of the sectors involved (National Health Service (NHS) organisations and Further Education (FE) Colleges) underwent massive re-organisations which greatly affected the progress of developing RLOs. In general, we can say that these two sectors that have slightly different educational purposes are more likely to experience both anticipatable and unanticipated obstacles during their participation. These needed to have been considered as much as possible at a very early stage in the project. Early involvement of line managers should have been sought to facilitate smooth progress. A further issue is that team members from the NHS and FE sectors are often under the management of two or three people, and so their managers' agreement and involvement was a prerequisite for their participation.

While collaboration was enthusiastically embraced by the staff involved from all three sectors, obstacles that arose at the level of the organisation. While the individuals (and, generally, their line managers) were keen to work together across institutions, the institutions themselves, pursuing divergent priorities, often caused problems for the team members. Some of these, such as major reorganisation were unavoidable. However, as the project progressed, senior managers became more aware of the project and its benefits for each collaborating institution. This often smoothed the path in terms of some of the obstacles that the team members encountered. A key recommendation is, therefore, to ensure that not only the direct managers, but also more senior managers, are aware of projects like LOLA and understand what benefits they can bring. We would concur with Jameson et al. (2006) that a collaborative management style within the LOLA project was a key factor in its success. Again, this was an explicit aim of the project, to build a community of practice (Wenger, 1998).

This opportunity was found to be beneficial to all members, because the exchange of perspectives led also to an exchange of knowledge. The development of an RLO became a learning process for all the members involved, including the Media Developer, whose role was to translate plans into digital media. To this extent, the teams did succeed in practicing interprofessional task-focussed team working, as members of different professions who communicated and learned from each other to achieve shared objectives.

However there was one emergent challenge: the centrality of the Media Developer was not always acknowledged: teams tended assume that if they put an RLO on paper, the Media Developer will be able to digitally construct the RLO, yet it was found in practice that they must be actively involved throughout the entire design process. This one-way non-interprofessional collaboration arguably reflected a 'blindness' towards the inadequacies of their own interprofessional team working, and one that such teams must be aware of in any future collaboration.

Therefore it is suggested that the Media Developer has an active role from the very first stages of the team formulation and the discussions around the development of specifications and contents of RLOs. This is not only for the team to be able to define what can be achieved in terms of technology per se but also what the selected media developer can create in terms of her experience and knowledge in the use of the relevant software packages. It is also desirable for the media developer to be able to understand and familiarise herself with the jargon (such as acronyms) used by members from other sectors, facilitating more effective communication within the time constraints. Lastly, this early involvement will offer other content authors the opportunity to expand their vision about a particular RLO in order to include advanced interactive features that might have not initially been thought of.

This collaboration could expand even more and include the exchange of teaching (for teachers) and learning (for students) techniques and styles among the involved sectors. By exchanging visits to sectors and classrooms, the members will be able to understand the different settings as well as the different styles and requirements of learning. In this way, the content authors and the media developers will be able to exchange feedback that could potential be invaluable for the development of RLOs since they would be more sensitive to particular learning styles.

## FUTURE RESEARCH DIRECTIONS

The case study reported in this chapter is relatively small scale, and the generalisability of any qualitative project are limited, however, it is clear that the issues of process, especially in terms of cross-sector working remain poorly understood. We would therefore recommend that further studies of this type are conducted alongside developments in e-learning, and perhaps this should become a standard for the evaluation of projects of this kind. It would also be interesting to see if these issues were reproduced at a much larger scale, or whether different organisational and managerial issues became more important.

## CONCLUSION

Team members drawn from three different educational sectors working interprofessionally to create RLOs felt that the experience had been a great opportunity to meet and exchange perceptions, ideas, thoughts and teaching materials about common areas of expertise. It also offers an opportunity for sectors (particularly FE) that do not have sufficient resources to proceed with such developments by taking advantage of those of other sectors which are more self-sufficient, especially in the area of media development.

However there was one emergent challenge to this interprofessional teamworking: teams tended assume that if they put an RLO on paper, the Media Developer will be able to digitally construct the RLO, yet it was found in practice that they must be actively involved throughout the entire design process. It is thus strongly recommended that interprofessional teams working to produce RLOs must reflect upon the inadequacies of their own interprofessional team working, and take steps to ensure that all team members are aware that Media Designers are not mere translators of ideas into a digital format, but have much expertise to offer on an equal footing with other team members. With the proliferation of digital media, it is likely that in future years, Media Developers will have gained their expertise during graduate and post-graduate studies of media production, and be certified academics in their own right.

## REFERENCES

Brady, D. (1997). Further and higher education partnerships: the future for collaboration. *Journal of Vocational Education and Training, 49*(1), 167–175. doi:10.1080/13636829700200292

Connolly, M., Jones, C., & Jones, N. (2007). Managing collaboration across further and higher education: a case in practice. *Journal of Further and Higher Education, 31*(2), 159–169. doi:10.1080/03098770701267630

Department for Education and Skills. (2003). *Towards a Unified e-Learning Strategy*. Consultation Document. Retrieved March 4, 2009, from http://publications.teachernet.gov.uk/eOrderingDownload/DfES-0455-2003.pdf

Eduserv Foundation. (2009). *Research and Innovation*. Retrieved March 4, 2009, from http://www.eduserv.org.uk/research

Fleming, T., Tammone, W., & Wahl, M. (2002). *E-Learning: Addressing the Challenges via Collaboration*. Washington, DC: US Department of Education. Retrieved February 14, 2009, from http://eric.ed.gov/ERICDocs/data/ericdocs2sql/content_storage_01/0000019b/80/1a/52/e9.pdf

Hanna, D. E. (1998). Higher education in an era of digital competition: Emerging organizational models. *Journal of Asynchronous Learning Networks, 2*(1), 66–96.

Hartnell-Young, E., Smallwood, A., Kingston, S., & Harley, P. (2006). Joining up the episodes of lifelong learning: A regional transition project. *British Journal of Educational Technology, 37*(6), 853–866. doi:10.1111/j.1467-8535.2006.00668.x

Harvey, B., & Beards, D. (2004). E-learning in Scottish further and higher education. *Education + Training, 46*(6/7), 353–360. doi:10.1108/00400910410555268

Jameson, J., & DeFreitas, S. (2006). Editorial: Collaborative e-support for lifelong learning. *British Journal of Educational Technology, 37*(6), 817–824. doi:10.1111/j.1467-8535.2006.00665.x

Jameson, J., Ferrell, G., Kelly, J., Walker, S., & Ryan, M. (2006). Building trust and shared knowledge in communities of e-learning practice: collaborative leadership in the JISC eLISA and CAMEL lifelong learning projects. *British Journal of Educational Technology*, *37*(6), 949–967. doi:10.1111/j.1467-8535.2006.00669.x

Mason, J., & Lefrere, P. (2003). Trust, collaboration, e-learning and organisational transformation. *International Journal of Training and Development*, *7*(4), 259–270. doi:10.1046/j.1360-3736.2003.00185.x

McConville, S. (2007). *Widening Participation: A Virtual Approach to F.E. Collaboration*. Wolverhampton, UK: University of Wolverhampton. Retrieved February 14, 2009, from http://wlv.openrepository.com/wlv/handle/2436/18076

Reid, I. C. (1999). Beyond models: Developing a university strategy for online instruction. *Journal of Asynchronous Learning Networks*, *3*(1), 19–31.

Scottish Centre for Information on Language Teaching and Research. (2008). *Cross sector collaborative activities to promote modern languages in Scotland*. Report to the Subject Centre for Languages, Linguistics and Area Studies.

Trim, P. (2001). An Analysis of a Partnership Arrangement between an Institution of Further Education and an Institution of Higher Education. *Journal of Further and Higher Education*, *25*(1), 107–116. doi:10.1080/03098770020030542

Universities Collaboration in E-Learning (UCEL). (2009). *UCEL Homepage*. Retrieved February 14, 2009, from http://www.ucel.ac.uk/

Wenger, E. (1998). *Communities of practice: learning, meaning, and identity*. Cambridge, UK: Cambridge University Press.

# Chapter 15
# Overcoming the Barriers to Promoting Online Interprofessional Education

**Dawn Forman**
*University of Chichester, UK & Curtin University, Australia & Auckland University of Technology (AUT), New Zealand*

**Marion Jones**
*Auckland University of Technology (AUT), New Zealand*

## ABSTRACT

*It is not always transparent how e-learning can facilitate the interaction necessary for individuals to learn with, from and about each other as advocated by CAIPE (2002). This chapter emphasises the critical importance of facilitators acquiring new techniques to enable learners to make optimum use of the opportunities provided by Electronic Interprofessional Education (EIPE). To understand how barriers to online interprofessional education can be overcome, it is necessary to understand approaches to the development of positive e-learning environments. With this understanding, e-learning in supporting Interprofessional Education can be implemented more smoothly for both academic staff and students.*

## INTRODUCTION

This chapter explores the cultural changes within higher education institutions that can facilitate the development of e-learning for interprofessional education. The chapter identifies processes essential to the development of e-learning materials that are flexible, user friendly and motivating for the user. Interprofessional working requires commitment and willingness to value diversity in working together to set goals and meet outcomes.

DOI: 10.4018/978-1-61520-889-0.ch015

In healthcare there is a long history of concern over poor communication and rivalry between different disciplines and competing professionals (Cleak & Williamson, 2007). Learning to work collaboratively requires education, adaptation, a creative attitude to change and a commitment to functioning effectively in teams. It is arguable that e-learning prepares professionals for practice that overcomes barriers perceived in relation to learning together. Specific considerations for developing e-learning resources for interprofessional education and overcoming the barriers to their use by both academic staff and students will be outlined. Attention is also given to the challenges of exploiting first e-learning and then EIPE in order to support today's learners, whether at undergraduate or postgraduate level.

## BACKGROUND

With the publication of the World Health Organisation's (2009) new Framework for Action on Interprofessional Education and Collaborative Practice, the world is again being encouraged to get to grips with interprofessional education and the ways it shapes practice. There is a need to provide convenient, accessible and flexible education that supports collaborative practice (MacDonald, Stodel, & Chambers, 2008). Any approach adopted must underpin the principles of teaching and learning so that, during the learning journey, the learner progresses at their own pace whether they are new to higher education, working on an undergraduate degree programme, or bringing experience from practice into a postgraduate learning context.

At the same time, the E-prefix is ever-present these days, from ecommerce to e-learning. Technologies such as Facebook, IPODs, Wickis, portals and virtual worlds are likely to be familiar to today's learners, so learning providers need a high level of awareness of them and to incorporate them where effective. Research by Huws, Jagger and Bates (2001) reports that the creation of 'E-Indicators' has resulted in a database covering 204 countries and including 171 variables. This proliferation reinforces the need to incorporate information and communication technologies in any education provision.

This learning journey needs therefore to bring together the benefits of interprofessional education and e-learning to address the needs of the health care student. First, however, the perceived barriers to IPEL must be understood and overcome. Biggs (2003, p. 213) stresses that the globalisation of education has coincided with the development of information technology, so there are enormous pressures on teachers to get to understand and harness this technology, not only to improve their teaching but also to enable their university to sell their learning wares across the world.

However, quality management systems must not be compromised by an assumption that e-learning is a default solution in every learning context. The alignment of course objectives, assessments and learning strategies is critical to engage learners in multiple learning activities. These must be interactive and lead to assessment and quality teaching that considers different learning styles. If this is ignored, the e-learning process itself can become a barrier to learning.

### Perceived Barriers

Pollard and Hillage (2001) outline the advantages of technology supported learning and show how this can be customised to suit learner needs. Their research highlights the following barriers to e-learning, which are similar to those identified earlier by Roberts (1998) who interviewed learners about their on line learning experience. Figure 1 outlines those similarities.

From the comparisons in Figure 1, it can be seen that Roberts focuses more on learning and teaching, while Pollard and Hillage focus more on change management, so the two complement each other. Change management is the focus of this chapter, which explores the conditions associated with the successful use of e-learning in interprofessional education. While e-learning is convenient, unfamiliarity with its methods and technologies can be a barrier to learning, especially where no consideration is taken of the fact that people learn differently and at different rates.

A study of online interprofessional learning by Miers et al. (2007) found that many students were apprehensive about online learning, concerned about their lack of IT expertise and anxious they would miss out on important learning. Professional cultures feed into the barriers especially

*FIgure 1. Perceived Barriers: Pollard and Hillage (2001) & Roberts (1998) compared*

| Pollard and Hillage (2001) | Roberts (1998) |
|---|---|
| **Safety of the infrastructure supporting Elearning.** One of the most challenging experiences for staff and students is when the system "goes down" and neither party can make contact. Support systems must be available if possible over 24 hours to assist the learner to maintain motivation to learn through this technology.<br><br>**Development costs (materials and staff development).** It should not be assumed academic staff have the skills to develop e-learning. This mode of study is also is daunting for new learners so they need to know they have a competent teacher available to them. The learning institution must provide resources for the development and implementation of processes that support individual creativity.<br><br>**Reluctance of facilitators to embrace the technology and modify their approach.** The challenge for the teacher is in overcoming barriers to creativity and innovation through awareness that they have become a facilitator of learning and not the deliverer of information (Magnussen, 2006).<br><br>**Learners' unfamiliarity with the E-learning method.** E-learning can be very daunting to the learner and, therefore, the processes offered must be user-friendly and sufficiently motivating to the learners. Effective communication processes are essential and need to value the diversity each discipline brings to learning.<br><br>**Change**. "The unknown" can be a barrier to learning and a threat to professional confidence. This must be addressed during planning and strategising as the programme is developed. E-learning tends to use constructivist approaches to learning. This requires learners to overcome the fear of using their own skill in developing learning which enhances their understanding as they progress (Peters, 2000). | **Technical issues.** Learners want to be able to access the on line learning when and where this suits them so 24-hour, 7-day a week technical support is necessary to ensure online materials are working appropriately<br><br>**Computer as a surrogate teacher.** Learners more familiar with learning with a tutor who is physically present in a classroom often have difficulty relating to even the most interactive of learning.<br><br>**Autonomy in e-learning.** Learners need to be self-motivated to access the online learning and to engage with the materials and the discussion groups provided.<br><br>**Communication.** This is almost identical to what Pollard and Hillage (2001) described as the learner's unfamiliarity with the learning method.<br><br>**Learning environment.** This relates to what Pollard and Hillage (2001) outline as the "change". |

for Postgraduate education, where students are qualified professionals. Therefore how values, beliefs, attitudes and socialisation process impact on learning, needs to be considered when developing learning activities that promote effective interprofessional teamwork (Jones, 2000).

The development of online interprofessional programmes has the added dimension of learning with people from different disciplines (CAIPE, 2002). It is clear, therefore, that the development of e-learning-based IPE needs to overcome the barrier of non-alignment of teaching and learning, while valuing the diversity different professions bring with them both in learning style and ways of learning. Therefore, it is important to explore the perceptions of each student group in relation to individual learning styles and professional cultural differences. Learning styles are affected by the context and the time the learner has to undertake an activity along with what the individual brings to the learning through their professional world-view. Understanding this context is important to motivate people to learn together through group activities and assessments. In addition, activities must be overtly relevant to their day-to-day work (Biggs, 2003).

A further concern is that the medium does not aid the interaction necessary for effective interprofessional education. A study by O'Neil (2008) indicates online structures were highly interactive. However, it is not suggested that interprofessional e-learning (IPEL) is the only method for delivering IPE, rather that EIPE is one of a range of tools and that it will appeal to the learning styles of some students more than others. Therefore, a consideration of learners' different learning styles may result in the application of a blended learning approach, making use of multiple learning activities and strategies.

To meet the challenge of helping people learn effectively, the infrastructure developed must support learning objectives and provide for the diversity of learning strategies to overcome or reduce perceived barriers to e-learning (Meads et al., 2009). Some students find e-learning a frustrating and lonely experience: they feel that getting to know other learners is much harder through online learning. Their learning experiences can be guarded in that, while students have time and space to work at their own pace, it can be daunting to critique the work of others and to give answers which may then be written about in the next online chat room. Therefore, it is critical for developers to develop exercises to foster trust and confidence among students. However, this may require a deep change in the way higher education institutions are organised.

## The Impact of an Infrastructure to Support EIPE

Infrastructure is a very important consideration for both the learner and the developer. Two models of e-learning infrastructure support which have emerged over the last 15 years are:

- The McKey (2000) Total Learner Experience Model, which indicates considerations necessary for a cost-effective infrastructure.
- The Salmon (2004) Five-step model for E-Moderators.

    These models are illustrated below. Figure 2 shows McKey's model, and Figure 3 shows Salmon's model.

Ways in which the systematic use of each model can help curriculum developers overcome barriers to e-learning will now be considered. Let us look first at McKey's model and the extent to which it addresses some of the barriers we have identified.

## McKey's Total Student Learning Experience Model

McKey's model is very much focused on a learner-centred experience, and considers the educational institution as a whole. The change is from the traditional model of learning, where an expert imparts information, to a process of facilitating learning by the learner, and at the learner's pace. Motivation for the learning therefore needs to come from the design of the learning activities which are created. Fink (2003) endorses this approach and urges facilitators to create learning-centred courses, as opposed to content-centred courses. The stress is on a high degree of interactivity. McKey's model (Figure 3) outlines four aspects critical for the e-learning infrastructure, and each of these will be discussed more fully.

### A. Administration: Can Create Barriers or Promote Learning

In this context administration covers the enrolment and performance monitoring of learners as they work towards their awards. An examination of how administrators perform the key role of communicating with all involved is necessarily complex, but McKey stresses the importance of this in the planning and development phase of e-learning. Only when these elements are available and the technical infrastructure can be guaranteed should any higher education institution consider offering total e-learning provision.

It therefore follows that any higher education institution embarking on the delivery of e-learning materials must consider almost every aspect of the learner's journey. They must ensure this is modified to meet the demands of both e-learning and the efficiency and effectiveness needs of the educational institution. Opportunities for non-

*Figure 2. The McKey (2000) total learner experience model (Adapted from McKey, 2000)*

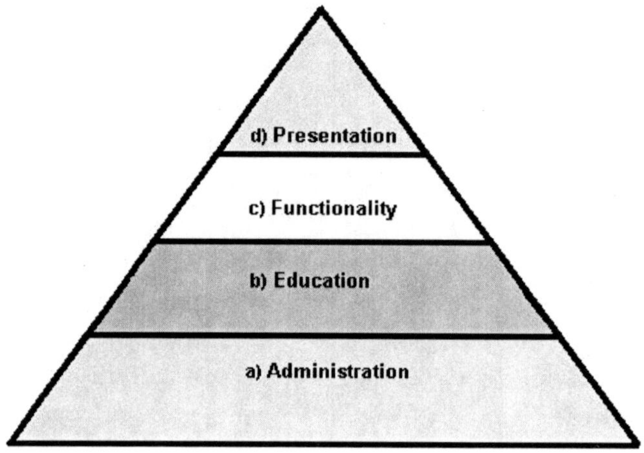


traditional learning activities must be made as user-friendly as possible: learners will come to e-learning with different expectations and capabilities. Therefore, the programme administrator needs to provide guidelines and a direct link to support (or the 'help desk') so all concerns can be addressed early in the process.

## B. Education

This section in McKey's model examines the design of e-learning materials. McKey makes it clear the learners will study in a different way from traditionally taught learners. The design of material will thus involve more than merely putting lecture notes into an electronic format. Once

*Figure 3. The Salmon (2004) five-step model for e-moderators*

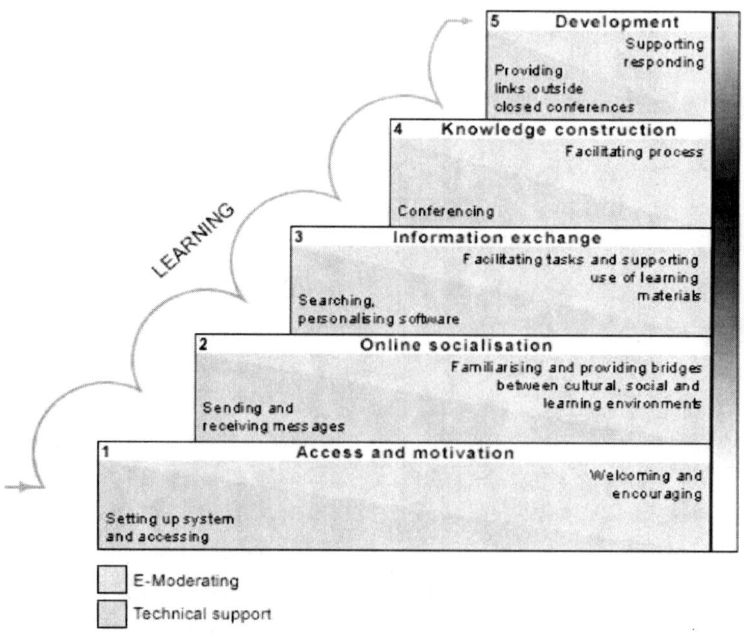

students overcome the fear of virtual learning they will develop skills to identify useful information, to interact constructively and to store information for later application. These activities are part of the educational strategy for students to develop confidence in learning. The emphasis is on their own problem solving and opportunities to tackle questions for themselves.

Curriculum designers are thus challenged in discarding traditional ways of portraying information and reconsidering learners' expectations and styles of study. Designers have generally found this to be a liberating experience. Indeed, this process seems to have re-ignited their motivation for teaching and learning differently (Quinney, 2008). In this way the expectations, feelings, and context of the student are considered, so as to promote learning and to create a positive learning environment.

It may be helpful to consider Ragan's (1999) three guiding principles and practices for the design and development of distance education:

- Learning goals and preferences must be clear and transparent.
- Interactions, including online socialisation, must be facilitated.
- Assessment and measurement (of learning outcomes) should be clear and appropriate to the method of study.

All parts of the design have equal importance in the total learning experience. As with any educational provision, the assessment is an important component of the learning experience and must harmonise with the method used. For interprofessional e-learning (IPEL) therefore, the assessment must be integrated with the learning outcomes. In the early days of e-learning, institutions tended to offer learners the choice of online or in the classroom (Ragan, 1999). This was satisfactory only as long as it enabled learners to choose their preferred learning style. In Ragan's example, the assessment method was common and therefore the outcomes of students' online learning could be compared directly.

A short cut to establishing essential e-learning resources is to use those which already exist. Walsh (2007) reports widening access to online materials originally developed for GPs so they can be used to support nurses and hospital doctors. This process seems to have been successful in a specific health context, but is more about sharing a learning experience than using IPEL. IPEL requires learning experiences which are not merely shared, but involve strong interaction and a constant exchange of ideas between participants from different disciplinary backgrounds.

Carter & Rukholm (2008) stress that online courses provide unique opportunities for critical thinking, teacher-student interaction and discipline-specific learning. Developing the skills of reflection is a further challenge with e-learning, but once trust and success have been experienced students can be encouraged to find a critical friend. Biggs (2003) outlines ways in which different people take on the role of critical friend, who may be a peer on the programme, a person at work or anyone that can relate to the student's own context. They must, of course, have the time and interest to reflect and discuss with the student learner.

The outcomes of two recent studies are of interest here. Campbell (2006) indicates the results of assessments were higher with online learning; and O'Neil and Fisher (2008) indicate results tended to be comparable. This validates online learning as a tool to aid learners to achieve the desired outcomes. As providing both in-classroom and online learning can be seen as a duplication of resources, many institutions now choose either the classroom or the online mode for IPEL.

However the technologies needed to deliver this in totally online form must be interactive, flexible and have a high level of usability. These characteristics are critical to providing the motivation to create learning that works towards the development of a collaborative model of practice for interprofessional work. O'Neil & Fisher (2008) found in a study of student nurses that learners were process oriented and highly interactive. The designing of programmes therefore needs to reflect the characteristics of the learners or there is the potential to create barriers unintentionally. The learner must feel that the learning experience makes them excited and keen to learn, not frustrated and de-motivated.

An additional consideration is that the wide variety of formative and summative assessment checks available on-line offer a more accurate way of determining learner success or failure in achieving learning outcomes. From an interprofessional perspective, assessment needs to develop interprofessional communication – debate and critique that promotes not only the students' confidence and professional integrity, but also collaboration skills and the ability to give constructive feedback. Some examples are group assignment, role play in chat rooms, visits to interprofessional practice areas and follow-up discussion of these experiences online.

## C. Functionality

This relates to the quality assurance necessary to ensure materials and methods of delivery are consistent and of a high standard. The whole infrastructure needs to meet learners' needs and to ensure all aspects of the learning journey, are perceived as seamless and usable by learners. Biggs (2003) differentiates between quality in standards and in meeting university requirements, and quality enhancement resulting from reflecting on the teaching, responding to student feedback and recognising that no one group of students is the same. It is clear that responding to difference is critical. Interprofessional learning is about difference, similarities, common purpose and vision: these factors should be the focus of the EIPE learning experience

Clearly, the incorporation of e-learning into any curriculum and overcoming of the barriers to interprofessional learning requires:

- An infrastructure (administrative and technical) to support the development of materials.
- Development of the facilitators' understanding of the experience of the interprofessional learner who will access the materials.
- Development supporting the designing of materials and activities appropriate to interprofessional e-learning.
- Development of learners to support them in maximising the opportunities of this learning environment and to enable them to recognise the importance of working with other disciplines and their facilitator.

In an interprofessional context, e-learning materials or learning objects have a hugely beneficial part to play in the learners' experience. They cannot fully replace the learning in the classroom, the clinical laboratory, the hospital or the community environment. But they can, if incorporated appropriately, provide a learning environment which both enhances the learning "with, from and about" (CAIPE 2002) of each of the professional groups and provides a means for learners from different professions (IPE learning set) to remain in contact even when on placement in different clinical environments.

Many higher education institutions have limited experience of using e-learning but have formed alliances with other universities in similar positions. In some cases they have outsourced the technical infrastructure in order to ensure 24-hour, 7-day a week technical support for learners. This partnership and outsourcing arrangement often provides access to markets which could not result from a higher education institution working on its own.

## D. Presentation

It may seem obvious that a learner will get used to how he or she interacts with the materials, and will therefore expect consistency in the presentation of all aspects of their journey. This may not, however, be simple to achieve. Often the administrators working to support the infrastructure, the designers of the materials and the facilitators will work in very different ways. Clear guidelines for everyone involved in facilitating this journey must be established, and a clear identification of the learners' needs should be paramount.

## Salmon's Five-Step Model for E-Moderators

Turning to Salmon's model (2004), this is somewhat different from McKey's model. The first step (see Figure 3) 'access and motivation' has a similar emphasis to McKey's 'infrastructure' in stating the need for all aspects of the learner's interaction with the process to be considered. The remaining steps are explored below.

## Online Socialisation

Salmon's second step, 'online socialisation' however, emphasises the importance of online socialisation. Lack of socialisation, particularly with EIPE, is seen as a major barrier. As Mancuso-Murphy (2007) indicates, socialisation comes to be seen as an important aspect of online learning. Learners value the convenience, accessibility and flexibility of e-learning. However, to succeed in online learning learners need to be motivated, self-directed and active participants in the learning process. Therefore, an efacilitator needs to take particular care in the design of both the e-learning object and in facilitating the learning process. Miers et al. (2007) indicate that learners are initially apprehensive about using technology and have differing views about its advantages and disadvantages. However, with support for learner interaction, learners report that they benefit from the peer review process.

## Information Exchange

Salmon's third step is termed 'Information Exchange'. This is the stage at which motivating and facilitating EIPE is considered. Early research into the skills needed to facilitate IPEL (Howkins & Bray, 2008) indicated:

- **Development of self awareness:** Howkins and Bray (2008) report that the panel reviewing a Delphi study were unanimous that self awareness is critical to the ability of the individual to be a facilitator.
- **Respect and valuing difference:** Appreciation of the huge range of factors which make each group different is key to the facilitator being able to undertake their role.
- **Managing power and hierarchy:** Groups often see this as a crucial concern and a facilitator must be able to understand the power dynamics in the group and indeed their own influence on the group.
- **Group processes:** The flexibility of the facilitator in allowing the group to develop their own way of working is a key aspect in developing interprofessional education.
- **Context and planning:** Gaining information in advance about the participants, their likely views and planning the session accordingly is crucial (Howkins and Bray (2008).

The development of the facilitator is therefore crucial if the barriers to EIPE are to be overcome, and this is promoted through staff development, through working in programme teams and through peer review.

Magnussen (2008) indicates that online learning uses adult learning theories and views the educator as a facilitator of learning. Collison et al. (2000) call deliverers of e-learning provision e-moderators and outline their experiences – as well as their fears and anxieties. For the purposes of this chapter, however, the term facilitator is used.

E-learning facilitation identifies the learner and facilitator relationships as crucial (Atack, 2002; Roberts, 1998). Furthermore, facilitators need to create an environment which encourages peer support between learners, supports the development of communicative dialogue and guides socialisation. Communication online is often referred to as "communication without voices": it is a form that can lead to feelings of abandonment and isolation (Mueller 2001). Dialogue is therefore critical (Atack, 2003; Robley et al., 2004) and the style of communication is the factor that makes interactions with learners gratifying (Hyde & Murray, 2005; Robley et al., 2004).

In considering the motivational aspects of IPEL, Miers et al. (2007) review the online interprofessional learning experience of people in seven professions: children's nurse; radiotherapist; diagnostic imager; mental health nurse; midwife; physiotherapist; and adult nurse. They indicate online socialisation is crucial to establishing relationships between learners and between learner and facilitator. Their research shows that, at times, there can appear to be gaps in communication if a question is posed and not immediately responded to. Furthermore, socialisation must include comments which are personal (e.g. congratulations on birthdays and weddings, and expressions of support). Forman's (2005) examination of the importance of the interpersonal aspects of IPE in building relationships between the learners reached similar conclusions.

Miers et al. (2007) indicate, however, that further development is needed to ensure learners are more comfortable with critiquing one another's work. Their evaluation of IPEL indicates successful interprofessional collaboration can be achieved online if attention is given to all aspects of the motivation and facilitation of the learners' experience. Early feedback on critique, and guidelines offering a framework for critique, help develop their skill and confidence.

Learners new to online programmes need support through the provision of an onsite troubleshooter experienced in computer technology. This prevents the frustration and de-motivation that can occur for both learners and facilitators. Magnussen (2008) suggests the success of implementing a web-based programme requires thoughtful planning and a committed faculty, with dedicated administrators to provide the technical support.

It is important to remember that any learning programme can be blended, using multiple learning strategies and interactions. While e-learning can be undertaken totally online, some can be delivered on campus, if all learners can access the delivery location. This can help motivate learners to capitalise on the ways of learning that they most enjoy.

## Knowledge Exchange

The fourth step of Salmon's model 'Knowledge exchange' emphasises the learning process, which has been described as involving the thoughts, feelings, commitments and actions of facilitators and learners within the learning situation. It also includes the interaction patterns and descriptions of the learning environment that result from that interaction (Bilinski, 2002; Harden, 2003; Huit, 1995). This view is endorsed by Carter and Rukholm (2008) who indicate that "online courses provide unique opportunities for critical thinking, facilitator interaction and discipline-specific writing".

One useful general taxonomy that can be adapted to e-learning is that proposed by Fink (2003). This taxonomy is based on six components necessary for significant learning: foundational knowledge; application; integration; human dimension; caring; and learning how to learn. They should be considered in relation to entry criteria for the programme being undertaken, and in relation to the level at which the programme is developed.

These aspects are re-emphasised in work by Forman et al. (2002) where aspects of these models had been systematically used to implement an e-learning infrastructure in a UK institution. Furthermore, Colla et al. (2008) cite the Demand Driven Learning Model, or learner-centred approach, as the most appropriate way to design and deliver online interprofessional learning. They outline four important factors for the development of e-learning:

- Introductory information for learners must be clear, concise and unambiguous with clarity around expectations of the interprofessional learner.
- Programming (or programme design) must facilitate ease of access and the saving of information.
- Organisation and navigation (user friendly instructions) must be employed and consistency established in offering choice, setting expectations and achieving course outcomes.
- Content should be designed to improve collaboration and interprofessional practice.

In summary, as Rosenberg (2008) suggests, the facilitator of demand-driven learning is key to delivering the knowledge base the learner needs. As Carter and Rukholm (2008) suggest, the role of the facilitator, the design of the learning materials and the support provided for the learner are critical factors in fostering critical thinking and professional growth in an online environment.

## Development of the Learner

Salmon's fifth step concerns learner development. Dealing with extensive, complex and varied student placement arrangements is a major challenge for higher education institutions. From the viewpoint of IPEL, once an effective infrastructure is in place and can be monitored, maintained and enhanced, e-learning becomes a popular way of meeting the challenge of dealing with large numbers with different placement arrangements: particularly when e-learning is combined with problem-based learning (increasingly known as PBeL). Further pedagogical values are also recognised and include:

- The ability of learners to engage with each other asynchronously.
- The ability of learners to consider and reflect on postings from other learners before adding to and posting their own contributions.
- The opportunity for less vocal or confident learners to contribute more equally.
- The scope for learners to record and (re) access posted feedback from facilitators as required.
- An increase of competence in computer usage and appreciation of how online engagement can be used for interprofessional interaction (Beck & Owens, 2007; Hughes et al., 2004; Kamin et al., 2006).

In these situations the PBeL for the interprofessional aspects of the programme often involves assigning an interprofessional group of learners to work with a facilitator for the period of a term, semester or year. Problem-based IPEL will, therefore, commence in the classroom when other academic aspects of the programme are being delivered. The learner will become familiar with aspects of problem-based learning and then move on to using this methodology with IPEL as indicated in Figure 4.

The evaluation of programmes developed and delivered in this way has been very positive, with learners enjoying the social aspects, even in an online environment, of getting to know learners

*Figure 4. The learner's progression from problem-based e-learning to problem-based interprofessional e-learning*

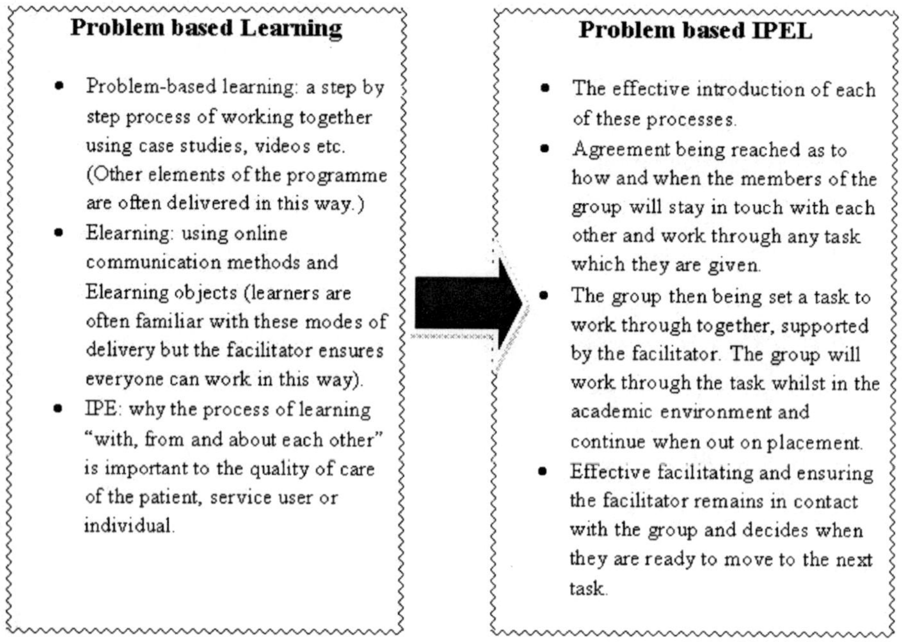


from other professions and working through the tasks as an interprofessional group (Beck & Owens, 2007; Hughes et al., 2004; Kamin et al., 2006).

## Discussion

Effective problem-based IPEL will be critically dependent on staff development. There is a need to combine the skills needed to provide the infrastructure as outlined earlier and the facilitation of IPEL or problem-based IPEL. Institutions must then ensure the learning objects are an appropriate pedagogical method for the learners: this may require considerable development work, often over many years. Sharing resources between institutions and networking is therefore seen as an effective way of learning and developing together.

Thus, any institutions considering E-IPE need to ensure appropriate time for staff development. Quinney. et al. (2008) review the interprofessional e-learning aspects of social work curricula, emphasise the need for ongoing staff development and emphasise the role of IPE in enhancing the individual professional identity. Jones (2000) asserts that interprofessional practice must be supported by facilitation skills which promote valuing diversity, providing effective communication, guidance on adaptation to change and team development. It seems, however, that this investment can lead to greater job satisfaction. Magnussen (2008) indicates that more time is needed in preparing online materials and in supporting staff development, but that educators enjoy the facilitation of learning in an online environment.

The World Health Organisation's Framework for Action (2009) looks at the development of IPE internationally and offers key guidance on its development. Learning to collaborate and work together requires education and adaptations to create change. It is by functioning effectively in diverse – and geographically extended – teams that we will overcome the barriers to IPE e-learning. Sharing information is critical in this framework.

*Figure 5. Global connections and relationships Forman (2008)*

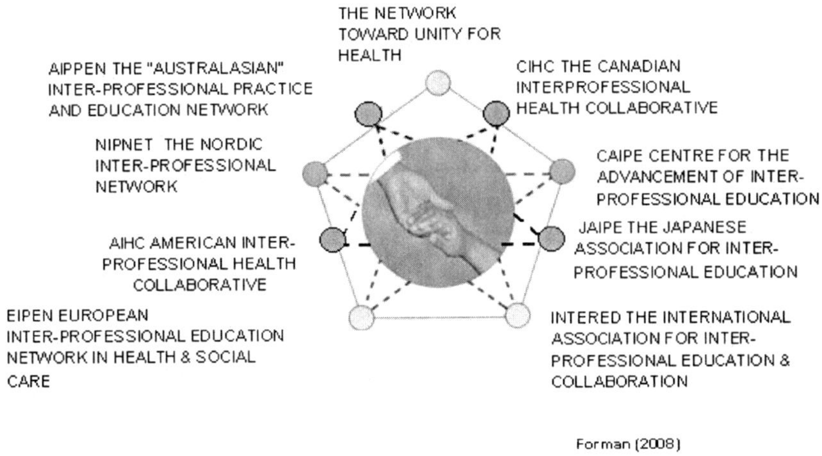


## Networking and Interprofessional Development

To facilitate development of learning objects and to undertake research into IPEL, some institutions are sharing resources: in some countries this has been actively encouraged by funding agencies. For example, in the UK, Centres of Excellence in Teaching and Learning (CETLs) have been funded to encourage these developments, in terms of both the underpinning research and the learning objects produced. In addition, many of the IPE Networks which exist internationally have begun working together to share resources both within their local network and internationally. All emphasise the need to concentrate on the benefits to the patient, individual and community. As Barr et al. (2005) asserts, we need to ensure we provide an effective IPE experience from the patient's perspective.

Figure 5 shows the international IPE bodies and the links which are now being encouraged between them to share good practice, and ways in which ideas and resources can be shared and barriers prevented. Along with these, more recently New Zealand has developed a National Centre for Interprofessional Education and Collaborative Practice in 2009.

## CONCLUSION

This chapter focuses on cultural changes needed within higher education institutions to support the development of effective IPEL. IPEL, if appropriately designed and facilitated, can overcome the perceived barriers and address issues of timetabling and maintaining provision, even when learners are on placement at different times and they come from several professional groups. The provisions of a virtual university, as once described by Teare et al. (1998), cannot be achieved in isolation. Neither can E-IPE. Just as the IPE philosophy is to "learn with, from and about each other" (CAIPE, 2002), institutions need to learn from each other in order to maximise the benefits of producing effective EIPE materials appropriate to today's and tomorrow's practitioners.

## REFERENCES

Atack, L. (2003). Becoming a Web-Based Learner: Registered Nurses' Experience. *Journal of Advanced Nursing*, *44*, 289–297. doi:10.1046/j.1365-2648.2003.02804.x

Atack, L., & Rankin, J. (2002). A Descriptive Study of Registered Nurses' Experiences with Web-Based Learning. *Journal of Advanced Nursing, 40*, 457–465. doi:10.1046/j.1365-2648.2002.02394.x

Atwal, A., & Caldwell, K. (2002). Do multidisciplinary integrated pathways improve interprofessional collaboration? *Scandinavian Journal of Caring Sciences*, 360–367. doi:10.1046/j.1471-6712.2002.00101.x

Barr, H., Koppel, I., Reeves, S., Hammick, M., & Freeth, D. (2005). *Effective Inter-professional Education*. Oxford, UK: CAIPE/Blackwell Publishing Ltd.

Biggs, J. (2003). *Teaching for Quality Learning at University* (2nd ed.). Berkshire, UK: Open University Press.

Bilinski, H. (2002). *Bates Guide to Physical Examination and History Taking* (8th ed.). Philadelphia: Lippincott.

Campbell, M., Gibson, W., Hall, A., Richards, D., & Callery, P. (2005). Online vs. Face to Face Discussion in a Web-Based Research Methods Course for Postgraduate Nursing Students: A Quasi-Experimental Study. *International Journal of Nursing Studies, 45*, 750–759. doi:10.1016/j.ijnurstu.2006.12.011

Carter, L. M. (2008). A Study of Critical Thinking, Teacher- Student Interaction, and Discipline – Specific Writing in an Online Educational Setting for Registered Nurses. *Journal of Continuing Education in Nursing, 39*(3), 133–138. doi:10.3928/00220124-20080301-03

Centre for the Advancement of Inter-professional Education (CAIPE). (2002). *Definition of Interprofessional Education*. Retrieved August 10, 2009, from http://www.caipe.org.uk

Cleak, H., & Williamson, D. (2007). Preparing health science students for interdisciplinary professional practice. *Journal of Allied Health, 36*(3), 141–149.

Collison, G., Elbaum, B., Haavind, S., & Tinker, R. (2000). *Facilitating Online Learning Effective Strategies for Moderators*. Madison, WI: Atwood Publishing.

Fink, L. D. (2003). *Creating Significant Learning Experiences: An Integrated Approach to Designing College Courses*. San Francisco: Jossey- Bass.

Forman, D. (2005). Recognising the Importance of the Interpersonal. In *The Theory Practice Relationship in Inter-professional Education* (pp. 55-60). Higher Education Academy Health Sciences and Practice Occasional Paper No. 7.

Forman, D. (2008). Networking IPE Networks. *The Network: Towards Unity for Health Newsletter, 26*.

Forman, D., Nyatanga, L., & Rich, T. (2002). E-learning and Educational Diversity. *Nurse Education Today, 22*, 1–7. doi:10.1054/nedt.2001.0740

Howkins, E., & Bray, J. (2008). *Preparing for inter-professional teaching: theory and practice*. Oxford, UK: Radcliffe Medical Press.

Huws, Jagger, & Bates. (2001). *Where the Butterfly Alights: the Global Location of E- work*. IES Report 480.

Hyde, A., & Murray, M. (2005). Nurses' Experiences of Distance Education Programmes. *Journal of Advanced Nursing, 49*, 87–95. doi:10.1111/j.1365-2648.2004.03267.x

Johns, C. (1995). Framing Learning Through Reflection within Carper's Fundamental Ways of Knowing In Nursing. *Advance for Nurse Practitioners, 22*(2), 226–234.

Jones, E. M. (2000). *Shaping team practice in the context of health reform: Challenges, tensions, benefits*. PhD Thesis, Flinders University of South Australia.

MacDonald, C. J., Stodel, E. J., & Chambers, L. W. (2008). An Online inter-professional Learning Resource for Physicians, Pharmacists, Nurse Practitioners, and Nurses in Long Term Care: Benefits, Barriers and Lessons Learned. *Informatics for Health & Social Care*, *33*(1), 21–38. doi:10.1080/14639230801886824

Magnussen, L. (2006). Applying the Principles of Significant Learning in the E–Learning Environment. *The Journal of Nursing Education*, 82–86.

Mancuso-Murphy, J. (2006). Distance Education in Nursing: An Integrated Review of Online Nursing Students' Experiences with Technology – Delivered Instruction. *The Journal of Nursing Education*, 252–260.

McKey, P. (2000). *Total Student Experience Model*. White Paper, NextEd.

Meads, G., Jones, I., Harrison, R., Forman, D. & Turner, W. (2009). Cultural Congruence between Higher Education and Health and Social Care Agencies. *Journal of Education and Work*.

Miers, M. E., Clarke, B. A., Pollard, K. C., Rickaby, C. E., Thomas, J., & Turtle, A. (2007). Online Interprofessional Learning: The Student Experience. *Journal of Interprofessional Care*, *21*(5), 529–542. doi:10.1080/13561820701585296

O'Neil, C., & Fisher, C. (2008). Should I take this Course Online? *The Journal of Nursing Education*, *47*(2), 53–58. doi:10.3928/01484834-20080201-04

Peters, M. (2000). Does constructivist epistemology have a place in nursing education? *The Journal of Nursing Education*, *39*, 166–172.

Pollard, & Hillage. (2000). *Exploring E-learning*. IES Report 376.

Quinney, A., Hutchings, M., & Scammell, J. (2008). Student and Staff Experiences of Using a Virtual Community, Wessex Bay, to Support Interprofessional Learning: Messages for Collaborative Practice. *Social Work Education*, *27*(6), 658–664. doi:10.1080/02615470802201754

Ragan, L. C. (1999). Good Teaching is Good Teaching: An Emerging Set of Guiding Principles and Practice for the Design and Development of Distance Education. *Cause/Effect*, *22*, 20-24.

Roberts, K. K. (1998). *A Naturalistic Study of Students' Experience in a Computer-Based Nursing Course*. Doctoral Dissertation University of Kansas, Dissertation Abstracts International 59 (12) 4352A (UMI No 9914 105).

Robley, L. R., Farnsworth, B. J., Flynn, J. B., & Horne, C. D. (2004). This New House: Building Knowledge Through Online Learning. *Journal of Professional Learning*, *20*, 333–343. doi:10.1016/j.profnurs.2004.07.012

Rosenberg, M. J. (2001). *E-Learning – Strategies for Delivering Knowledge in the Digital Age*.

Salmon, G. (2004). *E-moderating: the key to teaching and learning online* (2nd ed.). London: Routledge.

Teare, R., Davies, D., & Sandelands, E. (1998). *The Virtual University An Action Paradigm and Process for Workplace Learning*. London: Cassell.

Walsh, K. (2007). Inter-professional Education Online: The BMJ Learning Experience. *Journal of Interprofessional Care*, *21*(6), 691–693. doi:10.1080/13561820701436912

World Health Organisation. (in press). *A Framework for Action on Inter-professional Education and Collaborative Practice*. Belgium. *World Health Organisation.*

# Chapter 16
# The Utility of Disruptive Technologies in Interprofessional Education:
## Negotiating the Substance and Spaces of Blended Learning

**Maggie Hutchings**
*Bournemouth University, UK*

**Anne Quinney**
*Bournemouth University, UK*

**Janet Scammell**
*Bournemouth University, UK*

## ABSTRACT

*This chapter will consider the educational benefits and challenges of introducing e-learning objects within an interprofessional curriculum. It examines the tensions of curriculum development as content or process-driven in the context of facilitating interactive learning using blended learning strategies which combine online and face-to-face interactions. This chapter draws upon an evaluation of student and staff experiences of an interprofessional curriculum incorporating health and social care users and carers as case scenarios in a web-based simulated community, Wessex Bay, and highlights congruent and disruptive factors in negotiating transformative learning and cultural change. It draws conclusions and recommendations for informing practice in interprofessional education and suggests directions for future research to inform the substance (interprofessional case scenarios) and spaces (discussion boards, chat rooms, classroom) for collaborative learning in an interprofessional curriculum.*

DOI: 10.4018/978-1-61520-889-0.ch016

## INTRODUCTION

Interprofessional education (IPE) in health and social work has been strongly advocated to encourage effective partnership working in order to improve services and effect change (Barr, 2002; Barr & Ross, 2006; Whittington, 2003). The potential of online technologies in overcoming the constraints of place-situated and time-bounded classrooms is key here.

The potential for creativity and transformative learning are highlighted by Bruner in suggesting that "Knowing is a process not a product" (1966, p. 72). Curriculum development literature distinguishes between "curriculum as content and education as transmission, curriculum as product and education as instrumental and curriculum as process and education as development" (Kelly 2004, p. 46). The challenge is to encourage participants, both staff and students, to make the transition from perceiving the interprofessional case scenarios as ready-made e-learning products to a process of engaging creatively in the spaces for collaborative learning whether through online e-processes or face-to-face.

The case study reported in this chapter highlights congruent and disruptive factors in negotiating transformative learning and cultural change. It draws on evaluation of student and staff experiences of an interprofessional curriculum incorporating health and social care users and carers as case scenarios in a web-based simulated community, Wessex Bay. Finally, conclusions and recommendations for informing practice in interprofessional education and suggested directions for future research to inform the substance of e-products and spaces of e-processes in blended learning will be outlined.

## BACKGROUND

A new curriculum including interprofessional education (CAIPE, 1997) at each undergraduate level was introduced at Bournemouth University in 2005 for adult, child health, mental health and learning disability nursing disciplines, midwifery, occupational therapy, physiotherapy, operating department practice (ODP) and social work students. In common with many HEIs providing education in these subject areas, the challenge was to facilitate meaningful interprofessional learning whilst balancing a number of structural factors: the complexities of professional body requirements, the logistics of large student numbers and multi-site teaching across three sites in Dorset, Somerset, and Hampshire, and a wide range of practice placements.

The learning and teaching strategy adopted utilised a blended learning approach, providing a mix of online and face-to-face interactions (Sharpe et al., 2006) to provide interprofessional learning activities whilst supporting and engaging geographically dispersed student groups. The principal vehicles used to integrate IPE activities were an e-learning object known as 'Wessex Bay' and a Virtual Learning Environment (VLE), a commercial system named 'Blackboard'. The aspiration was to provide the technology affordances for promoting and supporting a virtual community of practice (Wenger, 1998) to enable the dispersed communities of practitioners, students and tutors, to collaborate and develop interprofessional practice through face-to-face and online interaction (Scammell et al., 2008).

Building on previous projects (Hutchings, 2002; Hutchings et al., 2007; Quinney, 2005; Mulholland et al., 2005) and drawing on pedagogic literature on both e-learning (Beetham & Sharpe, 2007; Laurillard, 2002; Miers et al., 2007; Salmon 2000, 2002) and interprofessional education and partnership working (Barrett et al., 2003; Clarke et al., 2007; Quinney, 2006), a simulated web-based community, 'Wessex Bay', was created to facilitate interprofessional collaboration and learning across a range of health and social work disciplines in an interprofessional curriculum. Its use was linked to the VLE, offering facilities for group announcements, discussion forums, chat and virtual classroom facilities. Together, they form an online learning environment which affords both the substance of digital content in the form of interprofessional case scenarios and digital spaces in the form of discussion boards, chat and virtual classroom facilities for collaborative learning.

The intention behind Wessex Bay was to facilitate learning and teaching strategies that promoted student enquiry through problem-based learning within a safe but challenging interprofessional environment for students to engage with 'live' individuals situated within an authentic and shared community (Brown et al., 1989; Jonassen & Land, 2000; Savin-Baden, 2000). The plan was for tutors to provide the necessary scaffolding for students through managing the progress of individual cases to encourage active student engagement in case analysis, discussion, debate, collaborative problem-solving and decision-making in a variety of interprofessional situations designed to achieve transformative learning (Dewey, 1933; Collins et al., 1989; Lave & Wenger, 1991; Vygotsky, 1978).

The development of Wessex Bay as a learning object was informed by evolving practice in the design of authentic activities situated in realistic learning environments (Herrington et al., 2004), acknowledging the importance of context and situated learning (Lave & Wenger, 1991). This approach was underpinned by constructivist educational theory (Mayes & Freitas, 2007) and experiential learning theory (Dewey, 1933, 1938). These underpinnings were supported by information and communication technologies, providing connectivity and opportunities for collaboration (JISC, 2004).

Herrington and colleagues (2004) argue that "the value of authentic activity is not constrained to learning in real-life locations and practice, but can be realized through careful design" (2004, p. 5) in both face-to-face and web-based learning environments. They distinguish ten characteristics critical in the design of authentic activities and provide useful evaluation criteria for e-learning objects:

Authentic activities:

- Have real world relevance.
- Are ill-defined, requiring students to define the tasks and subtasks needed to complete the activity.
- Comprise complex tasks to be investigated by students over a sustained period of time.
- Provide the opportunity for students to examine the task from different perspectives, using a variety of resources.
- Provide the opportunity to collaborate.
- Provide the opportunity to reflect.
- Can be integrated and applied across different subject areas and lead beyond domain-specific outcomes.
- Are seamlessly integrated with assessment.
- Create polished products valuable in their own right rather than as preparation for something else.
- Allow competing solutions and diversity of outcome (Herrington et al., 2004, p. 11-13).

To replicate the local environment, Wessex Bay comprised a 'virtual' seaside town and rural hinterland, with private homes and public facilities, including hospital and social services departments. Scenarios included information about service users and patients accessing the services as well as details concerning employed staff roles, with links to professional codes of practice, legislation and agency policies (Pulman et al., 2008).

Both static scenarios and scenarios that evolved over several weeks were used as triggers to engage students in blended learning activities relating to the residents. The activities were: to generate collaborative strategies for assessment and intervention; develop awareness of the service user perspectives; consider professional roles; develop awareness of services available; explore possible

*Figure 1. Wessex Bay home pages*

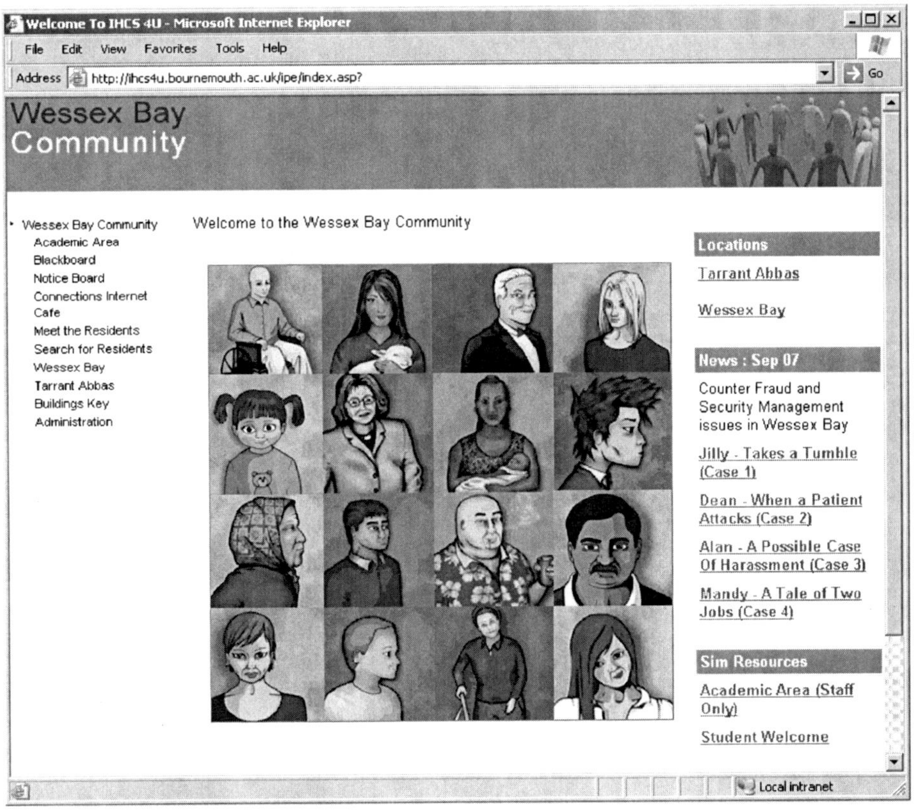


legal interventions and develop an understanding of the debates associated with these themes.

Wessex Bay is linked to the VLE, for group announcements, discussion forums, chat and virtual classroom facilities.

Online bulletin boards were used to update students on the service users' situation, with timed online meetings to explore the implications of 'breaking news' as well as the discussion forum facility to debate issues and share interprofessional perspectives.

## Wessex Bay: The E-Learning Object in Detail

A screenshot of the Wessex Bay home page can be seen in Figure 1. Students can search Wessex Bay for individual service users and patients by name, keyword, or map location, to identify their homes and find out about family structures, living situations and care needs. Health, social work and care facilities are located in this virtual community, including a health centre, care home, and day centres, with information about people accessing the services and the staff employed there (Scammell et al., 2008). Resources developed comply with accessibility guidelines (TechDis, 2002).

Wessex Bay is used for IPE teaching in Year 1 units of study on 'Communication' and 'Preparation for Professional Practice' and Year 2 units on 'Communication in Groups' and 'Risk Assessment & Risk Management'. A problem-based learning strategy (Savin-Baden, 2000) was adopted, in which community member case studies are selected as triggers for students working together in interprofessional groups. Students could, for example, consider the scenario of a 15 year old boy with a learning disability admitted to hospital

*Figure 2. Wessex Bay member profile and evolving scenario*

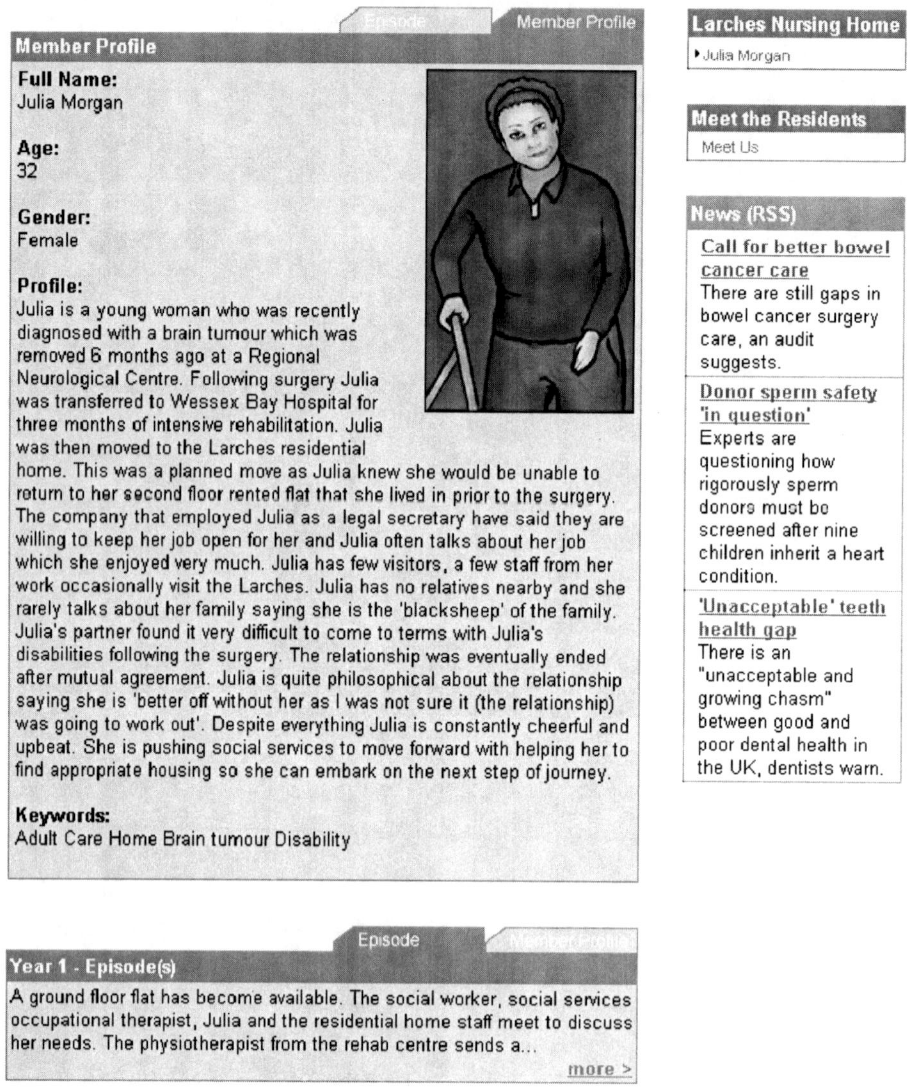


for a tooth extraction. An example of another such evolving scenario, involving a young woman diagnosed with a brain tumour, is demonstrated in Figure 2.

The Wessex Bay member profiles provide standardised brief case histories, including accommodation details and location, and links to family and community networks and can be developed through adding new episodes to identify critical issues for analysis and reflection by the students. Student engagement was facilitated by academic staff through announcements providing updates on the evolving situation for characters, and scheduled online chat meetings to explore the implications of developments. Discussion forums were used to debate issues, share interprofessional perspectives, and generate collaborative strategies for assessment, planning, intervention and review of care provision (Scammell et al., 2008). The blended learning activities form the basis of assessed unit learning outcomes.

## BENEFITS AND CHALLENGES FOR PRACTICE

The introduction of e-learning objects in the curriculum offers both benefits and challenges for practice. The evaluation study of Wessex Bay was funded by the Centre for Health Sciences and Practice Subject Centre of the Higher Education Academy (Scammell et al., 2008). There were approximately 600 student participants, of which the majority was female. There were 40 staff participants of which 30 were female and 10 male.

Data were collected through student focus groups (SFG), tutor focus groups and interviews (TFG), student (SQ) and tutor questionnaires (TQ) and student (ES) and tutor electronic surveys (TS) conducted over a two year period. Staff and student perspectives were compared, and the conclusions reached were supported with illustrative comments.

Key findings from the evaluation will now be considered in detail, so the substance and spaces of blended learning can be examined and implications for others developing technology enhanced learning in IPE can be identified. There is evidence that the aspirations, intentions and plans for implementing change were fulfilled. However, it was clear from the findings that there was some misalignment between the affordances of the technologies and the transferability of the learning for students and staff (Argyris & Schön, 1974).

### Paradoxes of Disruptive Influences

The benefits of realism and authenticity in facilitating IPE through Wessex Bay were clearly identified by students. They described the value of relating to real life characters and situations:

*It is easier to relate theory to practice when characters are created as they help bring problems and health issues to life (SQ1).*

Students also appreciated the timed release of episodes helping to maintain their interest and providing more depth and realism to cases:

*We would have triggers that could be in the form of referrals, it could be a telephone call, a letter, so in a way that was quite nice because that's a bit like real life in a way (SFG4).*

The IPE experience was valued by students and staff. Students identified benefits of working together within interprofessional groups, for gaining an understanding of different professions, sharing views, valuing different professional perspectives, and gaining confidence to work interprofessionally (Scammell et al., 2008). Student comments included:

*It really helped me to understand how important it is to understand other people's roles in order to get the best outcome for both patients and staff (SQ1).*

*Very good, it was about learning to respect other professionals' opinions and differences (SQ4).*

*I now feel I work as part of a multi disciplinary team, I can communicate better with different professions (SQ1).*

Yet some students challenged the potential of learning, identifying limitations of the case study information provided. They wanted fewer characters developed in more depth with greater interaction:

*You couldn't build a whole picture could you? It's like with the lad that I did, it didn't stress what the learning disability was, it just said that he was a teenager with a learning disability (SFG5).*

Similarly academic staff reported a wish for more sophistication in Wessex Bay, with added depth and interactivity of cases:

*Some of the case studies were a bit sketchy (TFG).*

*Very static community. Could be brought alive by more interaction (TS).*

Some students and staff also questioned if the focus of particular cases was sufficiently interprofessional, suggesting they were largely health oriented and did not represent the specialties of the different professions.

*I felt a little isolated as the only Learning Disability student nurse. The group mainly consisted of adult nurses which dominated the teaching of the unit. There were a few midwives and mental health nurses who felt that this was the case (SQ3).*

But some academic staff also acknowledged their responsibility in making the e-learning objects work more effectively:

*We could use our collective expertise and feed in and make a really good scenario. So maybe we need to build them collaboratively...maybe we need to build these cases interprofessionally as well (TFG).*

These findings provide evidence for improving the design of Wessex Bay as an e-learning object and meeting the expectations of students and staff about the substance of e-learning and have informed our recommendations. But such a technology-focused interpretation, concentrating as it does on the design of the e-learning object, could neglect the processes of transformative learning and cultural change for students and staff alike. We will argue that it could, in particular, miss some more fundamental indicators of disruptive influences at work in these processes.

Dewey's concept of experiential learning is a focused process, directed to solving a question or issue through purposeful thinking or reflection. Dewey's model provides the levers necessary for effective learning by identifying: "(1) a state of doubt, hesitation, perplexity, in which thinking originates, and (2) an act of searching, hunting, inquiring, to find material that will resolve the doubt" (1933, p. 12). It is in this sense that Dewey's concept encompasses both the space and potential substances for learning triggered by questions and issues. The Wessex Bay case scenarios were intended to act as levers for transformative learning by engaging students in purposeful disruptive processes where the case scenarios provided basic information and necessitated interprofessional collaborative action and reflection on practice (Dewey, 1933; Herrington et al., 2004).

However, acknowledging and working with scenarios in a process of engagement which may be ill-defined, complex and time-consuming is a significant challenge for introducing e-learning objects within an IPE curriculum. These findings reflect differences in assumptions and expectations about what the technology will be able to do, but also raise issues: concerns emerged around the learning and teaching strategies employed to encourage interprofessional learning and tensions were evident concerning the influence of profession-focused perspectives in interprofessional case scenarios and collaborative learning activities.

The double disruption of two educational innovations, "one predominantly related to curriculum as syllabus *content* (IPE), and the other to curriculum as *process* (blended learning), presented significant challenges for staff and students" (Scammell et al., 2008, p.8). But rather than suggesting the outcomes of the evaluation study were mainly related to the curriculum or to the technology, we want to argue that issues highlighted here are interconnected, and described as three zones of praxis (Scammell et al., 2008), namely the experiences and disruptions in engaging with technology enhanced learning, the learning and teaching strategies adopted, and the challenges of asserting professional identity while developing understanding and accommodating different world views.

The disruptive influences of the IPE curriculum and technology enhanced learning were felt in multiple and interlinked ways. Engagement with these innovations evoked strong emotions (Scammell et al., 2008). Students described feeling frustrated, anxious and confused:

*It was really stressful that week not knowing and being in control of things. Not being prepared. Even though it was new it made it really stressful. When you did get into it I thought it was really good (SFG1).*

Staff descriptions of working with mixed professional groups were more positive in expressing their experiences as 'enjoyable', 'challenging' and 'stimulating' (TQ), but they also reflected student emotions:

*That's what worried me that perhaps some people were isolated and not able to ask for help. I felt frustrated because I wouldn't have the skills to help anyway (TFG).*

Staff found that working with the IPE curriculum added to their own understanding of other professions and caused them to consider how to accommodate 'different world views' (TQ). It also helped them to develop a sense of 'commonality in purpose' (TQ) (Scammell et al., 2008, p. 15).

## Technology Preparedness

Differences in engagement with the IPE experience can be partially associated with variations in IT skills and requirements for familiarisation and support. Some students demonstrated more familiarity with the information and communication technologies and the VLE, through which Wessex Bay is accessed, than others and Year 2 students identified positive improvements in computer literacy skills over Year 1 (Scammell et al., 2008). Where students struggled with computers they suggested this affected learning and contributions to collaborative group work:

*We could have done group work by using Blackboard but it wasn't an option by the people who I was working with because they weren't interested in it or didn't know how to use it (SFG3).*

Students received an introduction to the VLE and IT systems. Some students felt they should be supported more but others found they could use the systems effectively following a period of practice and appreciated trial and error as a good way to master these systems. Students also became a learning resource helping each other learn the systems. Most students also agreed that when help was asked for, it was given and initial difficulties were soon resolved.

Staff revealed varying levels of engagement and preparedness, along with the need to engage in more staff development and self directed familiarisation with the technology. Staff described themselves as having a range of IT skills and expressed a variety of attitudes from enthusiasm to reluctance in engaging with technology enhanced learning:

*I think you've got to be computer-minded to like these things. I'll use it because I have to use it.... I don't like using it and it's a necessity of my work and it helps students but I can't say that I enjoy going onto it (TFG).*

The disruptive influence of introducing technology enhanced learning, both Wessex Bay and the VLE, were felt by students and staff alike. Students recognised that everyone was learning and that both they and some of their tutors had insufficient knowledge and skills to use the technology effectively.

*I think it was foreign to everyone at the beginning because it has only been in this year and even the tutors were struggling with Blackboard. They were struggling to get to grips with it the same as we were, who have never used it before (SFG2)*

## Teaching Interventions

Students identified learning benefits when academic staff added triggers for Wessex Bay case scenarios and discussion threads in the VLE. However some students felt they had not been informed about the features of Wessex Bay, that tutors had not developed the characters, and did not do enough to help students work with the cases. Although this finding may be partially associated with the level of IT skills possessed by staff, this is not sufficient to explain variations in engagement. It also represents different expectations of technology enhanced learning in providing content-rich environments (Jonassen & Land, 2000) for enhancing student learning and uncertainty about degrees of scaffolding (Vygotsky, 1978) and intervention required.

Collins, Brown and Newman (1989) when discussing the concept of cognitive apprenticeship, closely aligned with Lave and Wenger's (1991) legitimate peripheral participation, suggest the process of scaffolding:

*involves a kind of co-operative problem-solving effort by teacher and student in which the express intention is for the student to assume as much of the task on his own as possible, as soon as possible (Collins et a., 1989, p. 482).*

However Oliver (1999) points out that open and flexible learning often reduces opportunities for tutors to provide face-to-face teaching, and thus to scaffold learning hence proving disruptive and problematic for some staff:

*I felt frustrated because I wouldn't have had the (IT) skills to help anyway (TFG).*

There was evidence that scaffolding activities were taken on by some students in supporting and helping each other to use the technology (Scammell et al., 2008). This tends to demonstrate the importance of Vygotsky's emphasis on the significance of social interaction with more capable peers (1978).

Academic staff had opportunities to work collaboratively on the design of cases and for some units of study this worked very well. They were also encouraged to take ownership of particular cases, to develop episodes and manage the development of the character and family and carer networks. Some staff appreciated the value of Wessex Bay and could see its potential for further development but they also recognized the time required to take ownership of the cases:

*There is a wealth of potential in Wessex Bay/ MyBU but it requires careful pedagogy and time for academic staff (TS).*

The substance of the IPE curriculum was positively associated with realistic experiences in the form of case scenarios and related information provided as e-products. However, the spaces for collaborative teaching through facilitation and development of these scenarios and collaborative learning through analysis and deliberation by students within interprofessional groups proved more challenging and open to negotiation.

Undoubtedly some students benefited from successfully negotiating the spaces of blended learning. They were able to make the transition from externally validated e-product, the Wessex Bay case scenarios, to a process of engaging creatively with the substance of the case study in the spaces for collaborative learning. These spaces, whether facilitated online as e-processes or face-to-face, enabled students to interact, negotiate and construct shared interpretations:

*We tried to go through what the various problems were that that person would have, say for instance, if that person lived on their own, was heavily disabled, and what have you. So we each tried to pick up on something for particular professions. I had an occupational therapist, a physiotherapist, a couple of nurses, so they all tried to put their own input into what that patient would need from their own perspective, so we did it that way (SFG2/4)*

The concept of a problem based approach to IPE using realistic case scenarios as e-learning objects provides a basic, standardised and sustainable solution for interprofessional learning across a wide geographical area, involving large numbers of students and staff representing a wide range of professions. But expectations of what the technology can do may be at odds with this authentic problem-based approach to learning. Such an approach relies on staff "buying in" by providing resources, creating triggers and negotiating appropriate scaffolding for enriching interprofessional learning. It also relies on students not wanting to simply be instructed but to engage actively in creative problem solving using a social constructive approach, to review and plan interventions in care provision.

## DISCUSSION: SOLUTIONS AND RECOMMENDATIONS

The introduction on a large scale of e-learning approaches can be viewed as both innovative and problematic, with a range of pedagogic issues needing to be addressed for staff and students. There are challenges for the individual, interprofessional and uniprofessional programme team, student group, academic department and organisation, and their differing agendas have been considered. The concept of an all-encompassing e-learning object in which all the work of curriculum design and facilitation has been embodied for staff and students is juxtaposed with the realities on the ground, where processes of facilitation and learning can be difficult and challenging for students and academic staff alike.

Argyris and Schön's (1974) distinction between *espoused theory* and *theory-in-use* could help to explain the misalignment between perceived affordances of the technologies, practice experiences, and the transferability of learning. Argyris and Schön's theory is based on two distinct *theories of action*, the one guiding what people say the *espoused theory*, and the other what they do, the *theory in use*, leading to potential mismatches between the espousal of different individuals, groups, and organisations compared with their practices on the ground (1978, p. 11). We would therefore recommend that curriculum designers pay attention to the close integration of curricular content with learning processes.

Degrees of readiness *for* and levels of engagement *with* blended learning in the new curriculum were particularly pertinent themes emerging in the evaluation. The *double* disruption of changes to both curriculum content (the introduction of IPE) and curriculum process (the introduction of blended learning) was experienced and reported in both positive and negative terms. Whilst technologies are essentially disruptive and can be experienced as creative spaces they can also be experienced as uncomfortable and threatening spaces where static, familiar or defensive teaching and learning practices are used rather than opening up a safe space for transformative teaching and learning practices to emerge. An example of this is where staff simply printed off and distributed case study material from Wessex Bay to use in class, rather than enabling students to experience and interact with Wessex Bay scenarios online.

Other situations included ambitious and unrealistic expectations of the resource, imagining that it could or should function almost independently of the educator rather than being utilised to complement a range of learning activities, and some considered it to be less interactive than expected. In line with the existing literature in this area (Clark et al., 2007; Craddock et al., 2006), our evaluation indicated that 'buy-in' to the value of interprofessional education varied greatly, in both staff and student groups. This could also reflect both a degree of social distancing from the curriculum design and case scenario design by the staff facilitating the interprofessional education sessions, as has been discussed elsewhere (Scammell et al., 2008) and:

*a more optimistic and ambitious claim for learning technology applications to be able to release staff time and achieve interactivity through student-material interaction over student-student and student-tutor interaction (Scammell et al., 2008, p. 23).*

Focus group and online surveys contained student feedback relating to the varied level of ability and confidence with information and communication technologies to enhance teaching and learning (Quinney et al., 2008). However, when staff were positive about the utilisation of the resource to support and enhance interprofessional learning and realistic about the time investment in thoughtful pedagogy this, not surprisingly, had a positive impact. A student from a group containing social work students and a range of nursing students spoke positively about their tutor who demonstrated motivation and good preparation:

*she did really well giving examples for them and for us…she's actually done her homework (Quinney et al., 2008, p. 662)*

The interprofessional education literature demonstrates a growing awareness of the implications of e-learning for staff development (Clarke et al., 2007; Rees & Johnson, 2007). It was apparent in the case study reported in this chapter that people had different personal and professional agendas, and did not necessarily 'buy into' the one being promoted from 'the top'. Additionally as implementers of the strategy, we were associated with that top-down approach. The challenge for those implementing e-learning resources and for educators facilitating learning is to achieve 'optimum disruption', where the potential for creativity and transformation is seen as achievable and realistic, rather than experienced as too uncomfortable, too difficult or simply too unwelcome and therefore resisted or rejected. Strategies to these ends tend to advocate fostering a sense of 'ownership' among stakeholders in a blend of top-down and bottom-up approaches to change (Bromage, 2006: Elton, 1994: Berg and Östergren, 1977, 1979). This is one area where further research would be welcomed.

Experience with Wessex Bay as an e-learning object focused on IPE has raised a number of questions for further research, particularly with regard to exploring the relationship between lifeworld-led research and education (Todres et al., 2007). The development of blended learning approaches for health and social care education would benefit from further analysis of the relationships between different types of media and their combinations in learning objects and pedagogical strategies contributing to transformative learning. These could include lifeworld service user and carer stories in the form of short films, podcasts, poetry, drama and narrative case studies (Keen & Todres, 2007). The goal would be to outline a transferable holistic educational model that connects the three territories of praxis, technology enhanced learning, pedagogical strategies, and personal growth in achieving embodied relational understanding, critical judgement and ethical sensitivity for professional practice (Todres, 2008).

## CONCLUSION

There are positive benefits in introducing e-learning objects, not least being the disruptive utility that information and communication technologies can contribute to individual and organisational learning and cultural change. The experience of introducing the double disruption of an IPE curriculum and blended learning processes simultaneously has proved challenging within a complex organisation represented by differing individual, group and organisational agendas. The organisation, including individual academics but not necessarily uniprofessional academic groups, *espoused* (Argyris & Schön., 1974, 1978) views regarding the value of IPE and e-learning, however the *theories-in-use*, apparent in the findings, reflected some antipathy to both concepts and practice. This was mirrored by both individual students and groups of students. We strongly recommend that successful and continued negotiation of the substance and space of e-learning requires investment in staff development, while acknowledging that this presents a significant challenge to reach teaching staff who may be uncertain, unconvinced or indifferent to the efficacy of disruptive technologies in teaching and learning. The challenge is to achieve 'optimum disruption', where transformation is seen as achievable and realistic, rather than being experienced as too uncomfortable.

## REFERENCES

Argyris, C., & Schön, D. (1974). *Theory and practice: increasing professional effectiveness*. San Francisco: Jossey-Bass.

Argyris, C., & Schön, D. (1978). *Organizational learning: a theory of action perspective*. Reading, MA: Addison-Wesley.

Barr, H. (2002). *Interprofessional education: today, yesterday and tomorrow.* Occasional Paper No1. London: Centre for Health Sciences and Practice, LTSN/CAIPE.

Barr, H., & Ross, F. (2006). Mainstreaming interprofessional education in the United Kingdom: A position paper. *Journal of Interprofessional Care*, *20*(2), 96–104. doi:10.1080/13561820600649771

Barrett, G., Greenwood, R., & Ross, K. (2003). Integrating interprofessional education into 10 health and social care programmes. *Journal of Interprofessional Care*, *17*(3), 293–301. doi:10.1080/1356182031000122915

Beetham, H., & Sharpe, R. (Eds.). (2007). *Rethinking pedagogy for a digital age*. London: Routledge.

Berg, B., & Östergren, B. (1977). *Innovations and innovation processes in higher education*. Stockholm: UHÅ.

Berg, B., & Östergren, B. (1979). Innovation processes in higher education. *Studies in Higher Education*, *4*(2), 261–268. doi:10.1080/03075077912331377027

Bromage, A. (2006). The management of planned change: an interdisciplinary perspective. In Hunt, L., Bromage, A., & Tomkinson, B. (Eds.), *The Realities of Change in Higher Education: Interventions to Promote Learning and Teaching* (pp. 1–14). London: Routledge.

Brown, J., S., Collins, A. & Duguid, P. (1989). Situated cognition and the culture of learning. *Educational Researcher*, *18*(1), 32–41.

Bruner, J. S. (1966). *Towards a theory of instruction*. Cambridge, MA: Harvard University Press.

Centre for the Advancement of Interprofessional Education (CAIPE). (1997). Retrieved August 31, 2008, from http://www.caipe.org.uk/

Clarke, B.,A., Miers, M., Pollard, K. & Thomas, J. (2007). Complexities of learning together; students' experience of face to face interprofessional groups. *Learning in Health and Social Care*, *6*(4), 201–212. doi:10.1111/j.1473-6861.2007.00162.x

Collins, A., Brown, J. S., & Newman, S. E. (1989). Cognitive apprenticeship: teaching the craft of reading, writing and mathematics. In Resnick, L. B. (Ed.), *Knowing and learning: essays in honour of Robert Glaser* (pp. 453–494). Hillsdale, NJ: Erlbaum.

Craddock, D., O'Halloran, C., Borthwick, A., & McPherson, K. (2006). Interprofessional education in health and social care: fashion or informed practice? *Learning in Health and Social Care*, *5*(4), 220–242. doi:10.1111/j.1473-6861.2006.00135.x

Dewey, J. (1933). *How we think*. Boston, D.C.: Heath.

Dewey, J. (1938). *Experience and education*. New York: Simon & Schuster.

Elton. L. (1994). *Management of teaching and Learning: Towards change in universities*. London: CVCP.

Herrington, J., Reeves, T. C., Oliver, R., & Woo, Y. (2004). Designing authentic activities in web-based courses. *Journal of Computing in Higher Education*, *16*(1), 3–29. doi:10.1007/BF02960280

Hutchings, M. (2002). Enhancing learning with computer mediated communication: teacher intentions and student conceptions of online learning. In Rust, C. (Ed.), *Improving Student Learning Using Learning Technologies* (pp. 99–111). Oxford, UK: Oxford Centre for Staff Development.

Hutchings, M., Hadfield, M., Howarth, G., & Lewarne, S. (2007). Meeting the challenges of active learning in Web-based case studies for sustainable development. *Innovations in Education and Teaching International*, *44*(3), 331–343. doi:10.1080/14703290701486779

JISC. (2004). *Effective practice with e-learning*. Joint Information Systems Committee.

Jonassen, D. H., & Land, S. M. (Eds.). (2000). *Theoretical foundations of learning environments*. Mahwah, NJ: Lawrence Erlbaum.

Keen, S., & Todres, L. (2007). Strategies for disseminating qualitative research findings: three exemplars. *Forum Qualitative Sozial Forschung*, *8*(3). Retrieved from http://www.qualitative-research.net/fqs-texte/3-07/07-3-17-e.htm.

Kelly, A. (2004). *The curriculum. Theory and practice* (5th ed.). London: Sage.

Laurillard, D. (2002). *Rethinking university teaching: a framework for the effective use of educational technology* (2nd ed.). London: Routledge. doi:10.4324/9780203304846

Lave, J., & Wenger, E. (1991). *Situated learning: legitimate peripheral participation*. Cambridge, UK: Cambridge University Press.

Mayes, T., & de Freitas, S. (2007). Learning and e-learning: the role of theory. In Beetham, H., & Sharpe, R. (Eds.), *Rethinking pedagogy for a digital age* (pp. 13–25). London: Routledge.

Miers, M., Clarke, B., Pollard, C., Rickaby, C., Thomas, J., & Turtle, A. (2007). Online interprofessional learning: the student experience. *Journal of Interprofessional Care*, *21*(5), 529–542. doi:10.1080/13561820701585296

Mulholland, J., Mallik, M., Moran, P., Scammell, J., & Turnock, C. (2005). *Making practice-based learning work. An overview of the nature of the preparation of practice educators in five health care disciplines*. London: Higher Education Academy, Health Science and Practice Network.

Oliver, R. (1999). Exploring strategies for on-line teaching and learning. *Distance Education*, *20*(2), 240–254. doi:10.1080/0158791990200205

Pulman, A., Scammell, J., & Martin, M. (2008). Enabling interprofessional education: the role of technology to enhance learning. *Nurse Education Today*, *29*(2), 232–239. doi:10.1016/j.nedt.2008.08.012

Quinney, A. (2005). Placements online; student experiences of a website to support learning in practice settings. *Social Work Education*, *24*(4), 439–450. doi:10.1080/02615470500096951

Quinney, A. (2006). *Collaborative Social Work Practice*. Exeter, UK: Learning Matters.

Quinney, A., Hutchings, M., & Scammell, J. (2008). Student and staff experiences of using a virtual community, Wessex Bay, to support inter-professional learning: messages for collaborative practice. *Social Work Education, 27*(6), 658–664. doi:10.1080/02615470802201754

Rees, D., & Johnson, R. (2007). All together now? Staff views and experiences of a pre-qualifying interprofessional curriculum. *Journal of Interprofessional Care, 21*(5), 543–555. doi:10.1080/13561820701507878

Salmon, G. (2000). *E-moderating: the key to teaching and learning online*. London: Kogan Page.

Salmon, G. (2002). *E-tivities: the key to active online learning*. London: RoutledgeFalmer.

Savin-Baden, M. (2000). *Problem-based learning in higher education: untold stories*. Buckingham, UK: SRHE and Open University Press.

Scammell, J., Hutchings, M., & Quinney, A. (2008). *A virtual practice community for student learning and staff development in health and social work inter-professional education; changing practice through collaboration.* Bournemouth, UK: Bournemouth University, School of Health and Social Care. Retrieved December 30, 2008 from http://www.health.heacademy.ac.uk/publications/miniproject/scammell08.pdf

Sharpe, R., Benfield, G., Roberts, G., & Francis, R. (2006). *The undergraduate experience of blended e-learning: a review of UK literature and practice.* York, UK: Higher Education Academy. Retrieved August 20, 2009, from http://www.heacademy.ac.uk/assets/York/documents/ourwork/research/literature_reviews/blended_e-learning_full_review.pdf

TechDis. (2002). *The accessible curricula: good practice for all.* TechDis, LTSN Generic Centre and UWIC.

Todres, L. (2008). Being with that; the relevance of embodied understanding for practice. *Qualitative Health Research, 18*(11), 1566–1573. doi:10.1177/1049732308324249

Todres, L., Galvin, K., & Dahlberg, K. (2007). Lifeworld-led healthcare: revisiting a humanising philosophy that integrates emerging trends. *Medicine, Health Care, and Philosophy, 10*(1), 53–63. doi:10.1007/s11019-006-9012-8

Vygotsky, L. (1978). *Mind in society: the development of higher psychology processes*. Cambridge, MA: Harvard University Press.

Wenger, E. (1998). *Communities of practice: learning, meaning, and identity*. Cambridge, UK: Cambridge University Press.

Whittington, C. (2003). *Learning for collaborative practice with other professions and agencies*. London: Department of Health.

# Chapter 17
# Achieving Interprofessional Health Education Through the Use of E-Resources:
## Exploring the Experience of Two Different Undergraduate Professional Groups

**Karen Harrison**
*Coventry University, UK*

**Lorraine McFarland**
*Coventry University, UK*

## ABSTRACT

*This chapter presents a case-study of a large-scale programme of interprofessional health education delivered through a virtual learning environment (VLE) and reusable interprofessional learning objects. The project commenced in 2005/2006, and brought together 1,800 students from 14 different health and social care disciplines in an InterProfessional Learning Pathway (IPLP). Interprofessional e-learning pedagogy and resources were developed to permit asynchronous interactions between large numbers of students from different disciplines, operating in virtual learning sets based around a series of patient scenarios which had been developed with service-user involvement. After the IPLP had run for the first three years, the experience of two of the 14 participating professional groups, dietetics students and physiotherapy students, was explored in 2008 through the use of focus groups and questionnaires. Both student groups felt the IPLP learning experience had a positive impact on their awareness of other roles and interprofessional issues, and offered greater insight into the total clinical pathway of individual patients. Overall, Physiotherapy students found engagement in the IPLP a more useful learning experience than Dietetics students. A range of logistical issues, both positive and negative, were identified by both student groups in terms of the practical processes associated with the complex operation of the IPLP.*

DOI: 10.4018/978-1-61520-889-0.ch017

## INTRODUCTION

Interprofessional learning is a process through which students and practitioners have structured learning opportunities for shared learning, to enable them to acquire knowledge, skills and professional attitudes that they would not be able to acquire any other way (Funnell, 1995). However, there is evidence that indicates that whilst such interprofessional education provides a valuable learning experience for students (Barr, Koppel, Reeves, Hammick & Freeth, 2005), there are substantial logistical problems associated with the processes of seeking to bring large numbers of students from different professional courses together in a common learning environment (Thistlethwaite & Nisbet, 2007).

In order to seek to overcome these difficulties, in 2005 a novel programme of interprofessional education (IPE), delivered primarily through e-learning, was introduced in the Faculty of Health and Life Sciences, Coventry University. This was called the Interprofessional Learning Pathway (IPLP) and involved the engagement of 1,800 students from 14 different health and social care professions. The first students undertaking the whole three year IPLP completed their studies in 2008. The innovative nature of this programme demanded the development of several new e-pedagogies and e-resources; thus the experience was a novel one for learners and teachers alike (Clouder 2008).

Within this chapter, the impact of this educational intervention is explored with two of the fourteen participating student groups; BSc Physiotherapy and BSc Dietetics. The rationale for focusing in on specific groups was to consider how to improve their students' experience of the programme. This exploration considers both the learning that occurred and the practical issues, both positive and negative, that were associated with the new mode of delivery. There were several important differences between the physiotherapy and dietetics professional groups; these differences had an impact on how the IPLP learning experience was perceived by the two student groups.

## BACKGROUND

A healthcare professional from a single discipline would find it difficult to provide inclusive care that addresses all issues required to improve a patient's health. An interprofessional approach to treatment enables individuals from various disciplines to share their unique perspectives by integrating information and working as a team to enhance a patient's health status (Parsell & Bligh, 1998). Interprofessional learning aims to facilitate the communication and collaboration between healthcare professionals, which is required to address each aspect of a patient's rehabilitation (Lumague, Morgan, Mak, Hanna, Kwong, Cameron, Zener & Sinclair, 2006) by creating a holistic healthcare package through empowering a group of professionals to cope with issues that surpass the scope of any one profession (Barr, 2002).

Interprofessional learning was initially developed for improving the quality of patient care and safety. The importance of developing good interprofessional working relations has been highlighted as a consequence of incidents of patients put at risk or who came to harm (WHO, 1988). In the UK for example, the public inquiry into cardiac surgery at Bristol Royal Infirmary (Department of Health, 2001b) and the death of Victoria Climbié (Department of Health, 2003) both identified the need for radical reform of the education and training of professionals and the need to promote patient-focused collaborative working (Humphris & Hean, 2004). An integral part of the UK Department of Health's NHS Plan (DoH, 2000) and policy (DoH, 2001a, 2002, 2004; 2006, 2007, 2008; RCP, 2008) has been to improve interprofessional practice and enhance the quality of patient care and safety. Central to the policy is that pre-qualifying students 'learn with, from, and about one another to facilitate collaborative practice' (Barr, 2000, p. 3).

Hugh Barr suggests that interprofessional education was conceived to overcome ignorance and prejudice between health and social care professions. By understanding one another's craft, they would value what others brought to collaborative practice whilst setting aside negative stereotypes (Barr, 2002). Thus, collaborative competencies are strengthened to effect change or improvement in practice, thus the quality of patient care (Barr, 1998). In addition, interprofessional learning may enhance personal and professional confidence, promote mutual understanding between professions, and facilitate professional communication (Barr, 2000). Furthermore, interprofessional learning aims "to produce interdependent practitioners....able to know how and when to involve colleagues of other professions to achieve more concrete solutions" (Ohio State University, 1999, p. 2).

Systematic reviews of evaluations of interprofessional education (Barr, 2000; Freeth, Hammick, Koppel, Reeves, & Barr, 2002; Barr et al., 2005) find evidence suggesting that it can produce positive results. An evaluation of interprofessional education for nursing, occupational therapy, physiotherapy and social work students found that they enjoyed the experience and they had learnt about one another's professional roles (Banks & Janke, 1998). Research with team members based in general practice, showed that the team improved their collaborative working (Bond, 1997; Kristjanson, Dudgeon, Nelson, Henteleff & Balneaves, 1997). Falconer and colleagues (1993) report improved patient satisfaction and reduced length of patient stay in an interprofessional education evaluation. In addition, DePoy and colleagues (1997) reported improved attitudes towards other professional groups. However, Parsell and Bligh (1998) suggest it is not yet established whether learning together during basic training results in better 'working together' in practice.

## Contextualisating the (IPLP) Within Existing Provision

A series of previous cross-disciplinary initiatives at Faculty level meant that there was already a significant amount of 'shared teaching and learning' within the professional health and social care courses. This included the following two activities:

- Shared common core modules, in subject areas such as inequalities in health, research, evidence-based practice and management.
- Emphasis on interprofessional working on existing clinical placements, also reflected in the placement assessment requirements.

Although this represented a strategic move away from uniprofessional health care education, and to a large extent fulfilled Funnell's (1995) definition of interprofessional learning in providing 'structured learning opportunities for shared learning', the existing provision did not fulfil Barr's (2000) definition of enabling pre qualifying students to learn 'with, from and about one another to facilitate collaborative practice'.

The new (IPLP) was designed to address this perceived deficiency, and was to be delivered in addition to existing core shared modules (in health inequalities, research, evidence-based practice and management), and to existing enhanced interprofessional learning opportunities for students on clinical placements, (for instance on a stroke unit, paediatric placements, or specialised rehabilitation wards). These four key elements are represented schematically for the BSc Physiotherapy course in Figure 1.

The IPLP was launched at the beginning of year one, as part of the large 'Welfare, Health and Inequalities' module. Pathway elements were also linked to specific modules in years one, two and three of the course. Performance within the IPLP was formally assessed, and students could not progress from one year to another, or graduate

*Figure 1. Schematic representation of the place of the IPLP in the BSc Physiotherapy course*

<table>
<tr><td rowspan="3">Year One</td><td colspan="2">Interprofessional Learning Pathway</td></tr>
<tr><td colspan="2">Common core modules: Inequalities in Health, Research Methods</td></tr>
<tr><td colspan="2">Uni-professional modules</td></tr>
<tr><td rowspan="3">Year Two</td><td colspan="2">Interprofessional Learning Pathway</td></tr>
<tr><td>Common core module: Evidence Based Practice</td><td rowspan="2">Practice placements: Uni-professional and interprofessional working</td></tr>
<tr><td>Uni-professional modules</td></tr>
<tr><td rowspan="3">Year Three</td><td colspan="2">Interprofessional Learning Pathway</td></tr>
<tr><td rowspan="2">Practice placements: Uni-professional and interprofessional working</td><td>Common core module: Management in Health Care</td></tr>
<tr><td>Uni-professional modules</td></tr>
</table>

from individual programmes of study unless this element was passed.

The project commenced in the academic year 2005/2006, bringing together 1,800 students from 14 different disciplines in health and social care, utilising a Virtual learning Environment (VLE), initially WebCT, more recently Moodle. The following professional groups participated:

- Adult Nursing
- Child Nursing
- Clinical Psychology
- Dietetics
- Learning Disability Nursing
- Medicine
- Mental Health Nursing
- Midwifery
- Occupational Therapy
- Operating Department Practice
- Paramedicine
- Physiotherapy
- Social work
- Youth work

Before the IPLP was run for the first time, it was formalised through a large, multi-professional review, as the new learning pathway was part of the approved core programme for all recognised health and social care courses involved. New learning materials were designed, with associated novel assessment techniques. A specific, credited, training programme 'E-Spire', was developed totrain lecturers involved in delivering and assessing this new element. Development of these materials was co-ordinated and led through The Centre for Interprofessional e-learning (CIPeL), a collaborative Centre for Excellence in Teaching and Learning (CETL) between Coventry and Sheffield Hallam Universities in the United Kingdom, funded by The Higher Education Funding Council for England (HEFCE). A further collaboration involved the participation of medical students from nearby Warwick University Medical School.

Student learning was focused primarily on a series of 'patient-journeys': detailed accounts released in sequence to students over a series of weeks, as the individual patient and their family proceed along a typical clinical pathway, experiencing interventions from several healthcare professionals. Created utilising material based on 'real stories', the journeys involved substantial service-user engagement in their development. Students worked in action-learning sets of approximately 20 participants, drawn from a range of professional backgrounds which were facilitated by an E-spire trained lecturer. The intended learning outcomes were assessed using a range of activities, including a requirement to engage in making appropriate postings, and involvement

*Figure 2. Focus group discussion points*

**Focus: An exploration of the experience of participating in the inter-professional learning pathway as part of the BSc Dietetics/BSc Physiotherapy programme**

1. What is your key impression of the IPLP?
2. Did you find the on-line approach to IPLP effective (especially in year one)?
3. What do you think has been the main impact of participating in the IPLP as part of your course?
4. Did participating in the IPLP in the University make any difference to the way in which you worked on your clinical placements?
5. Could you comment on the way in which participating in the IPLP impacted upon the way that you see your own professional role, and the roles of other members of the health and social care team?
6. What has been your experience of inter-professional learning in the clinical environment?
7. What factors have supported your inter-professional learning experience?
8. What factors have hindered your inter-professional learning experience?
9. How do you feel that participating in both the University and the clinically based components of inter-professional education will impact upon your practice as a dietitian/physiotherapist when qualified?
10. How might the IPLP learning experience be improved?
11. On reflection, what is the single best and worst feature of the IPLP on your BSc Dietetics/BSc Physiotherapy course?

in group discussions responding to the release of online posting of information.

## Evaluating the Impact of the IPLP

In 2008, after the IPLP had run for the first three years, the experience of two of the 14 participating professional groups, dietetics students and physiotherapy students was explored. Two 'matched' research projects were commissioned by CIPeL to evaluate the impact of the IPLP on those students. Both projects involved collecting data through identical questionnaires and focus groups questions to permit a broad comparison of findings.

The two student groups differed in two ways. Firstly, physiotherapy was the larger group, the intake being 162 students in September 2005, compared with the dietetics intake of 45. Secondly, both professions operate differently on a daily basis, physiotherapists interacting in a wide range of different clinical situations with large numbers of different professional groups at many points along a given clinical pathway, a work pattern shared with other similarly large groups such as nursing, medicine and occupational therapy. In contrast, it might be argued that dieticians engage in less varied clinical interactions, often along a specific patient pathway which is typically very specialised, presenting fewer opportunities for multi-professional contact. This specialist mode of operation is typical of other professional groups, including speech and language therapists, midwives, paramedics, operating department practitioners and clinical psychologists.

The overall aim was to develop a clear picture of students' perceptions and experiences of participating in the IPLP, and more fundamentally, to consider the impact of participation on their learning, and their mode of working on the clinical placements that are integral to all health-care professional courses. Each project had complementary quantitative and qualitative elements that essentially covered the same areas of exploration. Data were collected using both focus groups, and a survey for each professional group. An open request for focus group participants was put to each cohort respectively. Eight physiotherapy students and six dietetics students responded and participated in two separate focus groups, one for each discipline. The discussion points listed in Figure 2, were considered within the focus groups.

The questions were designed to thoroughly probe the students' experience, and ascertain the perceived value of the IPLP as part of their larger programme. The impact of participation in the IPLP on later operation in the clinical placement element of the courses was explored particularly carefully, in order to address the observation of Parsell and Bligh (1998); that it was not yet established whether learning together during basic training results in better working together in practice.

The survey was administered conducted with the two cohorts at the end of a lead lecture. Questionnaires were completed by 97 out of 162 physiotherapy students and 22 out of 24 dietetics students. The questionnaire was designed to collect information on the following key areas:

- Satisfaction level of IPL modules (content, access, assessment, timing).
- Overall IPL experience within the University (enjoyment).
- The advantages and disadvantages of IPL learning modules, using open-ended questions (supported, hindered, suggestions for improvement).
- The importance, relevance and significance of IPL in the University and on clinical placement.
- The impact of IPL in terms of understanding, support and awareness of other professions.
- Demographic information of students: gender and age group.

There were a total of 20 rated items on the survey tool, and 5 sections for recording responses to open questions, which were subject to qualitative analysis. A substantial amount of information was collected from the survey (both quantitative and qualitative data) and the focus groups. The key messages and themes which emerge from the data overall are presented under the following headings:

- Impact of participation in the IPLP on learning.
- Practical issues related to the operation of the IPLP delivered through e-learning.
- A comparison of methodologies, and a comparison of groups.

## Impact of Participation in the IPLP on Learning

Students from both professions identified a range of benefits of participating in the programme which included:

- A better understanding of other professions and what they do.
- An appreciation of the value of team-working.
- Insight into the existence of a 'bigger picture', and the importance of thinking about it.
- Better insight into the way professionals need to work together in a given patient pathway.
- Insight into roles of others facilitating appropriate referrals.
- Pride in own profession, feeling valued.
- Confidence to initiate interprofessional interactions on later clinical placements.
- Increased learning about clinical conditions from the perspective of other healthcare professionals.
- Better preparation to actively engage with other professions in practice, and work supportively together.
- Ability to explore issues related to roles in a 'safe', non-clinical environment.
- Learnt to value the roles of others.

Overall, students describe a powerful learning experience that fulfils the definitions of both Barr (2002) and Funnell (1995). In addition, the stories told by students provide evidence that the experience of participating in the IPLP had a significant positive impact on interprofessional working in practice; an issue previously identified as unclear (Parsell & Bligh, 1998).

The learning outcomes in the IPLP required active participation of a wide range of health and social care professions, utilising e-learning resources that were strongly patient focussed, and were delivered in a phased manner, permitting the patient story to gradually unfold over time. Students claimed a greater understanding and knowledge of other professions and groups. At the same time, and somewhat paradoxically, their own professional identities were strengthened, through a dual process of engagement in interprofessional teamwork and promotion of their own profession, consistent with earlier findings (O'Neill, Wyness, McKinnon & Granger, 2000). For instance:

*[The IPLP has] given us a better understanding of other professionals.... It has also given us a chance to promote what we do... You come to appreciate why other professions do what they do.*

Poor perceptions of other professional groups are frequently linked to low levels of contact between these individual professional groups (Kennard, 2002). Communication between groups is an essential pre-requisite to foster understanding and appreciation of the roles of others. Freeth et al. (1998) described how participation in interprofessional education could reduce communication breakdown, increase morale and efficiency and avoid 'unhelpful protectionism'. Students described how participation in the IPLP promoted team working, removed barriers and reduced interprofessional rivalry, through improved perceptions of other professions roles and skills:

*IPLP breaks down some barriers...It promotes team working...I was a bit more receptive to the occupational therapists ideas and to what the nurses had to say, as opposed to isolating myself.*

Interprofessional learning may engender respect and understanding of other associated professions (McCroskey & Robertson, 1999). Students believed their professional practice would benefit from greater respect and appreciation for others. When on clinical placements, students described how senior clinicians, in both their own and other fields, appreciated their interprofessional learning experience:

*I think you'll have more appreciation and respect of somebody else and you have more respect for other professions.*

*People who've been qualified for a while, they're really quite keen on the fact that we have done this.*

Engagement in the IPLP meant that students gained insight into the bigger clinical picture, and hence into their role within the total patient scenario. This improved understanding of the roles of other professions and facilitated more effective patient referral.

*We'd say, 'this is what we would do' and wait for our part to come about... In that way you can see where your role fits in.*

*It helps you in referring a patient onto an appropriate person realising when you're not the right person for the job...*

Parsell and Bligh (1998) suggest that in order for interprofessional learning to be effective, professionals in training must feel themselves to be of equal standing with other professional groups and Barr (2000) argues that effective interprofessional learning may enhance personal and professional confidence. In fact, students felt that participation in the IPLP helped them to feel more confident in talking to other professionals, and that it facilitated communication and hence improved working relationships:

*it made me feel more confident with talking to other professionals because I knew what they were able to do.*

Participation in the IPLP elevated students' self-esteem. Having the confidence to say, 'this is what I would do' helped them to 'feel like the expert, valued, and proud' and gave them an authority they might not have otherwise experienced:

*Well I felt like the expert in the group... I valued my own profession more, thinking 'now I know what I'm talking about and I'm about to give you advice on what we do.*

Students gained the confidence to initiate interprofessional interactions and asked to shadow professionals in areas of specific interest, thus directing their own learning path:

*Rather than me being invited I just asked and said 'can I shadow you'. I think it may have been partly due to the IPLP experience.*

Students also emphasised how the IPLP contributed to the human aspects of their role; friendship, an interactive team, a nice atmosphere, the breaking down of barriers, and helping each other:

*It sets you up for becoming friends with everybody else... it's really important to work as a nice interactive team of friends.*

The existing body of research in this area suggests that interprofessional learning leads to increased knowledge of other disciplines, and generates opportunities for consultation with members of other professions (Geller, Rhyne, Hansbarger, Borrego, VanLeit & Scaletti, 2002). The outcomes of our study reflect previous findings, including the facilitation of communication between professions through mutual understanding and respect (Geissler, Freeth, Hammick, Koppel, Reeves & Barr, 2002) and increased personal and professional confidence (Sinclair, 2004; Parsell & Bligh, 1998). Students felt valued, appreciated their 'expert' status and increased pride in their role. Interprofessional learning encourages professionals to engage in reflective practice by examining their performance to make improvements (Geissler et al., 2002). This study showed that the Physiotherapy students recognised their dependency on the interprofessional team in order to enhance care and maximise patient outcomes. They recognised the importance of other roles and actively initiated situations where they could work with professionals from other areas.

## Practical Issues

When the IPLP was launched in September 2005, it was ambitious in both concept and implementation. Firstly, the large scale operation involved 1,800 students, 14 professional groups and two separate University sites (Coventry and Warwick). Additionally, it incorporated the development of new e-pedagogies, including the design and delivery of the 'E-spire' training course for facilitators, and the design and development of sophisticated re-usable learning objects, in the form of a series of 'patient-journeys' utilising service-user and clinician expertise to ensure authenticity.

As the programme moved into the second and third year of operation, new learning objects were generated, and an increasing number of e-learning facilitators undertook E-spire training. This course was subject to educational evaluation and external examiner scrutiny. As a result, adjustments were made to the mode of delivery. The students who took part in this evaluation described several problems, which may be expected in the delivery of any innovation running through a first cycle. Many of the issues raised were resolved as the delivery of the programme settled into an established routine in the first, second and third (physiotherapy only) years, and those outstanding are discussed below.

## Satisfaction Levels Associated with Participation in the IPLP

The combined results of both surveys showed that overall, students expressed high levels of satisfaction with access to the learning experience, moderate satisfaction with the timing and content, but only a third expressed satisfaction with the assessment of the module. Within these overall figures there was a marked difference between responses of the physiotherapy and dietetics students.

Generally, physiotherapy students were satisfied with their IPL experience. They recorded very high satisfaction with access to the IPE learning experience in all three years, although they were moderately satisfied with the IPL content, with a dip in the third year, but they expressed dissatisfaction towards the assessment of the IPL in all three years. In contrast, Dietetics students showed high dissatisfaction in their first year towards the content of the IPL module, the access, the assessment and the timing of the IPL learning experience. These proportions improved in their second year with respect to content and access but remained the same for assessment, and deteriorated for timing.

The content of the IPLP was viewed much more positively by Physiotherapy students than by Dietetics students, although by year three Physiotherapy students were less satisfied with the content. The focus group revealed that the dip in satisfaction with content in year three may have been due to students' feeling it had become predictable in the items presented. It seems they need to have more intellectual challenge utilising experience they had by then gained from working in clinical environments.

An important point to note is that both groups of students participated in the same IPLP, and hence were presented with the same learning materials, at the same point in time, with the same access to facilities, and undertaking the same assessment. However, the value of what was presented to them was subject to differential interpretation by the two groups, with Physiotherapists tending to value the experience and find it useful, but Dieticians defining the situation in a significantly less positive way.

## Benefits and Issues with Online IPL

The primary benefit offered by delivering a module using a VLE is pragmatic, it being impractical for such a large and diverse group to engage with one another in any other way. Thistlethwaite and Nisbet (2007) identified the key barriers to the success of interprofessional education; timetabling, rooms and resistance. Clouder (2008) describes how e-learning can address these practical problems, but identifies that issues such as 'differences in professional requirements and entrenched norms' pose long-term challenges. It is precisely these professional and social barriers that interprofessional learning, delivered in any mode, is designed to ameliorate.

At the beginning of the IPLP process, students described feelings of confusion as they struggled to determine how the new system operated, and what was required of them. The new development created operational problems, such as accessing the VLE whilst on clinical placement, clashes with assessment deadlines, and lack of timely posting by some students. However, these were generally quite quickly resolved, and mostly viewed as issues that might be expected with the initiation of any new system.

For both student groups these problems were largely resolved by the second year by which time they felt comfortable with the new system and valued the flexibility offered by a VLE that offered both space and time-independent learning. A dietetics student commented:

*I quite like the online because it's much more flexible with everyone on different courses having different workloads, it's not reliant on you meeting up at a particular time*

A similar comment arose in the physiotherapy focus group:

*The idea of online is quite good because I imagine it would be quite difficult to actually get that number of people together to talk about something, and if you were busy you didn't have a set time. You have to go on-line and contribute, if you were awake at midnight or whatever you could do it then, ... you could fit it around other things.*

However, the benefits of asynchronous learning were to some extent offset by learning-set participants not posting their profession-specific content until relatively late in the process:

*....towards the end where people haven't been organised and well, not had the time, you'd just have an influx of responses and you find that you thought 'well, if I'd had these responses before I'd probably have done a lot better in responding to them'....*

Additionally, although the convenience of an asynchronous posting facility was appreciated, students missed the face-to-face human element in online interactions. This was illustrated from the Physiotherapy focus group:

*You don't get how they're saying it do you across the Internet, so there's not non-verbal communication,.... it's just words on a page.*

Student Dietitians described similar feelings. However, they indicated that this was partially ameliorated where students participating in an action-learning set discussion forum were given an opportunity to meet prior to interacting on-line:

*I found it better because you knew who you were talking to because you had actually met them before, but I found that it wasn't as interesting as sitting in a room with a group of people where you could all have a discussion with each other, and it didn't have the same flow to it...*

As previously described, offering online IPL was essential; it was not possible to deliver interactive learning experiences to very large numbers of students from a wide variety of health and social care disciplines any other way. Recent studies have indicated that this mode of delivery is both popular and effective with adult learners in the health care field (Ellis, 2007; Scammell, Hurchings, Quinney, Hean, Dumbrell & Pulman, 2008; Walsh, 2007). These findings are generally supported in the detailed evidence provided from this study. On balance, students viewed this mode of delivery very favourably, feeling that the benefits substantially outweighed the costs.

## DISCUSSION

Each project had both a quantitative and qualitative element, these two components essentially covering the same areas for exploration. In most areas there was a strong consensus between evaluation findings ascertained using the two different approaches, but overall the findings from the focus groups were more positive towards IPL than those from the surveys.

This may be illustrated, for example, by Physiotherapy students' responses to the survey questions on the impact of IPE through e-learning, their understanding of their role, and their professional identity, Only a third (34) of students felt this was positively affected by engagement in IPE. In contrast, when the same question was asked in the focus group, participants gave numerous examples of the positive benefits of participation. A typical extract reads as follows:

*you've got more understanding of other peoples' roles means that you can talk to them more easily, and I think the people that I worked with anyway appreciated the fact that we'd taken the time to actually understand what it was that they did. So you got on much better with them from the start.*

It may be conjectured that the different nature of the two research tools is responsible for this apparent difference in emphasis. Survey responses are made fairly quickly on an individual basis, whereas focus group responses are generally more considered and discursive, and involve active listening and participation with other group members. However, it is also possible that as interprofessional education is generally accepted within the University as being of value, underlying social pressures would mean that students in a group discussion situation would not like to challenge the worth of such a well established practice, any more than they would wish to challenge evidence based-practice, or reflective practice.

Another key trend was that Physiotherapy students' experiences of IPL were overall more positively reported than those of Dietetics students in both questionnaire and focus groups. This could be due to the different nature of the two professions, Physiotherapy being a larger professional group having engagement with a wider range of patient pathways, whereas Dieticians are a smaller profession, with limited and specialist engagement with other professional groups. This was articulated by a participant in the dietetics focus group:

*we were one of the ones that contributed slightly less because there were quite a few parts of the scenario that we wouldn't have a great big role and say 'this is what we would do' and sort of waiting for our part to come about.*

However, it is important to note that these students did identify significant benefits in learning about the roles of others, and how they themselves fitted into that bigger picture. A dietetics student encapsulated this:

*oh I'm not just a dietitian, I fit into a bigger picture, and then there was an insight that the bigger picture existed, and you had all to be thinking about it.*

It has been suggested that some professional roles are so specific in their application that interprofessional working is less of an issue. Future research could examine how such groups can be made to feel fully included within the IPL process as well as ways that teaching, learning and assessment strategies might enhance their learning experience.

## SUMMARY AND RECOMMENDATIONS

This case study explores a large scale, complex and novel educational enterprise, the purpose of which was to permit large numbers of students from diverse professional backgrounds to work together around patient focussed scenarios. Facilitated online action learning sets provided an opportunity for students 'to learn with, from and about one another to facilitate collaborative practice' (Barr, 2000). It would be impossible for such large numbers of students to have undertaken such learning in any other way.

The evaluation indicated that the new system achieved the proposed educational objectives. Elements of the system were trialled in advance, but this was the first time the complete new system had been adopted. ed. Whenever a large-scale strategic change is introduced, there are likely to be initial operational problems that cannot be anticipated. Many of the issues raised in the evaluation were addressed through routine module evaluation and quality management.

However, there are areas where a benefit; such as the opportunity for flexible, asynchronous posting, has an associated cost; in this case the inability to maintain a normal dialogue or discussion between participants. Our evaluation suggests this can be improved if students meet face-to-face before engaging in online discussions, interestingly, similar conclusions have been reached regarding unprofessional online discussions (Bromage, 2004).

An unexpected finding was the difference in perception of participating in the IPLP of the two professional groups. This finding has not been highlighted in previous studies, probably because there has been no direct comparative analysis between participating groups. Whilst both groups found value in the experience, Physiotherapists demonstrated a substantially more positive view of IPLP compared to Dietetics students, as manifested in both quantitative and qualitative evidence. It may be possible to develop strategies to partially ameliorate the dietetics students' less positive perceptions, but the effect could be, at least in part, the inevitable consequence of the small scale and specialist nature of this profession. If so, these feelings might also be expected to be shared by other small and specialist groups, such as operating department practice or speech and language therapy.

The evaluation was undertaken at the end of the first full cycle of operation of the IPLP and so this also simultaneously represented the beginning of a new era of full integration of interprofessional education into all of the healthcare courses run within the Faculty of Health and Life Sciences. This new era demanded both that students learn to work together around patient-based scenarios, and that academics from different professional disciplines work together to develop and deliver this novel educational provision; a powerful learning experience for both staff and students.

*Coming together is a beginning; keeping together is progress; working together is success (Henry Ford).*

## ACKNOWLEDGMENT

A five-person team undertook the educational evaluation described within this chapter, two of whom were the authors. Grateful acknowledgement is made to Karen Davies, the questionnaire designer, Mollie Gilchrist who undertook the statistical analysis of both projects, and to Dr Jane Earland and Karen Davies who produced a final report for the BSc Dietetics and BSc Physiotherapy provision respectively.

The projects were funded through the Centre for Interprofessional e-learning (CIPeL), a collaborative Centre for Excellence in Teaching and Learning (CETL) between Coventry and Sheffield Hallam Universities in the United Kingdom funded by the Higher Education Funding Centre for England (HEFCE).

## REFERENCES

Banks, S., & Janke, K. (1998). Developing and implementing interprofessional learning in a Faculty of Health Professions. *Journal of Allied Health*, *27*, 132–136.

Barr, H. (1998). Competent to collaborate: towards a competency-based model for interprofessional education. *Journal of Interprofessional Care*, *12*(2). doi:10.3109/13561829809014104

Barr, H. (2000). *Interprofessional Education: 1997-2000. A Review* (5:03, p. 9). United Kingdom Central Council of Nursing, Midwifery and Health Visiting.

Barr, H. (2002). *Interprofessional education: Today, Yesterday and Tomorrow. A review Commissioned by HEA*. The UK Centre for the Advancement of Interprofessional Education.

Barr, H., Koppel, I., Reeves, S., Hammick, M., & Freeth, D. (2005). *Effective Interprofessional Education: Argument, Assumption and Evidence*. Oxford, UK: Blackwell Publishing. doi:10.1002/9780470776445

Bond, M. (1997). A learning team in the making. *Journal of Interprofessional Care, 11*, 89–98. doi:10.3109/13561829709040247

Bromage, A. (2004). Atavistic Avatars: Ontology, Education and 'Virtual Worlds. In Seth-Preston, D. (Ed.), *Virtual Learning and Higher Education* (pp. 133–150). Amsterdam: Rodopi.

Clouder, D. L. (2008). Technology enhanced learning – conquering barriers to interprofessional education. *The Clinical Teacher, 5*, 198–202. doi:10.1111/j.1743-498X.2008.00243.x

Department of Health. (2000). *A Health Service of all the Talents: report on the review of workforce planning*. London: Department of Health.

Department of Health. (2001a). *Working Together Learning Together. A Framework for Lifelong Learning for the NHS*. London: Department of Health.

Department of Health. (2001b). *Learning from Bristol: The Report of the Public Inquiry into children's heart surgery at the Bristol Royal Infirmary 1982-1995*. London: Department of Health.

Department of Health. (2002). *Shifting the Balance of the Balance of Power: Securing Delivery*. London: Department of Health.

Department of Health. (2003). *Audit of Victoria Climbie Inquiry recommendations. Social Services Inspectorate*. London: Department of Health.

Department of Health. (2004). *Delivering HR in the NHS*. London: Department of Health.

Department of Health. (2006). *Our Health, Our Care, Our Say: A New Direction for Community Services*. London: Department of Health.

Department of Health. (2007). *National Stroke Strategy*. London: Department of Health.

Department of Health. (2008). *High Quality Care for all; NHS Next Stage Review Final Report*. London: The Stationary Office.

DePoy, E., Wood, C., & Miller, M. (1997). Educating rural allied health professionals: an interdisciplinary effort. *Journal of Allied Health, 26*(3), 127–132.

Ellis, R. A., Goodyear, P., O'Hara, A., & Prosser, M. (2007). The university student experience of face-to-face and online discussions: coherence, reflection and meaning. *ALT-J. Research in Learning Technology, 15*(1), 83–97.

Falconer, J. A., Roth, E. J., Sutin, J. A., Strasser, D. C., & Chang, R. W. (1993). The critical path method in stroke rehabilitation: lessons from an experiment in cost containment and outcome improvement. *Quality Review Bulletin, 19*(1), 8–16.

Freeth, D., Hammick, M., Koppel, T., Reeves, S., & Barr, H. (2002). A Critical Review of Evaluations of Interprofessional Education. London: Learning and Teaching Support Network: Health Sciences and Practice.

Funnell, P. (1995). Exploring the value of interprofessional shared learning. In Soothill, K., Mackay, L., & Webb, C. (Eds.), *Interprofessional relations in healthcare* (pp. 163–171). London: Edward Arnold.

Geissler, C., Freeth, D., Hammick, M., Koppel, I., Reeves, S., & Barr, H. (2002). *A critical review of evaluations of interprofessional education*. Retrieved August 28, 2008, from http://www.health.ltsn.ac.uk/publications/occasionalpaper/occasionalpaper02.pdf

Geller, Z., Rhyne, R., Hansbarger, L., Borrego, M., VanLeit, B., & Scaletti, J. (2002). Interdisciplinary health professional education in rural New Mexico: A 10 year experience. *Learning in Health and Social Care, 1*, 33–46. doi:10.1046/j.1473-6861.2002.00003.x

Humphris, D., & Hean, S. (2004). Educating the future workforce: building the evidence about interprofessional learning. *Journal of Health Services Research & Policy*, *9*(1), 24–27. doi:10.1258/135581904322724103

Kennard, J. (2002). Illuminating the relationship between shared learning and the workplace. *Medical Teacher*, *24*, 379–384. doi:10.1080/01421590220145752

Kristjanson, L., Dudgeon, D., Nelson, F., Henteleff, P., & Balneaves, L. (1997). Evaluation of an interdisciplinary training program in palliative care: addressing the needs of rural and northern communities. *Journal of Palliative Care*, *13*(3), 5–12.

Lumague, M., Morgan, A., Mak, D., Hanna, M., Kwong, J., & Cameron, C. (2006). Interprofessional education: The student perspective. *Journal of Interprofessional Care*, *20*(3), 246–253. doi:10.1080/13561820600717891

McCroskey, J., & Robertson, P. J. (1999). Challenges and Benefits of Interprofessional Education: Evaluation of the Inter-Professional Initiative at the University of Southern California. *Teacher Education Quarterly*, 69–87.

O'Neill, B., Wyness, A., McKinnon, S., & Granger, P. (2000). *Partnership, Collaboration and Course Design: An emerging model of interprofessional education*. Retrieved November 8, 2008, from http://www.cstudies.ubc.ca/facdev/services/registry/pcacdaemoie.html

Ohio State University. (1999). *Assumptions for Interprofessional Education and Practice*. Ohio: The Interprofessional Commission of Ohio. Retrieved August 30, 2008, from http://www.osu.edu/ico/assumptn.html

Parsell, G., & Bligh, J. (1998). Interprofessional learning. *Postgraduate Medical Journal*, *74*(868), 89–95. doi:10.1136/pgmj.74.868.89

Parsell, G., Spalding, R., & Bligh, J. (1998). Shared goals, shared learning: evaluations of a multiprofessional course for undergraduate students. *Medical Education*, *32*, 304–311. doi:10.1046/j.1365-2923.1998.00213.x

Royal College of Physicians. (2008). *National Clinical Guidelines for Stroke* (3rd ed.). London: RCP.

Scammell, J., Hurchings, M., Quinney, A., Hean, S., Dumbrell, A., & Pulman, A. (2008). Summary - *A virtual practice community for student learning and staff development in health and social work Inter-professional education; changing practice through collaboration*. School of Health and Social Care, Bournemouth University/Higher Education Academy Health Science and Practice. Retreived October 12, 2009, from http://www.health.heacademy.ac.uk/publications/miniproject/summaryscammell08.pdf

Sinclair, L. (2004). Interprofessional education (IPE): New knowledge in need of translation? *Canadian Physiotherapy Association: Research News and Views Newsletter*, Spring, 11-13.

Thistlethwaite, J., & Nesbit, G. (2007). Interprofessional Education: what's the point and where we're at. *The Clinical Teacher*, *4*, 67–72. doi:10.1111/j.1743-498X.2007.00148.x

Walsh, K. (2007). Interprofessional Education Online. The BMJ experience. *Journal of Interprofessional Care*, *21*(6), 691–693. doi:10.1080/13561820701436912

World Health Organization (WHO). (1988). *Learning Together to Work Together for Health.* Technical Report Series 769. Geneva: World Health Organization.

# Section 4
# Creating and Managing Learning Resources

# Chapter 18

# A 'Video Trigger,' but no Silver Bullet:
## An Actor-Network Analysis of an E-Learning Tool in Health Science Education

**Marit Fougner**
*Oslo University College, Norway*

**Laurence Habib**
*Oslo University College, Norway*

## ABSTRACT

*This chapter analyses the design and implementation of a video trigger as a pedagogical tool to facilitate interprofessional understanding in several professional Bachelor courses at a faculty of Health Sciences. It uses actor-network theory as an analytical tool to uncover the main human and non-human actants that are involved in the processes of planning, designing, implementing and using the video trigger. It also attempts to bring to light the various types of interconnections that exist between those actants. The data that forms the basis of the analysis is qualitative, and includes meeting minutes and field notes. The analysis reveals the existence of a dichotomous relationship between technology and pedagogy, which is at the source of major challenges in the design and use of a video trigger as a pedagogical tool.*

## INTRODUCTION

This chapter reports on the issues associated with developing video triggers for case-based learning for use on an interprofessional module, in particular the difficult relationship between technological concerns and pedagogic concerns. It is based on a study of one of three compulsory interprofessional modules that are common to seven different bachelor programmes at the Faculty of Health Sciences in a Scandinavian institution of Higher Education. The study investigates the design and subsequent use of video triggers as a pedagogical tool aimed to further interprofessional understanding, using actor network theory as an analytical lens. The chapter provides an outline of the rationale behind using video triggers, a description of the processes of design and implementation of those triggers,

DOI: 10.4018/978-1-61520-889-0.ch018

and an overview of the main pitfalls encountered. It then presents an evaluation those processes in the light of the purposes intended by the Faculty and suggests some insights that may be useful to others who seek to develop similar initiatives.

The module in question, "Health clients and health workers: reflections on practice", was a compulsory module for all the students at the Faculty of Health Science, and was meant to foster interprofessional thinking and practice. In particular, it was to establish long-lasting bridges between the various professions that were represented at the Faculty of Health Science. To that end, it aimed to provide the students with insights into other health professions in order to make interprofessional cooperation a natural part of professional practice after graduation. The module offered the students a common platform for discussion around relevant themes, in particular issues related to communication and ethical dilemmas that arise within the realm of the interaction between health workers and their clients. About 450 students from 8 different bachelor programmes (medical laboratory science, dental technology, occupational therapy, pharmacy, physiotherapy, Mensendieck physiotherapy, prosthetics and orthotics, and radiography) took the module each year. Most of them were in their second year of a three-year bachelor programme, with a small number of third-year students in their ranks. All the students took on-campus programmes and all were based on the same physical site.

Case-based learning has a century-long tradition as a pedagogical approach for professional training (Herreid, 2006; Barnes, Christensen & Hansen, 1994). In recent years, case-based learning has been used to support interprofessional education (Lindqvist, Duncan, Shepstone, Watts, & Pearce, 2005). Case-based instruction aims to provide students with a way to bridge the gap between theory and practice (Heitzmann, 2008; Mayo, 2004; Wright, 1996). Cases are also used in order to allow students to grasp the complexity and the ill-structured nature of real-life situations (Choi & Lee, 2009; Silverman, Welty & Lyon, 1996; Boyce, 1992; Pomeroy & Philip, 1994), as well as the "contextual, local and situated" character of professional practice (Shulman, 1992, p. 28).

Although the case methodology has traditionally made use of written text as its main medium, newer technological advances have allowed for the development of so-called video cases or video triggers. Video cases have been described as providing rich and authentic insights into aspects of professional practice that are of direct relevance to the students of professional courses (Yadav & Koehler, 2007). They are also considered appropriate tools to promote the students' reflection and dialogue (Rowley & Hart, 1996). Much of the literature on video cases focuses on their use as a pedagogical tool. However, some works have also attempted to shed light on the very process of creating video cases (Kurz, Llama & Savenye, 2005).

Video cases are part of a wider array of online pedagogical tools that fall under a number of umbrella designations such as "e-learning", "online learning", "technology-supported learning" or "web-based learning". E-learning tools have been used to support interprofessional education. Typically, e-learning has been chosen in interprofessional education programmes to provide pedagogical flexibility, in particular in terms of time and space (Juntunen & Heikkinen, 2004). Also, the use e-learning platforms such as virtual learning environments can facilitate the inter-institutional delivery of courses (Skorga, 2002), and collaborative learning across geographically dispersed sites (Quinney, Hutchings & Scammell, 2008), both of which are often crucial elements in interprofessional education. In addition, online environments allowing for asynchronous communication have been described as bringing greater fluidity to the process of enquiry-based learning (Hughes, Ventura & Dando, 2004), which plays a significant part in many interprofessional courses.

## BACKGROUND OF THE STUDY

The methodological approach for this study is interpretative. The data gathered is qualitative and takes the form of meeting minutes and field notes based on observations made on a daily basis by one of the authors of this article throughout the study. The rationale behind using meeting minutes as an empirical source is that meetings were the primary arena of discussion and decision making for the project. Observations performed during the meetings attended were recorded systematically in a field diary. However, observations of the activities carried out by the lecturers involved in the project could not be comprehensive, as those activities were performed in various places and often simultaneously and the data gathering could only be carried out by one person. The researcher in charge of gathering the data proposed that team members would keep their own reflective diaries to obtain a more comprehensive overview of the various activities taking place throughout the project period. Nevertheless, none of the project members did keep a diary, presumably because they considered their involvement with the project as primarily focused on creating and quality assuring a video trigger, and may have regarded diary keeping as a secondary activity that was not worthy of the time investment it would involve.

### An ANT-Based Approach to Data Analysis

In this chapter, we use actor-network theory (ANT) as an analytical lens to describe and evaluate the whole "life cycle" of the video triggers. We start with the very idea of how the video triggers were originally formed, go on to investigate how they were designed and implemented, and explore how students used and evaluated their suitability as educational support.

ANT started as a branch of the sociology of science, and was pioneered by John Law (1987), Michel Callon (1986) and Bruno Latour (1987). ANT has proven useful in shedding light on situations where social and technical issues are intermingled, addressing their complexity without oversimplifying them or attempting to separate the technical from the social (Tatnall & Davey, 2005; Macome, 2008; Thompson and Cupples, 2008). Instead of portraying technology and society as dichotomic entities, ANT offers a conceptual framework that highlights the existence of heterogeneous relationships between humans and non-humans.

Since its early days in the 1980s, ANT has developed to become a rather complex body of knowledge, which is often difficult of access, and whereby different authors emphasise different concepts and areas of application. Some concepts are, however, recurrent throughout the ANT literature – in particular the concepts of actor/actant, network and enrolment. In this chapter, we also choose to make use of the concepts of inscription, translation and blackboxing that figure at the core of many of the major works in the ANT literature.

*Actors/actants* are defined as entities that act or make other actors act. One of the main characteristics of ANT (and one of its main controversial points) is that it considers both humans and non-humans as actors (or actants). Non-human actants may be elements of nature such as the ozone layer (Latour, 1993) and ocean currents (Callon, 1986), or man-made material or intellectual artefacts, such as technologies or metaphors. Those actors get together or enrol each other to form "*networks* of aligned interests". Such networks (or *actor-networks*) are characterised by the fact that all the actors/actants that form them have identical or similar interests, goals or purposes. They are often heterogeneous, i.e. they may include actants of very different kinds, such as humans, organizations, natural elements and man-made artefacts.

*Blackboxing* refers to a situation whereby an actor-network, or a part of this network, is so stable that it appears to the outside world as a given, a taken-for-granted entity that ceases to be the object of investigation. Focal actors may also become so central in the network that they end up being *obligatory points of passage*, i.e. they are rendered indispensable to other actors.

The notion of *inscription* is used particularly to refer to the design of non-human actants that are meant to be part of a larger network. In particular, many designers have a particular pattern of usage in mind when creating an artefact, and therefore "inscribe" this imagined pattern of action into the artefact itself. A famous example of inscription is given in Latour (1992): a bulky appendage to a hotel room key will be "inscribed" with the message "leave this key at the reception desk". *Translation* is a multi-faceted concept in ANT, but one of the uses of the term that is most relevant to this study is related to the way users of an artefact "translate" it so as to make it fit to their perceived needs. The notion of translation may here refer to situations whereby users exploit the artefact for a purpose that is different from what was originally intended by its designers, or physically modify it to harmonize it with their own particular context.

## ANAlysis of the Data from an ANT Perspective

### Identification of the Actors/Actants

This section defines the key actors and actants that were related to the video-trigger project described here. The actors and actants will then be discussed in more detail in the following sections. The project was initiated and carried out by a large group of lecturers that decided to divide into two teams: one "*technology group*" specialising in the technological aspects of designing and developing the video triggers, and one "*reference group*" that was meant to ensure that the produced videos would cover the curriculum requirements in terms of providing knowledge and insights into issues of interprofessionality and patient-centeredness.

The *students* at the Faculty were important actors in the process for a number of reasons. First of all, it was the student evaluation scores that prompted the academic management at the Faculty to look for an alternative and innovative learning form for the interprofessional module. Also, the students were to be the beneficiaries of the video triggers, which were meant to stimulate discussions about how to develop a patient-oriented rather than a discipline-oriented or profession-oriented approach to the diagnosis and treatment of patients.

The *academic management* of the Faculty provided the financial support of the project. At the time when the project was initiated, both the Faculty and the institution as a whole had defined technology use in learning and teaching activities as being a key strategic area for development. Concurrently, the institution had just invested in a commercial *virtual learning environment* (VLE) that was made available to all the faculties, and the general discourse throughout the institution was that teaching and learning forms that would make good use of the VLE were to be encouraged. In that sense, the VLE can also be identified as an actant that played a major role in the instigation and the continuation of the project. A certain *faith in technology*, a conviction that as long as the product was of high technological quality it would be useful to the students, emerged as another pivotal actant in the project.

The *video triggers* themselves, as well as the technical equipment necessary to develop them, became central actants in the project. In addition, those video triggers were an inherent part of a compulsory *assignment* for all the students at the Faculty. Within the realm of this assignment, the students were to carry out a number of tasks, including watching the video trigger, writing a text answer, publishing it on the VLE, reading other students' answers and providing written feedback on one or more of those answers. The teachers then provided their own feedback to each student answer. The assignment was thereby a significant actant, as it formed the pedagogical frame within which the video triggers could be used purposefully.

The *focus on interprofessionality* was an important actant at the outset of the project, as it originally was the major drive for the project. The analysis that follows will suggest that, although raising student awareness of interprofessional issues was among the foremost strategic objectives of the Faculty, it turned out to be tricky to implement.

## Analysis of the Processes of Enrolment/Alignment of Interest

### Enrolment of the Video Trigger by the Academic Management

The idea of using video triggers as pedagogical tools appeared against the backdrop of rather low scores for the interprofessional module in the student evaluation surveys of previous cohorts. The module was originally chiefly reliant on text-based case studies, a learning form that the students did not consider to be engaging. Some of the students in the previous cohort had also expressed discontent about the text-based case studies being of little relevance to their own profession and failing to shed light on the notion of interprofessionality. On average, the students from the programmes that were more heavily reliant on the hard science paradigm (for example medical laboratory sciences or dental technology) were more dissatisfied with the case studies than the students from programmes involving more of a soft science perspective (such as occupational therapy and physiotherapy). In other words, the text-based case studies did not bear a strong level of inscription of the interprofessional focus, and the students' translation of those case studies did not highlight interprofessionality either.

The data suggests that the use of video triggers had from the beginning been invested with a special status and with an ambitious mission at the Faculty. The management of the Faculty started therefore to look for alternative pedagogical practices that would stimulate interest among both the students and the teachers. As e-learning happened to be a buzzword both within the Faculty and throughout the institution as a whole and as the institution had invested in a new virtual learning environment (VLE) a couple of months earlier, it seemed to the module tutors both logical and timely to begin to use the VLE for the interprofessional module. Every student had access to the VLE, and could access any content published on the VLE independently of time and space. Content might take the form of text, sound files, pictures or videos. Because the module had traditionally used case-based assignments, and because the VLE provided new possibilities in terms of how the content of the cases could be presented, the management of the Faculty decided to sponsor the development of a video case that would trigger discussion on interprofessionality and patient-centeredness.

The general expectation was that using videos would significantly improve the learning outcome among students and, consequently, improve the scores attributed to the interprofessional module in the student evaluation exercise. First of all, it was hoped that using state-of-the-art technology to present the case studies would boost both the teachers' and the students' enthusiasm vis-à-vis the interprofessional module. The fact that a new and innovative learning form was introduced as part of the module was hoped to be a motivating factor for the students. Secondly, it was hoped that a video would be a more appropriate medium to convey the essence of interprofessionality and patient-centeredness than text-based cases. A video presenting a real person recounting a real story was anticipated to be a step forward in terms of authenticity compared to text-based cases.

Finally, both the video trigger and the VLE-based assignment that it related to were also meant to provide a solution to a logistical problem. In preceding years, it had been extremely difficult to arrange for students from different bachelor programmes to attend the lectures and seminars of the interprofessional module. This was partly due to the fact that different bachelor programmes have different schedules, in particular in terms of when students are on campus and when they are on work placement. It was hoped that having a common assignment on the VLE available independently of time and space would allow the students to form some sort of virtual community. It was expected that, within such a community, collaboration across courses would not be inhibited by the traditional logistical constraints of having to meet face to face. In particular, the fact that the VLE allowed for asynchronous discussions was seen as a major step in the direction of increased flexibility for the students. This new development was also expected to considerably reduce the logistical complexity that the course tutors would have to handle in order to make the module available to all.

The video trigger was thereby ascribed a focal function in the interprofessional module, as it was attributed the role of "saviour" of a so far somewhat unpopular module. Coupled with the new virtual learning environment and with the compulsory assignment that was based on it, the video trigger became a major actant within the interprofessional module.

## Faith in Technology and Pedagogical Interests as Irreconcilable Actants

It is apparent from the meeting minutes that the lecturers who were in charge of designing and producing the video triggers did not form a homogeneous entity and did not reach a common understanding of what the video trigger was meant to achieve.

*The greatest challenge here is to find a common understanding of issues and concepts, such as, e.g., user involvement and communication in such a way that everyone will be able to identify with them, while at the same time express their own particular perspective [Extract from meeting minutes].*

This situation was exacerbated by the fact that the group decided to split into two teams: a "reference group" which was to provide the content of the video trigger with a focus on interprofessionality, and an "e-learning group" which concentrated on the technical aspects of the video trigger. Although it might seem ironical that a team working on interprofessionality would split into separate specialist teams based upon their knowledge and experience, this decision was justified by practical and logistical concerns. The original rationale behind forming two teams was that team members would divide the workload among themselves according to their specialized knowledge and interest. It was also anticipated that having two smaller-size teams would ease the logistical problems related to the process of scheduling meetings involving a large number of persons.

However, having separate teams working on the same assignment had a number of unexpected consequences. One problematic area was the dynamics of the relationship between the two groups. It is apparent from the meeting minutes that each group was expecting the other to provide them with input about how to go about their part of the work. For a long while, the "reference group" was expecting the "technical group" to produce a prototype of the video trigger and present it to them, while the "technical group" was waiting for the "reference group" to provide them with specifications for the content of the video trigger.

*The e-learning group is requesting a speedier elucidation of what the content of the video trigger is supposed to be. The reference group needs to express their opinion as to how [a description of] a clinical situation will have to be described, and how it is going to relate to what it is supposed to illustrate: interprofessionality versus multiprofessionality [Extract from meeting minutes].*

It can be hypothesized that members of the "reference group" did not feel they had much to contribute to the discussion on video triggers because they did not have the same level of technological expertise as the "technology group".

Also, much of the data suggests that the two groups were treated differently as far as time management was concerned. Members of the "technology group" were formally granted a weekly number of hours that they were to use both to improve their technical skills and to work on the design and development of the video trigger. The members of the "reference group", on the other hand, were not granted any additional work time to be dedicated to the work they were to perform on video triggers. This work therefore had to be carried out in addition to their regular teaching and research activities, and could therefore not be prioritised in the same way as the work that the "technology group" was carrying out. The different treatment of the two groups as far as time allocation is concerned suggests that the academic management attributed relatively more status to the work performed by the "technology group" compared to the work performed by the "reference group". The former appeared to be considered to be more worthy of an investment than the latter, which seems to reflect a weightier concern with increasing the Faculty staff's technological knowledge than with providing room for a deep investigation of what is meant with interprofessionality and patient-centeredness.

In ANT terms, it can be said that the "technology group" and the academic management at the Faculty enrolled both the video trigger technology and a faith into technology to form a *stabilized network of interest*. Within this network, their efforts to make the technology as sophisticated and elegant as possible gained legitimacy and status. This phenomenon was probably intensified by the fact that some of the members of the "technology group" attended a video-trigger production course provided by a centre outside the Faculty. The course focused mainly on the technical aspects of the production of video triggers, and it can be suggested that it contributed to the technology group developing a special interest in achieving high technological quality in the production of video triggers. In that sense, the designers and tutors of the video-trigger production course can be seen as actors participating in and reinforcing the network of aligned interests whereby technology is at the centre and pedagogical issues are peripheral.

Conversely, the network that the "reference group" had formed by way of enrolling *the interprofessional and patient-centred focus* was relatively weak. It is interesting to note that although interprofessionality and patient-centeredness were supposed to be at the core of the video triggers, those two perspectives did not come through very clearly in the final video trigger products. Despite much effort, the notion of interprofessionality had remained a relatively empty concept. In that sense, it can be said that the designers failed to *inscribe* interprofessionality in the video triggers.

The data gathered in this study does not provide a clear answer as to why interprofessionality failed to gain a foothold in the video-trigger production process. One possible explanation may be that interprofessionality and patient-centeredness were themselves still relatively new ideas that did not always appear to be fundamental in the eyes of all the staff members in the Faculty. Another possible explanation is that many of the members of the academic staff have relatively limited clinical practice, and may therefore lack a hands-on knowledge of what interprofessionality means in practice. Finally, it may be suggested that the attempt at portraying interprofessionality in a common video trigger for a large array of professions turned out to be a rather artificial undertaking. In particular, the video trigger was meant to be equally relevant to students from a variety of professions that have different degrees of interprofessionality and that do not always collaborate with each other in real life professional contexts (e.g. biomedical laboratory scientists and occupational therapists).

## The Video Trigger as a Black Box and an Obligatory Passage Point

The video trigger was subjected to a pilot test in two phases. In the first phase, the video trigger was made available on the VLE, together with an assignment text. At this stage, the assignment text was purposely open-ended, merely enjoining the students to identify relevant interprofessional issues that emerged from the video and discuss them in the light of the theoretical concepts covered during the lectures. The reason behind having such an open assignment was to encourage the students to think for themselves, and let them come up spontaneously with reflections on issues of interprofessionality and patient-centeredness.

Before making the assignment available to the students, the members of the project team attempted to answer it themselves. In the light of this trial, they were concerned that the students would have difficulties identifying interprofessionality and patient-centeredness in their *translation* of the case study presented to them. The project team therefore decided that the assignment would have to be redesigned so as to provide the students with a more explicit focus on issues of interprofessionality and patient-centeredness. In order to do so, the video trigger was to be extensively reshaped and restructured.

The second phase in the trial process consisted, therefore, in redesigning the video trigger to make the interprofessional and patient-centeredness foci more manifest and relevant to the differing concerns of students from different professions. Such an endeavour proved to be rather arduous. At this stage in the project, time and resources had become scarce and it soon became clear that totally revamping the video trigger would have been an unrealistically time-consuming undertaking.

Instead, the "technology group" decided to simply create a number of additional video clips to address the shortcomings identified in the first phase of the pilot study. The main purpose of those new clips was to raise issues that would be relevant to students from different professions. Some of those new clips were to be appended at the end of the existing video trigger. Others were standalone add-ons to the main video trigger (and would appear as separate files on the VLE).

However, it soon became clear that any single clip was unlikely to achieve equal relevance across the large spectre of professions represented within the Faculty. In order to circumvent this problem, the "technology group" created several new clips, each of which highlighted a couple of issues that was considered to have a particular relevance for one of the professional programmes involved in the module. Overall, the changes and additions made to the video did not provide a clearer interprofessional focus. To palliate this problem, a new approach was taken, namely the implementation of changes to the assignment text accompanying the video.

The assignment text was reworked so as to be made more detailed and explicit. The new assignment text expressly presented questions guiding the students to reflect on how the notion of patient-centeredness was to be understood from the patient story presented in the video. It is to be noticed that the new questions related to the video focused more particularly on the issues of patient-centeredness, which was deemed to be a rather straightforward concept, rather than on issues of interprofessionality, which had shown to be more difficult to grasp from the video itself. In a subsequent phase, the new assignment text put forward a number of questions about interprofessionality, but those questions did not take the video trigger as a basis. Instead, the students were asked to gather a number of other sources (such as patient stories or situations they may have been witness to during their placement period), and use those as a basis for reflection on the issue of interprofessionality. In ANT terms, it may be purported that the assignment text lent itself to being *inscribed* with the focus of interprofessionality more easily than the original video trigger.

The additional video clips created in the second

phase of design can be considered to be new *actants* in the actor-network formed by the assignment, the designers and the rest of the project team. Those new actants played a crucial role in increasing the focus on multiprofessionality in the video trigger assignment and overshadowing the original focus on interprofessionality, as they were tailored to the interests of students from particular disciplines. The shift in focus from interprofessionality to multiprofessionality was both apparent and widely accepted throughout the Faculty as a way to deal pragmatically with the difficulties associated with the original focus on interprofessionality. As time went by, this shift in focus became so noticeable that the initial focus on interprofessionality had to be officially tuned down. This was done by formally changing the name of the module, from "interprofessional module" to "multiprofessional module" the year after the video trigger was first introduced.

It is interesting to see that despite being made aware of many serious shortcomings in the video trigger, the project team never proposed to formally discard it. Instead, they first attempted to have it redesigned, then modified the assignment text accompanying it. The fact that much energy, financial and human resources had been used to create the video was considered an adequate reason for holding on to it. From an ANT perspective, it can be suggested that infusing a strong faith in technology into the processes around the redesign of the interprofessional module has strengthened the position of the video trigger within the network formed around them. Because the creation and implementation of the video case had incurred a significant financial investment, the resulting video trigger became both a *black box* and an *obligatory passage point*.

The situation described above illustrates how a technological artefact such as the video trigger can become *blackboxed* when any attempt at changing its content would incur such a large amount of effort that it is deemed more reasonable to use it "as is" instead of trying to modify it. The video trigger became concurrently an *obligatory passage point* for all the students of the Faculty having to take the compulsory interprofessional (later multiprofessional) module. The blackboxing of the video trigger resulted in considerable modifications made to other actants in the network, namely the assignment text in the first place, and, later, in the very name of the module.

## CONCLUSION AND FUTURE RESEARCH DIRECTIONS

This chapter has related a story of a non-human actant - the video trigger - and how it came to pass that this technological development, rather than serving the needs of tutors, eventually dominated those needs, and forced them to change the very nature of their module to fit around the technology. In ANT terms, the video trigger has been made into an *obligatory passage point* for a particular module, and has been *enrolled* and *aligned* with the interests of a particular group. It has outlined how faith in technology has been a central non-human *actant*, playing a pervading role in the way other actors such as management staff and teaching staff related to the video trigger. Many of the actors involved in the process of sponsoring and creating the video trigger seem to have overestimated the benefits that could be drawn from technology use in the interprofessional module. It may be noted that such a phenomenon has been described in the literature on educational technologies as a recurrent problem (e.g., McAleese, 1996; Friesen, 2004).

The story told in this chapter is also one of a dichotomization between a technological outlook and an interprofessional and pedagogical perspective. In so doing, it illustrates one of the central challenges inherent to the work of designing educational technologies. Because educational technologies are still a rather new genre to some educators, they may know little about how to make the quest for a technologically elegant solution and the pursuit of pedagogical goals converge. The analysis presented in this chapter suggests that there is a critical need to go beyond this type of dichotomy and find ways to ***align*** the technological interests with the educational interests.

The key lesson that can be learned from the case presented here is that faculty who seek to develop reusable interprofessional e-learning objects may want to pay close attention to identifying what type of strategy may be deployed to avoid letting the focus on technology overshadow the pedagogical purpose of an educational tool. It may be useful to make the various actors in the design and use of the tool aware of the danger of having an excessively technology-oriented focus. It may also be valuable to remind the actors of the actual purpose of a pedagogical tool: to raise the students' knowledge or awareness level in an area that has been defined as relevant for their education. The case study illustrates an overreliance on technology, which resulted in pedagogy and content being relegated to a secondary place. A higher level of involvement of the whole academic team into the design of the technology ("learning by design"), as advocated by, for example, Koeler & Mishra (2005), and Mishra & Koeler (2006), may contribute to educational technological tools being more appropriate to the actual needs of the teacher and student users.

Other lessons that can be learned from this case relate to a number of project management issues. First, it is apparent from the data that the number of persons involved in the project (13 altogether) was too high, since it required splitting the project group into two teams. In addition, the senior management limited their involvement with the project to the planning stage, for example in designating who would work with the project, and in allocating additional time to members of the technical team. A heavier involvement in the project could have reduced the extent of the communication problems between the teams. In addition, the fact that academics working with technology were allocated more time to work with the project that those who "only" worked on issues of interprofessionality may have given the wrong signals as to whose work was most valued.

In the case described here, the focus was primarily on the final product (the video trigger) and not on the processes that came into play in order to achieve the products' educational goals. This may have played a part in the delays and lack of communication within the project group. This suggests that there is a need for the project manager to be kept abreast of the various activities taking place within the realm of the project. This may be achieved by requesting project members to provide regular field reports throughout the project period.

The notes that had been taken from the field in the form of diaries proved to be useful when looking back on the project and analyzing the chain of events that had led to the creation and the implementation of the video trigger. However, it is to be expected that diaries kept by only one person will have been colored by this person's background and preconceptions. In order to reduce the amount of bias stemming from the dairies, it might be useful to request that all project members keep a "project diary" and designate a certain amount of project hours to be used solely for diary keeping. Such diaries may also contribute to a wider and more democratic form of organizational learning, where each project member is given a voice in the process of creating collective information.

## REFERENCES

Barnes, L. B., Christensen, C. R., & Hansen, A. J. (1994). *Teaching and the case method: Text, cases, and readings* (3rd ed.). Boston: Harvard Business School Publishing Division.

Boyce, B. A. (1992). Making the case for the case-method approach in physical education pedagogy classes. *Journal of Physical Education, Recreation & Dance, 63*(6), 17–20.

Callon, M. (1986). Some elements of a sociology of translation: Domestication of the scallops and the fishermen of St Brieuc Bay. In Law, J. (Ed.), *Power, action and belief: A new sociology of knowledge?* (pp. 196–233). London: Routledge & Kegan Paul.

Choi, I., & Lee, K. (2009). Designing and implementing a case-based learning environment for enhancing ill-structured problem solving: Classroom management problems for prospective teachers. *Educational Technology Research and Development, 57*(1), 99–129. doi:10.1007/s11423-008-9089-2

Friesen, N. (2004). Three objections to learning objects. In McGreal, R. (Ed.), *Online Education Using Learning Objects* (pp. 59–70). London: Routledge/Falmer.

Heitzmann, R. (2008). Case study instruction in teacher education: Opportunity to develop students' critical thinking, school smarts and decision making. *Education, 128*(4), 523–542.

Herreid, C. F. (2006). *Start with a story: The case study method of teaching college science*. Arlington, VA: NSTA Press.

Hughes, M., Ventura, S., & Dando, M. (2004). Online interprofessional learning: Introducing constructivism through enquiry-based learning and peer review. *Journal of Interprofessional Care, 18*(3), 263–268. doi:10.1080/13561820410001731304

Juntunen, A., & Heikkinen, E. (2004). Lessons from interprofessional e-learning: Piloting a care of the elderly module. *Journal of Interprofessional Care, 18*(3), 269–278. doi:10.1080/13561820410001731313

Koeler, M. J., & Mishra, P. (2005). What happens when teachers design educational technology? The development of technological pedagogical content knowledge. *Journal of Educational Computing Research, 32*(2), 131–152. doi:10.2190/0EW7-01WB-BKHL-QDYV

Kurz, T. L., Llama, G., & Savenye, W. (2005). Issues and challenges of creating video cases to be used with preservice teachers. *TechTrends: Linking Research and Practice to Improve Learning, 49*(4), 67–73.

Latour, B. (1993). *We have never been modern*. New York: Harvester.

Law, J. (1987). Technology and heterogeneous engineering: The case of Portuguese expansion. In Bijker, W. E., Hughes, T. P., & Pinch, T. J. (Eds.), *The social construction of technological systems: New directions in the sociology and history of technology* (pp. 111–134). Cambridge, MA: MIT Press.

Lindqvist, S., Duncan, A., Shepstone, L., Watts, F., & Pearce, S. (2005). Case-based learning in cross-professional groups: The development of a pre-registration interprofessional learning programme. *Journal of Interprofessional Care, 19*(5), 509–520. doi:10.1080/13561820500126854

Macome, E. (2008). On implementation of an information system in the Mozambican context: the EDM case viewed through ANT lenses. *Information Technology for Development, 14*(2), 154–170. doi:10.1002/itdj.20063

Mayo, J. A. (2004). Using case-based instruction to bridge the gap between theory and practice. *Journal of Constructivist Psychology, 17*(2), 137–146. doi:10.1080/10720530490273917

McAleese, R. (1996). Technology in education to technology of education: concepts, conflicts and compromises. In Armstrong, S., Thompson, G., & Brown, S. (Eds.), *Facing up to radical changes in universities and colleges* (pp. 51–63). London: Kogan Page.

Mishra, P., & Koehler, M. J. (2006). Technological Pedagogical Content Knowledge: A framework for teacher knowledge. *Teachers College Record, 108*(6), 1017–1054. doi:10.1111/j.1467-9620.2006.00684.x

Pomeroy, V. M., & Philip, I. (1994). Healthcare teams: An interdisciplinary workshop for undergraduates. *Medical Teacher, 16*(4), 341–346. doi:10.3109/01421599409008271

Quinney, A., Hutchings, M., & Scammell, J. (2008). Student and staff experiences of using a virtual community, Wessex Bay, to support interprofessional learning: Messages for collaborative practice. *Social Work Education, 27*(6), 658–664. doi:10.1080/02615470802201754

Rowley, J. B., & Hart, P. M. (1996). How video case studies can promote reflective dialogue. *Educational Leadership, 53*, 28–29.

Shulman, L. S. (1992). Toward a pedagogy of cases. In Shulman, J. H. (Ed.), *Case methods in teacher education* (pp. 1–30). New York: Teachers College Press.

Silverman, R., Welty, W. M., & Lyon, S. (1996). *Case Studies for Teacher Problem Solving* (2nd ed.). New York: McGraw-Hill.

Skorga, P. (2002). Interdisciplinary and distance education in the Delta: the Delta Health Education Partnership. *Journal of Interprofessional Care, 16*(2), 149–157. doi:10.1080/13561820220124166

Tatnall, A., & Davey, B. (2005). *An actor-network approach to informing clients through portals.* Informing Science and Information Technology Education Joint Conference, Flagstaff, Arizona, USA, 16-19, 2005. Retrieved August 12, 2009, from http://informingscience.org/proceedings/InSITE2005/I60f75Tatn.pdf

Thompson, L., & Cupples, J. (2008). Seen and not heard? Text messaging and digital sociality. *Social & Cultural Geography, 9*(1), 95–108. doi:10.1080/14649360701789634

Yadav, A., & Koehler, M. J. (2007). The role of epistemological beliefs in preservice teachers' interpretation of video cases of early-grade literacy instruction. *Journal of Technology and Teacher Education, 15*(3), 335–361.

# Chapter 19
# Pedagogical Directions in Creating Interprofessional E-Learning Materials

**Frances Gordon**
*Sheffield Hallam University, UK*

**Karen Booth**
*Sheffield Hallam University, UK*

**Helen Bywater**
*Sheffield Hallam University, UK*

## ABSTRACT

*This chapter will provide new insights around the underpinning pedagogy of e-learning that supports the development of collaborative skills in health and social care. It will provide an articulation of practitioner epistemology concerning the development of e-learning materials designed to promote collaborative practice. To do this it will draw on a qualitative case study undertaken by the authors which aimed to uncover the working theories employed by academics when creating e-learning materials to support interprofessional education initiatives. The introduction to this chapter will set the scene by discussing how the initial drivers for developing e-learning materials are often logistics and efficient resource management. However, what this study uncovered were the underlying pedagogies lecturers used in developing e-learning materials. The introduction will signal discussions to be developed in the chapter of how these theoretical approaches to planning learning experiences are adapted and transferred from classroom based practice to the virtual environment.*

## INTRODUCTION

The Centre for Interprofessional e-learning (CIPeL) is a collaborative Centre for Excellence in Teaching and Learning (CETL) funded by the Higher Education Funding Council for England (HEFCE). The Centre for Interprofessional e-learning (CIPeL) has been engaged in developing e-learning materials which address some of the barriers to interprofessional education. This chapter will draw on our experience of combining these in the context of undergraduate, pre-

DOI: 10.4018/978-1-61520-889-0.ch019

qualifying courses for health and social care. This chapter aims to articulate pedagogical principles of interprofessional e-learning drawn from current evidence and experience. It will examine the main principles of creating and using e-learning materials in interprofessional education and discuss the challenges and suggest solutions to providing e-learning that supports collaborative practice.

## BACKGROUND

Interprofessional education (IPE) and e-learning are both high profile issues in the higher education sector. The policy focus on improving health and social care services emphasises the need for integrated services and collaborative working in the UK (DoH, 1999; DoH, 2001). These key ideas continue to be re-emphasised, underlining the continuing drive towards an agenda of improvement in health and social care services (Wanless, 2004; Department of Health, 2004a; Department of Health, 2004b; Department of Health, 2004c). The Climbié report focused attention on the need for improved communication between agencies and individuals involved in the education and welfare of children. The reports of disasters resulting from failures to safeguard individual children have provided a strong driver for identifying the knowledge and skills required for developing effective integrated services (Taylor, Whiting & Sharland, 2007).

The Bristol Enquiry Final Report (Kennedy, 2001) and the Victoria Climbié Report (Laming, 2003) represent significant breakdowns in service delivery in the UK and the UK Government has responded through policy initiatives. Successive policy initiatives around this area signal that effective collaborative working between agencies and with patients/service users must be supported by education and training that is genuinely multiprofessional (see The NHS Plan (DoH, 2000), Expert Patient (DoH, 2002) and the green paper Every Child Matters (DfES, 2003)). This has led to a requirement for interprofessional components within health and social care courses in Higher Education Institutions (DoH, 2000a, 2000b). This policy and the contractual requirements through commissioning processes, present higher education institutions providing prequalifying health and social care education with the challenge of ensuring that interprofessional learning is a feature of the courses they offer.

E-learning is a growing phenomenon within higher education and is seen as having the potential to support more learner-centred and personalised forms of learning (JISC, 2008). The Higher Education Funding Council for England (HEFCE) in its strategy for e-learning (2005, 2009) has indicated the need for the sector to embrace the use of the internet and new technologies. Prensky (2001) indicates that today's students are the first generation to grow up with new technology and as such think and process information differently from their predecessors. HEFCE recognises that today's students "are bringing new expectations of the power of technology into higher education" and states that the use of e-learning should become a normal or embedded part of lecturers' activities and that "staff require support so they can effectively exploit the potential of these new technologies" (HEFCE, March 2009, p. 5-6). E-learning, although perhaps to a varied extent across institutions, is already a feature of university life and promises to become a ubiquitous presence that will impact increasingly on learning and teaching.

## THE CHALLENGE OF INTERPROFESSIONAL E-LEARNING (IPEL)

The delivery of IPE, as suggested earlier, is a requirement of health and social care programmes. It is not, however, without problems. One of the most compelling of these is the logistical and organisational difficulties of bringing students from a wide range of courses together in order to "learn with, from and about each other" (Barr, 2002). We would argue that meeting the logistical challenge can be understood metaphorically through reference to Maslow's hierarchy of needs (Maslow, 1943), where the basic need of actually finding a way of bringing students together in the first place has to be met before the higher order needs of actualising effective learning can be considered.

Solving the problem of logistics brings with it, however, a new difficulty. Managing learning for the disparate groups of professions being brought together, with arguably unique cultural attitudes and customs, adds complexity. This diversity of professional culture brings richness to interprofessional learning and supports the aspiration of achieving positive interprofessional outcomes. However, if the learning environment is not managed effectively, negative stereotyping across the professions can deepen working against the building of co-operation and collaborative working (Carpenter, 1995a, 1995b; Hewstone & Brown, 1986; Hewstone, Carpenter, Rought & Franklyn-Stokes, 1994).

A related dilemma can also be identified within the widening participation agenda of Higher Education (DfES, 2003). Diversity, whilst being a welcome development to aid greater participation in higher education, also carries greater levels of attrition among students from non-traditional academic backgrounds (Stanton, 2008). The widening participation agenda is part of the British government's plans for increasing prosperity not only for the nation but for individuals, with learning being seen as the key to achieving these aspirations (DfEE, 1993). Policy initiatives to promote learning for all is seen in the discourses of both lifelong learning and the celebration of diversity through widening participation. The notion of life long learning implies that students need to develop ways of becoming increasingly independent in their learning, giving impetus to further discourse around the autonomous learner. We argue that those working in higher education, (and probably in all spheres of education), are challenged in facilitating contemporary, diverse student cohorts to become autonomous learners. This, we will go on to argue, becomes more challenging within interprofessional education.

Meeting the needs of an increasingly diverse range of learners is a recognised issue within universities generally. Students come from different educational and/or cultural backgrounds and with different life experiences, but it can be argued cohorts are homogenous at least with respect to the subject being studied. However, we would suggest that the challenge of diversity is exacerbated in interprofessional education. The additional aspects of difference found in interprofessional learning where the different professional orientations, the different knowledge bases and language of the 'home' course and the different ways of professional working, all influence attitudes and approaches to learning add to the difficulties of meeting the needs of all students.

Given these circumstances initially the major preoccupations of those embarking on, and maintaining, interprofessional education are the logistical issues of getting students together and subsequently providing a programme of learning that accommodates the diverse needs of students engaging in the learning. E-learning has been proposed as a solution to logistical issues and, as we argue in this chapter, can provide a way ahead to both the thorny problems of diversity and promoting learner autonomy. Santy et al. (2008) describe how the availability of on-line technology offers a flexible approach to interprofessional delivery in terms of time and place and how discussion boards, for example, are an effective way of bringing people together to work on issues, even though in reality they are some distance apart. Quinney et al. (2008) also indicate that online learning can facilitate learning across sites and enables the boundaries of time and space for dispersed interprofessional learners to be transcended.

However, the seemingly straightforward logistical solutions suggested in the literature are not without their cautions. Monahan et al. (2006) have warned that whilst these approaches have proved popular, they are often single user learning environments, employing asynchronous discussion that do little to promote interaction or stimulation for the student. Monteith and Smith (2001) report findings of an exploration of students' experiences of using a virtual learning environment where they consider that its contribution to learner autonomy was somewhat superficial, although students did use the available technology flexibly for communication and social aspects of learning. They argue that the superficial aspects of students' independence in learning are related to evidence within the literature that indicates a need for lecturers to mediate learning.

The popularity of employing e-learning strategies mentioned by Monahan et al is reflected by Deepwell and Malik (2008) who discuss how institutions have invested heavily in providing a robust and integrated technical infrastructure. This, they suggest, brings with it an increasing expectation for academic staff to use these institutional technologies to deliver teaching but they point to the uncertainties around the impact of these developments. Interestingly in the context of HEFCE's (2005) and JISC's (2008) notions that students expect to use technologies, Prenski's (2001) discussion about today's students being fundamentally different from their predecessors throws light on another problem. Prenski (2001, p.1) refers to contemporary students as "digital natives", referring to the current generation being the first to grow up in a world saturated by digital technology, they are digital "native speakers" – hence their expectation to use technology in learning. Their teachers on the other hand, are "digital immigrants" working in a system designed for yesterday's student with both teachers and the system trying hard, as Deepwell and Malik identify, to catch up.

Prenski's discussion articulates beautifully the problem faced: the digital native students are fluent in a language that their teachers, to a greater or lesser extent, struggle with. Students, however, need teachers and want to encounter them in their digital world. They have consistently rated support from lecturers as being the strongest influence on their on-line learning experiences (Alexander, 2001), with students revealing a "continuing desire for face to face contact" (Monteith & Smith, 2001, p. 126). This issue remains current, with Deepwell and Malik (2008) finding that although students were in favour of the use of technologies to aid learning, there was a significant need for direct guidance from tutors on how to use technology for learning more independently.

This, we suggest, may indicate that the actual technology plays a relatively minor part in what students perceive as positive about e-learning. They may expect to use technology in the same way that technology permeates their day to day living, but what they rate as important is the lecturer role in their learning, whether it is online or on-campus learning. The mismatch between digitally native and digitally immigrant experience can be seen where the technology becomes more important than the pedagogy. Barberra (2004, p. 14) reflects these issues when she reports Virtual Learning Environments (VLEs) as being useful for offering distance learning, but points to how VLEs tend to privilege "technological and aesthetic over educational criteria".

## PEDAGOGICAL DIRECTIONS

The use of technology for learning is expanding and e-learning is seen as a solution to a number of difficulties and challenges within higher education generally, and for interprofessional learning in particular. However, we suggest that these solutions cannot serve students well if implemented in isolation from educational theory. Muirhead (2007) reports an absence of research around the theories that support teaching and learning to be used in conjunction with e-learning in higher education. He raises concerns that well-established teaching styles utilising e-learning are being developed and used worldwide with little research to support the pedagogy underpinning it. We argue, therefore, that technology should not be the driver for designing the learning experience; rather that the pedagogical approaches to designing educational experiences should form the basis of promoting learning and how that learning may be enhanced by the planned use of technology. This section of the chapter presents aspects of the work we have conducted into exploring the working theories that lecturers employed when designing e-learning opportunities for their students.

The complexities around the drivers for interprofessional e-learning formed a backdrop to our interest in discovering how lecturers planned the development of e-learning materials to support students' collaborative working skills. Our institution faced a particular challenge due to a change of contract and a consequent large increase in student numbers. It was important to ensure that the quality of the interprofessional programme was maintained despite this increased demand on estate and human resource. Part of the solution was to develop a range of e-learning materials to support the programme. CIPeL provided funding for twenty one staff to be seconded to develop e-learning materials. This gave them time and space to plan the development outside their normal workloads.

## The Case Study

We conducted a study to explore the experiences of lecturers who had been involved in producing digital media for the interprofessional programme to support learning. The study was qualitative in approach employing semi-structured interviews to collect data that were analysed using thematic analysis processes. All twenty one of the lecturers involved in producing digital materials consented to participate in the study. Ethical approval and research governance was obtained through the Faculty Ethics Committee at our institution.

In this study we found that the initial drive towards developing interprofessional e-resources tended to be based on logistical and resource difficulties as identified earlier in the chapter. A starting point for some lecturers was the creation of e-lectures as a way of overcoming the organisational problems of very large numbers of students accessing lectures at the same time on-campus. Their strategy was that where the educational purpose was to transmit knowledge, providing the lectures on-line would be just as effective and would save resources.

*Well I suppose there was the sense that there needed to be some key e-lecture type materials to replace the lectures that used to happen. So that was sort of what drove me to e-lectures (lecturer 2).*

An example of such an e-lecture is one that deals with the principles of service improvement. Normally the background and key issues underlying this initiative would be given in a lecture hall to large numbers of students, perhaps followed up by a seminar. The e-lecture gives the same information but in an e-format enabling the information to be transmitted on- or off-campus. However, as the accounts unfolded it became clear that although the logistical issues were a starting point, the lecturers were engaging in deeper thinking about the pedagogy of the interprofessional e-learning they were designing.

The study aimed to gain understanding of the underlying pedagogical intent employed by lecturers when planning e-learning materials. The context of the learning was important and illuminates the issues discussed so far in the chapter. The resources were destined to be used in a blended programme of interprofessional learning for undergraduate health and social care students. Whilst experienced teachers do not necessarily consciously think about the learning theories they employ when planning their teaching activities, the lecturers had to think about quite complex issues of promoting collaborative learning and working using a digital environment. Their previous experience of face to face teaching with multiprofessional groups would have relied to some extent on the here and now dynamic within the classroom. When asked to reflect on promoting collaboration through e-learning, it became clear that they were reconceptualising their classroom experience by drawing on learning theories and how they applied them in that context.

*Interviewer: Do you think doing that project, undertaking that work has influenced you as an educator? Has it moved you anywhere?*

*Respondent: Yes, definitely. I think you have to think about the theories of learning and going back to basics and it's made me think very hard about that, increased my knowledge of learning theories in general (lecturer 11).*

Analysis of the participants' interviews uncovered pedagogical themes within their accounts. Those relevant to this chapter included: 'learner autonomy'; 'collaborative learning' and 'authenticity'.

## Learner Autonomy

When exploring the development of their e-lectures with the lecturers, it became clear that issues other than logistics were at play. There may be a difficulty in conceptualising the e-lectures as 'interprofessional' when relating to the "with, from and about" definition of interprofessional (CAIPE, 1997). What we argue is that these e-lectures drew to the attention of students, concepts that underpin interprofessional learning, which they would later utilise in processes of collaborative learning. The concepts are common to all the professions learning within our interprofessional programme and offer background knowledge to support learning 'with, from and about each other'.

Woo, Gosper, McNeill, Preston, Green & Philips (2008) discuss the fairly common practice of videotaping actual lectures and providing them on-line for students. This has been criticised for the negative effect on lecture attendance and subsequent diminishing of quality of the experience for students who do attend. In the context of this study, the provision of an e-lecture was not to provide materials for those who, for whatever reason, did not attend the 'real thing'; attendance was never the aim, the aim was to avoid the need for attendance. The lecturers did not see their e-lectures as merely replicating what they would have delivered in a live lecture.

Indeed, the participants' perceptions of the advantages of e-lectures over 'live' lectures seemed to reflect a move towards the promotion of student autonomy. They considered the students' ability to access this learning at times and places convenient to them and that they could repeat the lecture as often as required. The designs included additional resources within the e-lecture for students to use according to their own perceived needs. This approach gives the students control over the time, place and pace of learning that traditional lectures do not.

*This is about facilitating students' learning and trying to get them right at the start to be independent in their learning (Lecturer 13).*

From this perspective, the basic need to ensure that students could all access the learning, no matter where they were physically, was fairly easily solved. However, the medium of the e-lecture also allowed the lecturers to address some of the inherent diversity issues encountered in interprofessional learning. They were able to embed resources into the digital environment of the e-lecture in ways that cannot be achieved in 'live' lectures, and these resources could be targeted towards the interests of specific groups accessing the resource. Different groups of students could, in this way, utilise resources that met their professional and/or learning needs as appropriate, and use these resources to meet the learning outcomes of the educational opportunity.

The selection by students, with or without guidance by lecturers, towards specific resources that supported the content of the e-lecture, and how and when students used them gave some element of control in the learning. This is important in terms of learner autonomy and in meeting diverse needs arising both from a range of professional perspectives and the widening participation agenda. Pintrich (2003) asserts that students who have control over their learning are more likely to achieve positive learning outcomes. Students having control over pace of learning and of revisiting concepts they find challenging may go some way in promoting autonomous and successful learning.

## Collaborative Learning

The collaborative learning theme relates to how the lecturers reflected on their strategies for promoting interprofessional collaborative learning and working. The lecturers described how they thought about designing learning opportunities, using e-learning materials as part of the learning design, to encourage students to bring insights from their uni-professional learning and experience to the interprofessional context. It was planned to use these perspectives to generate insights that would aid the students' interprofessional development through learning with, from and about each other (CAIPE 1997)

Below, one lecturer discusses how a patient/client scenario would be used in an online component of a course. The intention was to develop collaborative skills in students, mediated through planning interactive activities in multidisciplinary groups.

*Yes, the students were required to work collaboratively in order to discuss the issues that were raised at each of the stages (of the resource), discuss how to proceed and then they were provided with a password that enabled them to carry on. So it was a case study that ran parallel to a discussion board, a discussion forum within that board that was facilitated by a tutor. So it wasn't just about the object we produced it was about ensuring that the student engaged in collaborative discussion online (lecturer 5).*

The quote below describes how the lecturer thought about ensuring interaction between students, and how these interactions were intended to promote collaborative team working. It also illuminates an intention for students to bring practice issues to the learning encounter, promoting relevance of the learning to the workplace.

*I thought it gave them a lot of scope as well to bring in any of their other experience from practice so we did really think about the kind of situations that were pretty common which students were going to meet in interprofessional collaborative learning and practice and we did it as well because I think that it really did try and mirror collaborative practice so that the way that we constructed it was, they had to work as a team ..... they couldn't really get away with not contributing (lecturer 9).*

This aspect of learning mirroring practice is related to the belief that students being able to collaborate on a practice orientated problem in the classroom, or online, equips them with the skills to collaborate on similar issues in real world practice. This belief can be challenged in that a mere mirroring of the practice world does not accommodate the complexity of practice which is almost impossible to replicate (Barton, McKellar & Maharg, 2005). We argue nevertheless that acquiring the skills of collaboration can be transferable to practice and that e-learning can offer a version of reality that reflects the ambiguities and uncertainties of the 'real world' (Jonassen, 1992).

Collaboration between the students was encouraged through the use of practice scenarios that contained issues which triggered discussion. They were designed to be used initially in the classroom and subsequently online. The scenarios produced by the lecturers were stories of plausible situations encountered in the practice context and centred on the perspective of the service user. These materials presented the service user in several ways including videos, photographs and illustrations. Curtis and Lawson (2001) indicate that collaborative activities and discussions can be achieved successfully in an online medium. They also report that the facility to challenge and explain in face to face interactions is more difficult to achieve successfully in asynchronous online discussions. It may be that such social learning can be promoted by students getting to know each other at the outset of these initiatives, through a context of familiarity with each other. Curtis and Lawson cite Cook (1997) as viewing social acknowledgements in discussion board posting as unproductive, but Curtis & Lawson considers such niceties promote collaboration.

The blended learning approach taken by the lecturers suggested that they viewed face to face contact between the students at the initial stages of the learning experience as important, reflecting ideas expressed by Borthick & Jones (2000). The issue of promoting social learning, and so assisting collaborative learning, by initial classroom based contact was only one element of the lecturers' thinking. The lecturer below indicates how working through the learning materials could enable the student to consider his or her development as a collaborative worker in practice.

*Well, one of things the students have to do at the end is to actually make an assessment of their performance as a collaborative worker in practice. So, the process that the learning object takes them through is about identifying themselves as such an individual and how others actually contribute to the general outcome for the patient or client. (lecturer 19).*

Monahan et al. (2008) consider that on-line systems can promote constructivist learning where a student learns from his/her own experience. This, they argue, results in a student-centred learning environment. It could be argued however, that whilst the lecturers may well have been thinking about the learning of students as individuals, they were also thinking about student-centredness in a collective way, and trying to encourage students to think about how collaborating together, in a co-constructionist fashion, produces positive learning outcomes. Encouraging collaboration led lecturers to think about what kinds of materials would motivate students to engage with the learning object and with each other. The lecturers believed that the achievement of this rested on the creation of situations or scenarios that had relevance to the practice arenas the students were preparing to be employed in.

## Authenticity

This theme of ensuring authenticity of practice for students was strong within the accounts. Considerations around providing students with learning opportunities that meet the precepts of adult learning theory were evident. The lecturers recognised that adults use their own experience as a base for learning and that adults' motivations to learn are around subjects that have immediate relevance to current problems they are engaged with (Knowles, 1984).

*I think that's one of the criticisms of interprofessional learning and online learning is that sometimes it's not real to life so I thought that if it was as genuine as it could be and that the range of material would be engaging with the students. So I think I thought if I could try to make it more real and more applied to practice ... (lecturer 9).*

However, providing relevant, practice based learning opportunities was not straightforward. The diversity challenge for interprofessional education has been identified earlier in this chapter. It was clear that the lecturers were drawing on their previous experience of grappling with the problem of achieving relevance for students when the cohort was richly varied in terms of professional backgrounds. Attempts to provide scenarios that meet the needs of a wide professional range of students have often, in themselves, added to perceptions of lacking relevance or even stretching credulity. This had been the experience of the lecturer quoted below.

*I think that students will really latch on to such scenarios where they see themselves ....... but elsewhere I've worked in modules where we've had all sorts of triggers and things, and when you get the disabled person with the brother with learning disabilities who's having problems....... it just becomes farcical as far as I'm concerned (lecturer 2).*

The lecturers described previous attempts to provide a 'one size fits all' situation that carried with it the often futile need to try and persuade students that even though the relevance to them was not obvious, they would at some point come to some sort of meaningful understanding of what the learning was about. Such approaches ignore the needs of adults to use their personal experience (Knowles, 1984). If the scenario presented cannot provide a scaffold of prior experience onto which new learning can be constructed, adults will tend to lose motivation.

*It's difficult, as usual, it's difficult to try and engage all the professional groups but we tried to make it clear to the students that if they couldn't immediately see the relevance of what they were discussing it didn't mean that it wasn't relevant to them. It's an old chestnut isn't it? (lecturer 5).*

The lecturers needed to find some solutions to these problems and their accounts featured a strong focus on developing learning materials that were authentic to practice. This, they knew from experience, was difficult and has been recognised elsewhere. Muirhead (2007, p. 181) has criticised the use of online scenarios as ending up being so carefully scripted that they "bear little resemblance to practice".

For the lecturers in this study, authenticity was conceived as ensuring that all students could participate in simulated real-life situations, with the use of digital formats offering a possible solution. However, there was a clear recognition that the sharing of one scenario across the total cohort did not meet the precepts of adult learning theory. The lecturers aimed to design a number of scenarios. The lecturer below describes two of the four that were eventually created. These materials featured story lines that were believable and contextually authentic and directed towards a group of up to five professions, thus ensuring every profession within the total cohort was featured in one or other of the scenarios.

*So we created these two on line enquiry based learning scenarios and we called them Ted and Jane. And we really used real life actors and we developed these stories essentially on people (lecturer 10).*

## Discussion

This chapter has presented lecturer perspectives on how they went about designing e-learning materials. The pedagogical insights gained through an analysis of their narratives reveal the complexity of task they were faced with. The initial driver was to resolve logistical problems. These were relatively easy to address through the technological support provided by the CETL. The educational imperative was to support interprofessional learning, to promote collaborative skills and offer a sound pedagogical base. Their accounts revealed an eclectic use of learning theories combined with digital technology to produce an effective e-learning experience for students.

## FUTURE RESEARCH DIRECTIONS

The study discussed in this chapter explored the experience of those who are experienced champions of interprofessional education. The wide scale implementation of interprofessional education necessarily carries the implication of a wider range of teachers becoming involved in supporting students' interprofessional learning. Payler, Meyer and Humphries (2008) raise the need for more work around articulating interprofessional pedagogies. Equally pressing is an exploration into the experience of those supporting students in a context of uncertainty and ambiguity. This calls for 'practice based evidence' (Margison, 2001) derived from the experience of those supporting interprofessional learning in both classroom and digital environments.

Clearly the issues discussed in this chapter give rise to many areas requiring further investigation, not least the need to find ways to evaluate e-learning resources (Tait, Tait, Thornton & Edwards, 2008). We are confident that the work of our colleagues in the Reusable Learning Object (RLO) CETL will take these perspectives forward. This chapter considered the issue of interprofessional e-pedagogy from the perspective of the lecturer participants only. The students' experiences of using the resources needs further exploration.

As Santy, Beadle and Needham (2009, p. 4) assert "More work needs to be done to illuminate the ways in which such activities do or do not foster learning in students …..". The need for further research is also necessary to investigate how the use of digital resources and e-learning strategies can address the aspirations of the lecturers. Learner autonomy for example, was something the lecturers wanted to promote though e-learning. However we contend that we are not much further in understanding how this plays out in keeping with assertions that the current lack of clarity about how far computer assisted learning can develop autonomy. Montheith & Smiths' (2001)

## CONCLUSION

Interprofessional e-learning pedagogy is situated within two problematic areas in higher education. The challenges to the wide scale implementation of IPE are well recognised and e-learning has been suggested as a solution for some of these difficulties. However, it seems that despite assumptions that technology can assist deep, independent learning, the evidence suggests, that students require guidance from tutors to achieve such autonomous learning. Prensky's (2001) characterisation of students as digital natives and tutors as digital immigrants provides a powerful metaphor to illustrate the potential language and practice mismatch between student and tutor. Support for both lecturers and students is needed to facilitate learning in the digital age.

The collision of interprofessional and e-learning 'problems' provide the context in which the participants of the study worked. Their interprofessional teaching practice was characterised by a concerted move away from what Paris (2000) has described the 'sage on the stage' to a 'guide on the side', promoting learner autonomy, and active, collaborative learning. Their challenge was to use e-learning within a pedagogy that valued students as adult learners and to harness students' motivation to learn to work effectively within multidisciplinary services.

*Yes there might be an instinctive reaction to say, of course, you'd do this, as a nurse, this is what you do. But then when they discussed it with (each other), when they explored the legal and social issues that perhaps social workers might have more knowledge about, they might change their minds. (lecturer 5).*

## REFERENCES


Alexander, S. (2001). E-learning developments and experiences. *Education + Training, 43*(4/5), 240-248.

Barberra, E. (2004). Quality in Virtual Learning Environment. *British Journal of Educational Technology, 33*(1), 13–20. doi:10.1111/j.1467-8535.2004.00364.x

Barnes, D., Carpenter, J., & Dickinson, C. (2000). Interprofessional education for community mental health: attitudes to community care and professional stereotypes. *Social Work Education, 19*(6), 565–583. doi:10.1080/02615470020002308

Barton, K., McKellar, P., & Maharg, P. (2005). *Authenticity in learning: transactional learning in virtual communities*. Draft paper submitted to the sixth International Clinical Conference. Accessed August 13, 2009, from http://www.uwex.edu/disted/conference/Resource_library/handouts/03Info_H03.pdf

Bielaczyc, K., & Collins, A. (1999). Learning communities in classrooms: a reconceptualisation of educational practice. In Reigeluth, C. (Ed.), *Instructional design theories and models. A new paradigm of instructional theory* (*Vol. 2*, pp. 269–293). Mahwah, NJ: Lawrence Erlbaum Associates.

CAIPE. (1997). Interprofessional education - a definition. *Caipe Bulletin, 13*, 19.

Carpenter, J. (1995a). Doctors and nurses: stereotypes and stereotype change in interprofessional education. *Journal of Interprofessional Care, 9*(2), 151–161. doi:10.3109/13561829509047849

Carpenter, J. (1995b). Interprofessional Education for medical and nursing students: evaluation of a programme. *Medical Education, 29*, 265–275. doi:10.1111/j.1365-2923.1995.tb02847.x

Cook, S. L. (1997). *Mindfulness and tolerance of ambiguity in online communication*. Unpublished PhD Thesis, University of Denver, Denver, CO.

Curtis, D. D., & Lawson, M. J. (2001). Exploring collaborative online learning. *JALN, 5*(1), 21–34.

Deepwell, F., & Malik, S. (2008). On campus, but out of class: an invesitgation into students' experiences of learning technologies in their self-directed study. *Research in Learning Technology, 16*(1), 4–14.

DfEE. (1993). *The learning age, Department for Education and Employment*. London: The Stationery Office.

DfES. (2003). *The future of higher education. White Paper cm 5735*. London: The Stationery Office.

DfES (Department for Education and Skills). (2003). Every Child Matters. Green Paper, Cm. 5860. London: The Stationery Office (TSO).

DoH. (1999). *National service frameworks for mental health: modern standards and service models*. London: Department of Health. The Stationery Office.

DoH. (2000a). *A health service of all the talents: developing the NHS workforce*. London: Department of Health. The Stationery Office.

DoH. (2000b). *The NHS Plan*. London: Department of Health. The Stationery Office.

DoH. (2001). *Valuing people a new strategy for learning disability for the 21st century a white paper*. London: Department of Health. The Stationery Office.

DoH. (2002). *The expert patient: A new approach to chronic disease management for the 21st Century*. London: Department of Health. The Stationery Office.

DoH. (2004a). *The NHS Improvement Plan: putting people at the heart of public services. Cm 6268*. London: Department of Health. The Stationery Office.

DoH. (2004b). *Choosing Health: making healthy choices easier*. London: Department of Health The Stationery Office.

DoH. (July 2004c.) *National Standards, Local Action: health and social care standards and planning framework 2005/06-2007/08*. London: Department of Health. The Stationery Office.

HEFCE. (2005). *Strategy for e-learning*. Retrieved July 5, 2009, from www.hefce.ac.uk

HEFCE. (2009). *Enhancing learning and teaching through the use of technology: A revised approach to HEFCE's strategy for e-learning*. Retrieved July 5, 2009, from www.hefce.ac.uk

Hewstone, M., & Brown, R. J. (1986). Contact is not Enough: an Inter-group Perspective on the "Contact Hypothesis". In Hewstone, M., & Brown, R. J. (Eds.), *Contact and Conflict in Intergroup Encounters*. Oxford, UK: Blackwell.

Hewstone, M., Carpenter, J., Rought, D., & Franklyn-Stokes, A. (1994). Intergroup contact between professional groups: two evaluation studies. *Journal of Community & Applied Social Psychology*, *4*, 347. doi:10.1002/casp.2450040504

JISC. (2008). *Effective practice with e-portfolios*. Bristol, UK: Joint Information Systems Council.

Kennedy, I. (2001). *The Bristol Royal Infirmary Inquiry, CM5207* (1). London: HMSO.

Knowles, M. (1984). *Andagogoy in action: applying modern principles to adult education*. San Francisco: Jossey Bass.

Laming, L. (2003). *The Victoria Climbié Inquiry*. Norwich, UK: HMSO.

Margison, F. (2001). Practice based evidence in psychotherapy. In Mace, C., Mooray, S., & Roberts, B. (Eds.), *Evidence in psychological therapies. A critical guide for practitioners*. Hove, UK: Brunner-Routledge.

Maslow, A. H. (1943). A theory of human motivation. *Psychological Review*, *50*(4), 370–396. doi:10.1037/h0054346

Monahan, T., McArdle, G., & Bertolotto, M. (2008). Virtual reality for collaborative e-learning. *Computers & Education*, *50*, 1339–1353. doi:10.1016/j.compedu.2006.12.008

Monteith, M., & Smith, J. (2001). Learning in a virtual campus: the pedagogical implications of students' experiences. *Innovations in Education and Teaching International*, *38*(2), 119–125. doi:10.1080/13558000010030176

Muirhead, R. J. (2007). E-learning: is this teaching at students or teaching with students? *Nursing Forum*, *42*(4), 178–184. doi:10.1111/j.1744-6198.2007.00085.x

Paris, D. C. (2000). Is there a professor in the class? In R. A. Cole, (Ed.), (2000), Issues in web-based pedagogy: A critical primer. London: Greenwood Press.

Payler, J., Meyer, E., & Humphries, D. (2008). Pedagogy for Interprofessional education – what do we know and how can we evaluate it? *Learning in Health and Social Care*, *7*(2), 64–68. doi:10.1111/j.1473-6861.2008.00175.x

Prensky, M. (2001). Digital natives, digital immigrants part 1. *Horizon*, *9*(5), 1–6. doi:10.1108/10748120110424816

Quinney, A., Hutchings, M., & Scammell, J. (2008). Student and staff experiences of using a virtual community, Wessex Bay, to support interprofessional learning: messages for collaborative practice. *Social Work Education*, *27*(6), 658–664. doi:10.1080/02615470802201754

Santy, J., Beadle, M. & Needham, Y. (2009). Using an online case conference to facilitate interprofessional learning. *Nurse Education in Practice*, doi:10.1016/j.nepr. 2008.11.04

Stanton, G. (2008). Access to higher education in England: its position within the wider education ecology. In *Unfinished Business in Widening Participation* (pp. 9–26). London: The Learning and Skills Network.

Stew, G. (2005). Learning together in practice: a survey of Interprofessional education in clinical settings in south-east London. *Journal of Interprofessional Care*, *19*, 223–235. doi:10.1080/13561820500138685

Tait, M., Tait, D., Thorton, F., & Edwards, M. (2008). Development and evaluation of a critical care e-learning scenario. *Nurse Education Today*, *28*, 970–980. doi:10.1016/j.nedt.2008.05.016

Wanless, D. (2004). *Securing Good Health for the Whole Population: final report. Commissioned by the Department of Health*. London: Department of Health.

West, M. A., Borrill, C., Dawson, J., Scully, J., Carter, M., & Anelay, S. (2002). The link between management of employees and patient mortality in acute hospitals. *International Journal of Human Resource Management*, *13*(8), 1299–1310. doi:10.1080/09585190210156521

Woo, K., Gosper, M., McNeill, M., Preston, G., Green, D., & Philips, R. (2008). Web-based lecture technologies: blurring the boundaries between face-to-face and distance learning. *Research in Learning Technology*, *16*(2), 81–93.

# Chapter 20
# Reusable Learning Objects in Health Care Education

**Richard Windle**
*University of Nottingham, UK*

**Heather Wharrad**
*University of Nottingham, UK*

## ABSTRACT

*This chapter will review the definition, development and characteristics of reusable learning objects (RLOs) and outline examples of how these resources are meeting the challenges of interprofessional learning. It will discuss the ways in which pedagogy is developed and expressed within RLOs and how this may impact on interprofessionality.*

## INTRODUCTION

Reusable learning objects (RLOs) are small, granular e-learning resources. They generally utilize multimedia elements to engage the learner in a visual and interactive learning experience. They are mostly web-based and increasingly are being offered as open-education resources, which can easily be accessed and used. Screen-shots from some health-related RLOs which have been used with interprofessional groups are shown in Figure 1, and "live" examples can be found at the following websites/repositories SONET (University of Nottingham, 2009), CIPEL (2009) and UCEL (2009).

DOI: 10.4018/978-1-61520-889-0.ch020

As the name suggests, sharing and reusability are important drivers for the RLO model of e-learning. This philosophy has obvious resonance with interprofessional learning. At its best, it influences every aspect of the design, creation and delivery of the resources. This includes the types of repositories used, compliance with technical standards to allow interoperability between IT systems and the licensing/copyright models adopted. A robust learning object economy based on sharing of resources between diverse stakeholder groups has long been the vision of many in this field (Gunn, Woodgate & O'Grady, 2005; Weller, 2004). Such an economy offsets the production-costs of high quality, media-rich learning resources, by the number of times they can be reused; by different cohorts on the same

*Figure 1. Screenshots taken from examples of RLOs used for interprofessional learning*

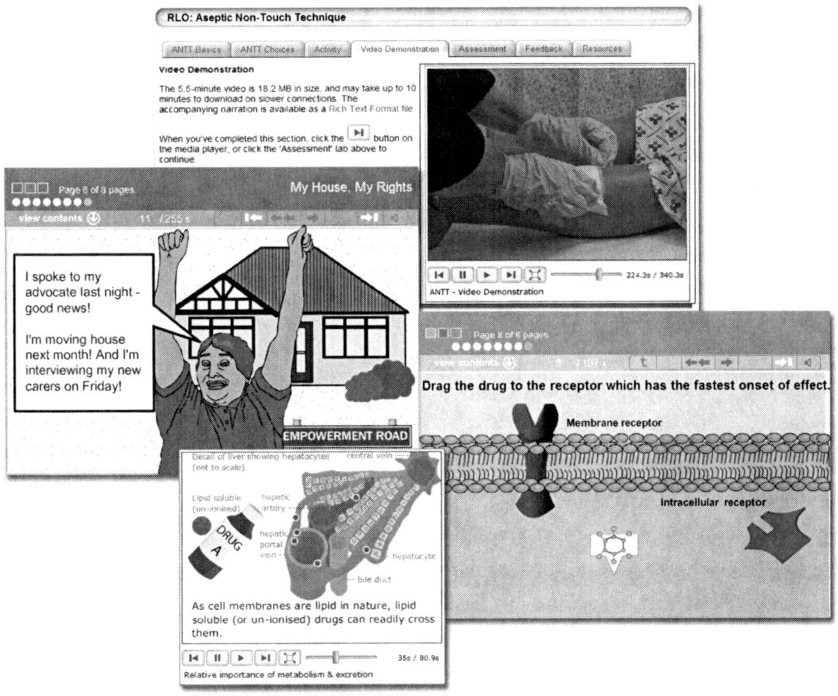


course (vertical reuse), on different courses within the same institution (horizontal reuse) and between institutions, making the final cost per student a miniscule amount.

Although this model has resonance in many areas of higher and further education, it has particular potential in the area of interprofessional health and social care education given the large numbers of students involved. This is multiplied by the wide variety of health and social care courses offered by any one institution, the number of institutions offering such courses and the emphasis on continuing or life-long learning in this area. Taken together these factors mean that certain subject areas are revisited many times. For example, think for a moment how often hand-washing is taught within a given institution; between cohorts, on different courses and for different health and social care disciplines. When this is multiplied by the number of institutions teaching these courses in one form or another, the true potential for sharing becomes apparent.

Indeed, RLOs have been taken up extensively in health and social care education and have proved highly effective in meeting the educational challenges that these curricula face. For example, a series of chemistry RLOs were developed for an interprofessional nursing and midwifery programme to address the difficulties that these students experienced in this area. These proved to be highly popular and effective for these groups, both reducing anxiety and improving knowledge (Windle et al., 2007b). Moreover, these RLOs have also been widely reused by other health care professional students throughout the world. This suggests that that the RLO approach is effective at addressing similar areas of difficulty across the he care professions.

*Figure 2. The reusable learning object continuum, demonstrating examples of the range of resources that have been defined under this umbrella term*

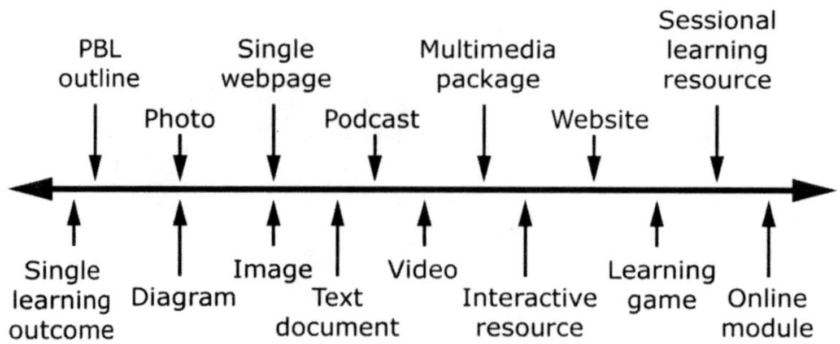


Despite examples like the one given above, and those that appear in a chapter by Wharrad and Windle in this volume, examples of widespread cross-institutional reuse are rare. Take the example of hand washing again. A quick internet search reveals the multitude of virtually identical learning objects produced by different organisations in this area. There does not appear to be an unwillingness to share resources on the part of those developing them (Currier & Campbell, 2002), but rather an unwillingness, or inability to reuse resources produced by others. This so-called "not invented here" syndrome is complex, but is partly related to a perceived lack of ownership of the resources by the potential reuser (Bond, Ingram & Ryan, 2008). Therefore, there is a need to design RLOs to overcome this barrier in order to optimize their potential for reuse.

Although the drivers and barriers to sharing and reuse do not necessarily equate directly with those that influence interprofessional effectiveness, there are obvious similarities. It could be argued that a resource that is optimized for reuse and which empowers the reuser will have many qualities that make it suitable for an interprofessional audience. This chapter will explore aspects of the RLO philosophy that make it successful in modern health and social care education and the extent to which these support or hinder sharing in an interprofessional environment. In particular it will:

- Explore the definitions of RLOs;
- Discuss the community of practice approach that is so often at the heart of RLO development;
- Build a model of the pedagogical attributes that are encompassed by RLOs to investigate their relevance to interprofessional use and sharing.

## BACKGROUND

### Exploring the Definition of Reusable Learning Objects

The term reusable learning object is used widely and can mean very different things depending on the context of its use, or the user. Perhaps the most widely accepted definition of an RLO is that proposed by Wiley as "*any digital resource that can be reused to support learning*" (Wiley, 2000). Although this is solid and objective, it is a very broad definition and can be applied to a wide array of resources that have traditionally be given the label "RLO." Some examples of materials that have been included within the RLO umbrella are shown in Figure 2. The term can be seen to encompass everything from a couple of lines of text, to an entire multimedia module.

However, this broad interpretation of what constitutes an RLO is problematic for a number of reasons. Firstly, learning object repositories are often composed of a bewildering assortment of resources. This can hinder sharing and reuse as any potential user must navigate through an eclectic jumble of resources with variable pedagogical qualities and value. Not only does this make searching for the appropriate resource time consuming and difficult, but it may also undermine confidence in the quality of the resources being offered. Secondly, it is the materials at either end of the continuum (Figure 2) that are most open to criticism. At one extreme, materials are criticized for being so small that they have little or no educational value as learning resources, being meaningless out of the context for which they were created. At the other extreme, materials are too large to be seriously reused in any other context (Friesen, 2003; Polsani, 2003). Therefore, it may be argued that a more focused, less inclusive understanding of what constitutes an RLO is needed to provide confidence in their use and reuse. The very name reusable learning object itself provides a good starting point for this more focused understanding.

## Reusable

Reusable implies that the content is appropriate for learners, outwith the immediate group that it was intended for. Thus the content being covered must have meaning and importance for different groups of learners. It also means that any contextual references included must not preclude its use by other groups. The reduction in contextual reference points can provide a tension between use and reuse (Freisen, 2003; Nurmi & Jaakkola, 2005). Generally, materials designed for use with a particular group benefit from rich contextualization in the form of familiar examples and so on. However, reuse is best served by context-neutral materials. This tension has obvious resonance for the potential use of resources in an interprofessional sphere. Take an RLO on an aspect of pharmacokinetics for example. It is clear that in order to explain this concept clearly, learning facilitators would prefer to use examples of drugs that are familiar to the learning group they are working with, but that these examples may be unfamiliar to practitioners in a different practice-discipline, who may wish to reuse the resource. However, with careful design it is possible to balance this issue (see Wharrad and Windle's chapter in this volume) if consideration is given to the range of examples of drugs included and to their context of use. In some cases generic or hypothetical examples are used to illustrate the learning goal being addressed.

Reusability also requires that a resource is platform-independent. Resources requiring specialized software for delivery should be excluded, as should resources that are developed for, or locked away within, proprietary learning management systems or web environments. Formats such as HTML are most appropriate. Formats that have gained a high level of ubiquity in distribution such as Adobe Flash (player) are also appropriate. However, caution should be applied when downloadable drivers or driver-updates are required. Even if these are freely available, the presence of prohibitive security safeguards on networked computing environments is common in most health-care environments and may hinder reuse.

Resources should also be excluded if they are released with restrictive or prohibitive licensing agreements. Many suitable licensing models are available to support reuse, such as the Creative Commons Licensing model (Creative Commons, 2009). This allows the owner of the material to distribute them freely for use and repurposing whilst retaining the copyright. The distributor of the RLO must have rights to use and distribute the media assets included within any resource, paying particular attention to materials where the copyright may be owned by a third party.

Case-studies, simulations and scenarios such as the RLO "Gateways to Health" (UCEL, 2009) are popular approaches to the development of RLOs for an interprofessional audience. Authenticity is an important aspect of interprofessional learning, but in order to allow resources to be reusable careful attention must be paid to any potential issues of confidentiality for an individual or institution. Likewise, ethical and cultural issues must be respected in the design of such resources,

Finally, for a resource to be considered reusable, it must be available for distribution in a form that is easy to locate and embed. Materials should be labeled with key terms that are searchable and meaningful to the different interprofessional groups who may access it. Materials should also be offered in a format that is standards-compliant, such as an IMS content (IMS, 2005) or SCORM package (SCORM, 2004), so that it can be easily incorporated into learning systems.

## Learning

For the term "*learning*" to be applied to a resource requires that it is more than merely a knowledge or content object, but rather that it has an inherent pedagogy (Boyle and Cook, 2003). The lack of attention to pedagogy in some resources, or the fact that they are instructivist rather than constructivist in nature, has invited further criticism of RLOs in general (Nurmi & Jaakkola, 2005), as has the fact that their design is sometimes driven more by technical considerations than real learners' needs (Friesen, 2003). However, in recent years there have been attempts to address these concerns with an increased emphasis on pedagogy. Therefore, it can be argued that for any resource to be considered an RLO it should have an inherent pedagogy which engages the learner in an interactive learning experience. It might require them to develop, consider, manipulate, reflect upon, measure or analyse knowledge. Examples of this interactive approach can be seen in many of the RLOs in the collections outlined above. Another chapter in this volume by Wharrad and Windle outlines how this approach has been used in order to develop a family of objects that inform and challenge health care staff from a range of professions about improvement culture within their organizations. It should be noted that adherence to these guidelines may considerably limit the amount of resources falling within the RLO category, but will allow the user or learning facilitator to approach materials bearing the RLO title with far more confidence.

Confidence in the learning quality of a resource can also be fostered by more robust attention to quality control to ensure the validity of the content and the pedagogical approaches adopted. These quality control processes should be apparent and transparent. It is perhaps going too far to suggest that in order to be called an RLO a resource should have undergone such a procedure, but it does represent best practice. The current metadata schemes that have been designed for learning objects to not include provision for recording such processes, but this could easily be addressed. The issue of quality assurance is of particular relevance to resources used in relation to health and social care where the accuracy of the content within a resource may be of paramount importance. A quality assurance process also allows the materials to be reviewed by members of different professional groups, ensuring that that everyone can have confidence in its ability to support h.

## Object

The term "*object*" when applied to an RLO encompasses two important aspects of its construction. Firstly, it implies a level of portability and unity. For a resource to be called an RLO, it should be complete within itself and self-contained, adhering to the decoupling and cohesion model proposed by Boyle (2002). It should require no external dependencies or links to other materials in order to function as a learning resource. Secondly, the term implies something that is small in size, or granular in nature. Evaluation with health care students has shown that small packages of e-learning are often their preferred format (Wharrad, Kent, Allcock & Wood, 2001).

However, granularity is very hard to define. What is small, what is granular? Lack of clarity or direction in relation to this parameter, perhaps more than any other, has led to the diversity of resources encompassed within Wiley's definition (Wiley, 2000). So how should we begin to define a level of granularity that constitutes an RLO? It may be possible to determine this pedagogically. In general a learning object should only cover a single concept or learning goal.

Alternatively, it may be possible to define a suitable level of granularity from a behavioral perspective, by suggesting that a single RLO should represent about 5-20 minutes worth of learning activity for the learner. However, this too is problematic for obvious reasons, as it supports an instructivist approach to learning as the time is defined by the RLO author rather than the learner. A pragmatic approach to this question based on the analogy of RLOs as "learning atoms" (Young & Morrison, 2002) may be helpful. In general an RLO should represent the highest level of granularity, or smallest divisible unit of learning, that it is possible to achieve whilst retaining the purpose and holism of a single learning task.

It is recognized that a reductionist approach to e-learning has again been a major source of criticism of RLOs, but it should be noted that despite their individuality, RLOs often work synergistically with one another, especially when combined and contextualized as part of a well planned learning facilitation. Modern tools are being developed that support this combining of resources, whilst allowing each individual resource to retain its independence.

## QUALITIES OF RLOS IN RELATION TO THEIR USE IN HEALTH CARE EDUCATION

### Development Processes

To appreciate the full potential of RLOs for inter-professional learning, it is important to take a step back from the resources themselves and consider their developmental framework. This discussion will draw upon experiences from the Centre for Excellence in Teaching and learning for Reusable Learning Objects (RLO-CETL, 2009), Centre, but has resonance with many contemporary RLO-based development initiatives. The RLO-CETL was set up and funded by the Higher Education Funding Council for England in 2005 in an attempt to seed improvements in learning and teaching within higher education. The RLO-CETL is based at three institutions, London Metropolitan University, the University of Cambridge and the University of Nottingham. Its aim is to develop, research, implement and disseminate best practice is RLO development and design.

The detail of the development process that has been developed by the RLO-CETL has been published previously (Boyle et al., 2006). It begins with scoping workshops in which teams are brought together. These workshops have the functions of team building, exploring the nature and potential of RLOs and beginning to scope the outlines of the resource. At this point, the process remains technology-neutral and initial planning is done using large wipable storyboards. This is important as it prevents disenfranchisement of those who do not feel comfortable with technology and also prevents the constraint of creativity by the confines of the technology. Then the process of iterative development begins. Initially, the team work to develop a detailed storyboard, or specification. Before the specification enters the media development phase it undergoes a series of quality control audits or reviews. It is reviewed by an independent content specialist, representatives from the target learning group and also undergoes pedagogical and technical review if required. A series of tools have been constructed for this purpose. Only on successful completion of these reviews and necessary revisions do the materials enter the media development phase. Developers are free to use whatever software and media are appropriate and best meet the pedagogical needs outlined, so long as they remain within the guidelines outlined above.

Following completion of this phase, the review stage outlined above is repeated with a set of tools that focus on the media representations of the pedagogy. Finally, following further revision, the RLO is released. Generally it will be released for use with the target group of learners first. Extensive feedback and evaluation is collected from this group in the initial phase of release. Details of the evaluation toolkit that has been created for this purpose, together with examples of evaluations of projects where RLOs have been used for interprofessional learning within Higher Education, Further Education and work-based learning are discussed in a chapter by Wharrad and Windle in this volume. Following evaluation and any necessary revision, the RLO will be released for open access as a content package labeled with metadata in the form of keywords, description and information regarding provenance, educational and technical format. Evaluation of the resource and is reuse continues through the inclusion of an online feedback form.

Traditionally many e-learning developments have been seen as "top-down" in nature with too great an emphasis on technical standards rather than pedagogy. This can lead to a mismatch between the developed resources and real learner-requirements. Conversely, RLO development initiatives are often seen as "bottom-up" approaches that seek to empower key stakeholders to participate in and manage the e-learning development process (Cook, Holley, Smith, Haynes & Bradley, 2006). The RLO-CETL, like many other projects, has a development framework based on Wenger's community of practice model of organization (Wenger, 1998). The communities of practice fostered by the RLO-CETL consist primarily of lecturers and students from various disciplines who together develop small families of RLOs to address learning needs that they have identified. The communities are supported by instructional designers and multi-media developers. Whilst the model adopted does not represent a perfect adherence to the community of practice model, it does illustrate three elements that Wenger suggests should be found in a functional community of practice.

Firstly, members of a community of practice should have a sense of connectedness, effective communication and belonging. Extensive research using network analysis has been conducted within the RLO-CETL's communities of practice. This has demonstrated a high degree of connection between the members of the RLO-CETL's communities and also shown them to be dynamic and adaptive, mirroring the "life cycle" model of community outlined by Wenger (Morales & Carmichael, 2007).

Secondly, members should have an active and meaningful role. The actions they undertake should have purpose and value and be suited to the skills and needs of the individual providing them. It is certainly the case that tutors and learners are best placed to understand real learning needs and how these can be addressed within a given field. Other members of the team help in the formation of successful pedagogical strategies and delivery. This may include may include the translation of a pedagogical interaction from one format, for example a seminar, into a form that is appropriate for delivery within an RLO. During network analysis the different stakeholder groups within the RLO-CETL's communities were able to articulate their roles and were able to see the value of their contribution to the finished product (Morales & Carmichael, 2006).

The third characteristic of a community of practice is that members are empowered. Very often tutors feel that their role in e-learning development to is to hand over their content at the technologist's door and leave. However, the community of practice model adopted here, together with the iterative development process that will be outlined shortly, aims to empower tutors and students to participate and influence all aspects of the development process. This is important as tutors and students not only understand the content that needs to be delivered, but will also have a clear idea of the types of pedagogical approaches that will, or will not work, in addressing these. Network analysis within the RLO-CETL demonstrates empowerment at all levels within these communities of practice. For example, this quote taken from a recent study shows the sense of empowerment that student members of the RLO-CETL have articulated:

*Student (2007): Being part of something that is designed to help us. The RLO-CETL really wants us to contribute rather than just saying 'yes, we have students'. But we are actually taking part and getting involved (Windle et al., 2008).*

This empowerment has many ramifications. Many tutors and students are encouraged to develop further resources or to experiment with other areas of technology enhanced learning. Perhaps the greatest benefit from this empowerment, however, is the sense of ownership of the RLOs that it engenders (Windle et al., 2007b). This sense of ownership means that tutors and students are more likely to use the resources, undertake research and evaluation around their use, and also, crucially, be more willing to share the resource and actively recommend them to others, thus providing a springboard for reuse. The role that students have developed as advocates for the resources through social networking sites is one of the surprising, unintended consequences of the RLO-CETL project. For example, another student from the recent study states:

*Student (2007): I would like to help students reusing the RLOs and developing them for their own courses - that will be fantastic! (Windle et al, 2008).*

It has been suggested that communities of practice such as those envisaged by Wenger cannot be created; they either exist or they do not (Wenger, 1998). However, our experience is that communities of practice can be facilitated by providing resources, time, spaces, reward and recognition for individuals and groups (Morales, Carmichael, Leeder, Wharrad & Windle, 2007).

## Benefits of Community of Practice Approach

Although this approach to e-learning development can be time consuming, there are a number of clear benefits, especially for the development of materials for interprofessional health and social care education. Firstly, placing those on the "pedagogical front-line", namely tutors and students, at the centre of the development process should ensure that the resources produced are highly aligned to real-world learning needs. Data from projects, such as those outlined in this book, suggest that these RLOs have a high degree of alignment with the needs of the learners. Learners consistently rated them very highly in terms of the appropriateness of level, content and pedagogical approaches taken. Consistently they rate questions such as "*this RLO supported my learning*", "*was pitched at the right level for me*", "*introduced new concepts clearly*" or "*was appropriate for m course*" with approval ratings of over 90%. Moreover, these ratings are seen consistently across a wide range of subject areas, despite differences in the design of the RLOs, suggesting that tutors and students are indeed developing materials to meet the specific requirements of their own learner groups (Windle et al., 2007).

Secondly, RLOs developed by a community of practice approach have proved to be a highly

*Figure 3. The growth of RLOs within the University of Nottingham School of Nursing, showing the exponential increase in the number of RLOs and number of RLO authors.*

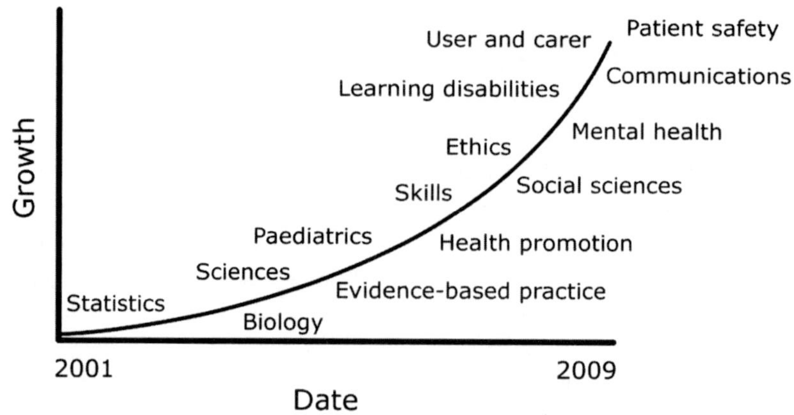


effective way of unlocking content and encouraging many into the e-learning development process for the first time. There are a number of reasons for this. The sense of empowerment, control and ownership undoubtedly play a large part, but it is also the granular nature of the resources themselves that helps. It is far easier to commit to developing content for a ten-minute learning package, perhaps involving a few days work, than a more traditional e-learning resource taking several months.

The effectiveness of the RLO development process in enabling many into the e-learning process can be clearly seen by the exponential growth in RLO development within the School of Nursing at the University of Nottingham from 2001-2009 (see Figure 3). The diagram shows an exponential increase in the number of RLOs being developed over this time and the number of tutors becoming involved in the process, from a handful of enthusiasts in the early days, to a much broader spread including many who would not consider themselves comfortable with technology. It is estimated that over 40% of school staff have been involved in RLO development by 2009. Similarly, the subject areas being covered by RLOs has widened over this time (see Figure 3). To start with the subjects that had been traditionally receptive to e-learning, such as biological sciences and pharmacology were quick to experiment with this new approach, but over time RLO development has spread into areas of the curriculum that were seen as less conducive to e-learning, such as learning disabilities, mental health and social sciences. This has particular relevance to the broad curricula that make up many health and social care education programmes.

By encouraging an increased involvement of staff in e-learning development, the RLO initiative has been seen as a model for institutional change across the RLO-CETL (Cook et al., 2007). The model suggests that there are tipping points beyond which the initiative's further rate of adoption becomes self-sustaining. These tipping points can be seen in the development of a critical mass of RLOs for delivery in programmes and also in the number of individuals becoming involved in the RLO development process. Thus, as a model of institutional change, the RLO development process may have much to offer the growth of interprofessional learning initiatives within the UK and elsewhere.

Perhaps the biggest advantage of the community of practice RLO development model for interprofessional learning is the inclusive nature of the approach. The multidisciplinary community allows different individuals from different fields or areas of expertise to contribute equally to the development process. These individuals work together to develop a learning resource that reflects the experience and knowledge of each. In doing so each discipline represented can gain a sense of ownership. Moreover, other stakeholder groups can also contribute their knowledge and experience to the learning resource. The involvement of students has already been noted. Indeed, in some project areas, the vocational experience of students working with individuals with a learning disability, for example, has formed the very core of resources aimed at supporting an interprofessional approach to working with this client group and has tackled issues such as communication and person-centered planning (Windle et al., 2008). This methodology has also allowed us to engage with practitioners, practice-managers, service users and carers in the development process. Obviously this is of crucial importance in modern health care education and provides a greater range of learning resources for the students.

## Pedagogical Attributes of Learning Objects

When considering best practice for the design of RLOs for interprofessional learning, it is helpful to look at the pedagogical attributes that RLOs display. The facts that the community of practice approach necessitates a move away from high levels of standardization means that designs often reflects the pedagogical preferences of the content authors. This has led to a rich diversity of RLOs designs. Therefore, analysis of existing RLO collections enables us to investigate how different aspects of pedagogy influence factors such as reuse and interprofessionality. However, learning designs within RLOs are rarely explicit. Earlier attempts to classify RLOs tended to focus on their technical attributes or formats, such as the eCornell "*Periodic Table of Online Learning Elements*" (eCornell, 2003), or take a far broader view of what constitutes an RLO as discussed previously (Wiley 2001).

However, as part of an analysis of the ways in which pedagogy affects reusability of RLOs, we have constructed and validated a tool that reviews the pedagogical attributes of RLOs. The tools, known as the learning object attribute metric (LOAM) tool (LOAM, 2007), consist of twelve pedagogical attributes. An example of an RLO scored using the tool is shown in Figure 4. The attributes themselves are derived from the IMS learning design framework (Windle, Wharrad, Leeder & Morales, 2007). IMS Learning Design provides a framework for codifying any instance of learning, from a whole module, to a single learning activity and has been widely adopted by projects that seek to capture learning design, such as the Learning Activity Management System (LAMS 2002).

At its principle level IMS learning design defines a learning instantiation under three major headings: the environment in which the learning takes place, the roles played by those involved in the learning process and the activities undertaken (IMS, 2005). Although the twelve attributes of the LOAM tool were identified with the explicit intention to measure their effect on reuse, many of the drivers and barriers to reusability may equally apply to their interprofessional use. Therefore these attributes will be discussed in relation to the design or selecting of RLOs for use in health care in interprofessional learning.

The attribute called "*objective*" describes the scope and focus of the learning objective addressed by the RLO and the extent to which it is explicit to the learner. The objective can range from small and highly focused to a broader, more diffuse goal. In general, for reuse, a highly focused goal is more appropriate. This tends to make the RLO more portable and clearer from the user's perspective. Broader objectives tend to lose definition and to be more poorly aligned with the content of the RLO. The same guidelines may apply to the use of resources for interprofessional learning.

*Figure 4. An example of the pedagogical footprint generated by the LOAM tool for the RLO "Levels of Measurement" that has been widely used in interprofessional learning.*

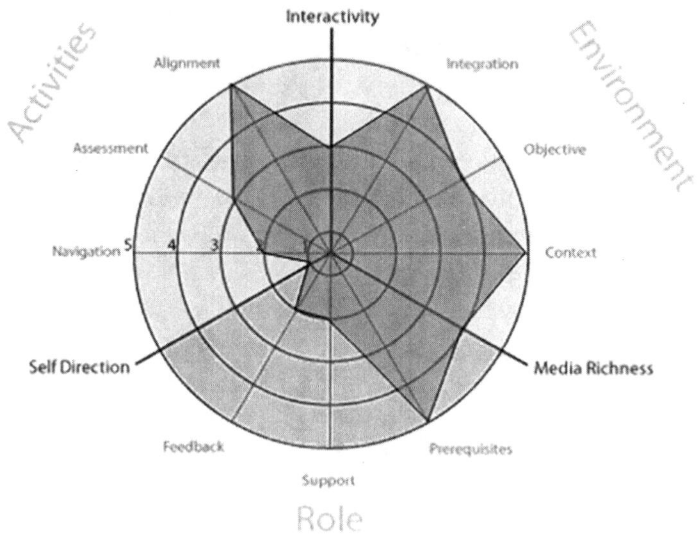


Here especially, a tight focus is required so that the learning goal can be interpreted by learners coming to the RLO from different backgrounds and cultures. There is a recognized tension in that different professional groups may wish to append different learning outcomes to the same resource to meet the needs of their learners, but the ability of a group to coalesce around a single, jointly agreed objective may well be an important first step in designing truly interprofessional materials.

The attribute called "*context*" describes the degree to which the materials are contexualised for a particular group of learners. This attribute represents a major source of conflict between use and reuse, as outlined above, but in general, reuse is best served by materials that are as context neutral as possible. The same may well apply to interprofessional learning. Here the degree of contexualisation must be minimized below the level at which it would bias or alienate one professional group or another through the use of cultural references, language or terminology. The example of pharmacology has already been discussed, but another example involves RLOs that have been developed in the area of patient safety and communication for use with all health practitioners. Here, it was important to include examples from a range of professions and ensure that no professional group was used as an example of poor communication.

"*Media richness*" describes the standard and variety of media elements used within an RLO. In general, learners respond well to a resource that contains a rich mix of media elements, with visual and audio components known to support the learning process for health care students (Lymn, Bath-Hextall & Wharrad, 2008). A high standard of media elements also gives the resource a level of professionalism and credibility that impacts on learning. However, care must be taken to ensure that complex media elements do not detract from learning by over burdening it with an unnecessary cognitive load or become too complex to reduce reusability. The production of an authentic environment through the judicial use of media elements is likely to support interprofessional learning. RLOs that take the learner on a journey through simulated living or care environments have proved particularly useful for exploring issues such as safety, and person-centered planning with an interprofessional audience.

The attribute "*integration*" describes the appropriateness with which different media elements have been have employed to address specific pedagogical issues and the extent to which they have been combined so that the whole is greater than the sum of the parts. The careful selection of media elements is of particular importance in health and social care. Photographs and videos may be appropriate to illustrate certain techniques, or conditions, whereas in other situations the flexibility or "*distance*" afforded by animations may be more useful. Interprofessionality must recognize that social constructs within media representation may have different meanings for different groups. A cartoon that one professional group might find amusing may be seen as offensive or a barrier to learning in another.

"*Interactivity*" describes the extent to which the learner can engage with the RLO. In order to have an active and engaging learning experience, the learner should be able to engage interactively with the learning object. Linear (or didactic) constructions limit interactivity but may be necessary for effective communication of knowledge. In general, simple but engaging interactions spaced throughout the resource are most appropriate. Where interactivity is included it should be designed to engage all professional groups, or to allow different professional groups to step into the shoes of another. RLOs such as Gateways to Health (UCEL, 2009) and Responses to Sexuality in People with a Learning Disability (University of Nottingham, 2009) are good examples of this approach.

"*Navigation*" describes the extent to which navigation forms an integral aspect of the learning design. Linear or directed navigation can be repetitive and tedious, but more open, non-linear constructs with multiple pathways can be confusing and disorienting. The related attribute of "*self-direction*" describes the extent to which the learner has choice in how they work through the resource and its activities, and whether they can choose to undertake particular sections. Branching navigation and a high level of self-direction is often selected as an option when materials are developed for diverse groups of learners, such as those within an interprofessional setting, but their use should be carefully considered, as they can have the effect of hiding information from the learner and may actually act to reinforce interprofessional boundaries.

The attribute "*assessment*" defines the level of self-assessment available to the learner, whilst the related attribute "*alignment*" describes the extent to which the assessment elements measure attainment of the learning objective. In general self-assessment is a good aspect of a learning design within an RLO, but this must be gauged. Self-assessment elements have been shown to engender a sense of ownership of the learning process in health care students, especially when grappling with complex multifaceted curricula (Childs, Blenkinsopp, Hall, & Walton, 2005). However, self-assessment may not always be appropriate, if an RLO is used to explore an ethical dilemma or to challenge perceptions for example. An example of such an RLO is a resource entitled "*Should Sarah Smack her Child*". This interactive tool is aimed at interprofessional groups studying early years education courses. It invites the learner to listen to a range of stakeholders as they give Sarah their own advice and to contribute to the debate (University of Nottingham, 2009).

The attribute "*pre-requisite*" is another attribute that has a major impact on reuse and is similarly relevant to interprofessional learning. It defines the level of knowledge required by the learner prior to undertaking the RLO. Obviously the pre-requisite level of knowledge in a particular subject area will vary between the different professional groups and the learning facilitator must be mindful of this in preparation. Tools such as glossaries may also help to bridge gaps between different professional audiences.

The last two attributes are "*support*" and "*feedback*." Support describes the amount of help and support provided to the learner by the content author within the RLO, including help menus, glossaries navigational support, on screen advice etc, whilst the attribute "*feedback*" describes the level of information provided to the learner in response to the learning tasks and assessments undertaken. Generally, the provision of support and feedback is quite low within RLOs despite the fact that a higher level of support is likely to support reuse and in the context of interprofessional learning; customizable help and support may be appropriate to meet the needs of the different learning groups involved.

Attributes described here were designed to review the reusability of learning objects. Other attributes may need to be considered when designing or choosing materials for interprofessional learning. Indeed, colleagues from the Centre for Interprofessional e-learning (CIPeL) CETL are currently developing a series of attributes to meet this need. The attributes identified include authenticity, multiprofessionalism, problem solving, valuing-others and person-centeredness.

## FUTURE RESEARCH DIRECTIONS

It can be seen from the brief discussion here, that many challenges still exist in the construction of RLOs for reuse and interprofessionality, including the ability to produce granular, context-neutral materials, but at the same time provide integrated, contextualized materials. A number of tools are being developed that address these conflicts. As well as commercial tools, some, such as the Umbrella Learning Object (ULO) creation tools are being developed as open source, or open- license software that will be freely available to learning facilitators and tutors (Taylor, Windle & Wharrad, 2008). This allows individual RLOs to be integrated to form a broader learning resource and also allows the addition of contextual elements to personalize resources for particular groups of learners. Other tools are being developed that allow the adaptation of core RLOs. These so called generative learning object (GLO) templates allow the modification of content to meet the needs of specific learning groups (Boyle, 2006). Finally there is a growing array of rapid e-learning development tools that allow the almost immediate creation of learning objects from existing resource. These include Powerpoint-flash conversion tools and XML driven templating environments such as XERTE (Taylor et al, 2007).

Therefore, there are a number of options available to the tutor to customize materials for interprofessional use. However, whichever of these tools is chosen, it is essential that the pedagogical imperatives outlined above are addressed if the materials created are to be effective in supporting learning and reuse. The future is also likely to see a divergence between RLOs with Web 2.0 enabled functionality which will increase the collaborative nature of learning with RLOs; this will surely benefit the interprofessional audience and will require an extension of the current pedagogical approaches displayed by RLOs.

## CONCLUSION

In conclusion, RLOs represent an e-learning format with a great potential for use within the interprofessional setting. A community of practice model provides an approach that has the potential to harness and empower a range of stakeholders in the creation of learning resources. It also has the potential to develop the collaborative sense of ownership of resources, which is central to sharing and reuse. Reusable learning objects themselves, by the very fact that they are designed for reuse also exhibit many characteristics that suit them for use in an interprofessional setting. However, it is the role of the learning facilitator to contextualize these resources and provide the higher levels of learning designs that is required in the complex environment of interprofessional pedagogy.

## REFERENCES

315-320). Retrieved May 15, 2009, from http://users.tpg.com.au/adslfrcf/lo/whoseLO.pdf

Bond, S. T., Ingram, C., & Ryan, S. (2008). Reuse, repurposing and learning design – lessons from the DART project. *Computers & Education, 50*, 601–612. doi:10.1016/j.compedu.2007.09.019

Boyle, T. (2002). Design Principles For Authoring Dynamic, Reusable Learning Objects. In A Williamson, C. Gunn, A. Young & A. Clear (Eds.), *Winds of change in the sea of learning, 19th annual conference of the Australasian Society for Computers in Learning in Tertiary Education:* Retrieved May 15, 2009, from http://www.unitec.ac.nz/ascilite/proceedings/programme.html

Boyle, T. (2006). The design and development of second generation learning objects. In E. Pearson & P. Bohman (Eds.), *Proceedings of Ed-Media 2006 World Conference on Educational Multimedia, Hypermedia & Telecommunications* (pp. 2-11), June 26-30, 2006; Orlando, Florida.

Boyle, T., & Cook, J. (2003). Learning Objects, Pedagogy and Reuse. In Searle, J. (Ed.), *Learning Technology in Transition. From Individual Enthusiasm to Institutional Implementation* (pp. 31–44). Lisse, The Netherlands: Swets & Zeitlinger B.V.

Boyle, T., Cook, J., Windle, R., Wharrad, H. J., Leeder, D., & Alton, R. (2007). An Agile Method for Developing Learning Objects. In L. Markauskaite, P. Goodyear, & P. Reimann. (Eds.), *23rd annual conference of the Australasian Society for Computers in Learning in Tertiary Education: Who's learning, whose technology?* Retrieved May 15, 2009, from http://papaya.edfac.usyd.edu.au/ascilite_papers/p64.pdf

Campbell, L. M. (2003). Engaging with the learning object economy. In A. LittleJohn (Ed.), Reusing online resources (pp. 35-45). London: Kogan-Page.

Childs, S., Blenkinsopp, E., Hall, A., & Walton, G. (2005). Effective e-learning for health professionals and students – barriers and their solutions. A systematic review of the literature – findings from the HeXL project. *Health Information and Libraries Journal, 22*(Suppl 2), 20–32. doi:10.1111/j.1470-3327.2005.00614.x

CIPEL. (2009). *Learning Objects*. Retrieved August 14, 2009, from http://www.cipel.ac.uk/learning_objects/learning_objects.htm

Cook, J., Holley, D., Smith, C., Haynes, R., & Bradley, C. (2006). Team Enhanced Creativity: An approach to Designing User-Centred Reusable Learning Objects. In Méndez-Vilas, A., Solano Martín, A., Mesa González, J. A., & Mesa González, J. (Eds.), *Current Developments in Technology-Assisted Education (2006)* (*Vol. 3*). Badajoz, Spain: FORMATEX.

Cook, J., Wharrad, H. J., Windle, R. J., Leeder, D., Morales, R., Boyle, T., & Alton, R. (2007). Implementions, change management and evaluation: A case study of the centre for excellence in teaching and learning in reusable learning objects. *Journal of Organisational Transformation and Social Change, 4*(1), 57–73. doi:10.1386/jots.4.1.57_1

Creative Commons. (2009). *Homepage*. Retrieved 15 May 2009, from http://creativecommons.org/

Currier, S., & Campbell, L. M. (2002). Evaluating learning resources for reusability: the DNER and learning objects study. *Winds of change in the sea of learning, 19th annual conference of the Australasian Society for Computers in Learning in Tertiary Education.* Retrieved May 15, 2009 http://www.ascilite.org.au/conferences/auckland02/proceedings/papers/059.pdf

eCornell. (2003). Learning Molecules. In: *eCornell's Approach to Designing Learning Systems*. Retrieved May 15, 2009 from http://www.ecornell.com/about/approach/whitepaperv2_1.pdf

Friesen, N. (2004). Three objections to learning objects. In McGreal, R. (Ed.), *On-line education using learning objects* (pp. 59–70). London: Routledge.

Gunn, C., Woodgate, S., & O'Grady, W. (2005). Repurposing learning objects: a sustainable alternative? *ALT-J Research in Learning Technology*, *13*(3), 189–200.

IMS. (2005). *IMS Global E-learning Consortium*. Retrieved May 15, 2009, from http://www.imsglobal.org/

Ip, A., Young, A., & Morrison, I. (2002). Learning Objects - Whose are they? In *Proceedings of the 15th Annual Conference of the National Advisory Committee on Computing Qualifications* (pp.

Learning Activity Management System. (2002). *Background to LAMS*. Retrieved May 15, 2009, from http://www.lamsinternational.com/about

LOAM. (2007). *The learning object attribute metric tool*. Retrieved May 15, 2009, from http://www.nottingham.ac.uk/nursing/sonet/projects/loam

Lymn, J., Bath-Hextall, F., & Wharrad, H. J. (2008). Pharmacology education for nurse prescribing students - a lesson in reusable learning objects. *Biomedical Central Nursing*, *7*(1), 2.

Morales, R., & Carmichael, P. (2007). Mapping Academic Collaboration Networks: Perspectives from the First Year of the Reusable Learning Objects CETL. *Journal of Universal Computer Science*, *13*(7), 1033–1041.

Morales, R., Carmichael, P., Leeder, D., Wharrad, H., & Windle, R. (2007). *Wolfson reward programme: A boundary zone for development of activity systems in the RLO-CETL through dialogue*. Paper presented at CAL 2007, March 2007, Trinity College Dublin.

Muzio, J. A., Heins, T., & Mundell, R. (2002). Experiences with reusable E-learning objects from m theory to practice. *The Internet and Higher Education*, *5*, 21–34. doi:10.1016/S1096-7516(01)00078-1

Nurmi, S., & Jaakkola, T. (2005). Problems Underlying the Learning Object Approach. *International Journal of Instructional Technology and Distance Learning*, *2*(11), 61–66.

Polsani, P. R. (2003). Use and abuse of reusable learning objects. *Journal of Digital Information*, *3*(4), 164–173.

RLO-CETL. (2009). *Resusable Learning Objects Centre for Excellence in Teaching and Learning*. Retrieved May 15, 2009, from http:www.rlo-cetl.ac.uk

SCORM. (2004). *ADL SCORM*. Retrieved May 15, 2009, from http://www.adlnet.org/Technologies/scorm/default.aspx

Taylor, M., Windle, R., Wharrad, H. J., Rai, H., Tenney, J., Beggan, A., et al. (2007). Supporting the development of flexible learning object development to support reuse in health science education. In: L. Gómez Chova, D. Martí Belenguer & I. Candel Torres, (Eds.), *International Technology, Education and Development Conference*. Valencia, Spain: INTED Proceedings.

Taylor, M., Windle, R. J., & Wharrad, H. J. (2008). *Reuse of Learning Objects. – Introducing the ULO*. Paper presented at Online Educa Berlin: ICWA.

UCEL. (2009). *Reusable Learning objects*. Retrieved August 14, 2009, from http://www.ucel.ac.uk/rlos/

University of Nottingham. (2009). *SONET: RLOs available for use*. Retrieved August 14, 2009, from http://www.nottingham.ac.uk/nmp/sonet/rlos/rlolist.php

Weller, M. (2004). Learning objects and the e-learning cost dilemma. *Open Learning*, *19*(3), 293–302. doi:10.1080/0268051042000280147

Wenger, E. (1998). *Communities of Practice learning, meaning and identity*. Cambridge, UK: Cambridge University Press.

Wharrad, H. J., Kent, C., Allcock, N., & Wood, B. (2001). A comparison of CAL with a conventional method of delivery of cell biology to undergraduate nursing students using an experimental design. *Nurse Education Today*, *21*(7), 579–588. doi:10.1054/nedt.2001.0602

Wiley, D. A. (2000). Connecting learning objects to instructional design theory: a definition, a metaphor, and a taxonomy. In D. A. Wiley (Ed.), *The Instructional Use of Learning Objects*. Retrieved May 13, 2009, from http://reusability.org/read/chapters/wiley.doc

Windle, R., Wharrad, H. J., Laverty, H., Keay, C., Hallawell, B., & Williams, A. (2008). *Student generated content. Permanent footprints or shifting sands*. Paper presented at Online Educa Berlin: ICWA.

Windle, R., Wharrad, H. J., Leeder, D., & Morales, R. (2007a). Analysis of the Pedagogical Attributes of Learning Objects in an attempt to identify Reusable Designs. In C. Montgomerie, & J. Seale, (Eds), *Proceedings of World Conference on Educational Multimedia, Hypermedia and Telecommunications* (pp. 2676-2685). Chesapeake, VA: AACE.

Windle, R., Wharrad, H. J., McCormick, D., Dandrea, J., Bath-Hextall, F., Leighton, B., et al. (2007b). Does a community of practice approach to the development of learning objects support reuse of e-learning materials in health science education. In L. Gómez Chova, D. Martí Belenguer & I. Candel Torres, I. (Eds.), *International Technology, Education and Development Conference*. Valencia, Spain: INTED Proceedings.

# Chapter 21
# Case Studies of Creating Reusable Inter Professional E-Learning Objects

**Heather Wharrad**
*University of Nottingham, UK*

**Richard Windle**
*University of Nottingham, UK*

## ABSTRACT

*Reusable learning objects can play an important part in enhancing interprofessional learning. They provide flexible support to students of health care and provide an opportunity during the creation process, for interprofessional educators to share knowledge and understand more about each other's roles. When creating learning objects, a development and evaluation framework including technical expertise and quality control at critical stages is important, however it is the interprofessional community brought together at workshops at the start of the development cycle and the underlying pedagogical design principles that ensure the materials are fit for purpose and guarantee reuse across professional groups.*

## INTRODUCTION

Within the context of interprofessional e-learning, this chapter will describe three case studies illustrating the use of a development and evaluation model for creation of reusable learning objects (RLOs). A range of issues around RLO design, repositories for interprofessional learning (IPL) and collaborative approaches will be discussed and some recommendations made for future research and development. The four objectives of the chapter are:

DOI: 10.4018/978-1-61520-889-0.ch021

- To provide a working definition of RLOs and explain the relevance and importance of the RLO approach for interprofessional learning.
- To show a generalised model for the creation and evaluation of interprofessional RLOs in health and social care (HSC) and discuss the variants of this in the case studies described.
- To show how educational and content creation issues have been addressed in three case studies of RLO development and use involving interprofessional teams in health care education:

- **Case study 1:** Pharmacology RLOs to support interprofessional learning.
- **Case study 2:** Cross sector development of RLOs for interprofessional learning in health care (LOLA).
- **Case study 3:** RLOs for service improvement in the UK National Health Service (NHS).

- To discuss issues arising from the case studies and make recommendations for further work and research.

## BACKGROUND

Current e-learning practice has moved away from putting whole courses online towards the creation of smaller self contained 'chunks' of learning, recently termed reusable learning objects (RLOs) (Wiley, 2000; Harden & Hart, 2002). RLOs present a number of educational advantages; they are stand-alone units of learning, which can be used in many different ways and across interprofessional disciplines (Windle et al., 2007a). This makes them extremely flexible and cost-effective. Material can be kept up to date more readily; it is much easier to update a single resource than an entire course. Students and teachers alike have access to these resources at any time or place through a standard web-browser. Teachers can combine various RLOs to form the basis for their own custom-made courses (Mason, Pegler & Weller, 2005) or they can direct students to individual RLOs to support or explain particular concepts or processes as part of a blended learning approach (Lymn, Bath-Hextall & Wharrad, 2008). Critics of RLOs would argue that reusability is a myth, and any lecturer will always deliver a subject in their own style reflecting their own slant even in the most concrete of disciplines. In his paper on 'Learning objects: weapons of mass instruction', Butson (2003) says "The overwhelming acceptance of learning objects is baffling given that they represent a decline of learning into a form of reductionism" (p. 667). If learning objects are used like Lego bricks to build courses comprising simply of content, then most educational practitioners would agree with Butson. This chapter will refute this, by demonstrating in three case studies, how pedagogically designed RLOs, created by interprofessional groups are making a real difference to learners.

Wiley (2000) first proposed the definition of an RLO to be 'a digital resource that can be reused to facilitate learning' and since then there have been a plethora of definitions in the literature and this is discussed in more detail in a chapter in this volume by Windle & Wharrad. One of the more pragmatic definitions (Leeder et al., 2002) is relevant to the RLOs described in this chapter: A reusable learning object is a web-based multimedia digital resource based on a single learning objective or goal, comprising a stand-alone collection of four components:

- **Presentation:** a presentation of the concept, fact, process, principle or procedure to be understood by the learner in order to support the learning goal;
- **Activity:** something the learner must do to engage with the content in order to better understand it;
- **Self-assessment:** a way in which the learner can apply their understanding and test their mastery of the content;
- **Links and resources:** external resources to reinforce the taught concept and support the learning goal (Figure 1).

Both the process of developing RLOs and the utilisation of RLOs as resources within health and social care (HSC) education can contribute to the fostering of IPL in a number of ways. The collaborative development of RLOs, comprising a face to face workshop followed by further follow up team working, is a key element of our approach (described below). Such development provides an opportunity for interprofessional

*Figure 1. Components of an RLO*

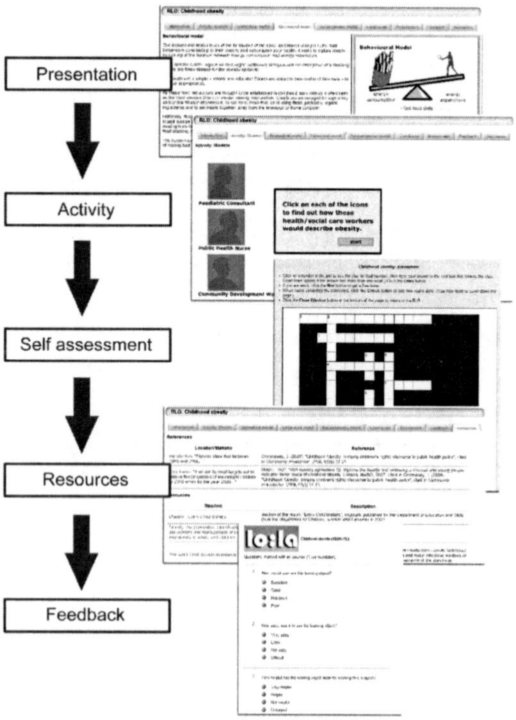


educators and practitioners to combine intellectual effort and expertise towards a common goal (the production of a high quality RLO) in a 'neutral' setting, meaning that the individuals involved are usually novices at RLO development and, in this context, the learners as members of the group become equals in the task. These features of this collaborative process, the common goal, neutrality and equality, provide an ideal opportunity for (inter)professional development via exchange of knowledge and skills (Miller, Ross & Freeman, 2001), out with the pressures and constraints of the usual work setting. Gaining a better understanding of, and respect for, the different professional roles in health and social care (HSC) is an important positive outcome of this collaborative, community based approach (Craddock, O'Halloran, Borthwick & McPherson, 2006).

Recognition of the value of online learning resources in interprofessional education is growing and there are many examples in the literature of projects seeking to capitalise on the affordances of technology for this purpose (Pulman, 2007; Miers et al, 2007; Cooper, Spencer-Dawe & Mclean, 2005; Juntunen & Heikkinnen, 2004). The benefits of RLOs as resources to support interprofessional learning within HSC education will be discussed within each of the case studies described.

## RLO CREATION AND EVALUATION

### Interprofessional Collaboration in RLO Creation: A Model of the Development

The RLO development methodology used in the case studies is schematised in Figure 2. This approach and the associated tools and templates were developed and validated over a number of years, firstly by the Universities Collaboration in e-learning (UCEL www.ucel.ac.uk) (Leeder et al., 2002) and more recently by the Centre for Excellence in Teaching & Learning in Reusable Learning Objects (RLO-CETL www.rlo-cetl.ac.uk) (Boyle et al., 2006).

The process begins with an 'unlocking content' workshop. Developing a community around the development process initiated via workshops is a crucial feature of the methodology. Our and others' research shows this community of practice approach provides a forum for critical debate around the content creation leading to relevant, reusable and high quality materials (Windle et al., 2007a; O'Keeffe, O'Regan & Cashman, 2008). The workshops foster a sense of ownership of the process and the outputs, leading not only to the use of the learning objects themselves but a growing community of use as networks are extended via a range of dissemination routes. This, inevitably, is a labour intensive and expensive process, yet it is recognised that rich multimedia coupled with high production values have enormous power to engage the learner and aid understanding (Edelson

*Figure 2. Schematic diagram showing the developmental framework for the production and quality assurance of RLOs*

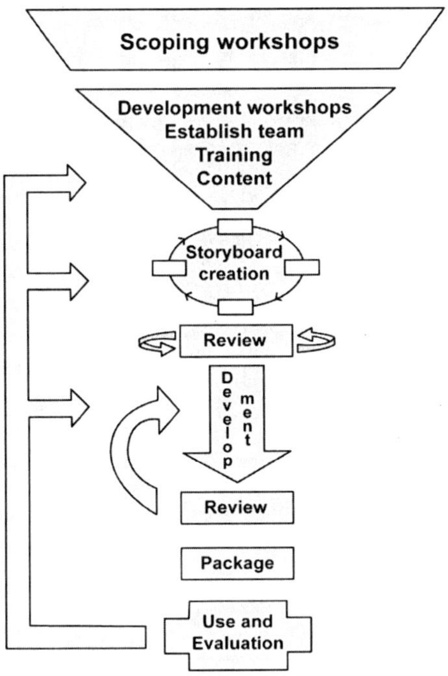


& Pittman, 2001). Outputs from the workshop are in the form of A0 size (1 square meter) laminated posters (Figure 3); these are digitised and stored as an archive forming the basis of the next stage which is the written specification or storyboard.

Storyboards are written through an iterative development cycle overseen by an instructional designer and learning technologist. Completed storyboards are peer reviewed; quality assurance is extremely important to the validity of any e-learning development, especially in the area of HSC. Reviewed storyboards then begin the media production process. A second peer review phase is undertaken on completed media. A set of standard forms available as Microsoft Word documents are available for each of the stages (specification template, peer reviews 1 and 2 forms and evaluation questionnaires). Following appropriate amendments the completed RLOs and their assets are packaged, metadata tagged and added to a repository.

All the RLOs produced are freely available for educational purposes under the terms and conditions outlined by the Creative Commons Licence (http://creativecommons.org/licenses/by-nc/2.0/uk/legalcode) and are available at www.nottingham.ac.uk/nursing/sonet. The recent policy announcements around open educational resources by the UK government (Lipsett, 2008) and various initiatives led by the UK Joint Information Services Committee (JISC) (www.JISC.ac.uk) suggests that open content initiatives will become an increasing focus for higher education (HE) and further education (FE) sectors in the UK.

## Pedagogical Design

The pedagogical design underpinning the RLOs in each of the case studies to be described was based on IMS learning design principles (IMS Global Consortium, 2003) with the emphasis on the environment in which the learning occurs, the roles played by the learner and "RLO-author" and the activities undertaken. Designs ensure that the most appropriate multi-media environment is created, so that learners take active roles within the learning process and are supported by help and feedback (Windle et al., 2007b). Activities and self assessments in the RLOs are aligned with the learning goal (Biggs, 2003) and are important because users must be actively engaged in the process of learning (Laurillard, 2002) and need feedback from self assessments to determine whether they have successfully achieved the learning goal. A variety of activities including crosswords, drag and drop, open text, annotating diagrams, multiple choice quizzes were incorporated. High granularity, partly to facilitate reuse and "just-in-time learning", but also as our research (Wharrad, Kent, Allcock & Wood, 2001) shows, small online learning units of between five and 20 minutes are the most valued and effective for the learner.

*Figure 3. Examples of A0 laminated storyboards – the outputs from the RLO creation workshops*

## Evaluation Framework and Toolkit

The framework and tools used to evaluate the process and outputs of RLO creation have been devised by the RLO-CETL. The theoretical framework for the RLO-CETL evaluation strategy informed by Cultural-Historical Activity Theory has been reported elsewhere (Morales, Carmichael, Wharrad, Bradley & Windle, 2007). The primary purpose of the evaluation is to ensure the pedagogical effectiveness of the RLOs, but we also determine the tutors' views on the usefulness of the RLOs and an understanding of the various teaching and learning contexts in which the RLOs are situated, so for example whether the context is interprofessional or multiprofessional. The evaluation tool kit includes: technical evaluations of the new content; assessments of changes in teaching and learning environments and practices; formative evaluation by teachers and students; and summative measures of use.

The key elements of the toolkit used in the case studies are listed below (the toolkit can be downloaded from www.rlo-cetl.ac.uk):

- A short online survey added to each RLO comprising 10 questions about usability and quality (collated using Zoomerang software www.zoomerang.com).

- A more detailed questionnaire completed by students who have used one or more RLOs in their module.
- Tutors' forms that collect information about the module and the context for the RLOs' use. The tutor's evaluation of learning effectiveness form is based on the JISC case study template (http://www.e-learning.ac.uk/effprac/html/cs_template.htm); this is a tool designed to encourage tutors to reflect on the RLOs' use.
- Peer review forms comprised of fixed and open response questions to determine from independent experts in the field whether the RLO content accurately represents the subject area..
- Learning Object Attribute Metric (LOAM) tool. This allows RLOs to be classified according to various attributes such as media richness, interactivity etc and is described in more detail in another chapter in this volume by Windle & Wharrad.

## CASE STUDIES

### Case Study 1: Interprofessional Development of Pharmacology RLOs

Students across all health professions find it difficult to relate pharmacology concepts to clinical experience. As more nurses, pharmacists and allied health care professionals seek training in prescribing, the levels of academic knowledge of trainee prescribers are likely to become more diverse. Lymn, Bath-Hextall & Wharrad (2008) stated that almost half of students attending their prescribing course had no more than a GCSE in a biological science subject. There is also a lack of classroom time available for teaching pharmacology (Wharrad, Allcock & Chapple, 1994; Latter, Rycroft-Malone, Yerrell & Shaw, 2001; Latter, Maben, Myall & Young, 2007; Avery & Pringle, 2005), and students state that they would benefit from preparatory material covering basic prin-

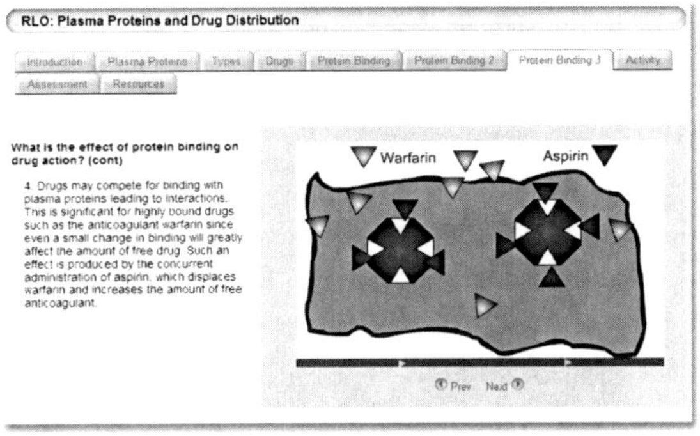


*Figure 4. Screenshot from RLOs for Case Study 1*

ciples to develop familiarity with pharmacological terminology (Bradley, Bradshaw & Nolan, 2006). Perhaps part of the problem lies with the nature of pharmacology education which appears to utilise traditional teaching methodologies almost exclusively (Banning, 2004). While lectures result in delivery of information they do not necessarily engender learning and understanding, which may be better supported by blended or more applied teaching methodologies such as RLOs.

The first workshop to explore the possibility of developing pharmacology RLOs (Figure 4) for healthcare courses was held in 2003 and was attended by seven lecturers responsible for teaching pharmacology to nurses, midwives and medical students (on both traditional and graduate entry courses) and two learning technologists. The different academic and professional requirements were debated and draft specifications were agreed and drawn up for seven RLOs covering bioavailability, half life, first pass metabolism, plasma proteins, drug-receptor interaction, understanding the synapse and the lock and key hypothesis. Each lecturer produced a more detailed specification for one of the RLOs including sketching ideas for images, animations, activities, self assessment and glossary. Animations were produced in Macromedia Flash and embedded in HTML web pages. Although funding was available to buy out lecturers from their teaching, in fact they did not or could not arrange this. No existing templates were available so these had to be developed from scratch and copyright free images were difficult to find and most had to be drawn. Peer review at the written specification and prototype RLO stages, lengthened the process (this was similar to journal article reviewing) so it took over a year to complete all the RLOs.

Since their release in 2004, thousands of students from medicine, nursing, midwifery, physiotherapy, dentistry, podiatry and veterinary science have now been used interprofessionally and there is wide global reuse of the RLOs by a range of healthcare learners. This success has led to further workshops involving other lecturers from nursing, medicine and allied health professions. The number of pharmacology RLOs released to date is eighteen and evaluations are almost exclusively positive. Access to RLOs statistically significantly increased an interprofessional group of post registration students' perceived understanding of pharmacology in successive cohorts and had an impact on their clinical practice (Lymn et al., 2008). The ability to visualise processes as computerised animations or videos compared to trying to understand from static text appears to enhance learning and increase test scores (Thatcher, 2007; Chew, Stiles, Joseph & Whitley, 1994). Qualitatively, student comments from our online evaluation feedback reinforces this claim,

*For me it was the visual aspect, actually seeing the concept visually was a huge bonus for me because it just made things click. We'd had key lectures and I'd read about things, but I think for me just to see how things worked visually was what I needed to put the whole picture together (Non medical prescribing student, University of Nottingham).*

*The simplicity and the fact that it had text and diagrams to support the explanation. I also thought that the activity at the end was a good idea as it reinforced learning (Podiatry student, University of Northampton).*

The value of these pharmacology RLOs may be partly because they are, by their very nature, different to traditional e-learning tools – they require no more than around 15 minutes to complete thus they do not require a time-intensive input making them more flexible for students to use at work or home. Many students commented on being able to repeatedly use particular RLOs until they understood a concept. Similarly the visual, audio and interactive nature of these RLOs means that they have an appeal for visual, auditory and kinaesthetic learners an important issue bearing in mind data which suggests that learning style is important in web-based e-learning (Manochehr, 2006).

## Case Study 2: Cross Sector Development of RLOs for Lifelong Learning in Health Care Education

Despite many UK government policy recommendations (DFES, 2003; 2005) cross sector education collaboration involving UK HE and FE are difficult to initiate and sustain for a whole variety of reasons (Weiss, 1987); however, more support is needed for learners moving from one educational sector to another. This case study (LOLA - Learning Objects, Lifelong Application), a three year project funded by the UK Eduserv Foundation in 2005, describes a cross sector collaboration involving a UK HEI (post 18 years university), FE College (post 16 years education) and an NHS Trust. The aim was to collaboratively develop a set of interprofessional RLOs in health spanning the lifelong learning continuum. The RLOs would be catalogued and housed in an 'open' repository along with all their constituent media components.

Along with a project director, learning technologist and researcher, the project team consisted of two lecturers from the HE sector (pharmacologist and infection control specialist),

*Figure 5. Screenshot from the LOLA website, Case Study 2*

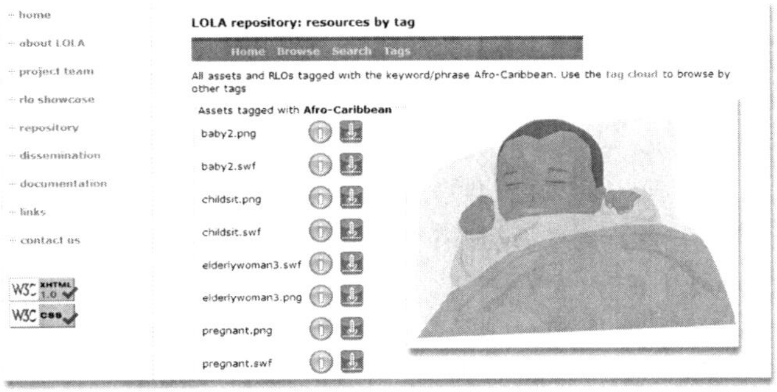


two NHS employees (a senior pharmacist and a senior infection control nurse both delivered professional development sessions to NHS staff and medical students) and two lecturers from a local FE College (a health visitor and a biologist). These members were content experts in the areas of infection control and medicines management and taught interprofessional groups of students. RLOs were produced following the process outlined in Figure 2.

Fifteen RLOs were produced on: aseptic technique, glove use, personal protective equipment, childhood obesity, body mass index, general adaptation syndrome, aminoglycoside dosing, volume of distribution, inflammatory response, bacteria and viruses, sizes and scales, prokaryotes and eukaryotes, introduction to clearance and clinical consequences of changes in clearance and home hazards. All the RLOs with one exception (Home Hazards) used a custom built 'media player' which housed the text, audio commentary and Flash and video components of the RLO and provided the navigational control buttons. The design template and functionality was informed by feedback from users of previous RLO designs (Windle, McCormick, Dandrea & Wharrad, 2009). RLOs were packaged using RELOAD and catalogued with the UK LOM Core Metadata Schema (this schema has fields relating to instructional design and pedagogy). A hundred and eighty media assets were catalogued using the simpler Dublin Core schema (Figure 5).

Online feedback data shows that the LOLA project and its outputs have benefited the wider community well beyond the immediate project team (Wharrad et al., 2008). The RLOs have been embedded into many modules particularly for the HEI courses; the partner sector use is not as great, for the NHS sector this is because they do not deliver as many sessions; for the FE college the lack of IT facilities, lack of IT skills of tutors and students, and a concern that despite input and peer review from FE lecturers, the RLOs were too difficult for FE students not necessarily in terms of the subject content but the level of the language used in the text and narration. Whilst the individual RLOs produced in this project are benefiting many healthcare students, the repository represents an example of a bank of health resources available for different professional groups.

Chapter 14 in this volume by Timmons et al. describes further research findings from this project and the project website and repository contains the outputs and documentation from the LOLA project (www.nottingham.ac.uk/nursing/lola).

## Case Study 3. RLOs for Health Professionals and Staff on Service Improvement in the NHS

The final case study describes a commissioned project funded in 2008 by the UK NHS Institute for Improvement and Innovation (NHSI), which took five months to complete. The project remit was to produce one hour of e-learning on service improvement in the NHS in the form of learning objects to be used by any health professional or other NHS employee either as part of an induction or training programme. The overarching aim of the improvement project is 'to develop a future workforce that has the mindset to challenge and be challenged in the way that healthcare is delivered, together with the skills to make the necessary improvements' (NHS Institute for Innovation and Improvement, 2008). This is to be achieved by threading the message into professional education and training that everyone, whatever discipline or grade, has a contribution to make to provide better, safer healthcare. Fifteen text based Leader Guides on improvement theory and practices (Laher & Penny, 2005) had already been produced and included for example, the plan, do, study act (PDSA) cycle, process mapping and producing run charts. For us there were two key challenges: firstly how to select the most desirable content for the RLOs from the extensive text based resources already available; and secondly how to design them in a way that would be appropriate for this diverse user group.

At the initial workshop twenty NHS interprofessional improvement trainers and educators who worked in NHS Trusts across the UK scoped ideas and content outlines for an RLO, each taking one of the broad Leader Guide themes. Using their expertise the key messages to convey in the RLOs were established along with some analogies that could be developed into animations to illustrate the ideas.

The first of three RLOs is on 'Improving Care'. The key message is that improvement is everyone's responsibility and that little improvements can have a big impact. The user is asked to think about an activity they have organised recently, maybe a family event such as a party, they then work through a model for improvement that introduces an animated version of the Plan, Do, Study, Act cycle (Langley, Nolan, Nolan, Norman & Provost, 2009). 'Planning for Improvement' is the second RLO which takes the form of an interactive patient journey. The user identifies areas for improvement along this journey, and then uses the PDSA cycle to show how improvements could be planned. The third RLO challenges the user to think about their practice and how this relates to the six core dimensions of the Knowledge and Skills Framework (Department of Health, 2009). A portfolio tool allows learners to record and print information as they work through the package (Figure 6).

In this case study, there were variations to the standard development framework (Figure 2). The project team (two e-learning health science academics, two NHSI academics and two media developers) developed the RLO content using the workshop storyboards as a guide and a 'distributed' peer review process, implemented by putting links and online forms onto a closed project website, was used to allow the workshop attendees to feedback on the prototype RLOs. The RLOs are in the NHS e-learning object repository (http://www.e-learningrepository.nhs.uk/) and are also available at www.nottingham.ac.uk/nursing/sonet. The RLOs are being incorporated into NHS induction programmes and HEI pre-registration interprofessional health courses, with positive evaluations.

*Figure 6. Screenshot from RLOs for Case study 3. (© 2008, NHS Institute for Improvement and Innovation. Used with permission).*

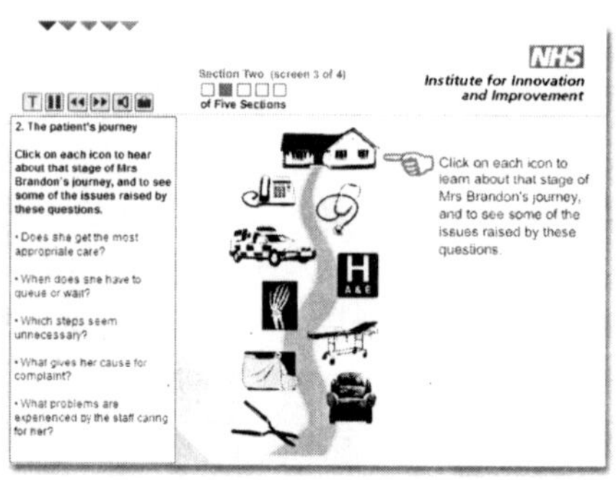

## DISCUSSION AND FUTURE RESEARCH DIRECTIONS

This collaborative process for interprofessional RLO creation based on learning design principles has addressed some of the criticisms identified in previous investigations of the effectiveness of e-learning technologies for health professionals, among them cost, poorly designed packages, lack of skills, need for a component of face-to-face teaching, time intensive nature of e-learning (Wharrad et al, 2001) and computer anxiety (Wilkinson, Forbes, Bloomfield & Gee, 2004; Childs, Blenkinsopp, Hall & Walton, 2005). Whilst the RLOs have been exclusively rated by users as high quality and having an impact on learning (Lymn et al., 2008; Windle et al., 2007; Wharrad et al., 2008; Windle et al., 2009), each case study project brought with it a different challenge in terms of the creation of RLOs: devising appropriate visual representations of complex scientific processes and reaching agreement on the core concepts in the first case study; in the second, the challenge of cross sector working and building a repository and the third – condensing a large amount of material into learning objects that would be suitable (not too difficult on the one hand or patronising on the other) for a very diverse range of users. A further synopsis of some key issues will follow with a focus on future directions and research.

### Repositories of Interprofessional RLOs

A Gartner report (Yanosky, Harris & Zastrocky, 2004) claimed that e-learning repositories have the potential to be transformational in the way HE delivers education yet many repositories are initially successful but have been difficult to sustain. Expensive knowledge repositories have been set up in a 'top down' manner and not used by lecturers for a range of technical, cultural and attitudinal reasons (Tate & Hoshek, 2009). Even community driven 'bottom up' repositories are difficult to sustain (Brosnan, 2005). In order for interprofessional users to share knowledge residing in a repository some key factors that we have identified, driving success are: ensuring that there is a critical mass of RLOs to make the repository worth visiting and searching; ensuring that the metadata describing the RLOs is effective so that lecturers and learners can access content in focused ways (Kwan & Balasubramanian, 2004; Nash, 2005); the need for enough context to allow evaluation of the content (Weiss, Capozzi & Prusak, 2004); providing guidance and support for contributors about copyright issues and continually adding to the RLO bank by seeking funding for new projects and partners. There is a difficult tradeoff between ensuring quality and control over the depositing of materials into a repository and the necessary flexibility and agility demanded by contributors who do not have a lot of time. Further research is needed to understand more about how lecturers and students are using repositories especially when federated searches and harvesting of RLOs from multiple interprofessional repositories becomes possible.

### Granularity and Context

Despite the success of collaborative workshops in producing fit for purpose interprofessional RLOs, granularity remains a key factor in ensuring reuse. The focus on a distinct learning goal is relevant to the use of RLOs within an interprofessional learning setting. Whilst a defined focus ensures that the RLO has a limited size or granularity, this can be a problem in interdisciplinary development teams when each discipline seeks to retain its own specific context. As shown in the case studies, by means of negotiation and cooperation, a satisfactory compromise can be reached and this process in itself is valuable for understanding more about the culture and practices of different professional groups.

## Quality and Workflow

There is a tension between ensuring RLO quality, by adhering to the quality assurance steps, limited funding (Pulman, 2007) and deadlines for release of the materials or delivery to student groups. A major consideration is managing the process illustrated in Figure 2. In our experience the lecturer-centric approach is crucial to ensure materials are fit for purpose and reuse. This impacts on the complexity of project management since RLO creation has to be fitted into an already busy lecturer's and clinician's schedule and project managers have to ensure work flow to media developers is not interrupted. As RLO communities of practice diversify to include more interprofessional groups, the logistics of project management become more complex. In the case studies described the time spent on this aspect of the project is hidden but it does add up to many person hours; however, the bite-sized RLO approach does have advantages over many technology development projects where loss of funding and momentum leads to unfinished products.

## Collaborative Approaches

Creating RLOs collaboratively using the standardised framework and tools has allowed lecturers and clinicians new to, and sometimes sceptical of, e-learning to become involved. Creators of RLOs are required to analyse their perspectives and teaching materials on the topics they deliver and often be more critical and creative; the workshops always draw debate on what the focus of the RLOs should be; this is a new type of role for lecturers, lesson preparation is more often done in isolation and therefore much less transparent and open to scrutiny. Web 2.0 philosophies, that of students, users, carers and client groups contributing their own content in the form of stories and reflections using various technological platforms, (Pulman, 2007) is becoming commonplace and we too are now involving these groups in RLO development (Windle et al., 2008; Hallawell, Windle, Wharrad, Laverty & Williams, 2009). Further research might address how an interprofessional sharing culture is influenced by the process of RLO creation and how reuse is influenced by different content creators.

# CONCLUSION

This chapter has illustrated via three case studies how an interprofessional collaborative method for creating RLOs has resulted in a widely, flexibly used and growing bank of high quality materials. The collaborative process of creation itself forces us to think in different ways. It demands that we take the wider view, broaden our horizons. Interprofessional collaboration enriches our perspectives, makes us more aware of others; this can only benefit healthcare outcomes in the long term. Interprofessional lecturers and educators, engaged in creating RLOs discover the need to reflect on the very processes of teaching and of learning at a fundamental level; this informs their teaching practice in many other situations beyond the world of e-learning whilst also creating valuable high quality resources available for present and future generations of students. RLOs can thus be seen as a vanguard for interprofessional educational change.

# REFERENCES

Avery, A., & Pringle, M. (2005). Extended prescribing by UK nurses and pharmacists. *British Medical Journal*, *331*, 1154–1155. doi:10.1136/bmj.331.7526.1154

Banning, M. (2004). Nurse prescribing, nurse education and related research in the United Kingdom: a review of the literature. *Nurse Education Today*, *24*, 420–427. doi:10.1016/j.nedt.2004.05.002

Biggs, J. B. (2003). *Teaching for quality learning at university* (2nd ed.). Buckingham, UK: Open University Press/Society for Research into Higher Education.

Boyle, T., Cook, J., Windle, R., Wharrad, H. J., Leeder, D., & Alton, R. (2006). An Agile Method for Developing Learning Objects. In L. Markauskaite, P. Goodyear & P. Reimann. (Eds.), *23rd annual conference of the Australasian Society for Computers in Learning in Tertiary Education: Who's learning, whose technology?* Retrieved April, 30, 2009, from http://papaya.edfac.usyd.edu.au/ascilite_papers/p64.pdf

Bradley, E., Bradshaw, C., & Nolan, P. (2006). Nurse lecturers' observations on aspects of nurse prescribing training. *Nurse Education Today, 26*, 538–544. doi:10.1016/j.nedt.2006.01.008

Brosnan, K. (2005). *Developing and sustaining a national learning object sharing network:A social capital theory perspective.* Retrieved October 28, 2008, from http://www.ascilite.org.au/conferences/brisbane05/blogs/proceedings/12_Brosnan.pdf

Butson, R. (2003). Learning objects: weapons of mass instruction. *British Journal of Educational Technology, 34*, 667–669. doi:10.1046/j.0007-1013.2003.00359.x

Chew, F. S., Stiles, R. S., Joseph, E., & Whitley, M. D. (1994). Computer-assisted instruction with interactive videodisc versus textbook for teaching radiology. *Academic Radiology, 1*, 326–331.

Childs, S., Blenkinsopp, E., Hall, A., & Walton, G. (2005). Effective e-learning for health professionals and students – barriers and their solutions. A systematic review of the literature – findings from the HeXL project. *Health Information and Libraries Journal, 22*(Suppl 2), 20–32. doi:10.1111/j.1470-3327.2005.00614.x

Cooper, H., Spencer-Dawe, E., & Mclean, E. (2005). Beginning the process of teamwork: Design, implementation and evaluation of an inter-professional education intervention for first year undergraduate students. *Journal of Interprofessional Care, 19*(5), 492-508. Retrieved May 5, 2009, from http://www.informaworld.com/smpp/title~content=t713431856~db=all~tab=issueslist~branches=19 - v19

Craddock, D., O'Halloran, C., Borthwick, A., & McPherson, K. (2006). Interprofessional education in health and social care: fashion or informed practice. *Learning in Health and Social Care, 5*(4), 220–242. doi:10.1111/j.1473-6861.2006.00135.x

Department of Families. Education & Schools. (2003). *Towards a Unified e-Learning Strategy Consultation Document.* Retrieved October 1, 2008, from http://www.dcsf.gov.uk/consultations/downloadableDocs/towards%20a%20unified%20e-learning%20strategy.doc

Department of Families. Education & Schools. (2005). *Harnessing Technology – Transforming Learning and Children's Services.* Retrieved October 1, 2008, from http://www.dcsf.gov.uk/publications/e-strategy/

Department of Health. (2009). *The NHS Knowledge and Skills Framework (NHS KSF) and the Development Review Process.* Retrieved May 5, 2009, from http://www.dh.gov.uk/en/Publicationsandstatistics/Publications/PublicationsPolicyAndGuidance/DH_4090843

Edelson, P. J., & Pittman, V. V. (2001). *E-Learning in the United States: New Directions and Opportunities for University Continuing Education.* Retrieved May 5, 2009, from http://www.sunysb.edu/spd/dean_papers/newdelhi.pdf

Hallawell, B., Windle, R., Wharrad, H. J., Laverty, H., & Williams, A. (2009). *Student generated e-learning within Learning/Intellectual Disabilities.* Paper presented at 4th International conference on e-learning. Toronto, Canada.

Harden, R. M., & Hart, I. R. (2002). An international virtual medical school (IVIMEDS): the future for medical education? *Medical Teacher*, *24*(3), 261–267. doi:10.1080/01421590220141008

IMS Global Learning Consortium. (2003). IMS *Learning Design information model version 1.0 final specification.* Retrieved March 14, 2009, from http://www.imsglobal.org/learningdesign/ldv1p0/imsld_infov1p0.html

Juntunen, A., & Heikkinen, E. (2004). Lessons from interprofessional e-learning: piloting a care of the elderly module. *Journal of Interprofessional Care*, 18(3), 269-278. Retrieved Auguest 10, 2009, from http://www.informaworld.com/smpp/title~content=t713431856~db=all~tab=issueslist~branches=18 - v18

Kwan, M. M., & Balasubramanian, P. (2003). KnowledgeScope: Managing knowledge in context. *Decision Support Systems*, *35*(4), 467. doi:10.1016/S0167-9236(02)00126-4

Laher, A., & Penny, J. (2005). Service improvement. In E. Pack (Ed.), Organisational Development In Healthcare: Approaches, innovations, achievements. Oxford, UK: Radcliffe publishing.

Langley, G. L., Nolan, K. M., Nolan, T. W., Norman, C. L., & Provost, L. P. (2009). *The Improvement Guide: A Practical Approach to Enhancing Organizational Performance* (2nd ed.). San Francisco, CA: Jossey-Bass Publishers.

Latter, S., Maben, J., Myall, M., & Young, A. (2007). Evaluating nurse prescribers' education and continuing professional development for independent prescribing practice: Findings from a national survey in England. *Nurse Education Today*, *27*, 685–696. doi:10.1016/j.nedt.2006.10.002

Latter, S., Rycroft-Malone, J., Yerrell, P., & Shaw, D. (2001). Nurses' educational preparation for a medication education role: Findings from a national survey. *Nurse Education Today*, *21*(2), 143–154. doi:10.1054/nedt.2000.0528

Laurillard, D. (2002). *Rethinking University Teaching: A Conversational Framework for the Effective Use of Learning Technologies*. London: RoutledgeFalmer. doi:10.4324/9780203304846

Leeder, D., McLachlan, J. C., Rodrigues, V., Stephens, N., Wharrad, H. J., & McElduff, P. (2004). Universities' Collaboration in e-learning (UCeL): a virtual community of practice in health professional education. In P. Kommers, P. Isaias & M. B. Nunes (Eds.), Lisbon, Portugal IADIS Web-based communities, (pp. 386-393).

Lipsett, A. (2008). UK universities should take online lead - Calls for revival of e-learning to secure future of higher education. *The Guardian*. Retrieved May, 5, 2009, from http://www.guardian.co.uk/education/2008/nov/11/e-learning-university

Lymn, J., Bath-Hextall, F., & Wharrad, H. J. (2008). Pharmacology education for nurse prescribing students - a lesson in reusable learning objects. *Biomed Central Nursing*, *7*(1), 2.

Manochehr, N.-N. (2006). The influence of learning styles on learners in e-learning environments: An empirical study. *Computers in Higher Education Economics Review*, *18*, 10–14.

Mason, R., Pegler, C., & Weller, M. (2005). A learning object success story. *Journal of Asynchronous Learning Networks*, *9*(1), 97–105.

Miers, M., Clarke, B., Pollard, C., Rickaby, C., Thomas, J., & Turtle, A. (2007). Online interprofessional learning: the student experience. *Journal of Interprofessional Care*, *21*(5), 529–542. doi:10.1080/13561820701585296

Miller, C., Ross, N., & Freeman, M. (2001). *Interprofessional education in health and social care*. London: Arnold Publications.

Morales, R., Carmichael, P., Wharrad, H. J., Bradley, C., & Windle, R. (2006). *Developing a Multi-Method Evaluation Strategy for Reusable Learning Objects: an approach informed by Cultural-Historical Activity Theory.* Presented at the 1st European Practice-based and Practitioner Research Conference - Improving quality in teaching and learning: Developmental work and Implementation challenges, University of Leuven, Belgium.

Nash, S. S. (2005). Learning objects, learning object repositories, and learning theory: Preliminary best practices for online courses. *Interdisciplinary Journal of Knowledge and Learning Objects, 1*, 217–228.

NHS Institute for Innovation & Improvement. (2008). *Improvement in Pre-Registration Education for better, safer healthcare: An NHS Institute initiative to introduce pre-registration health and social care students to improvement*. Retrieved May 7, 2008, from http://www.institute.nhs.uk/images//documents/BuildingCapability/Building%20 improvement%20capability/Preregistration%20 training/51427%20Pre-registration%20leaflet.pdf

O'Keeffe, M., O'Regan, L., & Cashman, D. (2008). *Supporting the development of Communities of Practice: informal versus formal communities*. Paper presented at the Association for Learning Technology conference 'Rethinking the Digital Divide' University of Leeds, UK.

Pulman, A. (2007). On the road to Virtual Europe – Redux. *The Electronic Journal of E-learning, 5*(4), 297–304.

Tate, M., & Hoshek, D. (2009). A model for the effective management of reusable learning objects (RLOs): lessons from a case study. *Interdisciplinary Journal of E-learning & Learning Objects, 5*, 51–72.

Thatcher, J. D. (2006). Computer Animation and Improved Student Comprehension of Basic Science Concepts. *The Journal of the American Osteopathic Association, 106*(1), 9–14.

Weiss, J. (1987). Pathways to cooperation among public agencies. *Journal of Policy Analysis and Management, 7*(1), 94–117. doi:10.2307/3323353

Weiss, L., Capozzi, M. & Prusak, L. (2004). Learning from the internet giants. *MIT Sloan Management Review, Summer,* 79-84.

Wharrad, H. J., Allcock, N., & Chapple, M. (1994). A survey of the teaching of biological sciences on undergraduate nursing courses. *Nurse Education Today, 14*, 436–442. doi:10.1016/0260-6917(94)90004-3

Wharrad, H. J., Herman, L., Timmons, S., Randle, J., Lymn, J., Shaw, W., et al. (2008). *How easy is cross sector collaboration? A case study of the collaborative development of e-learning in infection control, medicines and prescribing*. Proceedings of International Conference Education Research & Innovation, Madrid, Spain.

Wharrad, H. J., Kent, C., Allcock, N., & Wood, B. (2001). A comparison of CAL with a conventional method of delivery of cell biology to undergraduate nursing students using an experimental design. *Nurse Education Today, 21*(7), 579–588. doi:10.1054/nedt.2001.0602

Wiley, D. A. (2000). Connecting learning objects to instructional design theory: a definition, a metaphor, and a taxonomy. In D. A. Wiley (Ed.), *The Instructional Use of Learning Objects*. Retrieved May 5, 2009, from http://reusability.org/read/chapters/wiley.doc

Wilkinson, A., Forbes, A., Bloomfield, J., & Gee, C. F. (2004). An exploration of four web-based open and flexible learning modules in post-registration nurse education. *International Journal of Nursing Studies, 41*, 411–424. doi:10.1016/j.ijnurstu.2003.11.001

Windle, R., McCormick, D., Dandrea, J. & Wharrad, H.J. (2009). The use and reuse of reusable learning objects for addressing challenging areas of the curriculum. *Nurse Education Today* (submitted for publication).

Windle, R., Wharrad, H. J., Laverty, H., Keay, C., Hallawell, B., & Williams, A. (2008). *Student generated content. Permanent footprints or shifting sands*. Paper presented at Online Educa Berlin: ICWA.

Windle, R., Wharrad, H. J., Leeder, D., & Morales, R. (2007b). *Analysis of the Pedagogical Attributes of Learning Objects in an attempt to identify Reusable Designs.* In C. Montgomerie & J. Seale (Eds.), *Proceedings of World Conference on Educational Multimedia, Hypermedia and Telecommunications* (pp. 2676-2685). Chesapeake, VA: AACE.

Windle, R., Wharrad, H. J., McCormick, D., Dandrea, J., Bath-Hextall, F., Leighton, B., et al. (2007a). Does A Community Of Practice Approach To The Development of Learning Objects Support Reuse of E-Learning Materials In Health Science Education. In L. Gómez Chova, D. Martí Belenguer, & I. Candel Torres (Eds), *International Technology, Education and Development Conference*. Valencia, Spain: INTED Proceedings.

Yanosky, R., Harris, M., & Zastrocky, M. (2004). *Hype cycle for Higher education Gartner report G00121162*. Retrieved September 28, 2008, from http://www.gartner.com/it/section.jsp?type=press_releases&format=archive&year=2004

# Chapter 22
# Customisation and the Interprofessional Application of E-Learning Objects

**Helen M. Lynch**
*Canberra Institute of Technology, Australia*

**Kerry Trabinger**
*Canberra Institute of Technology, Australia*

## ABSTRACT

*Toolbox learning objects are a class of pedagogically rich, sophisticated e-learning objects created for the Australian vocational education and training system (VET). Their richness makes them very attractive to teachers and trainers working across a range of learning contexts but at the same time makes them difficult to reuse. While these e-learning objects have been designed to be customised and are often repurposed for use within one vocational context, an approach is emerging that sees them increasingly customised for reuse across a range of intervocational or interprofessional contexts. This chapter describes this approach, focusing on the tools and techniques of customisation, and presents a model of reuse that can be implemented elsewhere with any pedagogically rich web based e-learning object in intervocational and interprofessional settings. Toolbox learning objects are freely available to anyone with internet access from the Toolbox Learning Object Repository website. The Repository is fully searchable and objects can be previewed from the Repository website and downloaded without charge for educational use. This chapter will be of value to teachers, trainers and academics who are exploring the reuse of pedagogically rich web based e-learning resources for interprofessional or intervocational education.*

## INTRODUCTION

Definitions of e-learning objects, reusable learning objects or just plain learning objects (Ashley, Davis & Pinsent, 2008, p.13) are as many and varied as the names by which they are known. Kay and Knaack (2005), after an extensive literature survey, settled on a definition incorporating perspectives from the field of object orientated programming and teaching and learning. They defined learning objects as "reusable, interactive web-based tools that support the learning of specific concepts by

DOI: 10.4018/978-1-61520-889-0.ch022

enhancing, amplifying, and guiding the cognitive processes of learners" (Kay & Knaack, 2005, p. 230). In providing this functional, working definition from the literature the authors identified a quality of learning objects that has been greatly debated: that of reusability (Elliott and Sweeney, 2008; Ashley, Davis & Pinsent, 2008; Paris, 2006).

Emerging from the debate is a suggestion that the capacity of practitioners to reuse learning objects depends on the development of an understanding of reuse that addresses it in three important contexts. First, conceptually, a clarification of the idea of reuse and its "defining properties" (McDonald, 2006, p. 542) is needed. Second, practically, practitioners require "better tools ... to create and edit their own and others' online learning activities" (Fill, 2006, Section 6). Finally, what is also needed is a practical and holistic approach to the implementation of pedagogically coherent learning objects that will guide teachers and trainers in their efforts to repurpose them for use across diverse professional and vocational contexts (Elliot & Sweeney, 2008).

In the Australian VET system practitioners are already working with a holistic approach using tools and techniques that support the reuse of a class of e-learning objects called Toolbox learning objects. These are derived from an e-learning resource called Flexible Learning Toolboxes created for Australian VET practitioners through Australia's national VET e-learning strategy. The work discussed in this chapter has the potential to be implemented with many types of pedagogically rich web based e-learning objects outside the Australian VET system in ways that enhance and support the use of e-learning objects in interprofessional education.

## BACKGROUND

Definitions of e-learning objects have generally focused on specifying their attributes. Polsani (2003) suggests that there is general agreement that e-learning objects should have technical qualities that make them accessible, interoperable and reusable. Further definitions suggest that e-learning objects should not just support learning but be instructionally designed to ensure they are credible educational resources (Harvey, 2005). Pedagogically rich, highly granular e-learning objects amply meet these criteria. However their size and sophistication can make them difficult to reuse because of the depth at which the learning or subject specific context is often addressed in such objects. This difficulty highlights the fundamental tension between pedagogically rich, high granularity objects and their reuse.

Despite this, practitioners recognise and are excited by e-learning objects that are well designed, instructionally sound, and rich with activities, explanations and information. They immediately see how such objects could be customised to support reuse with their student group or teaching and learning context. What stops them from undertaking such activity is a lack of technical skill and appropriate, freely available, easy to use tools and techniques for repurposing (Fill, Leung, DiBiase & Nelson, 2006; McDonald, 2006). They are further hampered by trying to work with e-learning objects where little attention has been paid in their design, presentation and creation to the ways in which they might be reused (Paris, 2003).

## An Approach to Customising E-learning Objects

### Flexible Learning Toolboxes

The Australian VET system has access to a group of web based teaching and learning resources whose development is funded through the Government's e-learning strategy: The Australian Flexible Learning Framework (The Framework). Toolbox learning objects (TLO) are one type of resource in this group. They are highly granular (consisting of many segments or pages), peda-

*Figure 1. Working with clients – a scenario for healthcare workers*

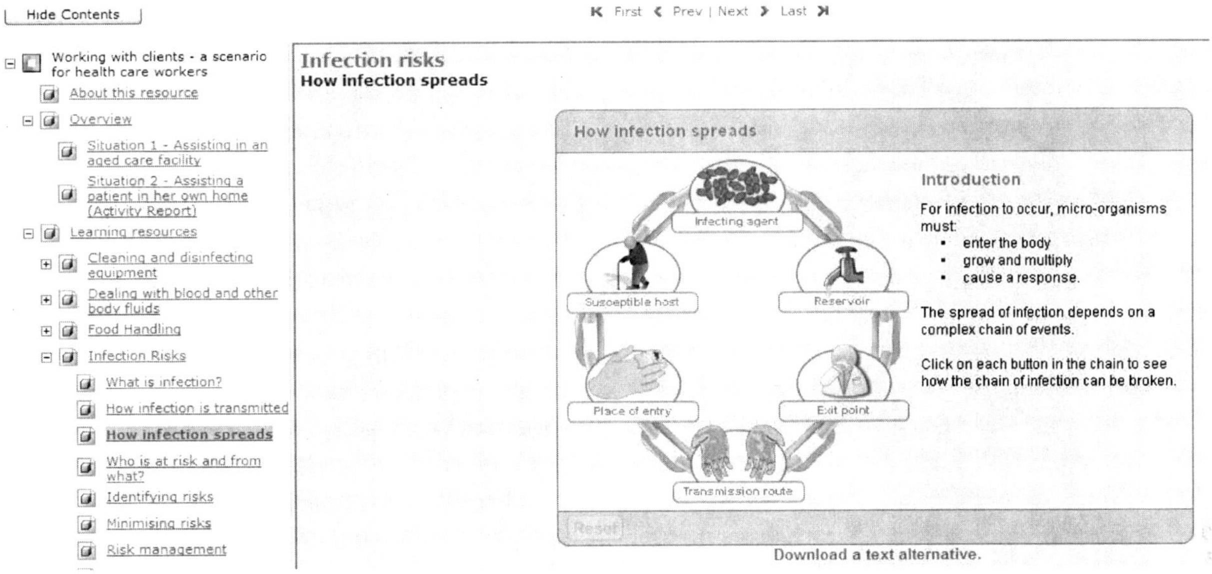


gogically sophisticated and have been designed to be disaggregated, and customised (Thompson & Kimberley, 2005) to enhance their reuse.

## Toolbox Learning Objects

TLO are created in a particular format. They consist of material designed to support learning developed in a series of web pages sequenced and packaged for online delivery. Each web page has teaching and learning content such as activities, knowledge resources and learner supports (instructions, schedules diagrams, or explanations). Pages are ordered to form a learning sequence. Each page in the learning object, for example, may be a single page containing text, images, links and embedded video or audio resources. Each page itself might also contain a link to a further learning sequence (a linked series of web pages), such as a complex activity, explanation, or quiz. The learning sequence is instructionally designed to meet learning goals and will typically contain learning design elements in the form of tasks, resources and supports (Oliver & Herrington, 2001).

The page shown in this learning object window (Figure 1) is an individual page with images, text, and additional pages that open when a computer mouse is hovered over an image in the infection cycle. The page is the third in a learning sequence called "Infection Risk". This learning object can be found, previewed and freely downloaded from the Toolbox Learning Object Repository.

These learning objects are created to meet the requirements of IMS content packaging standards. The IMS Content Packaging Specification is an international technical standard for creating e-learning resources which are interoperable and accessible. IMS is the name of the international corporation that publishes these standards. IMS content packages can be deployed in different learning or content management systems (like Moodle™) without having to be converted to different formats. They can be stored, described, searched for, discovered and downloaded from digital repositories and they will function on different computer systems.

It is in this form that Toolbox learning objects are made freely available over the internet from the Toolbox Learning Object Repository. The IMS package format means the learning objects have a metadata file which describes them and allows them to be stored and retrieved from a learning object repository. They have a manifest file that records their organisation so that a content package player or learning management system can display them correctly (keeping the pages in the right order). Learning objects downloaded from the Repository contain the Framework Content Package Player™ so they can be viewed easily in a web browser without the need for learning or content management system. Free and open source software such as Reload Editor™ or the VET Reload Tool™ can be used to create these IMS compliant learning objects by sequencing and packaging individual web pages in any desired order.

## What is Customisation?

Customisation is the technical process for repurposing pedagogically rich e-learning objects for reuse with learner groups and learning contexts for which they were not specifically designed. Toolbox learning objects have been specifically designed to enable customisation in order to increase this opportunity for reuse. Reuse of Toolbox learning objects through customisation is supported by the inclusion of adaptable, generic learning designs and broad learning objectives. Technically, customisation is supported by the use of web pages and resources that are easily edited, replaced or recreated in other forms.

For example, such repurposing means that an e-learning object with a problem based learning design, resources and activities for trainee nursing assistants learning about teamwork, could be extended and used by trainee allied health professionals (naturopaths and trainee mental health or Indigenous primary health care workers). The shared use of such a customised resource has the potential to facilitate collaboration by a range of trainee health care workers both in the classsroom and in their future workplaces. The options for customising in this way are only limited by the relevance of the e-learning object's learning design and the adaptability of the learning objectives addressed by the e-learning object. Generic design and the use of broad rather than narrow learning objectives enhance reusability on this level. This is generally one of the goals of e-learning object development (De Salas & Ellis, 2006). While much work still needs to be done on e-learning objects, learning design and learning theory, there is evidence of an emerging body of work that is exploring these issues, and moving forward with reusable learning designs to support e-learning object use (Oliver and McMahon, 2006).

There are a number of customisations that can be undertaken on web page based e-learning objects. Customisation most commonly completed by teachers and trainers working with students online requires a relatively low level of technical skill. The easiest customisations consist of replacing text documents that have been included in the e-learning object. This basic type of customisation includes becoming familiar with the structure of an e-learning object; its folders, files and pages. A Toolbox learning object typically contains assessments, short answer or practical workplace exercises as downloadable text documents. These documents are easily replaced with documents custom created by the teacher or trainer.

A slightly more technical customisation includes using a What You See Is What You Get (WYSIWYG) web editor to replace text or images, to add links or embed Web 2.0 applications such as a YouTube™ video (a video stored on the website, www.youtube.com, which allows for free video sharing over the internet) or Really Simple Syndication (RSS), a type of web page on a website that allows frequently updated content to be published and sent instantly to a user who has subscribed to the site. Highly technical customisations include using source computer code (java script or action script) to alter interactions or animations in order to change the way they behave in response to a user's action. One example of this type of interaction is where the user selects and drags an element (a phrase, image or icon) with the computer mouse to a new location on the screen which prompts feedback on the user's selection. This type of customisation requires more technical skill than most teachers or trainers could be expected to acquire.

## An Approach to Reusing E-Learning Objects Through Customisation

The benefits that accrue to learners, teachers, trainers and training organisations from e-learning object use, as well as those gained by other flexible approaches to teaching and learning, are well documented. E-learning objects can offer more individualized learning, and are more accessible. They facilitate the sharing of knowledge across disciplines and they provide opportunities for the reuse and repurposing of content in new ways (De Salas & Ellis, 2006). E-learning objects increase the level of resources available to learners and increases teacher's resource choices (Oliver & McMahon, 2006). Further benefits include convenience of use and ease with which currency of content can be maintained and student's learning experiences enriched (Recker et al., 2005). By far the most significant benefit is one delivered by the very nature of e-learning objects as reusable resources. If a technically skilled teacher or trainer can envisage how a learning activity (in an e-learning object) can be enhanced through customisation to facilitate intervocational or interprofessional learning they can create an enriching learning experience which has been tailored specifically to meet their student's learning needs.

As practitioners become more familiar with online pedagogies and the e-learning object format and develop the basic technical skills of web page editing, the emergence of an approach to reusing pedagogically rich e-learning objects becomes evident. This emerging approach comprises customisation techniques and software applications that begin to address one of the major impediments to reusing or repurposing e-learning objects at any technical level and for any purpose: the lack of technical skill and appropriate, freely available, easy to use tools and techniques to make customisation achievable (Fill, Leung, DiBiase & Nelson, 2006; McDonald, 2006).

The discovery and use of free, open source software available for download from the internet is one of the first steps in preparing to undertake e-learning object customisation. WYSIWYG web editing software such as Kompozer™, image editing software like Irfanview™ and audio editing software such as Audacity™ are freely accessible to anyone with internet access. These tools support changing text on a web page, replacing an image in a web page and creating an audio file as an alternative or supplement to text. Content packaging tools such as The VET Reload Tool™ and Framework Content Package Player™ that help to create and maintain standards compliant e-learning objects, are also freely available on the internet. These tools support the packaging of e-learning objects and their use on compact disk or personal computer as well as in learning or content management systems. These free open source applications minimize the technical skill and the financial investment in software needed to undertake such work. Experience suggests that free and open source tools are generally easier to use than proprietary software simply because they have fewer functions and offer a less sophisticated approach to the task at hand.

Tools and techniques for editing web pages that are easily applied by teachers and trainers with limited technical skills are highly valued. With these tools and a willingness to apply or acquire a very basic understanding of how web pages are created and edited, a practitioner (teacher or trainer) can:

- Add, remove, edit or replace text files incorporated in an e-learning object.
- Add, remove and edit content in a web page.

- Add, remove or replace whole web pages.

Adding or editing the content of an e-learning object web page can include:

- Replacing images.
- Adding, or replacing audio files.
- Replacing flash action script or java script interactions with alternatives such as images and audio files.
- Embedding or linking to video and other such multimedia files.
- Embedding or linking to other web resources.

The approach also facilitates adding interactivity and dynamic content to e-learning objects by linking to or embedding Web 2.0 applications. Web 2.0 is often referred to as social software because it offers such opportunities for socially engaging online activities emphasising communication, collaboration and interaction. Examples of social software include Flickr™ (a photo sharing website) and Scribblar™ (a collaborative site with text chat and shared whiteboards). Many other Web 2.0 applications such as wikis (simple web pages that groups can edit together over the internet) and blogs (a weblog, a type of website used by an individual to regularly make commentary that can be read over the internet by other people) can also be incorporated in an e-learning object webpage in the same way.

## Undertaking Basic Customisation

Teachers and trainers using e-learning objects developed from web pages can very simply undertake customisation to adapt them either to the needs of their learner group or to a shared vocational or professional context in one of two ways:

- Editing or adding to the content of web pages in an object.
- Changing the format in which that content is presented.

Adherence to a few basic rules designed to maintain the technical integrity of the e-learning object, such as not altering its folder and file structure is required and, once understood, allows for both types of customisation to be accomplished with only a minimum level of technical skill.

Customisations at this basic level can take any number of forms. Some examples might include:

- Replacing selected images with specific images more relevant to learners' workplaces or professional contexts such as clinical settings.
- Using images of people, places and equipment with which the learners are familiar to replace generic images.
- Duplicating text content with content in audio form for students who would rather listen than read or to learning situations where listening is more appropriate.
- Replacing icons used in instructional sequences so they are more appropriate to the learning context;
- Replacing text documents, consisting of support material or assessment activities, with new custom designed text documents.
- Replacing or reorganising text on pages to change the learning context or learning design.
- Adding to the text on pages providing students with direction, instruction and explanation more suited to their learning context.

With the exercise of a little more technical skill pages can be removed or completely new pages added. However removing or adding pages changes the way the object is structured so it must be repackaged to allow changes to be recorded in the e-learning object's manifest. It is worth noting that changing the content on individualweb pages in an e-learning object does not require repackaging to occur.

Repackaging can be completed with the VET Reload Tool™ but this kind of customisation has not been well known or widely practiced until recently and is regarded as a little more advanced than text or image editing. However with the publicised availability of this tool from the Australian Flexible Learning Framework website practitioners are experimenting with making their own e-learning objects and attempting more sophisticated customisation work with existing e-learning objects.

## A Customisation Scenario for Reuse in Health Care

A Toolbox learning object called '*Infection Risk*', freely downloadable from the Toolbox Learning Object Repository, addresses the topic of infection transmission and risk management in health care settings. This is an eight page learning sequence with very generalized factual information about its topic, containing a link to an authoritative website detailing formal infection control guidelines. It has a very simple interactive drop, drag and click activity where the learner identifies infection risks in a series of scenes. The overall design and content are very generic and straightforward. The materials in the learning object could be suitable for medical, dental, nursing, allied health, community and social care students in clinical and welfare settings. It should be noted that there are a number of versions of the *Infection Risk* learning object in the Repository, this is the least sophisticated of them.

### A Scenario

Management, staff and regular health care practitioners at a local aged care facility are concerned about recent incidents which suggest health practitioners (assistant nurses, personal care assistants and regular visiting allied health practitioners such massage therapists, dental hygienists) may not be as vigilant about infection risk as they should be when attending to patient needs. After general consultation with those concerned, management has decided to deliver refresher training on infection control. Managers hope to provide opportunities for the different health care practitioners to learn collaboratively in the hope that this will improve the overall management of infection control. Their key learning activities will be a number of scenarios about infection control based on issues faced by all practitioners and some which might be unique to the activities of each separate area of health care practice. This should raise cross discipline awareness of the range of professional health care activities in an aged care setting that have an infection risk. The learning activity will require practitioners to share examples of recognize/analyse/solve risk scenarios from their own experience and provide solutions to scenarios posed by other practitioners. It is expected that participants will be working asynchronously and, for some activities, in pairs with a practitioner who has a different professional background.

The refresher course is to be completed using a learning object and a community of practice site (Australian educators have Edna Groups™) offering discussion forums and collaborative spaces. The learning object can be placed on every computer in the facility and made available to all practitioners on CDRom or USB drive or from a server or web site. Study, preparation and the learning activities can be completed at times convenient to each practitioner either at home or while at the facility. Customisation opportunities could include enhancing the learning object to:

Emphasise the shared workplace and collaborative and interprofessional nature of the refresher training:

- Name and logo of the aged care facility is included on pages where appropriate.

- Images of common work areas and equipment are added to emphasize a familiar context for learning.
- Text can be added to introduce the activity and explain the need for training and what it will achieve.
- Infection control problem/solutions exemplars can be added from different practitioner perspectives to model desired infection control risk management approaches.

Expand the factual information about infection control by including:

- Links to more authoritative web sites and more sophisticated knowledge resources.
- Annotated lists of links to documents, reports or research findings (sourced from the web or the workplace server).
- Linked (or embedded) audio visual resources from a variety of sources.

Create a stronger more supportive learning design by re-ordering materials over the eight pages of the learning object:

- Cut, copying and paste content to move it between pages.
- Add more instruction/learner support such as voice files from the training facilitator, diagrams, and mind maps.
- Use icons to show or draw attention to learning design elements such as additional readings or exemplars.
- Include instructions about the learning activity, scheduling, how to contact a training facilitator, how to participate in a learning community (or community of practice).

## The Tools of Customisation

Software and techniques of customisation do not have to be complex or difficult to learn. Teachers and trainers customising the *Infection Control* learning object discussed can use freely available open sources software to complete the task. The techniques applied required only the basic technical skill of webpage text editing, linking and the simplest form of image manipulation. Tutorials in a learning object format detailing these tools and techniques are available for download and viewing in the Toolbox Learning Object Repository. Basic customisation techniques are easy to learn and apply; software is easily accessible and freely available. For example:

- Irfanview™ for image editing, from http://www.irfanview.com/
- Komposer™ for web page editing, from http://www.kompozer.net/

Both software packages can even be run from a USB drive (often called a memory sticks or thumb drive) so a practitioner can work on their e-learning object from any computer without having to install software on that computer.

The availability of two other key pieces of software as free downloads from the Framework web site means there is now a suite of tools to fully support low to medium levels of customisation. Learning objects are downloaded from the Toolbox Learning Object Repository already containing the Framework Content Package Player™ which allows them to be viewed with a web browser. This makes the learning objects easy to examine and evaluate for use. Toolbox learning objects are packaged with the VET Reload Tool™.

The VET Reload Tool™ has only been available to Australian practitioners since 2008. It offers exciting possibilities for more advanced customisation activities. While Reload Editor™ on which it is based has been available for much longer it appears not to have been widely known or used in the Australian VET system. Reload Editor™ in the form of the VET Reload Tool™ enables individual pages of content to be organized or structured in a form that can be "read" inside a learning management system (LMS). The tool allows for tremendous flexibility in manipulating e-learning object pages. Other file formats such a text documents, image, video and audio files can be packaged alongside web pages. The user can create nested sequences from individual pages, determine what pages appear and in what order. Pages can be hidden from view in the e-learning object and easily reinstated. A large e-learning object with many pages or sequences can be easily reduced to just one or two pages. Sequences can be enriched with the addition of new web pages or other types of resources such as video.

The main advantage of the software is that a user does not have to create navigation between the pages in a learning sequence. Creating navigation on a web page requires web development skills that are generally beyond the skills of most teachers and trainers. With the VET Reload Tool™ a teacher can open an e-learning object and remove a page from the organisation or change the place it appears in the e-learning object without creating web page navigation. A teacher can download a template from the web, use WYSIWYG web editing software to add or change text or images and simply package the separate pages together as an e-learning object.

The Framework's promotion of the VET Reload Tool™ and content package player has been well received. Practitioners quickly recognized the flexibility that the tool offers both for customising Toolbox learning objects and for making their own e-learning objects from the resources they have already created.

## FUTURE RESEARCH DIRECTIONS

The value of using and customising e-learning objects for interprofessional or intervocational education, as this practice grows, needs to be explored, particularly in relation to how learning designs and objectives facilitate customisation. Work also needs to be done to discover how best to raise the awareness of teachers and trainers to customisation opportunities with the potential to bring together learners from different vocational contexts in the same fields of study with the ultimate aim of improving collaborative outcomes in the workplaces. Future directions for research lies also in documenting and analysing the Australian experience of Toolbox learning object users and their learners with respect to customising e-learning objects for use with various learner groups and across vocational contexts.

Further work is needed to study the mechanisms of professional development that will best support teachers and trainers to customise e-learning objects and to make their own using the VET Reload Tool™ or its equivalent. Research should identify what further tools might be needed to support customisation and deal with new tools and applications (the emerging Web 3.0) that increase ways to add interactivity and dynamic, personalised content.

## CONCLUSION

There are obviously a number of factors that contribute to building a workable approach to repurposing pedagogically rich e-learning objects for use in interprofessional education and training. Some of these are directly related to the technical qualities of e-learning objects and how they are designed and developed while others relate to the tools and techniques of customisation and packaging practiced and shared by educators. Case studies conducted by Oliver and McMahon (2006) on the use of Toolbox learning objects certainly suggest that strategies incorporating learning object design and development: systems, and tools and techniques to support the practitioners who use them, are critical if the use of e-learning objects is to become more mainstream. But most importantly individual practitioners and training organisations should be encouraged to identify and create high quality, pedagogically rich e-learning objects that offer opportunities for reuse and customisation in situations where students who train for different vocations and professions can learn with and from each other. The Australian VET experience with e-learning objects and the creation of sophisticated online learning resources is certainly worth considering as a model of implementing e-learning strategies based on reusable resources.

## REFERENCES

Ashley, K., Davis, R., & Pinsent, E. (2008). *Significant properties of e-learning objects (SPeLOs).* Draft Version, 0.92. A Report commissioned by the JISC Information Environment Committee under the Information Environment Strategic Theme. Retrieved February 16, 2009, from http://www.jisc.ac.uk/media/documents/programmes/preservation/spelos_report.pdf

De Salas, K., & Ellis, L. (2006). The development and implementation of learning objects in a higher education setting. *Interdisciplinary Journal of E-learning and Learning Objects*, (2), 1-22. Retrieved July 10, 2009, from http://ijklo.org/volume2.html

Elliot, K., & Sweeney, K. (2008). Quantifying the reuse of learning objects. *Australasian Journal of Educational Technology, 24*(2), 137-142. Retrieved February 23, 2009, from http://www.ascilite.org.au/ajet/ajet24/elliott.pdf

Fill, K., Leung, S., DiBiase, D., & Nelson, A. (2006). Repurposing a learning activity on academic integrity: the experience of three universities. *Journal of Interactive Media in Education, 1*. Retrieved February 15, 2009, from http://jime.open.ac.uk/2006/01/

Harvey, B. (2005). Learning Objects and instructional Design. *The International Review of Research in Open and Distance Learning, 6*(2). Retrieved February 25, 2009, from http://www.irrodl.org/index.php/irrodl/article/view/227/310

Kay, R., & Knaack, L. (2005). Developing Learning objects for secondary school students: A multi-component model. *Interdisciplinary Journal of Knowledge and Learning Objects, 1*(1), 229–254. Retrieved March 30, 2009, from http://ijklo.org/Volume1/v1p229-254Kay_Knaack.pdf

McDonald, J. (2006). Learning object: A new definition, a case study and an argument for change. *Who's learning? Whose technology?: Proceedings of the 23rd Annual ASCILITE Conference,* Sydney, Australia. Retrieved December 30, 2008, from http://www.ascilite.org.au/conferences/sydney06/proceeding/pdf_papers/p99.pdf

Oliver, R., & Herrington, J. (2001). Online learning design for dummies: professional development strategies for beginning online designers. In P. Barker & S. Rebelsky (Eds.), *Proceedings of ED-MEDIA 2002, World Conference on Educational Multimedia, Hypermedia and Telecommunications*. Norfolk, VA: AACE (p. 1500-1505). Retrieved September 20, 2009, from http://elrond.scam.ecu.edu.au/oliver/

Oliver, R., & McMahon, M. (2006). *Trials of Learning Objects (Series 7).* Retrieved February 16, 2009, from http://toolboxes.flexible-learning.net.au/documents/pdfs/lotrials_report06.pdf

Paris, M. (2003). Reuse in Practice: Learning objects and software development. *Interact, Integrate, Impact: Proceedings of the 20th Annual ASCLITE Conference*. Adelaide, Australia. Retrieved March 12, 2009, from http://www.ascilite.org.au/conferences/adelaide03/docs/pdf/679.pdf

Polsani, R. P. (2003). Use and Abuse of Reusable Learning Objects. *Journal of Digital Information, 3*(4). Retrieved January 15, 2009, from http://jodi.tamu.edu/Articles/v03/i04/Polsani/

Recker, M., Dorward, J., & Dawson, D. Xin Mao, Ye Liu, Palmer, B. et al. (2005). Teaching, Designing, and Sharing: A Context for Learning Objects. *Interdisciplinary Journal of Knowledge and Learning Objects,* (1), 197-216. Retrieved July 10, 2009, from http://ijklo.org/Volume1/v1p197-216Recker.pdf

Thompson, L., & Kimberley, S. (2006). *Audit of Toolbox usage* (Final Report). Australian Flexible Learning Framework: Department of Education, Science and Training, Commonwealth of Australia. Retrieved March 16, 2009, from http://toolboxes.flexible-learning.net.au/documents/pdfs/NationalAuditOfToolboxes05.pdf

## WEBSITES

Audacity™ (free audio editing software): http://audacity.sourceforge.net/

Edna groups; community spaces for Australian educators: http://www.groups.edna.edu.au/

Guide to Social E-Learning. http://sociale-learning.flexible-learning.net.au/index.htm

Irfanview™ (free image editing software): http://www.irfanview.com

Kompozer™ (free WYSIWYG web editing software): http://kompozer.net/

Reload™ (Reusable e-learning Object Authoring and Delivery, and Learning Design Player™): http://www.reload.ac.uk/

The Australian Flexible Learning Framework. http://www.flexible-learning.net.au/

The VET Reload Tool™ and Content Package Player™: http://e-standards.flexible-learning.net.au/tools.htm

Web 2.0 applications index: http://www.go2web20.net/

# Chapter 23
# A Case Study Exploring a Multi-Disciplinary Collaborative Initiative to Use E-Learning to Meet the Professional Learning Needs of Health and Social Care Practitioners

**Karen Ousey**
*University of Huddersfield, UK*

**Stephen White**
*University of Huddersfield, UK*

## ABSTRACT

*This chapter explores the early development stages of an interactive interprofessional online learning package that updates and supports health and social care professionals who mentor students in practice settings. The package aims to present content that is relevant and useful to fourteen different disciplines accessing it. A benefit of online content is that learning can be undertaken when convenient for the mentor, 24 hours a day-7 days a week, with the facility to stop and restart as needed. Additionally the package is constructed so both individuals and groups can use it; this both meets a regulatory body's requirement for having a face-to-face update every year, and provides support for interprofessional learning between mentors from different disciplines.*

## INTRODUCTION

This chapter outlines the process of developing an interactive and interprofessional online e-learning package that updates and supports health and social care professionals who mentor students in practice settings. The concepts of e-learning are explored, including the potential benefits, the importance of ensuring it meets the learners' needs, and how barriers to developing this tool were overcome. Finally the package is presented; identifying the

DOI: 10.4018/978-1-61520-889-0.ch023

steps mentors need to progress through to ensure they have developed the knowledge necessary to be effective in the role, to meet the standards set by professional bodies.

## BACKGROUND

Health and social care students undertake placements within the workplace because practice environments are excellent for gaining experience and building confidence and competence (Benner, 1984; Schön, 1987). Qualified members of staff are expected to support them on placement, facilitate learning, and in some disciplines assess them, to ensure the student has the necessary psychomotor skills and underpinning knowledge base required for autonomous practice (Cope et al., 2000; Jones at al., 2001). Certain professional disciplines require these members of staff to regularly undergo mandatory updating (e.g. nursing, midwifery and operating department practitioners); for other professional groups this is a possible aspiration for the future.

Traditionally, for professions having mandatory requirements, update sessions are timetabled to be available many times throughout the year in a face-to-face event held in the placement areas. The purpose is to inform the clinical practitioners of curricula amendments, any changes within the Higher Education Institution (HEI) that may influence students' learning, as well as offering a forum to discuss issues and to ask questions. Each event is facilitated by a member of academic staff (a health or social care professional) from the university placing students within this area. Whilst sessions may include content that may seem relevant to more than one profession, particularly with generic content such as the role of staff in supporting learners, they tend to focus on one profession's content for curriculum and course structure. Therefore, in a placement area that supports students from more than one HEI or discipline, multiple sessions tend to be held, all of which must be attended if staff wish to be up-to-date.

This way of delivering the updates is very resource-hungry for academic staff, however, the staff presenters consider it a 'necessary evil' in order to provide sufficient opportunities for the vast numbers of practitioners to access all the necessary information. The academic staff would rather utilise this time to visit placement areas and discuss activities that are taking place; indeed, both Pulsford et al. (2002) and Hutchings et al. (2005) determined that members of staff supporting students wanted more contact with the HEI, which matched findings from previous research (Atkins & Williams, 1995). In addition, this delivery mode made it extremely difficult to exploit the opportunity that the update sessions offer for interprofessional learning between mentors from different disciplines and institutions.

Despite the many available opportunities to attend, increased workloads and limited staffing resources are making it progressively more difficult for healthcare professionals to attend any activities requiring them to leave the practice environment. Pulsford et al. (2002) found that fewer than half of the members of staff surveyed had attended an update session in the past year, and 20 per cent had never attended one. When further questioned on what prevented them from attending, "Staff Shortages" was the most significant barrier, with "Lack of Prior Notification" and "Inconvenient Times" being also cited as problems. A positive note was that respondents appeared committed to the principle of updating, with only 2% citing lack of interest as a reason for non-attendance. However, this last point was contradicted by Joshua-Amadi (2002) who believed that staff often lacked incentive, motivation, encouragement and leadership to attend training.

Whilst concurring with Pulsford et al. (2002) that the staff members desired to provide support, Wilkes (2006) found similar problems within the working environment, such as lack of resources or time. More recently Myall et al. (2008) identified that whilst most staff members were aware of the significance of their role in facilitating the development of students' clinical skills and acting as a source of support, staff shortages and an increasing demand on placement capacity contributed to an increased workload, often leading to a lack of time to carry out the role; such constraints were identified as preventing staff from attending regular updates. The problem may be exacerbated by the employers' themselves. Jones (2005) identified that many Trusts failed to recognise supporting students as an investment priority, making it difficult for staff to undertake relevant training. In an attempt to both overcome structural barriers to participation, and to exploit the opportunity for interprofessional learning between mentors from diverse disciplines and institutions, it was decided to investigate the possibilities offered by e-learning for updating these members of staff, and reducing the burden on attendance and delivery.

## Why Online?

Online learning has been reported to overcome some drawbacks inherent in traditional classroom teaching, especially its lack of flexibility (Rosenlund & Damark-Bembenek, 1999). Sigulem (2001), Neville (2003) and Cader and McGovern (2003) all reported effective education of health professionals using information and communication technologies, whilst Fox et al. (2001), Atak and Rankin (2002) and Huckstadt and Hayes (2005) demonstrated its positive use for post-registration Continual Professional Development education. In many cases, the authors focussed on the benefits to the organisation providing the courses, including widening accessibility of resources.

Others, such as O'Malley and McGraw (1999) report that students found online courses fitted in better with their schedules and saved time; a view concurred by Sit et al. (2005), with approximately 30% of their respondents commenting that online learning reduced competing demands from work and study. In online learning, attending lectures is replaced by remote access of e-learning materials, which provides 'on demand' education at the students' convenience. The most frequently reported satisfying aspect of online learning is its flexibility in learning, where students can study at their own pace and at convenient times (Rouse, 2000; Heidari & Galvin, 2002; Bentley, at al., 2003).

In healthcare education this is particularly important. Kozlowski (2002) argues that in nursing, the shift from 'teaching' to 'learning'; from a teacher-centred approach emphasising instructing and lecturing, leads to a student-centred approach that aims to help students organise and sift through information. Peacock and Hooper (2007), investigating physiotherapy education, found this can encourage students to take personal responsibility for learning, and become independent learners. This was further demonstrated by Sit et al. (2005), who found that over 80% of their respondents agreed that through online learning they learnt to take responsibility for their own study and work at their own pace. Furthermore Jonassen (2000) argued that using technologies in this way helps learners elaborate on what they are thinking and to engage in meaningful learning. This can be achieved through interactive learning experiences, to increase stimulus and interest for students (Rosenlund & Damark-Bembenek, 1999).

The difficulty of producing an e-learning tool that facilitates the participation of members of different professions in interprofessional learning was recognised; indeed, the health and social care professional bodies themselves sought to understand the role of staff members who support students, to meet the needs of a multi-skilled workforce (Marshall, et al., 2004). One early example of the problems encountered throughout the project concerned terminology. For example, there was extensive discussion on the title of the project, as this was influenced by the title given to members of staff who carry out this role, which differs significantly across the professions and had other connotations in some disciplines. This phenomenon is not new; as far back as 1995 Wilson-Barnett et al. identified the potential for confusion caused by varying descriptions of this role, as, more recently, did Mulholland et al.'s (2005) exploration of five health care disciplines and the preparation of placement support staff. Ultimately the update working group agreed to adopt the term 'Mentor' (Association of Operating Department Practitioners [AODP], 2006; Nursing and Midwifery Council [NMC], 2008). This term will be used in the remainder of this chapter.

As this term is not used across all the disciplines involved, it was agreed that the context of 'Mentor' needed to be clearly defined within the e-learning package to ensure those who do not use the term understood that it still applied to them. To further embed this thinking, definitions from relevant professional bodies of what this role entails within each discipline are presented in the first section of the package. This allows the mentors to consider similarities and differences of how their role is implemented across professions.

## Mentorship

Russell and Adams (1997, p. 2) define mentorship as '*an intense interpersonal exchange between a senior experienced colleague (mentor) and a less experienced junior colleague (protégé) in which the mentor provides support, direction, and feedback regarding career plans and personal development*' This view is supported in health-care contexts by Dorsey and Baker (2004) who suggest that mentorship is a planned relationship between an experienced person and one who has less experience to achieve identified outcomes; in doing this, the mentor needs to create a safe place for the protégé to expose their vulnerability and be coached into confidence- and growth-promoting situations (Davis et al., 1977). Indeed, Barker (2005) states that mentoring involves bridging the gap between the educational process and the real-world experience; in doing this, mentors inevitably undertake clinical teaching, supervise students' work, ensure professional socialization, and assess students' practice (AODP 2006; Chow & Suen, 2001; Jones, et al., 2001; NMC 2002, 2007; Spouse, 2000).

Mentors play a crucial part in engendering a positive learning environment (Myall et al., 2008). Well-prepared mentors are vital to the success of practice placements (English National Board/Department of Health, 2001); they provide direct support in terms of the development of clinical skills, encourage critical thinking and reflection on practice (Cope et al., 2000), and promote interprofessional working. Watson (2000) suggests that mentors must be provided with education, training and ongoing facilitation to function effectively in this demanding role; Andrews and Chilton (2000) also noted that a mentor has to be familiar with a range of information in order to be effective, including the programme of study, the documentation and the terms used in it.

Both the AODP (2006) and NMC (2007) suggest how mentors should be integrated into students' learning. Mentors should have knowledge of the student's programme to identify current learning needs; create and develop opportunities for students to meet their learning needs towards achieving learning outcomes; and be able to facilitate integration of learning from practice and educational settings. Pellatt (2006, p. 33) argues that the '*role of the mentor in the preparation of practitioners who are fit for practice is paramount*', but that '*better training, support and evaluation of their performance*' and an increase in their status, are needed.

## Developing the E-Learning Package

The concept initially evolved at the University of Huddersfield in September 2007 to unify the updates for three professional groups; nurses, midwives and operating department practitioners (ODPs). There were sufficient commonalities between mentor standards of these disciplines to make this a realistic challenge. A multi-professional working party, with staff representation from both academia and clinical areas was thus established; even at this early stage it was felt that involving key stakeholders was vital to promote ownership and ensure the package addressed the needs of all concerned.

A basic concept was conceived, whereby mentors of all three professions would access the update package online. It was structured as a linear branching pathway, which ensured that the update had to be completed, but in doing so it made the material relevant for everyone accessing it. This was achieved by presenting an initial generic section that reinforced the roles of the mentor and the mentee. Subsequently the user chose a route leading to profession-specific content, such as course structure, placement dates etc. Afterwards, they could go back and undertake the specific content for another discipline; this was particularly relevant for those who mentored more than one profession, for example, staff in the operating theatre. These early stages are further described by Ousey & White (2009).

## An Expanded Remit

The package was almost ready for dissemination when regional initiatives influenced change. In May 2008, the regional Strategic Health Authority (NHS Yorkshire and the Humber) decided that the concept of a multidisciplinary updating process deserved wider consideration, so funding was provided to establish a working group. This initially consisted of at least one health and social care academic from each university within the region, as well as Practice Learning Facilitators (PLFs) to represent staff from practice; these are very similar to the Practice Education Facilitators (PEFs) described by Carlisle et al. (2009). The following nine Higher Education Institutions (HEIs) participated in the project:

- University of Bradford
- University of Huddersfield
- University of Hull
- University of Leeds
- Leeds Metropolitan University
- The Open University
- Sheffield Hallam University
- University of York
- York St. John University

Initially this group reviewed several existing models of update package, both e-learning and paper versions, prior to choosing the system that the University of Huddersfield had developed, as the template for the new package. This meant the new package would need to provide content for fourteen health and social care disciplines being delivered across nine universities, as shown below:

- Audiology
- Clinical Physiology
- Diagnostic Radiography
- Dietetics
- Midwifery
- Nursing
- Occupational Therapy
- Operating Department Practice
- Paramedics
- Physiotherapy
- Podiatry
- Radiotherapy
- Social Work
- Speech and Language Therapy

## Collaborative Working Towards a Common Goal

Miccolo & Spanier (1993, p. 446) describe working collaboratively as a partnership *'where mutual goal setting occurs, where authority and responsibility for actions belong to individual partners'*. The importance of professional groups working in a multidisciplinary fashion was identified by the UK Department of Health [DH] (2000) who presented their National Health Service

(NHS) workforce strategy calling for "genuinely multi-professional" education and training that promotes:

- Teamwork
- Partnership and collaboration between professions, between agencies and with patients
- Skill mix and flexible working between professions
- Opportunities to switch training pathways to expedite career progression
- New types of workers

The International Council of Nurses code (2000) states that a co-operative relationship with co-workers in other fields is necessary, whilst the NMC in its publication 'Standard to Support Learning and Assessment in Practice', actively encourages interprofessional working and learning (NMC, 2008, p. 13, Section 1.3). Further to these, and particularly relevant to this project, Mallik and McGowan (2007) recommended standardisation of mentor programmes to promote an interprofessional practice education framework. To achieve this, the members of the working group needed to move away from independence, to interdependence, where strength and power are derived through working in partnership this required them to respect and trust each other, whilst sharing the vision, the planning, goal setting, work, successes and failures (Sullivan, 1998).

## Maintaining Integrity

Although online learning may have seen a genesis of novel delivery, Barker (2004) stresses that sound web-based courses tend to be structured similarly to traditional classroom courses. Therefore, in developing the update package, it was important to embed the foundation established through years of presenting it by traditional methods. As Knowles et al. (1984) suggest, adults are most interested in learning about subjects that have immediate relevance to their job or personal life; and students' learning can be influenced by satisfaction with their learning experience (Ramsden, 1991; Espeland & Indrehus, 2003). Therefore, emphasis was placed on ensuring that the online package content provided information relevant to the role of supporting students in practice. It was further noted that specific elements were required in order to meet professional body mandates (NMC, 2007; AODP, 2006).

It was also important to consider what mentors themselves believed they needed. To determine this, mentors were asked their opinions during informal lunchtime meetings and one to one discussions. Their responses emulated much of Mallik and McGowan's (2007) findings; their multi-professional study into what mentors wanted, identified that the most frequently agreed subject areas were: roles, responsibilities and accountability of the mentor; learning contracts; setting student tasks; and monitoring students' progress.

Mallik and McGowan (2007) also found that, in addition to the areas listed above, mentors indicated that they themselves should also be receiving supervision/mentoring. To promote the facility for supervision, the online update was developed to allow either individuals or groups to undertake it. This provides the opportunity for the group discussions to lead to interprofessional learning, where group members are from more than one discipline. One final benefit to the group facility is that updates carried out this way meet a requirement of the Nursing and Midwifery Council (2008, p. 14), for annual updating activities '*to include an opportunity to meet and explore assessment and supervision issues with other mentors/ practice teachers*'.

## E-learning Package Structure

The package is designed with a linear structure, requiring mentors to work from start to finish before being designated as having completed the update. This 'forced completion' mode was necessary in order to meet regulatory body requirements; during inspection visits evidence will be required that mentors have undertaken the update and whilst no-one is suggesting that these healthcare professionals would omit areas of content if able to do so, assurances would be needed that this is impossible. This does not, however, mean every mentor carries out the same activity, due to branching options at certain stages, as will be explained later.

## Supporting Interprofessional Interaction

It was imperative that the package is not merely a repository of content. Chapman (2000) argues that philosophies of adult education, including Liberal, Behaviourist and Humanistic, for example, must still apply when information and communications technologies are used; e-learning should not merely refer to disseminating lecture information and content through the Internet. The questions and scenarios presented throughout the package provide the interactive nature of the package. Users are unable to progress through the sections without responding, so are thus being encouraged to consider their own practice. Responses denoting 'best practice' are offered after each questions and scenario, so users can reflect on and reconsider their initial responses.

Integral to the package are audio recordings; 'soundbytes' based on experiences of 'real' mentors and students. These introduce content at intervals throughout the package and allow the mentor to hear others' thoughts around areas such as '*what being a mentor means to a registered practitioner*' and '*what a student perceives are the qualities that makes a good mentor*'. For mentors who have hearing difficulties, or whose computing facilities do not have sound, a text version of this content is provided. The inclusion of video or animation was also considered, so a test website was created for members of the working group to access at placement sites within the region. Unfortunately, due to security and firewall settings of computers within these sites, no single video format or player could be viewed at all placements. This meant video and animation were ruled out, as content would not be available to all users.

Finally, a 'chat' facility was included, providing opportunities for users to participate in synchronous discussions with others undertaking the update. Chat, or Instant Messaging, is a popular means of communication in virtual communities; its informality and interactive, instant support for negotiation of meaning has the many characteristics needed to provide help and information where needed (Nardi at al., 2000). On entering this area, the mentor can see the current conversation, as well as names and disciplines of all mentors who are currently online. They can then either join the current discussion, or send a private message to another user. Conversations can take place in the 'open chat' area, or within their profession-specific area.

## Introduction and Generic Section

Figure 1 displays the opening page introducing the package and putting the update into context; it asks for the mentor's details, to ensure the correct person is recorded as having completed the update. The mentor enters their choice of username and password, which they are reminded of both on-screen and by automated email. When logging back in, the mentor is asked to confirm that their personal details are unchanged, and then they are given the opportunity to re-enter the package at any point they have previously viewed; this allows for content to be reviewed and reconsidered.

A section of generic mentor content that all users work through is then introduced. This material was derived from lengthy discussions by the working group, to ascertain what information was relevant to all disciplines. The section enables the mentor to review their current knowledge of mentorship; allowing them to reflect on their

*Figure 1. The first stage of the update*

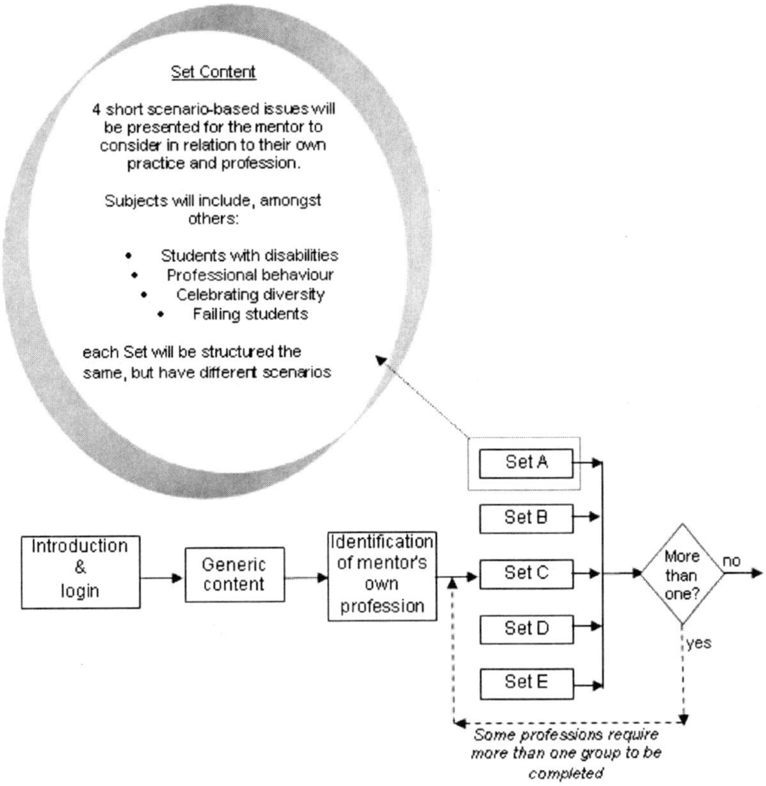


experiences undertaking the role; and updating their knowledge and awareness of contemporary mentorship. The intention is that they will address the objectives identified in Figure 2.

Next, the mentor has to choose from one of five Sets; each Set includes four problem based learning scenarios, based around issues of 'Students with Disabilities', 'Professional Behaviour', 'Celebrating Diversity' and 'Failing Students', with differing scenarios in each Set. In each scenario the mentor is asked how they would react to a particular situation, after which they are presented with suggested resolutions against which they can evaluate their own response. It is anticipated that

*Figure 2. Update objectives*

- Review the principles of supporting learning and assessment in practice
- Identify available support mechanisms
- Acknowledge the multi-professional dimensions of supporting learners and assessment in practice
- Recognise the accountability and responsibilities of supporting learning and assessment in practice
- Consider how to improve the quality of the learner's experience
- Contribute to a Standard for those supporting learners in practice from the Yorkshire and the Humber region

*Figure 3. The second stage of the update*

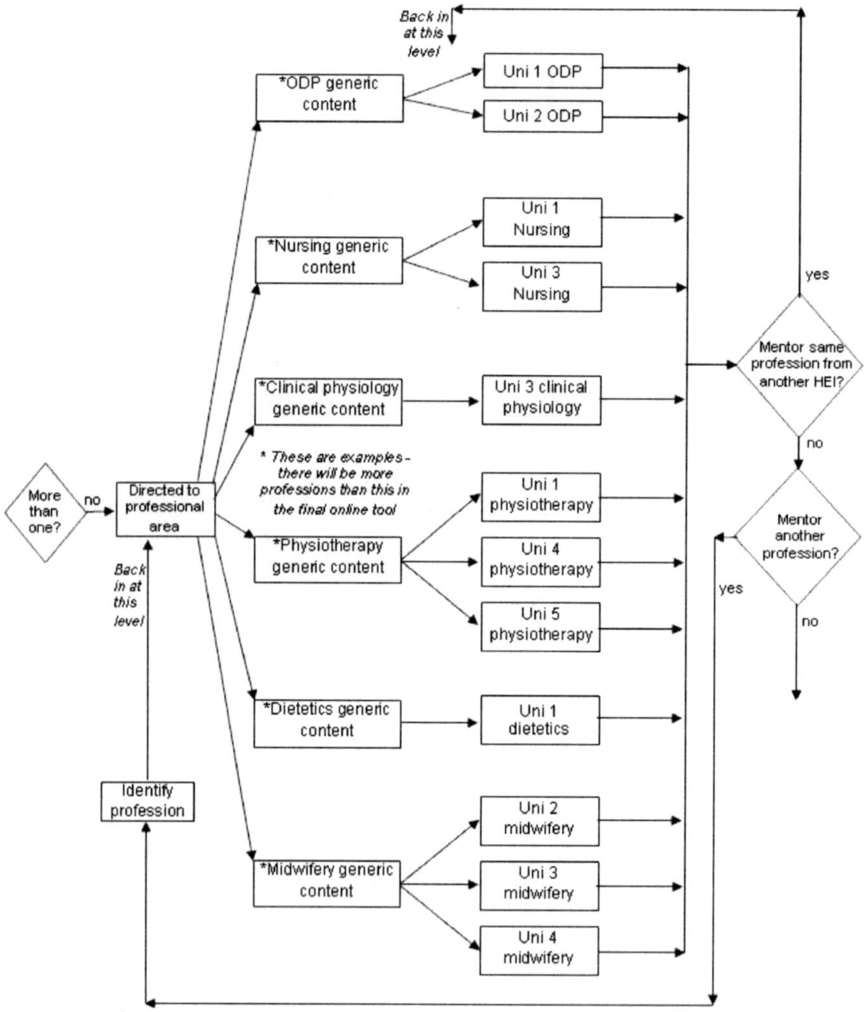


the mentor will explore and challenge the way they would have previously dealt with the situations. This would provide a meaningful way in which the participants could identify the importance of understanding the mentoring process.

Silen (2001) maintains that meaningful learning entails constantly making choices and being able to view the knowledge in relation to other fields and areas of application. Being able to decide on the relevance of information and determine when it is necessary to study something in greater depth and why, are also important for developing self-directed learning skills. The minimum to be completed by the mentor is one Set (four scenarios); they then progress to stage two of the update. Some disciplines have a greater expectation of their mentors, so more than one Set has to be completed before they can progress.

## Generic Professional Section

As shown in Figure 3, each mentor is then presented with content specific to their own discipline, regardless of which university their students attend; for example, this may be national requirements, or regulatory and statutory guidance. Some

*Figure 4. The final stage of the update*

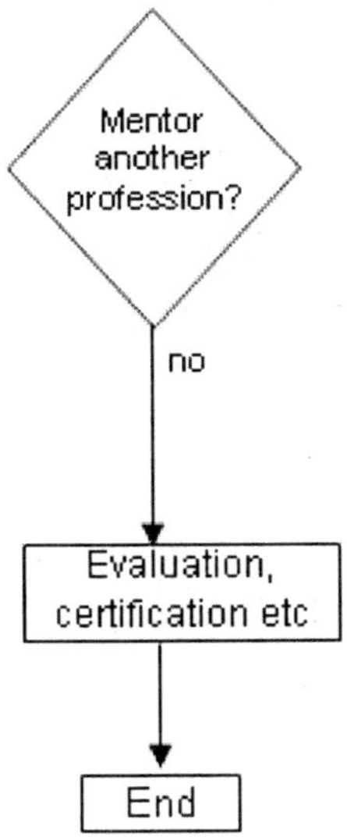


disciplines also chose to add extra developmental content, due to the 'infancy' of their mentorship education and update processes; for these groups, this package would also be utilised as a teaching tool, for trainee mentors.

The mentor then identifies which university their mentees come from and course-specific content is made available; for example, hours and patterns of work that the student is expected to complete while in the placement areas; an overview of academic assignments required by the students in each term/semester and so on. Hyperlinks within these sections allow the mentor to access content and documents on the institution's own website, allowing them to view the most up to date information.

Following completion of this section some mentors will have completed their update and will progress to the evaluation and certification stage. However, where the placement is utilised by a range of students, the facility exists for the mentor to explore content for the same profession, but from another university, or material from another discipline entirely; a process that can be repeated as many times as necessary. Although this will extend the time it takes to complete the activity, it precludes the mentor from having to attend multiple events as they used to for traditional face to face updates.

## Evaluation and Certification Section

Figure 4 exemplifies that once the mentor has completed their update, they evaluate the package, which collects quantitative data through a set of closed questions (Figure 5). Each question allows the mentor to elaborate by providing a qualitative comment. The questions aim to establish whether the objectives have been met, and to determine any improvements that may be needed to the package.

On completion, an email is sent to the mentor's email address, providing a web address linking to their certificate of completion. In addition to confirming they have carried out the activity, and the date of completion, the certificate features logos for each university that the mentor accessed the content of; providing evidence that they have been updated for all institutions that their mentees originate from. On the second page are a short series of reflective questions, allowing the mentor to review the update activity and how it has affected their practice; this provides evidence for their annual appraisals or professional portfolios. On subsequent pages the responses they gave to questions asked during the update are provided; this can be used for review and personal development.

## FUTURE RESEARCH DIRECTIONS

The online mentor package provides an alternative to the traditional 'face to face' sessions and allows multi-professional groups to learn and work together. This provides the opportunity for participants to discuss and explore various mentoring styles and engage in discussion. The package is now ready to undergo heuristic evaluation, as described below, followed by pilot implementation.

*Figure 5. Evaluation questions*

- **Do you now know where to access information relating to the mentorship of students? If not, please comment.**
- **Do you feel you have an improved understanding of how the mentor role can function in a multi-professional way? If not, please comment.**
- **Do you feel you have been given all the relevant information to assist you in your mentor role? If not, please comment.**
- **Do you feel that your ability to mentor has been enhanced through carrying out this activity? If not, please comment.**
- **There are accountability and responsibility issues associated with mentoring – has this update improved your understanding of them? If not, please comment.**
- **Has your understanding of how to improve the learner's experience been enhanced? If not, please comment.**
- **Has this activity developed you professionally? If not, please comment.**
- **Was the update package easy to navigate? If not, please comment.**
- **Was carrying out the update an enjoyable experience? If not, please comment.**
- **Is there any information that you would like added to this package? If so, please comment.**
- **Overall, has the package met your needs? If not, please comment.**
- **Do you have any further comments?**

Heuristic evaluation will be used as it has been recognised as an effective method for the formative evaluation of educational software (Quinn, 1996; Squires & Preece, 1999; Albion, 1999). A small team of independent evaluators, representative of the professions taking part and who are knowledgeable in the field of mentor updates, undertake the package whilst considering a set of usability guidelines; the "heuristics." Where problems are encountered they are graded in terms of severity against a Likert scale; with 1 being a minor issue, through to 5 meaning it would make the package unusable. This process was used in the initial development work carried out at the University of Huddersfield and proved very effective in identifying problems which could then be resolved before the pilot process.

The emerging trend within health and social care is that practitioners do not have enough time to leave their clinical areas to attend educational sessions and are missing out on information that is not only vital to the support of learners, but also to their own personal and professional development; this may have repercussions on the quality of the care that is being delivered. The underpinning ethos behind the development of this package is to provide health and social care practitioners with the opportunity to develop their support of learners in practice, at a time that best suits their professional and personal schedules. The inclusion of learning sets and scenarios supports the philosophy of problem-based learning; where the learning needs of users are related to the real-life issues they may encounter when supporting learners in practice. The benefits of presenting the content in this way can be used by others who are engaged in supporting learners. Indeed, the template of this package could be adapted for a variety of subjects.

An evaluation of the package will be undertaken following a six month trial period to ascertain if it is meeting the educational needs of the mentors; to establish if it develops the knowledge base of mentors at least as effectively as attending the traditional face to face sessions and if it promotes the problem solving skills of mentors. An action research approach will be used, allowing for users of the update package to be involved in its future development and to ensure that it meets the needs of its multidisciplinary audience.

## CONCLUSION

The e-learning package described in this chapter was a multidisciplinary collaborative project to support thousands of staff who mentor health and social care students across the Yorkshire and the Humber region. Effective interprofessional learning is attainable if common ground can be found for each profession to discuss and learn. The sharing of good practice is paramount if the health and social care professions are to work and learn together in a seamless fashion.

The package provides content relevant and useful for all participants, regardless of their profession, and the opportunity to learn and work together. The availability of the package, which can be accessed at a time convenient to the mentors either within or outside of the health care areas, alleviates the barriers of being 'too busy' to attend the traditional face to face sessions. In turn, staff will be able to utilise this time in visiting the placement areas to provide further support, advice and feedback on their performance in the role.

## REFERENCES


Albion, P. R. (1999). *Heuristic evaluation of educational multimedia: from theory to practice. 16th Annual conference of the Australasian Society for Computers in Learning in Tertiary Education*. ASCILITE.

Andrews, M., & Chilton, F. (2000). Student and mentor perceptions of mentoring effectiveness. *Nurse Education Today*, *20*, 555–562. doi:10.1054/nedt.2000.0464

AODP. (2006). *Qualifications Framework for Mentors Supporting Learners in Practice: Standards and Guidance for Mentors and Practice Placements in Support of Pre-registration Diploma of Higher Education in Operating Department Practice Provision*. Cheshire, UK: Association of Operating Department Practitioners.

Atack, L., & Rankin, J. (2002). A descriptive study of registered nurses' experiences with web-based learning. *Journal of Advanced Nursing*, *40*(4), 457–465. doi:10.1046/j.1365-2648.2002.02394.x

Barker, A. (2004). Using online learning technology to develop nurse leaders. *Nurse Leader*, *2*(5), 32–35. doi:10.1016/j.mnl.2004.07.011

Barker, E. (2005). Mentoring—A complex relationship. *Journal of the American Academy of Nurse Practitioners*, *18*(2), 56–61. doi:10.1111/j.1745-7599.2006.00102.x

Benner, P. (1984). *From Novice to Expert: Excellence and Power in Clinical Nursing Practice*. Reading, MA: Addison-Wesley.

Bentley, G. W., Cook, P. P., Davis, K., Murphy, M. J., & Berding, C. B. (2003). RN to BSN program: transition from traditional to online delivery. *Nurse Educator*, *28*(3), 121–126. doi:10.1097/00006223-200305000-00005

Cader, R., & McGovern, M. (2003). Introducing Blackboard: an electronic learning platform. *Nursing Times*, *99*(32), 24–25.

Carlisle, C., Calman, L., & Ibbotson, T. (2009). Practice-based learning: The role of practice education facilitators in supporting mentors. *Nurse Education Today*, *29*(7), 715–721. doi:10.1016/j.nedt.2009.02.018

Chapman, L. (2000). Distance learning for post-registered nursing: the facts. *Nursing Standard*, *14*(18), 33–36.

Chow, F. L. W., & Suen, L. K. P. (2001). Clinical staff as mentors in pre-registration undergraduate nursing education: students' perceptions of the mentors' roles and responsibilities. *Nurse Education Today, 21*, 350–358. doi:10.1054/nedt.2001.0569

Cope, P., Cuthbertson, P., & Stoddart, B. (2000). Situated learning in the practice placement. *Journal of Advanced Nursing, 31*(4), 850–856. doi:10.1046/j.1365-2648.2000.01343.x

Davis, L., Little, M., & Thornton, W. (1977). The art and angst of the mentoring relationship. *Academic Psychiatry, 21*, 61–67.

Department of Health (DH). (2000). *A Health Service of all the talents: developing the NHS workforce*. London: Department of Health.

Dorsey, L., & Baker, C. (2004). Mentoring undergraduate nursing students. *Nurse Educator, 29*, 260–265. doi:10.1097/00006223-200411000-00013

English National Board/Department of Health. (2001). *Placements in Focus: Guidance for Education in Practice for Health Care Professions*. London: ENB/DoH.

Espeland, V., & Indrehus, O. (2003). Evaluation of students' satisfaction with nursing education in Norway. *Journal of Advanced Nursing, 42*(3), 226–236. doi:10.1046/j.1365-2648.2003.02611.x

Fox, N., O'Rourke, A., Roberts, C., & Walker, J. (2001). Change management in primary care: design and evaluation of an internet-delivered course. *Medical Education, 35*(8), 803–805. doi:10.1046/j.1365-2923.2001.00974.x

Heidari, F., & Galvin, K. (2002). The role of open learning in nurse education. Does it have a place? *Nurse Education Today, 22*, 617–623. doi:10.1016/S0260-6917(02)00133-8

Huckstadt, A., & Hayes, K. (2005). Evaluation of interactive online courses for advanced practice nurses. *Journal of the American Academy of Nurse Practitioners, 17*(3), 85–89. doi:10.1111/j.1041-2972.2005.0015.x

Hutchings, A., Williamson, G. R., & Humphreys, A. (2005). Supporting learners in clinical practice: capacity issues. *Journal of Clinical Nursing, 14*, 945–955. doi:10.1111/j.1365-2702.2005.01239.x

International Council of Nurses. (2000). *The ICN Code of Ethics for Nurses*. Geneva: ICN.

Jonassen, D. H. (2000). Transforming learning with technology: Beyond modernism and post-modernism or whoever controls the technology creates the reality. *Educational Technology, 40*(2), 21–25.

Jones, M. L., Walters, S., & Akehurst, R. (2001). The implications of contact with the mentor for pre-registration nursing and midwifery students. *Journal of Advanced Nursing, 35*, 151–160. doi:10.1046/j.1365-2648.2001.01832.x

Jones, R. A. (2005). Prioritising to safeguard the integrity of nurse education. *Nursing Times, 101*(34), 26–27.

Joshua-Amadi, M. (2002). Recruitment and retention: a study in motivation. *Nursing Management, 9*, 7–21.

Knowles, M. (1984). *Andragogy in action: Applying modern principles of adult education*. San Francisco: Jossey-Bass.

Knowlton, D. S. (2000). A theoretical framework for the online classroom: A defense and delineation of a student-centered pedagogy. *New Directions for Teaching and Learning, 84*, 5–13. doi:10.1002/tl.841

Kozlowski, D. (2002). Using online learning in a traditional face-to-face environment. *Computers in Nursing, 20*(1), 23–30.

Mallik, M., & McGowan, B. (2007). Issues in practice based learning in nursing in the United Kingdom and the Republic of Ireland: Results from a multi professional scoping exercise. *Nurse Education Today*, *27*(1), 52–59. doi:10.1016/j.nedt.2006.02.007

Miccolo, M. A., & Spanier, A. H. (1993). Critical care management in the 1990s: making collaborative practice work. *Critical Care Clinics*, *9*, 443–453.

Mulholland, J., Mallik, M., & Moran, P. (2005). *An Overview of the Nature of the Preparation of Practice Educators in Five Health Care Disciplines*. 6th Occasional Paper. Higher Education Academy, London: Higher Education Academy: Health Sciences and Practice. Accessed August 10, 2009, from http://www.health.heacademy.ac.uk/publications/occasionalpaper/occp6.pdf

Myall, M., Levett-Jones, T., & Lathlean, J. (2008). Mentorship in contemporary practice: the experiences of nursing students and practice mentors. *Journal of Clinical Nursing*, *17*(14), 1834–1842. doi:10.1111/j.1365-2702.2007.02233.x

Nardi, B., Whittaker, S., & Bradner, E. (2000). Interaction and Outeraction: Instant Messaging in Action. *Proceedings of ACM Conference on Computer Supported Cooperative Work (CSCW '00)*, (pp. 79-88). New York: ACM Press.

Neville, M. W. (2003). Testing techniques. Blackboard allows students to take tests on the go. *Nurse Educator*, *28*(5), 207–209. doi:10.1097/00006223-200309000-00005

Nursing and Midwifery Council (NMC). (2002). *Standards for the preparation of teachers of nursing and midwifery*. London: Nursing and Midwifery Council.

Nursing and Midwifery Council (NMC). (2007). *Standards for the preparation of teachers of nursing and midwifery*. London: Nursing and Midwifery Council.

Nursing and Midwifery Council (NMC). (2008). *Standards to support learning and assessment in practice: NMC standards for mentors, practice teachers and teachers* (2nd ed.). London: Nursing and Midwifery Council.

O'Malley, J., & McGraw, H. (1999). Students perceptions of distance learning, online learning and the traditional classroom. *Online Journal of Distance Learning Administration*, *2*(4).

Ousey, K. J., & White, S. A. (2009). Developing a Multi-professional e-Learning Mentor Update Package. In Proceedings *of the International Conference on Mobile, Hybrid, and On-line Learning, elml,* (pp.103-107).

Peacock, S., & Hooper, J. (2007). E-learning in physiotherapy education. *Physiotherapy*, *93*(3), 218–228. doi:10.1016/j.physio.2006.11.009

Pellatt, G. C. (2006). The role of mentors in supporting preregistration nursing students. *British Journal of Nursing (Mark Allen Publishing)*, *15*(6), 336–340.

Pulsford, D., Boit, K., & Owen, S. (2002). Are mentors ready to make a difference? A survey of mentors' attitudes towards nurse education. *Nurse Education Today*, *22*(6), 439–446.

Quinn, C. N. (1996). *Pragmatic evaluation: Lessons from usability*. Paper presented at the 13th Annual Conference of the Australasian Society for Computers in Learning in Tertiary Education.

Ramsden, P. (1991). A performance indicator of teaching quality in higher education: the course experience questionnaire. *Studies in Higher Education*, *16*, 129–150. doi:10.1080/03075079112331382944

Rosenlund, C. H., & Damark-Bembenek, B. (1999). Assessing the effectiveness of an online program. *Nurse Educator*, *24*(1), 5–6. doi:10.1097/00006223-199901000-00001

Rouse, D. P. (2000). The effectiveness of computer-assisted instruction in teaching nursing students about congenital health disease. *Computers in Nursing, 18*(6), 282–287.

Russell, J. E. A., & Adams, D. M. (1997). The changing nature of mentoring in organizations: An introduction to the special issue on mentoring in organizations. *Journal of Vocational Behavior, 51*, 1–14. doi:10.1006/jvbe.1997.1602

Schön, D. A. (1987). *Educating the Reflective Practitioner*. San Francisco, CA: Jossey-Bass.

Sigulem, D. M. (2001). A Web-Based distance education course in nutrition in public health: case study. *Journal of Medical Internet Research, 3*(2), E16. doi:10.2196/jmir.3.2.e16

Silen, C. (2001). *Between chaos and cosmos-a driving force for responsibility and independence in learning, The Power of Problem Based Learning. PROBLARC, The 3rd Asia Pacific conference on PBL, 2001*. Australia: The University of Newcastle.

Sit, J. W. H., Chung, J. W. Y., Chow, M. C. M., & Wong, T. K. S. (2005). Experiences of online learning: students' perspective. *Nurse Education Today, 25*, 140–147. doi:10.1016/j.nedt.2004.11.004

Spouse, J. (2000). Supervision of Clinical Practice: The Nature of Professional Development. In Spouse, J., & Redfern, L. (Eds.), *Successful Supervision in Health Care Practice* (pp. 126–154). Oxford, UK: Blackwell Science.

Squires, D., & Preece, J. (1999). Predicting quality in educational software: Evaluating for learning, usability and the synergy between them. *Interacting with Computers, 11*, 467–483. doi:10.1016/S0953-5438(98)00063-0

Sullivan, T. J. (1998). *Collaboration – A Health Care Imperative*. London: McGraw Hill Company.

Watson, S. (2000). The support that mentors receive in the clinical setting. *Nurse Education Today, 20*(7), 585–592. doi:10.1054/nedt.2000.0463

Wilkes, Z. (2006). The student-mentor relationship: a review of the literature. *Nursing Standard, 20*(37), 42–47.

Wilson-Barnett, J., Butterworth, T., & White, E. (1995). Clinical support and the Project 2000 nursing student: Factors influencing this process. *Journal of Advanced Nursing, 21*, 1152–1158. doi:10.1046/j.1365-2648.1995.21061152.x

# Chapter 24
# Mixing Students Mixing Methods:
## Evaluating a Blended Approach to Interprofessional Education

**Christine Dearnley**
*University of Bradford, UK*

**Melissa Owens**
*University of Bradford, UK*

**Pete Greasley**
*University of Bradford, UK*

**Caroline Plews**
*University of Bradford, UK*

## ABSTRACT

*This chapter describes an innovative interprofessional education (IPE) module delivered with a blended approach, incorporating face-face teaching and online problem based e-learning (PBeL) activities. This has been evaluated using qualitative and quantitative methods. The qualitative work provided an overall perspective on how students experienced IPE. The quantitative work demonstrated significant differences in how different professions experienced the same curriculum; it also provided a useful insight into attitudinal changes over the period of learning. The evaluation findings are presented here with discussion of their implications for future delivery of IPE. Key issues include: maximising student engagement by optimising the timing of the module, the development of IT skills and managing the professional ethnocentrism and differences in professional epistemology and philosophy, which impact on students expectations of, engagement with, and subsequently the delivery of IPE.*

DOI: 10.4018/978-1-61520-889-0.ch024

## INTRODUCTION

This chapter presents an account of an innovative interprofessional education (IPE) module developed in the School of Health Studies (SoHS), University of Bradford, in the North of England. It is a case study from which we can offer critical guidance on key IPE issues. A blended approach was taken to the module delivery, utilising face-face learning and e-learning with problem based (PBL) processes, thereby creating problem based interprofessional e-learning (PBIPEL). An evaluation, drawing upon both qualitative and quantitative data was undertaken and the outcomes have informed development of the programme. The findings are presented here with discussion related to their implications for the future delivery of IPE.

## BACKGROUND

In September 2005 a new module was introduced within the SoHS enabling approximately 350 pre-registration students from Midwifery (MW), Nursing (Nu), Occupational Therapy (OT), Physiotherapy (PT) and Radiography (RT) to come together to learn with, from and about each other (CAIPE, 2002). This module was the first of two specific IPE modules that students were to take and was delivered in semester one, year one, of their respective programmes.

Moving towards curricula that includes the principles of IPE has required a paradigm shift in thinking, away from the traditional health and social care pre-registration programmes (Colyer et al., 2005). Traditionally, emphasis has been on creating individuals able to practice in their individual, chosen profession. Students would become immersed in the particular customs and traits associated with that profession without appreciation, or examination of, what those traditions actually were (Dall'Alba, 2009).

Today's health service, however, requires individuals capable of working with colleagues across different professional backgrounds. This requires them to have an appreciation of other professionals' skills, values and knowledge base; to be capable of team working and to be able to work collaboratively in order to provide a joined-up package of care. Pre-registration health care curricula therefore, need to re-focus and enable students to learn skills that will enable them to work in this way. These include the ability to reflect on their own working practices, as well as having an appreciation of others (Martin, 2005).

The definition of IPE most commonly cited, for example by Barr et al. (2005), Colyer et al. (2005) and Freeth et al. (2005), is that of CAIPE (2002): 'occasions when two or more professions learn with, from and about each other to improve collaboration and the quality of care'. Fundamental to this is the underpinning aim to break down negative stereotypes and promote an ethic of practice that is both patient-centred and mutually supportive of (interprofessional) team members (Hammick et al., 2009). IPE therefore is very different to multi-professional learning (MPL), whereby students are simply placed together, perhaps in a large lecture theatre, with little focus on interaction.

Student professionals are known to hold potentially harmful stereotypes of other professional groups even prior to commencing their professional programmes (Piertroni, 1991); the aims of IPE therefore, must be to eradicate these stereotypical images and create a respect for others' skills and knowledge (Hammick et al., 2009). Whilst MPL does not necessarily prohibit IPE, which can occur informally, for example, when sharing coffee outside of the formal learning environment (Owens et al., 2007), IPE is considered most effective when it is specifically planned, supports reflectivity, is group focused and based on principles of adult learning theories (Barr et al., 2005).

Achieving this can be both complex and challenging. Providing IPE within the university setting generally involves large numbers of students with often large differences in numbers of students undertaking different professional programmes. Facilitating small group-work, therefore, is very resource-heavy in terms of both facilities and staff, and complex in achieving a reasonable 'mix' of students from different professions in each group without one profession (generally nursing students) being overly dominant. Working across different curricular to find times when all students can be brought together in the classroom setting provides an additional challenge to the co-ordination of any IPE activity (Freeth, 2001; Freeth et al., 2005; Hughes et al., 2004; Kamin et al., 2006; Owens et al., 2007).

An increasingly popular way of meeting the challenges outlined above is to combine e-learning initiatives with PBL processes, thereby creating problem based e-learning (PBeL) (Hughes et al., 2004). Traditional PBL places the student at the centre of the learning process. Using a 'problem' as a trigger to learning, students are encouraged to explore whole issues based on clinical practice. Using a six step process, students identify and set group learning objects, which they then work together, over an agreed time period, to achieve (Alavi, 1995).

The additional pedagogical values of PBeL, however, are becoming recognised, particularly in relation to IPE initiatives. It allows asynchronous, flexible debate, enabling students to access and engage at a time most suitable to them (Beck & Owens, 2007). Furthermore, whilst initial IPE activities may occur in the university setting, it is likely that students will be required to continue to engage beyond this point, when they are dispersed in geographically distanced (practice) settings and working different and anti-social hours. Online PBeL enables students to continue to engage despite conflicting demands on time and space (Kamin et al., 2006; Beck & Owens, 2007). Additionally, PBeL allows students to be reflective in their responses: reading and considering others' responses prior to undertaking additional research and adding to debate by posting their own contributions in a cyclical manner (Hughes et al., 2004). In this way, the PBL process becomes more flexible and fluid and potentially encourages deeper learning than traditional face-face methods where each stage of the process is completed discreetly (Alavi, 1995).

## IPE and E-Learning Provision at the University

IPE at the university was developed within a framework of PBeL. Two 'PBIPEL' modules were created, to be delivered within years one and three of the curricula, with an additional IPE activity used within year two. Those pre-registration professions involved in the IPE activities are shown in Figure 1. The study described here relates to the year one module.

A developmental approach was used, beginning with an introduction to interprofessional working in semester one, year one. The module aim was for students to develop knowledge and understanding of factors integral to effective interprofessional team working. This module was a foundation to provide some insight into issues that could impact on effective interprofessional working, prior to clinical practice taking place. Students were brought together using a blended approach of face-face contact and on-line group work, through the web based virtual learning environment (VLE) Blackboard (BB)©. Becta

*Figure 1. Overview of IPE provision in the school of health studies*

| Year One | Year Two | Year Three |
|---|---|---|
| Midwifery, Nursing, Occupational Therapy, Physiotherapy, Radiography, | Midwifery, Nursing, Occupational Therapy, Physiotherapy, Radiography, Social Work, | Midwifery, Nursing, Occupational Therapy, Physiotherapy, Radiography, |

(2007) described VLE's as a combination of some or all of the following features; most of which were available on BB:

- Communication tools: email, bulletin boards and chat rooms.
- Collaboration tools: online forums, intranets, electronic diaries and calendars.
- Online assessment and marking.
- Controlled access to curriculum resources.

## Overview of Programme

Day One:

- Introductory lecture and a computer based VLE training session.
- Overview of module content and principles behind IPE and PBL.
- Students socialised into using the VLE with peers (Salmon, 2003) and given a practical opportunity to use its interactive elements

Day Two:

- Lecture regarding the rationale for interprofessional working, based on the Victoria Climbie case (Laming, 2003).
- Time in small, face-face, interprofessional groups - introduced to the principles of PBL and began work on their first PBL problems.

Day Three:

- Individual research around the learning objectives and posting onto group VLE discussion board, co-ordinated and collated by student Chair and posted on the VLE at the start of day four.

Day Four:

- Feedback of findings and evaluation in face-face groups, setting additional ground rules, based on evaluations, for the two further PBL problems.
- Lecture on the role of different health care professionals.

Day Five:

- Students given the problems for two further PBL scenarios and asked to set learning objectives for these.
- Lecture on diversity issues pertaining to interprofessional working.

Students were then given eight weeks to work together via their group's on-line discussion boards, to research, debate and collate their findings for these two problems. The first PBL/PBeL problem required them to gain increased understanding of organisational structures and interprofessional working at government policy level (see Figure 2).

*Figure 2. Problem one*

The Bradford Bugle
10th January 2005
1st Edition 32p
LOCAL HOSPITAL ACHIEVES FOUNDATION STATUS

The Bradford Bugle
2nd Edition 32p
12th June 2005
LOCAL HOSPITAL STRUGGLES UNDER BURDEN OF FOUNDATION STATUS

The Bradford Bugle
Late Edition 32p
5th September 2005
LOCAL PRIMARY CARE TRUST STATES 'DEBT NOT OUR RESPONSIBILITY'

*Figure 3. Example of questions from on-line examination*

| Question Asked | Area Assessing |
|---|---|
| A member of your student group is not contributing a fair proportion of the work. How would you best, initially, deal with this? | PBeL group work |
| Tuckman (1965) identified 4 stages of group development. Which of the following describes the stages? | PBeL group work |
| Which one of the following was a recommendation of the Victoria Climbié Inquiry? | Interprofessional working |
| Physiotherapy is best described as a healthcare profession concerned with: | Interprofessional working |

By providing time for students to work through the whole PBL/PBeL process and then discuss and redress difficulties experienced face-face, it was anticipated that students would be better equipped to engage effectively within the online study element for the remainder of the module (Salmon, 2003). Additionally, students were given a 'Netiquette' guide advising them how they were expected to engage via their online discussion board. This included advice regarding the need to critique work, rather than just posting and pasting, as well as responding in a constructive way to others' postings.

Each group had its own staff facilitator to oversee the PBL and PBeL process and ensure the students remained focused on the set task. In face-face PBL, staff facilitators support their student group and encourage them to stay 'on track'. However, this can often be achieved by physical presence alone with students given time and 'permission' to work through issues themselves (Alavi, 1995). During PBeL, the facilitator is invisible and therefore must show their (virtual) presence more frequently by regularly posting words of encouragement to their student group (Kamin et al., 2006), to encourage actual interprofessional interaction rather than parallel postings of information. In addition they were required to provide extended feedback when the final collated response had been provided: this could be used as a revision guide for the final exam.

On completing the module, students were assessed via an on-line multiple-choice questionnaire. In developing the module, the staff team had been keen to use a mode of assessment that would assess both knowledge gained and level of inteprofessional engagement. However, due to the volume of students involved and the resources available, this was considered unrealistic. In creating the assessment however, questions were set which tested two elements: knowledge gained by completing the module and engaging in the PBeL group work and issues that impacted on effective interprofessional working. Example questions are shown in Figure 3.

## EVALUATION OF THE MODULE

Funding was obtained from the Higher Education Funding Council for England (HEFC) e-learning funds to evaluate the IPE provision with an overarching aim of improving and developing the programme. The evaluation was undertaken in two parts, firstly a qualitative study based on uniprofessional focus groups, and secondly a quantitative evaluation based on a survey. The former gives an insight into students' perceptions and experiences of interprofessional e-learning whilst the second elucidates the impact of the intervention on different professional groups, allowing comparisons to be made.

*Figure 4. Key categories and subcategories*

| |
|---|
| **Developing knowledge of other professions (category)** |
| **Timing of the module (category)**<br>-Being new;<br>-Placements; |
| **Mode of delivery processes and implications (category)**<br>-IT skills;<br>-Content;<br>-Communication/participation;<br>-PBL;<br>-Promoting learning; |

## Qualitative Evaluation

A Preliminary evaluation was conducted using the standard module evaluation form immediately after the module had been delivered. This identified notable variations in response according to the student's professional group. Physiotherapy students, for example, tended to give unenthusiastic feedback, whereas nursing students generally remained positive.

In order to explore and understand these differences more fully it was decided to carry out uniprofessional focus groups. Seven focus groups, each with between ten and six students participants (N=59) were conducted between June and November 2006. Students from learning disability and mental health nursing were grouped together due to low response rates.

Semi-structured questions with supporting probe questions covered two main areas, (1) whether students perceived the module had increased their knowledge and understanding of other professions and in what way these elements of the module could be strengthened, and (2) how students had experienced the e-learning aspect of module delivery. The interview data were transcribed and analysed, with the initial coding of recurrent themes carried out independently. We then discussed the findings in detail in order to provide a validity check for the data and to develop the final map of categories and sub categories arising from the data set (see Figure 4). These will be briefly discussed.

## Knowledge of Other Professions

Whilst there was some agreement across groups that the module had not increased their knowledge of other professional roles, it did appear to have provided a useful precursor to interprofessional working by enhancing general communication and team working skills.

*so I learnt a little bit more how to tactfully draw people that are outspoken back to the group's purpose.*

## Timing of Module

Generally students would have preferred the module to occur later in their own programme of study. They reported feeling uneasy so early in the course about their own professional identities and limited clinical knowledge that they could share. This appeared to be linked to students viewing IPE as separate, or in addition to, their uniprofessional programmes rather than integral.

*I didn't feel confident speaking out because I ……… was basically going to say (what), I thought, whereas …… later on I would say I actually know this because I've done it*

## Mode of Delivery: Processes and Implications

### Information & Communication Technologies (ICT) Skills

The preferences for placing the IPE module later in the academic year were generally based on issues related to cultural identity or exposure to professional practice. However, some students felt that they didn't have, at this stage, the ICT skills required to undertake the module. Not all had access to a home computer and not all were sufficiently skilled in their use. The demands of contributing to an on-line module therefore were significant and students reported that it created extra stress for them at that time. As one student stated:

*If I had to do it now it wouldn't be a problem but I found blackboard overwhelming.*

**Content**

Identifying 'problems' or scenarios of equal relevance to all professions will always prove challenging. Most groups suggested it would have been useful to have at least one 'problem' focused on an episode of patient care that would allow exploration of each particular professional role. Interest was expressed by several groups for an insight into how their own professional role fitted in with other roles within the UK National Health Service (NHS). We were able to build on this finding when developing future online PBeL scenarios. When preparing scenarios for groups comprising several professions, it is advisable to have several short scenarios that together cover all professionals, rather than a single scenario covering 5-6 different professions working together.

**Communication/ Participation**

Students liked the flexibility of being able to post their comments on the VLE discussion board, but it was suggested that a real time 'chat room' similar in function to MSN would have been useful. Some groups reported that there was little student collaboration within the group in terms of producing work. Student outcomes may have depended very much on the type of group they found themselves in. Those who had had negative experiences expressed the need for more face-face contact as a motivator and reported poor engagement by some students whose only contribution appeared to be copy and pasting from websites.

There seemed also to be problems with 'Netiquette'. Some students apparently did not know the communication style appropriate to emails and e-learning, and this caused tensions. This is perhaps related to levels of IT skills at this stage in the course. For example:

*People just didn't stick to the rules because they just, they weren't used to this format, the, the discussion ... thing and they were putting things like, 'I'm off home now'.*

Other students reported that getting to know people on-line was difficult:

*our only real meeting of these people was done on-line and to be able to work with people in a multi- professional team you need to build a foundation and a foundation can't be done I don't believe across Blackboards.*

### Problem Based E-Learning

While there was a general acceptance that PBL was an effective teaching and learning method, not all students experienced it this way. The internet was the chief source of information and similar search patterns resulted in duplicated information, exacerbated by a lack of knowledge of how to access e-journals. Some reported having little engagement with the material that they researched.

Whilst some groups had understood the problem solving approach, several issues also arose regarding the facilitation of PBL. Where academic staff were engaged with their group this was acknowledged as extremely useful. All students would have liked more feedback on their group work, in particular whether or not the issues identified by the group were similar to those identified by academic staff.

Within the university, the whole of the midwifery curriculum is delivered by PBL. Some midwifery students, when reflecting on the IPE experience, were able to offer significant insight into some of the key issues that needed to be addressed. As one student midwife noted:

*It's taken us a whole year to build PBL groups that work in our own course and you can't really do it in a week.*

Another recognized the different learning styles adopted by the different professions:

*I actually got some comments back when I started talking about personal experiences and they couldn't grasp the concept that that's what PBL works like that you bring, that you can bring in personal experiences and that to help your learning you know but it's an aspect of PBL but just putting that forward was really difficult.*

### Promoting Learning

Several challenges to the delivery mode of IPE have been identified above. Crucially, these have included:

- ICT skills.
- Different learning styles and expectations across professional groups.
- Appropriate content.

However there was evidence that the module had promoted learning and encouraged engagement with IPE. Students liked the flexibility offered by the e-learning aspect of this module, which was more substantive than in their profession-specific modules. They liked not having to travel into university and recognized its strength as a tool for developing the skills of independent learning.

## The Quantitative Evaluation

The quantitative study involved developing a three part questionnaire in an on-line format, using the software package 'Questionmark Perception' (QMP), issued to students undertaking the year one IPE module. The questionnaire comprised of three parts. Part one was the university standard unit/module evaluation form. Part two focused specifically on the evaluation of the on-line aspect of the module and was created by the project team, which is discussed in detail below. Part three was the validated 'Readiness for Inter-professional Learning' (RIPLs) questionnaire (Parsell & Bligh, 1999) which comprised of nineteen questions and again using a likert-type scale. For the purpose of this chapter we will discuss the findings from parts two and three.

Part two of the questionnaire was created by the project team and focused specifically on evaluating the on-line elements of the learning experience. Current tools to evaluate on-line learning were examined and cross-referenced against the unit evaluation form in order to avoid duplication. From this, four themes were identified as follows:

- Preparation for on-line engagement.
- Accessibility and usability of on-line materials.
- Quality of information and relevance of on-line materials.
- Computer Skills.

*Figure 5. Questions forming part 2 of the questionnaire*

| **Part 2 of the questionnaire:** |
|---|
| I was provided with sufficient information to prepare me to use the online materials within this module. |
| The layout of the site enabled me to navigate easily around the online materials within this module. |
| The way in which the online materials were presented distracted me from their content. |
| I was provided with sufficient time in order to complete the on-line materials for this module. |
| A disabled person is defined in the Disability Discrimination Act as someone with a physical or mental impairment that has a substantial and long-term impact on their ability to carry out day-to-day activities. This excludes situations where sight can be corrected by glasses or contact lenses. Having read this do you consider yourself to have a disability? |
| The on-line learning materials prepared me for the assessment of this module. |
| The online materials stimulated my interest to learn. |
| I liked the fact that this module used online materials. |
| Were you required to complete an on-line assessment for this module? |
| The requirements for completing the online assessment for this module were clear. |

Using these themes, ten questions were created for part two of the questionnaire (See Figure 5). To enhance usability, a five-point likert scale indicating levels of agreement/disagreement with the statements was used for participant responses. Where this was inappropriate, for example when establishing whether students considered themselves to have a disability, a yes/no response was required.

Students completed the RIPLS questionnaire prior to commencing the IPE year one module. They then completed all three parts of the questionnaire, including the RIPLS questionnaire, on completing the module. This allowed students' attitudes towards the module to be evaluated via parts one and two of the questionnaire and attitudinal change, in terms of readiness for interprofessional learning, to be measured through the RIPLS questionnaire.

*Figure 6. Professional groups of students: original sample and completed questionnaires*

| **Profession** | **% Original Sample** | **% Questionnaires returned** |
|---|---|---|
| Adult Nursing | 41.9 | 45.7 |
| Radiography | 12.9 | 9.6 |
| Physiotherapy | 12.1 | 13.5 |
| Mental health | 11.2 | 8.3 |
| Midwifery | 8.4 | 8.3 |
| Occupational Therapy | 5.6 | 7.0 |
| Child Branch (Nu) | 5.3 | 6.1 |
| Learning Disability Branch (Nu) | 2.5 | 1.7 |
| Total | 100.0 | 100.0 |

*Figure 7. On-line evaluation % responses (N=225-238)*

| Evaluation Criteria | Agree % | Neutral % | Disagree % |
|---|---|---|---|
| Sufficient time to complete online materials | 81 | 10 | 9 |
| Requirements for completing online assessment were clear | 81 | 14 | 5 |
| Layout of site enabled easy navigation around materials | 75 | 15 | 10 |
| Provided with sufficient information to use online materials | 71 | 19 | 10 |
| Online materials prepared me for the assessment | 65 | 21 | 14 |
| Presentation (did not) distract from content* | 60 | 30 | 10 |
| Liked the fact that module used online materials | 53 | 35 | 12 |
| Online materials stimulated my interest to learn | 44 | 37 | 19 |

## Results from Part 2 of the Questionnaire

Of the 356 students who took the IPE year one module, 240 (67%) completed and submitted questionnaires at the end of the module. Figure 6 provides the relative percentages for each professional group, which demonstrates that similar proportions responded to the questionnaires.

### Evaluation of the On-line Learning Elements

Figure 7 shows the results of the on-line evaluation. For simplicity, the results of the five-point likert scale indicting levels of agreement/disagreement were re-coded into the following three categories: agree – neutral – disagree. The figures in the table show that the majority of students were positive about time-scales/requirements for completing on-line materials/assessments, and the website layout. However, only 53% liked using on-line materials on the module and only 44% felt they stimulated their interest to learn.

Interestingly, cross-tabulation across professional groups for these two areas of concern revealed that levels of disagreement resided in particular professional groups. Hence, from the data provided in Figure 8, we can see that the occupational therapists in particular, did not like using online materials.

From this data it is also evident that differences existed in levels of disagreement across professions. Whereas only 10% of nurses disagreed that 'online materials stimulated their interest to learn' for example, this rose to 23% for radiographers, 27% for occupational therapists, 32% for Midwifery and 48% for physiotherapists.

## Evaluation of the RIPLS Questionnaire

The RIPLS questionnaire was administered online to students both prior to the module and shortly after its completion. 355 students completed the questionnaire before the module and 221 after the module. Median ratings for each question were compared before and after the module using the non-parametric Mann-Whitney statistical test. From Figure 9 it is clear that there was an overall change of attitudes towards IPE amongst the students, from a negative prior to the undertaking the module, to a positive after the module.

Examination of Figure 9 demonstrates that there were some remarkable changes in attitude towards teamwork and collaboration. Prior to the

*Figure 8. On-line evaluation: % disagreement across professions**

| *Number of student nurses, all branches = 105; Number of other professions ranged from 19-31. | **Liked fact that module used online materials %** | **Online materials stimulated interest to learn %** |
|---|---|---|
| Nursing (Nu), | 11 | 10 |
| Midwifery (MW), | 11 | 32 |
| Occupational Therapy (OT), | 40 | 27 |
| Physiotherapy (PT), | 10 | 48 |
| Radiography (RT), | 5 | 23 |

start of the module, most students disagreed with the following statements (numbered to match their position in Figure 9):

1. Learning with other students will help me become a more effective member of a health care team.

*Figure 9. Median ratings on RIPLS question before (N=355) and after module (N=221)*

| **RIPLS Question** | Median Before Module | Median After Module | Mann-Whitney U | Z | Sig. (2-tailed) |
|---|---|---|---|---|---|
| **Teamwork & Collaboration** | | | | | |
| 1: Learning with other students will help me become a more effective member of a health care team | D | A | 9400 | -15.8 | 0.001 |
| 2: Patients would ultimately benefit if health care students worked together to solve patient problems | D | A | 5698 | -17.8 | 0.001 |
| 3: Shared learning with other health care students will increase my ability to understand clinical problems | D | A | 11265 | -14.9 | 0.001 |
| 4: Learning with health care students before qualification would improve relationships after qualification | A | A | 29131 | -5.7 | 0.001 |
| 5: Communication skills should be learned with other health care students | A | A | 38794 | -0.2 | 0.853 |
| 6: Shared learning will help me to think positively about other professionals | D | A | 7288 | -17.0 | 0.001 |
| 7: For small group learning to work, students need to trust & respect each other | A | A | 36013 | -1.8 | 0.068 |
| 8: Team-working skills are essential for all health care students to learn | D | A | 2209 | -19.7 | 0.001 |
| 9: Shared learning will help me to understand my own limitations | A | A | 28637 | -6.1 | 0.001 |
| **Professional Identity** | | | | | |
| 10: I don't want to waste my time learning with other health care students | A | D | 3149 | -19.0 | 0.001 |
| 11: It is not necessary for undergraduate health care students to learn together | A | D | 3240 | -19.2 | 0.001 |
| 12: Clinical problems solving skills can only be learned with students from my own department | A | D | 3647 | -18.9 | 0.001 |
| 13: Shared learning with other health care students will help me to communicate better with patients and other professionals | A | A | 37644 | -0.9 | 0.367 |
| 14: I would welcome the opportunity to work on small group projects with other health care students | A | A | 30191 | -5.2 | 0.001 |
| 15: Shared learning will help to clarify the nature of patient problems | SA | A | 22054 | -9.7 | 0.001 |
| 16: Shared learning before qualification will help me become a better team worker | A | A | 30617 | -4.8 | 0.001 |
| **Roles & Responsibilities** | | | | | |
| 17: The function of nurses & therapists is mainly to provide support for doctors | A | D | 3288 | -19.1 | 0.001 |
| 18: I'm not sure what my professional role will be | SA | D | 1496 | -20.2 | .001 |
| 19: I have to acquire much more knowledge & skills than other health care students | D | D | 28835 | -5.8 | .001 |

**Key for Figure:** 1= Strongly Disagree (SD); 2= Disagree (D); 3 = Neither (N); 4 = Agree (A);
5 = Strongly Agree (SA). Note: For ease of interpretation, abbreviated labels are used rather than raw numbers

2. Patients would ultimately benefit if health care students worked together to solve patient problems.
3. Shared learning with other health care students will increase my ability to understand clinical problems.
4. Shared learning will help me to think positively about other professionals.
5. Team-working skills are essential for all health care students to learn.

After the module, however, most students agreed with these statements, thus showing a change in attitude from negative to positive towards the importance of teamwork and collaboration. It is interesting to note that while there was no change in the median values for the following two statements (agreement before and after the module), statistical analysis showed a highly significant difference:

1. Learning with health care students before qualification would improve relationships after qualification.
2. Shared learning will help me to understand my own limitations.

## DISCUSSION

Based on the outcomes of this study, we would suggest that there are three key challenges when delivering IPE; maximizing student interaction, overcoming ethnocentrism (Nyatanga, 2005) and recognizing and acknowledging differences in profession epistemology and philosophy. Students in this study showed that they were not highly motivated to interact with other professions. Although the module was delivered early in their respective programmes, they had clearly formed an immediate and intuitive bond to their own, uniprofessional group.

This study showed that many students commenced the programme with a negative attitude for interprofessional learning, and changing those attitudes was both challenging and rewarding. Similarly notable differences existed in values and beliefs related to knowledge construction. For example, not all students valued learning that arose from examining personal experiences. These issues are fundamental to planning effective IPE and will therefore be examined further below.

### Maximizing Student Interaction

Whilst PBeL can be an effective way of delivering IPE (Hughes et al., 2004), problems were identified in the facilitation processes, establishing interprofessional interaction and scenario development. Scenarios need to have substance and relevance, to the extent that it is possible, to all professions. The challenge of developing mutually relevant clinical based scenarios is that they can seem unauthentic, and balance has to be achieved.

In terms of interprofessional interaction, all students interacted to some extent on the VLE, but some students reported that getting to know each other was more difficult on-line. Miers et al. (2007) report similar findings. Miers et al. (2007) report third year students in their similar study admitting to never having used the university web pages or a VLE before being compelled to do so for the IPE module. They found that only 53% of their students liked using the on-line materials and only 44% felt that they stimulated their interest to learn.

Creating e-learning materials and processes that are sustainable, interactive and mutually beneficial across professions is a challenging activity. However, advances in available technology, such as webcams enabling online face-face discussions and social networks such as NING that provide personal pages, complete with photos and other audio/visual images, could offer interesting opportunities to support this.

## Ethnocentrism

Nyatanga (2005) defined ethnocentrism as the view that one's own group is the centre of everything (the in-group), whilst other groups (out-groups) are scaled and rated in relation to one's own group. He describes 'archetypes' as an inherited pattern of emotion, thought and behaviour and states that "*human groups intuitively share a collective unconscious and use archytypes to effect in-group and out-group behavior*" (Nyatanga, 2005, p. 70). Therefore, as we found, students are likely to experience an immediate and intuitive bond to their professional group. This theory underpins any decisions related to timing of the module.

In our study, careful planning ensured that IPE was delivered early in respective programmes, giving less time for students to develop professional 'tribalism'. Yet this was problematic because students viewed the beginning of their professional courses as overwhelming in their own right and the addition of IPE was perceived as an extra burden. This, supported by the findings from the RIPLS survey undertaken by students before the module started, suggests that the relevance of the module was not evident initially. At this stage (the beginning of their professional programme) students said that what they really wanted to learn about was others' roles in interprofessional teams, but paradoxically felt they had not yet anything from which others could learn.

Nyatanga (2005) discusses the work of Sherif and Sherif, who as far back as 1964, were investigating the impact of the in-group/ out-group phenomenon. They resolved the differences of their two experimental groups by creating goals that were common to both groups and were only achievable by collaboration. This is central to IPE delivery. Arguably therefore, the timing of the module is not as important as the content the groups are required to engage with and the ways in which they are required to interact.

## Acknowledging Differences in Professional Epistemology and Philosophy

We would argue that within schools of health studies, there remain differences in epistemology and philosophy across the professions which impact on IPE when students are exposed to ways of teaching and learning that are different to their previous experiences and are thereby perceived as less important or even 'wrong'. For example, a student midwife tried to explain PBL to students from another profession who questioned the validity of using personal experiences. The survey data provided further evidence of different expectations of the professional groups in relation to e-learning, as is illustrated in Figure 7.

Epistemology describes a system of values and beliefs in knowledge construction. It defines a philosophy on learning; how to learn and how to teach. A constructivist epistemology, the belief that truth is socially and historically situated, underpins constructivist approaches to teaching; methods that encourage students to problem solve and develop their own knowledge and understandings based on a unique blend of experience, evidence and reflection. A positivist philosophy however, supports the notion of single truths and often underpins the technical, the testable and the objective and favours learning that is more absolute and testable. Schön (1984) recognized that different schools held different beliefs and conflicting views as he called for a united epistemology of practice, based on reflection in and on practice, and we suggest that this is the way forward.

## FUTURE RESEARCH DIRECTIONS

There is a clear link between the advances in technology proposed as a solution to delivering IPE in this chapter and the use of mobile devices in practice settings to make IPE accessible any time anywhere. This School of Health Studies is a partner in the Assessment & Learning in Practice Settings (ALPS) Centre for Excellence in Teaching and Learning (CETL), which has developed a suite of interprofessional assessment processes to be delivered and accessed using mobile technology. We are currently embedding ALPS outcomes in our IPE curricular and impact evaluation will be an essential element of this process. There is also further work to be done in exploring shared resources (across and within institutions) such as reusable IPE learning objects (available electronically in web-based and mobile formats) to overcome the challenges of content and delivery.

In relation to the module evaluation described here, further amendments to its delivery and assessment have been made, based on the findings of this study. Whilst no changes to the timing of the delivery have yet been made, alterations to both the focus of the problems used in the PBL scenarios and the assessment have taken place. An initial 'problem' focuses on principles of team building with the subsequent two 'problems' based on a clinical issue involving a wide number of agencies, and user involvement. For the assessment, students continue to undertake an MCQ exam, but are additionally peer-assessed on their involvement in group work, both face-face and on-line. Whilst much is written about the benefits and processes of IPE, there is considerably less about interprofessional assessment (IPA). A key response to the findings discussed in this chapter is an exploration of issues related to IPA and professional epistemologies that is currently being undertaken by two authors of this chapter.

## CONCLUSION

The case study presented in this chapter highlights that there will be in-group /out-group behaviours during interprofessional learning that can be managed by setting shared goals requiring interprofessional collaboration to achieve them. We would suggest that students need to have some socialisation into their own profession and even some practice based experience before the delivery of an IPE module. We identify that different professions have different epistemologies, which need further exploration in relation to the delivery of IPE and IPA. Our findings also suggest that IPE should be attempting to achieve different things at different times in the students' programme, with students learning about each other before they learn with and from each other.

## REFERENCES

Alavi, C. (Ed.). (1995). *Problem-based learning in a health science curriculum*. London: Routledge. doi:10.4324/9780203428023

Barr, H., Koppel, I., Reeves, S., Hammick, M., & Freeth, D. (2005). *Effective Interprofessional Education: arguments, assumption & evidence*. Oxford, UK: Blackwell. doi:10.1002/9780470776445

Beck, J., & Owens, M. (2007). Monitoring Student's Level's of Online Engagement in a Pre-registration, Interprofessional Education Module. *Proceedings of the EIPEN Conference*, September 2007, Poland.

Becta. (2007). *Virtual learning Environments (VLE's)*. Retrieved July 2, 2009, from http://partners.becta.org.uk/index.php?section=rh&rid=13640

CAIPE. (2002). Interprofessional Education- a Definition. CAIPE Bulletin. London: Centre for Advancement of Interprofessional Education (CAIPE).

Colyer, H., Helme, M., & Jones, I. (2005). Pragmatic approaches and the theory-practice relationship. In *Interprofessional Education*. Higher Education Academy Health Sciences and Practice Occasional Paper no. 7.

Dall'Alba, G. (2009). Learning professional ways of being: ambiguities of becoming. *Educational Philosophy and Theory*, *41*(1), 34–45. doi:10.1111/j.1469-5812.2008.00475.x

Freeth, D. (2001). Sustaining Interprofessional Collaboration. *Journal of Interprofessional Care*, *15*(1), 37–46. doi:10.1080/13561820020022864

Freeth, D., Hammick, M., Reeves, S., Koppel, I., & Barr, H. (2005). *Effective Interprofessional Education Development Delivery and Evaluation*. Oxford, UK: Blackwell Publishing. doi:10.1002/9780470776438

Hughes, M., Ventura, S., & Dando, M. (2004). On-line interprofessional learning: introducing constructivism through enquiry-based learning and peer review. *Journal of Interprofessional Care*, *18*(3), 263–268. doi:10.1080/135618204 10001731304

Kamin, C., O'Sullivan, P., Deterding, R., Younge, M., & Wade, T. (2006). A case study of teaching presence in virtual problem-based learning groups. *Medical Teacher*, *28*(5), 425–428. doi:10.1080/01421590600628241

Miers, M., Clarke, B., Pollard, K., Rickaby, C., Thomas, J., & Turtle, A. (2007). Online interprofessional learning: The student experience. *Journal of Interprofessional Care*, *21*(5), 529–542. doi:10.1080/13561820701585296

Nyatanga, L. (2005.) The archetypal roots of ethnocentrism. In *Interprofessional Education*. Higher Education Academy Health Sciences and Practice Occasional Paper no. 7.

Owens, M., Allen, L., Hutchinson, J. & Pearson, D. (2007). *Final report on: Promoting interprofessional learning across West Yorkshire*. Report presented to the West Yorkshire Strategic Health Authority.

Parsell, G., & Bligh, H. (1999). Interprofessional learning. *Postgraduate Medical Journal*, *74*, 89–95. doi:10.1136/pgmj.74.868.89

Perry, W. (1970). *Forms of intellectual and academic developments in the college years*. New York: Holt.

Piertroni, P. (1991). Stereotypes or archetypes? A study of perceptions of health-care students. *Journal of Social Work Practice*, *5*(1), 61–69.

Reason, P. (1996). Power and conflict in multidisciplinary collaboration. In Piertroni, P., & Piertroni, C. (Eds.), *Innovation in community care and primary health: the Marylebone experiment*. New York: Churchill Livingstone.

Salmon, G. (2003). e-Moderating: The Key to Teaching and Learning Online, (2nd Ed). London: Routledge Falmer.

Schön, D. A. (1992). The crises of professional knowledge and the pursuit of an epistemology of practice. *Journal of Interprofessional Care*, *6*(1), 49–63. doi:10.3109/13561829209049595

World Health Organisation. (in press). World Health Organization Framework for Action on Interprofessional Education & Collaborative Practice. *WHO Library Cataloguing*.

# Chapter 25
# Interdisciplinary Studies in Built Environment Education:
## A Case Study

**Gerard Wood**
*University of Salford, UK*

**Song Wu**
*University of Salford, UK*

## ABSTRACT

*The School of the Built Environment at the University of Salford redesigned its undergraduate programmes to include multidisciplinary project work at all 3 levels. This chapter provides a case study of the development and implementation of the interdisciplinary module at final level catering for students from five different disciplines. Overall, students responded positively to the module and academic tutors and visiting practitioners were also positive about student performance, but thought insufficient time had been allocated for module delivery and management, which was demanding than the traditional lecture/ tutorial pattern. The use of a dedicated website for communications was seen as a useful co-ordinating and cohesive device although the use of ICT could be significantly expanded. The greatest challenges concern operational difficulties associated with managing large numbers of students in teams, and composing clear requirements with associated assessment criteria.*

## INTRODUCTION

Although promoted by several movements in the USA since the 1930s (Chettiparamb, 2007) it is only over the past 20 years that interdisciplinary approaches to problem solving within the built environment has had any real impact. This corresponds with a similar pattern in academic disciplines as diverse as healthcare, music, engineering, law, history, computer science, and biotechnology for example. Education policy in the UK has over recent years accepted the importance of interdisciplinarity / multidisciplinary collaboration and it is now encouraged within both teaching (Higher Education Academy, 2009) and research (EPSRC, 2009).

In response to these agendas, the School of the Built Environment at the University of Salford has redesigned its undergraduate programmes to include multidisciplinary and interdisciplin-

DOI: 10.4018/978-1-61520-889-0.ch025

ary project work at all 3 levels of study. This chapter reports on a case study of the design, development and implementation of the 2008/09 interdisciplinary module at final level. There were approximately 200 students across five disciplines: Architectural Design & Technology, Building Surveying, Construction Project Management, Quantity Surveying and Property Management & Investment.

## BACKGROUND

### Disciplines and Professions Within the Built Environment

As Adam Smith predicted, the enormous expansion of economic activity in the 18[th] and 19[th] centuries encouraged a tendency towards increasing specialization, and with the rise of technology and occupational expertise, many groups began to claim *professional* status: architects, engineers and surveyors them. This fragmentation of disciplines within construction and real estate provided a degree of efficiency in the performance of the various tasks. However, by the second-half of the 20[th] century, the UK industry was plagued by conflict and characterised by a win-lose mentality, thereby becoming increasingly ineffective. Attempts to improve the management and co-ordination of projects had only a marginal impact on performance.

As a result several recent reports consistently and heavily criticised the industry as being unable to satisfy its clients, citing the relatively divisive nature of the construction industry in the UK when compared to Japan, USA and other European countries (Collier et al., 1991). A recurring conclusion was the need for greater collaboration amongst professionals. Attempts to encourage an interdisciplinary approach appeared to gather momentum through reports published by the Royal Institute of British Architects (Burton, 1992), the Construction Industry Council (Andrews & Derbyshire, 1993), Latham (1994) and conferences at the University of Cambridge (1991) and the University of Central England (1995).

In parallel, the significant growth in the use of *partnering* or *alliancing* systems of procurement also created a focus on: the identification of mutual objectives; robust problem resolution techniques; and systems to monitor continuous improvement in performance (Bennett & Jayes, 1998). All three would require substantially improved levels of co-operation between disciplines and project participants if partnering approaches were to be successfully implemented (Wood, 2005).

### The Nature of Interdisciplinary Activity

Arguments for interdisciplinarity generally stem from debates surrounding disciplinarity. According to Chettiparamb (2007) two main threads can be found. The first argues for interdisciplinarity normatively, positioning it either in terms of filling the interstitial gaps left between the disciplines, or in terms of collaborative transcendence surpassing what disciplinarity can achieve independently. A perspective summed up by Sir Alan Cockshaw, "What we can do together that we cannot do apart is truly amazing" (Spence et al., 2001, p. xii). The second type of argument for interdisciplinarity is more phenomenological, in that it emanates from observations of practice. This view proposes that interdisciplinarity already exists within disciplines.

Cooper (2002) makes the distinction between multidisciplinary work where two or more disciplines work together without stepping outside their traditional knowledge domains, and interdisciplinary work where there is a shared perspective which transcends boundaries. Newell and Swan (2000) use the jigsaw, kaleidoscope analogy: multidisciplinary being the former where pieces are combined but not changed to create the overall solution; interdisciplinary being the latter where components of the picture are shaken and stirred beyond individual recognition. There are obvious synergies here with Chettiparamb's first thread of normative interdisciplinarity and it is this position that is accepted here as that most representative of the built environment context.

It is not only the fragmentation of roles and responsibilities already described which has had an impact.

*Well-building hath three conditions: commoditie, firmness and delight (Wotton, 1624 (adapted from Vitruvius)).*

The common interpretation of Vitruvius' 1st century BC definition as good buildings being those that work well, are built well and give pleasure to users and visitors arguably reflects the intention of most built environment professionals. Vitruvius expanded his third aim into six separate categories - order, arrangement, eurhythmy, symmetry, propriety and economy - to which he proceeded to subordinate most other architectural aims (Scruton, 1979). This multi-faceted nature of design and production in the built environment becomes ever more complex as the centuries unfold and as materials and technologies progress.

Increasing project complexity, more demanding clients, and the blurring of boundaries between disciplines are often cited as important issues. As a typical example, interdisciplinary and team-working skills have been identified as serious deficiencies for urban regeneration practitioners (Bailey, 2005). More recently, the surge of interest in environmental considerations has led some to call for an interdisciplinary forum to "widen and deepen the debate on the ethical dimensions of building. .. from a variety of disciplinary perspectives", (Fox, 2000, p. 4). In a similar vein, Dowden (2007) highlights the important role of interdisciplinary partnerships in responding to climate change adaptation. Furthermore, the fast moving world of information and communications technologies is seen as both a driver for, and enabler of, greater collaboration (Kolarevic et al., 2000; Cheng, 2003). Whichever imperative is preferred, there is a consensus that built environment projects are facing complex challenges, and the observation by Nobel laureate Gunner Myrdal that, "Problems do not come in disciplines", (Gann & Salter, 1999, p. 18) is germane.

## Interdisciplinary Activity Within Higher Education

Chettiparamb's (2007)*Interdisciplinarity: a literature review* provides an excellent and detailed consideration of the diverse interpretations of the concept of interdisciplinarity. However, to retain focus, it is necessary to capture the essence of the industry reports and conferences cited earlier in this respect. Wood (1999) suggests that they appear consistent in recognising that higher education plays an important role:

*Future developments [of courses] should encourage the view that students of related disciplines benefit from working and learning together and that collaborative working is a positive and important component in an education programme. (Inter-disciplinary Studies for the Built Environment, Collier et al., 1991).*

Following Latham's (1994) lead, the Construction Industry Board Working Group 9 report '*Educating the Professional Team*' (1996) recommended that all undergraduate built environment courses should provide a common outcomes under the headings of Communication, Group dynamics and Professional awareness. It did not suggest when, how and why the outcomes were to be achieved. Wood (1999) speculated that such minor changes in approaches to learning, i.e. students achieving a common platform of skills reached independently rather than through active collaboration, would only result in minor changes in outcomes. In any event, there is no evidence to suggest any broad acceptance or implementation of these proposals within university departments or schools. According to Canning (2007) discussions of interdisciplinary learning and teaching are largely characterised by concerns for 'barriers' impeding the student and staff experience in terms of learning styles, assessment, and quality assurance.

Nevertheless, a number of universities have adopted coordinated approaches at undergraduate levels over recent years, but with mixed results; for example: UCE (now Birmingham City University), Leeds Metropolitan University, De Montfort in the UK and the Universities of New South Wales and Western Sydney in Australia (Wood, 1999). Specific case studies have been published through the Centre for Education in the Built Environment including initiatives at Heriot-Watt (Higgins, 2004), Greenwich (Clelford & Hopkins, 2009), and Groningen (Nunnington, 2009). Others report on similar work at Glasgow (Macrea, 2003), Leeds Metropolitan (Howes, 2003), and Heriot-Watt (Haniff, Kaka, Barron & Banfill, 2003). In contrast some institutions, notably the University of Cambridge, concluded that interdisciplinary skills are best taught on postgraduate course to students who already have work experience (Oliver et al., 1994), and Cambridge continue to offer '*Interdisciplinary Studies for the Built Environment*' as a Masters Level programme, now in its fifteenth year.

Although the above studies provide useful information and general reflections, with the exception of Higgins (2004), there is limited data exploring the experience and performance of students in such learning environments. This chapter seeks to address that lack of data by exploring the views of students undertaking a final year interdisciplinary module in the School of the Built Environment at the University of Salford.

## FINAL LEVEL INTERDISCIPLINARY MODULE: A CASE STUDY

### Interdisciplinary Studies at Salford School of the Built Environment

Senior academics at Salford accepted that the drivers for interdisciplinary working in the built environment were compelling. It was therefore necessary to design into the educational process opportunities for students to develop the ability to participate effectively in multidisciplinary teams. In approaching this task, Salford was guided by Wood's (1999) suggestion that, in an educational context, the development of interdisciplinary capabilities would require:

- (Some) common foundation of knowledge
    - a proper understanding of discipline role and knowledge
    - awareness of important issues facing the built environment
- Collaborative practice
    - ability to contribute discipline expertise effectively
    - co-ordination of design and production solutions
    - understanding of, and respect for, the expertise of other disciplines
- Reflective practice
    - comprehensive approaches to complex problems
    - challenging current norms
    - preparation for professional change

This selection (not exhaustive) gives an indication of the breadth of capabilities and levels of competence required. At Salford, it was decided to develop multi/interdisciplinary modules at all three consecutive levels of undergraduate study (years one, two and three) in an attempt to address the above outcomes.

*Figure 1. Location plan*

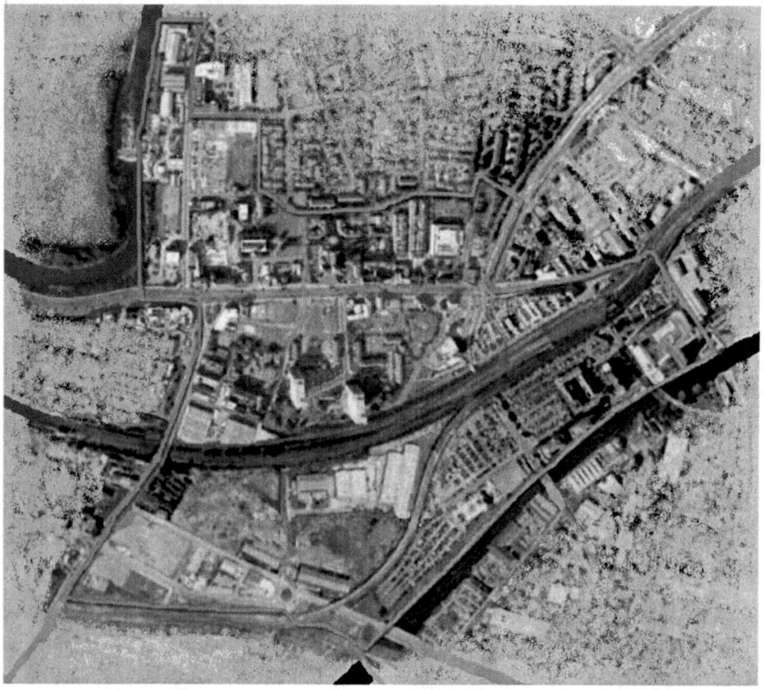

## Level 3 Interdisciplinary Project: Extract from Student Briefing

The overall aim of the interdisciplinary project at Level 3 is to tackle a complex re/development problem that challenges students to explore creative technical, economic and managerial solutions. It builds on existing experience of effective and productive group work in a multidisciplinary team and takes it to new levels of creative interactivity. The students are first presented with an urban regeneration scenario, outlined in brief below.

## Scenario

The Central Salford Urban Regeneration Company seeks to stimulate the next phase of regeneration of the Chapel Street area of the City of Salford. Chapel Street formed the old commercial and administrative core of the c. It still contains a number of business interests, the magistrates courts and residential areas, and there are several examples of new interventions aimed at reversing the previous economic and physical decline, including housing and landscape improvements, the refurbishment of Central Salford railway station, a major office scheme for the Inland Revenue and the 5-star Lowry Hotel adjacent to the new Calatrava footbridge over the River Irwell.

The ambition is to address the problem areas in the next ten years. Two separate masterplans have recently been commissioned – one for the area immediately around Chapel Street itself and one for the area between Salford Central Station, the railway viaduct and the River Irwell. Details of the regeneration framework and masterplans are provided. More details about the plans can be obtained from the Salford City and Regeneration Company websites:

- http://www.centralsalford.com/admin/assets/uploads/Documents/sc_devtfrwk_consultation%20draft.pdf

- and Salford Central Planning Guidance - Salford City Council.

The students were split into teams, each comprising approximately 6 students from different disciplines, and each student should assume his or her own discipline, i.e., as project manager, surveyor, designer, cost specialist, etc. Each team is given two tasks.

## Task 1

Your team is to critically review one of the current masterplan proposals. The review must consider how the plan could be delivered as a whole or, possibly more likely, as a series of block/sites to be offered for redevelopment over time by a range of separate developers. It should analyse the feasibility of the proposals in technical, economic and social terms. It should then identify the range of development opportunities and rate them as an aid to maximising return on investment. This needs to address both tangible and intangible returns. From this analysis the review should suggest any improvements to the masterplan necessary to address the current market situation.

## Task 2

Select one of the key redevelopment sites in the masterplan and develop an outline proposal in the form of a detailed feasibility study including an outline design, a cost plan and a building programme.

## Virtual Learning Environment

Good communications are essential to all of the facets of successful project working. In order to facilitate this, a dedicated project management website has been set up via an online Virtual Learning Environment (VLE), a commercial system named 'Blackboard'. It is anticipated that the student teams will set up internal communications through regular meetings and email to support effective team-working and for exchanging information or asking questions. Teams with part-time student members will rely more on remote communications, and for this type of distance working to be effective, regular face-to-face meetings are necessary to build trust and confidence. It is intended that all relevant documents will be distributed via the VLE. Requests for clarification and additional briefing meetings should be addressed via the dedicated email address on the website and the project coordinator will then direct them to the regeneration company and other subject specialists as necessary.

## Assessment

Learning Agreements are used to establish the detailed objectives of each group, the assessment criteria, marks to be allocated to each and the weighting of sections of work. The mode of assessment reflects the need to assess both the end product and the process through which that product was developed. As a general guide the marks should be allocated on the basis of 50% for the end product and 50% for process, as set out below. These proportions may be adjusted in the Learning Agreement to suit each team's objectives and targets.

- Product 50%:
  1. Critical Review of the Masterplan 15%
  2. Detailed Building Development Proposal 25%
  3. Final Team Presentation 10%
- Process 50%:
  1. Individual Competency Audit 10%
  2. Individual Critical Appraisal 20%
  3. Team Effectiveness Audit 20%

*Figure 2. Survey sample*

| Discipline | Full time | Part time | Total |
|---|---|---|---|
| Quantity Surveying (QS) | 43 | 50 | 93 |
| Building Surveying (BS) | 26 | 10 | 36 |
| Property Management and Investment (PMI) | 18 | 10 | 28 |
| Construction Project Management (CPM) | 18 | 12 | 30 |
| Architectural Design and Technology (ADT) | 5 | 3 | 8 |
| Total | 110 | 85 | 195 |

## THE MODULE EVALUATION PROCESS

The evaluation elicited both academic staff and students' responses to the module. The five senior module tutors were interviewed and key issues of the module contents, assessment and feedback were discussed. A brief SWOT analysis was also conducted to identify the key strength, weakness, opportunities and threats. A voluntary and anonymous student survey was administered at the end of the module during the debriefing process. It was intended to reveal the level of student satisfaction with module contents, assessment and feedback. A self-assessment of module learning outcomes was also included in the survey questionnaire; this would enable a comparison with the actual results of the module as assessed by staff.

The questions were divided into 3 sections:

- Learning experience
- Assessment and feedback
- Learning outcomes

A five-point balanced Likert scale (1 = strongly disagree, 2 = disagree, 3 = neutral, 4 = agree, 5 = strongly agree), commonly used in attitude assessment, was adopted to ascertain the respondent's level of agreement or disagreement to statements in the survey sections of learning experience, assessment feedback and general questions in the learning outcomes section. The self assessment of learning outcomes used a numerical scale (1= poor, 10=excellent). Students were asked to make a self-assessment of the learning outcomes based on the module specification. The questionnaire was piloted with 5 randomly selected students. This was to check for clarity, terminology and consistency of the questions, and appropriate improvements were made to the survey questionnaire.

### Student Participants

The finalised questionnaire was given to all 195 students who undertook the module (Figure 2) from Quantity Surveying (QS), Building Surveying (BS), Property Management and Investment (PMI), Construction Project Management (CPM) and Architectural Design and Technology (ADT). The combination of five disciplines and both full-time and part-time students meant that this was a mixed group with varying levels of relevant occupational experience in construction and property industries (i.e. from none for many full-time students to 5+ years for many part-time students). In addition, the group contained significant numbers of overseas students (approximately 20%). It was made clear to the students that responses would be anonymous and not compulsory.

Figure 3 illustrates the number and percentage of the sample population who were sent, and subsequently responded to, the survey by programme and study mode. In total, 170 out of a total population of 195 students (87.2%) completed to the questionnaire. The overall response rate is quite high because the questionnaire was administered after the module debriefing process, for which students' attendance was required.

*Figure 3. Response rates to survey questionnaire by programme and study mode*

| | Nr of FT in sample | Nr of FT respondents | % | Nr of PT in sample | Nr of PT respondents | % |
|---|---|---|---|---|---|---|
| **ADT** | 5 | 5 | 100 | 3 | 3 | 100 |
| **BS** | 26 | 23 | 88.5 | 10 | 8 | 80 |
| **CPM** | 18 | 16 | 88.9 | 12 | 10 | 83.3 |
| **PMI** | 18 | 15 | 83.3 | 10 | 7 | 70 |
| **QS** | 43 | 38 | 79.2 | 50 | 45 | 90 |
| **Total** | 110 | 97 | 88.2 | 85 | 73 | 86.9 |

## SUMMARY OF RESULTS

The data from the survey questionnaire was entered into SPSS and descriptive statistics were compiled for every question. This incorporated a frequency count of the possible responses, the percentages, mean, mode, median, standard deviation, standard error, range, confidence level and sample variance. The results are summarised below.

### Section 1: Multi-Disciplinary Experience-Learning Experience (Q3-Q8)

Students were invited to rate their learning experience in respect of team working skills, real life experience, and the importance of the multi-disciplinary project in the degree programme as follows.

- For each statement, please rate the extent of your agreement or disagreement. (1 = Strongly Disagree, 2 = Disagree, 3 = Neither disagree or agree, 4 = Agree, 5 = Strongly Agree)
- Learning experience
  - 3: I work better in a team environment
  - 4: My team working skill has been enhanced by the multi-disciplinary project
  - 5: My discipline based key skills are improved by working with students from other disciplines
  - 6: The multi-disciplinary project helps me understand the complexity of a real life design and construction project
  - 7: The multi-disciplinary project makes my study more relevant to real life
  - 8: The multi-disciplinary project modules are an important part of my degree course

*Figure 4. Descriptive statistics of learning experience questions*

| | N | Minimum | Maximum | Mean | Std. Deviation |
|---|---|---|---|---|---|
| **q3** | 170 | 1 | 5 | 3.68 | 0.817 |
| **q4** | 169 | 2 | 5 | 4.01 | 0.694 |
| **q5** | 169 | 1 | 5 | 3.89 | 0.862 |
| **q6** | 170 | 1 | 5 | 3.91 | 0.975 |
| **q7** | 169 | 1 | 5 | 3.73 | 0.948 |
| **q8** | 169 | 1 | 5 | 3.51 | 1.036 |
| Valid N (listwise) | 166 | | | | |

*Figure 5. Descriptive statistics of academic support questions*

| | N | Minimum | Maximum | Mean | Std. Deviation |
|---|---|---|---|---|---|
| **q9** | 170 | 1 | 5 | 3.56 | 1.002 |
| **q10** | 169 | 1 | 5 | 3.20 | 0.984 |
| **q11** | 169 | 1 | 5 | 3.61 | 0.900 |
| **q12** | 170 | 1 | 5 | 3.45 | 1.141 |
| Valid N (listwise) | 168 | | | | |

The results are illustrated in Figure 4. The mean response to all questions in this section fell between the 'neutral' and 'strongly agree' criterion. It is clear that all student groups generally agreed that the learning experience was positive.

## Section 1: Multi-Disciplinary Experience-Academic Support (Q9–Q12)

Students were asked to indicate the satisfaction level of the academic support in the project as follows:

- 9: I have received sufficient advice and support from tutors of my discipline
- 10: I have received sufficient advice and support from tutors of other disciplines
- 11: I have been able to contact tutors when I needed to
- 12: Learning agreement is well executed by the team tutor

The mean response fell between the neutral and strongly agree criterion (see Figure 5). It suggests the students generally agreed that the academic support was sufficient. However, the standard deviations of questions 9 and 12 are higher than others. This suggests that there are different opinions among the student groups. The detailed results of questions 9 and 12 are illustrated in Figures 6 and 7.

The results in Figure 6 suggest that Quantity Surveying (QS) students showed a tendency to responded 'disagree' or 'strongly disagree' in question 9. They felt there was not enough support from the QS tutors. The QS group is by far the largest student group (93 QS students) and only three QS tutors were allocated for academic support. This may have caused the dissatisfaction, if

*Figure 6. Question 9 results by student discipline*

| Q9 | Response by students of each discipline | | | | | |
|---|---|---|---|---|---|---|
| I have received sufficient advice and support from tutor of my discipline | ADT | BS | CPM | PMI | QS | Total |
| Strongly disagree | 0 | 1 | 0 | 0 | 3 | 4 |
| Disagree | 1 | 3 | 3 | 4 | 18 | 29 |
| Neither Disagree or Agree | 1 | 2 | 5 | 3 | 16 | 27 |
| Agree | 5 | 21 | 14 | 12 | 35 | 87 |
| Strongly Agree | 1 | 4 | 4 | 3 | 11 | 23 |
| Total | 8 | 31 | 26 | 22 | 83 | 170 |

*Figure 7. Question 12 results by student discipline*

| Q12<br>Learning agreement is well executed by the team tutor | Response by students of each discipline | | | | | Total |
|---|---|---|---|---|---|---|
| | ADT | BS | CPM | PMI | QS | |
| Strongly disagree | 1 | 2 | 1 | 2 | 6 | 12 |
| Disagree | 0 | 2 | 8 | 3 | 14 | 27 |
| Neither Disagree or Agree | 1 | 6 | 3 | 2 | 18 | 30 |
| Agree | 5 | 17 | 10 | 10 | 33 | 75 |
| Strongly Agree | 1 | 4 | 4 | 5 | 12 | 26 |
| Total | 8 | 31 | 26 | 22 | 83 | 170 |

the QS tutors didn't have enough time to support all QS students.

The results in Figure 7 suggest that a higher proportion of CPM students (9) and QS students (20) responded 'disagree' or 'strongly disagree' in question 12. They felt that the learning agreement was not well executed by their team tutor. The learning agreement was a new assessment method for both students and tutors. The quality of guidance on the learning agreement provided by tutors might differ between tutors from different professional backgrounds. The lack of consistency was felt by the students, especially by student project leaders, who were responsible for preparing the learning agreement.

## Section 2: Assessment and Feedback (Q13-Q18)

Students were asked to indicate their satisfaction with the assessment and feedback by rating the extent of their agreement with the following statements.

- 13: The project brief provides sufficient material to complete the project
- 14: The assessment criteria used in marking have been clear in advance
- 15: Group assessment arrangements and marking have been fair
- 16: The assessment arrangement can reflect the individual performance fairly in the team

*Figure 8. Descriptive statistics of assessment and feedback questions*

| | N | Minimum | Maximum | Mean | Std. Deviation |
|---|---|---|---|---|---|
| **q13** | 170 | 1 | 5 | 2.51 | 1.067 |
| **q14** | 170 | 1 | 5 | 2.84 | 1.062 |
| **q15** | 169 | 1 | 5 | 3.33 | 0.955 |
| **q16** | 169 | 1 | 5 | 3.24 | 0.985 |
| **q17** | 168 | 1 | 5 | 3.39 | 0.922 |
| **q18** | 169 | 1 | 5 | 3.40 | 1.098 |
| Valid N (listwise) | 167 | | | | |

*Figure 9. Questions 13 results by student discipline*

| Q13 The project brief provides sufficient material to complete the project | Response by students of each discipline | | | | | |
|---|---|---|---|---|---|---|
| | ADT | BS | CPM | PMI | QS | Total |
| Strongly disagree | 3 | 5 | 4 | 7 | 12 | 31 |
| Disagree | 0 | 12 | 11 | 9 | 32 | 64 |
| Neither Disagree or Agree | 2 | 5 | 7 | 3 | 18 | 35 |
| Agree | 2 | 9 | 4 | 3 | 20 | 38 |
| Strongly Agree | 1 | 0 | 0 | 0 | 1 | 2 |
| Total | 8 | 31 | 26 | 22 | 83 | 170 |

- 17: Feedback on the tasks has helped me clarify things I did not understand
- 18: Learning agreement is useful to co-ordinate the assessment requirements with the team's skill set

The results are illustrated in Figure 8. Apart from question 13 and 14, the mean response fell between 'neutral' and 'strongly agree' responses. The results indicate clear dissatisfaction with the project brief and assessment criteria.

The detailed results of the question 13 are illustrated in Figure 9. The results indicated that the project brief didn't provide sufficient material to complete the project. More than half the respondents (55.9%) responded 'strongly disagree' or 'disagree' to this question. There were confusions at early stages of the project in terms of the scope of the development and project requirement. The project brief was revised several times during the year, and in consequence, many students felt that time was lost at the beginning of the project. This problem was further confirmed by staff tutors during their interviews.

The results for question 14 are illustrated in Figure 10. The students' responses were affected by the same problem as the project brief. The assessment criteria weren't set out clearly enough at the early stages of the project. Although revised later, this caused the students some confusion. The results indicated that 115 out of 170 responses fell between 'neutral' and 'strongly disagree'.

*Figure 10. Questions 14 results by student discipline*

| Q14 The assessment criteria used in marking have been clear in advance | Response by students of each discipline | | | | | |
|---|---|---|---|---|---|---|
| | ADT | BS | CPM | PMI | QS | Total |
| Strongly disagree | 3 | 1 | 3 | 3 | 9 | 19 |
| Disagree | 1 | 11 | 7 | 10 | 20 | 49 |
| Neither Disagree or Agree | 2 | 7 | 7 | 7 | 24 | 47 |
| Agree | 1 | 11 | 9 | 2 | 27 | 50 |
| Strongly Agree | 1 | 1 | 0 | 0 | 3 | 5 |
| Total | 8 | 31 | 26 | 22 | 83 | 170 |

*Figure 11. Descriptive statistics of learning outcome questions*

| | N | Minimum | Maximum | Mean | Std. Deviation |
|---|---|---|---|---|---|
| **q19** | 165 | 2 | 10 | 7.03 | 1.420 |
| **q20** | 163 | 1 | 10 | 6.31 | 1.745 |
| **q21** | 165 | 1 | 10 | 7.15 | 1.559 |
| **q22** | 165 | 2 | 10 | 7.64 | 1.623 |
| **q23** | 165 | 2 | 10 | 6.92 | 1.567 |
| **q24** | 165 | 0 | 10 | 7.56 | 1.555 |
| **q25** | 164 | 0 | 10 | 7.20 | 1.574 |
| **q26** | 165 | 3 | 10 | 7.10 | 1.385 |
| **q27** | 165 | 3 | 10 | 7.93 | 1.457 |
| **q28** | 165 | 0 | 10 | 7.42 | 1.570 |
| **q29** | 165 | 1 | 10 | 7.36 | 1.600 |
| **q30** | 165 | 0 | 10 | 7.38 | 1.632 |
| Valid N (listwise) | 161 | | | | |

## Section 3: Learning Outcomes (Q19-Q30)

Respondents were asked to complete a self-assessment of the learning outcomes. The learning outcomes were extracted from the programme module specification as follows.

- Knowledge and understanding:
- On completion of this module, I am able to:
  - 19: Apply core discipline knowledge and skills in a multidisciplinary project environment
  - 20: Use a multi dimensional product model to support design and production
  - 21: Effectively demonstrate their own disciplinary role and integrate it with the roles of other built environment professionals responding to a problem solving scenario
  - 22: Use ICT to a competent professional standard in the preparation and presentation of a project submission and team presentation
  - 23: Implement realistic personal development plans that identify learning deficiencies and develop strategies for implementation into their career development

    Transferable/Key skills:

    On completion of this module, I am able to:
  - 24: Develop team working and leadership skills
  - 25: Use a range of technological systems and techniques in the searching for, use and presentation of information
  - 26: Read and interpret project related data presented in differing formats
  - 27: Work collaboratively within a team
  - 28: Make written and oral presentations which are ethical, coherent, cogent and logically structured to a standard appropriate for a built environment professional

*Figure 12. Comparison between actual module results with the question results*

| | | Average Result (Q19 –Q30) | Actual Result | Difference |
|---|---|---|---|---|
| ADT | PT | 7.04 | 6.98 | 0.88% |
| | FT | 7.36 | 7.17 | 2.05% |
| BS | PT | 7.30 | 6.92 | 5.23% |
| | FT | 6.83 | 6.77 | 0.94% |
| CPM | PT | 7.57 | 7.12 | 5.89% |
| | FT | 7.75 | 6.57 | 15.23% |
| PMI | PT | 7.11 | 7.01 | 1.37% |
| | FT | 7.31 | 6.81 | 6.82% |
| QS | PT | 7.27 | 7.17 | 1.36% |
| | FT | 7.17 | 6.69 | 6.74% |

- 29: Access and use a range of primary and secondary sources of information
- 30: Solve practical project related problems appropriate to this level of study

The results are illustrated in Figure 11. This dataset was also used to compare with actual final module results (Figure 12). The results suggested that the students' self-assessments were quite close to the actual module results. Only the CPM full time students' self-assessments are significant higher than the actual results, probably because most of them were acting as project leader and their role and extra contribution were not directly reflected in their final results. Also 50% of their final module mark was based on the group, not individual, marks, but the questionnaire results are based on the individual assessment.

## General Questions (Q31-Q34)

Four general questions were included in the end of the questionnaire where students were asked to indicate their overall view of the multi-disciplinary project.

- 31: I developed my abilities to collaborate effectively with other team members
- 32: I developed an understanding of the roles of other team members from different disciplines
- 33: I developed trust in other students' knowledge
- 34: Overall, I would recommend this module to a friend who is interested in the programme

The results in Figure 13 show that the mean response fell between the 'neutral' and 'strongly agree' criterion. The students showed a consistent positive attitude towards these four questions.

*Figure 13. Descriptive statistics of general questions*

| | N | Minimum | Maximum | Mean | Std. Deviation |
|---|---|---|---|---|---|
| q31 | 168 | 1 | 5 | 3.96 | .712 |
| q32 | 167 | 1 | 5 | 3.98 | .685 |
| q33 | 168 | 1 | 5 | 3.92 | .844 |
| q34 | 168 | 1 | 5 | 3.43 | 1.081 |
| Valid N (listwise) | 167 | | | | |

## Student Evaluation: Summary on Interdisciplinary Working

Overall, the results of the student survey can be interpreted as positive. The learning experience and academic support were satisfactory (Q3-12) and the students assessed themselves as good/very good in terms of achieving the module learning outcomes (Q19-30) with scores ranging from 6.31 to 7.93. However, it must be recognised that the assessment and feedback issues may need to be addressed (Q13-18). There is particularly a dip in the scores related to the clarity of the project scenario / briefing and the assessment criteria. It is arguable that this reflects the complex nature of a real-life tangled, multi-faceted urban regeneration problem, especially at the strategic level - and that there is inevitably some discomfort for all disciplines when first addressing such challenges.

There may well be some truth in this view; it will be important for the academic team to reflect on both the nature of the problem presented and the students' response to it when devising future scenarios.

It seems clear therefore, that the biggest challenges with interdisciplinary working are not related to convincing students of its value (see Q31-33), but more to do with operational difficulties associated with managing large numbers of students in teams, and composing clear requirements with associated assessment criteria.

## Staff Evaluation

Semi-structured interviews with five module tutors (responsible for 60% of the students who undertook the module) were conducted, to explore their perceptions of the students' reactions and their own personal experiences. All interviewees were lecturers or senior lecturers with experience of large group teaching on project based assignments prior to this particular module. Transcripts were coded to identify categories of subject matter and then broken into meaningful text segments. The text segments were then de-contextualised from the narrative of the interview and copied into coded folders. This allows some quasi-quantification of agreement/disagreement using the constant comparison method that Glaser and Strauss (1967) advocate. Manifest items, i.e. those physically present in the content (Robson, 2002), were also identified to avoid any reliance on latent content which is a matter for inference or interpretation. The responses can be organised into three themes of student experience, staff experience, potential improvements and the more effective use of ICT.

## Staff Views on the Student Experience

All staff tutors felt the students had concerns in terms of the marking of this final year module, particularly students allocated in a perceived 'weak' team. The students were concerned that their own mark would be affected by non-performing students, with a subsequent negative impact upon their degree classification. Tutors also felt the assessment mechanism didn't always achieve a satisfactory balance between the group mark and the individual mark.

The tutors agreed that the overall student experience was good. Most students appreciated the team working experience; particularly the complex nature of real world regeneration problems and the need for, and exposure to, the knowledge of other disciplines.

Tutors observed many students struggling at the beginning of the project, mainly because the complex brief and the new mode of working. Gradually, the majority settled into their teams and were able to work effectively as a team by the end of the project. The tutors needed to provide significant levels of support in order to increase students' confidence in this respect. It was thought that this problem could be eased if the brief was clearer and less complex, and team working was introduced earlier in the degree programme.

Tutors reported that students complained about the assessment method. Although there was a genuine attempt to balance the group and individual marks through the assessment mechanism, students did not all agree that this worked effectively. It was suggested that the weighting of the individual component of the assessment needs to be increased to allow for a more 'personalised' module mark. The tutors agreed that it was essential to fully co-ordinate the marking, and that the process must allow for internal moderation of marks and some normalisation across assessing tutors.

## Staff Experience

Overall, the tutors had positive experiences from the module. They felt the size of the student teams was manageable, but the ideal number of students for each tutor should be no more than 25-30. The demand on tutors' time was felt to be higher than the conventional class room based teaching module.

The tutors found the assessment method was over-complicated and didn't balance well between group and individual marks. It was suggested that the assessment should be simplified, and for both students and tutors to be better informed earlier in the module.

The tutors also agreed that the module needs a core team (minimum of 5-6 tutors) and a strong module leader. The module leader must coordinate tutors to ensure consistent communication between tutors and students. Tutors need to have a common understanding of the assessment criteria, to ensure consistent marking across the student teams.

The module needs to be better resourced in terms of staff time, and tutors should be allocated sufficient workload to provide adequate support for the students. It was thought that in this instance, too many staff had been involved, and consequently the workload allocation was spread too thinly. As a result, it was difficult for some staff to spend the time that students demanded, and it too often relied on staff's good will to allocate extra hours. This led in some cases to inconsistencies in the student experience between tutor groups.

The grouping of students into teams had in some cases been unsuccessful in achieving a suitable mixture of disciplines. It was felt that this must be addressed in future, since it defeats the purpose of a 'multi-disciplinary' project. It was thought that this was mainly due to the large number of one particular discipline (quantity surveyors) undertaking the module. However, it could be avoided through better planning and organisation.

## Potential Improvements and More Effective Use of ICT

A SWOT analysis was also conducted as part of the interview. The aim of the SWOT analysis is to plan and improve this project for next academic year. The results are shown in the Figure 14.

*Figure 14. SWOT analysis*

| Strengths | Weaknesses | Opportunities | Threats |
|---|---|---|---|
| • Team working experience<br>• Real life project experience | • Complex brief and assessment method<br>• Inconsistent assessment<br>• Coordination among tutors<br>• Better student grouping | • Exposure to different discipline<br>• Flexible project material<br>• Extra skill set for the students<br>• Industry involvement | • Student complaints,<br>• High demand of staff time<br>• Communication with part time students |

With regard to ICT use in the module, the Virtual Learning Environment (VLE) 'Blackboard' was deployed in the project. The latest project and module information (of various kinds) was continually uploaded onto the VLE by tutors during the module. In turn, students were required to submit both their individual project and group work on the VLE for assessment. The original intention had been to also use the VLE extensively to facilitate group collaboration and communication. Unfortunately, this was not effectively implemented, and the system was primarily used as an information portal to distribute project materials and collect assignments.

The main reason was thought to be a lack of training for both students and tutors. Blackboard provides collaboration tools such as discussion groups, Wiki blogs, etc. Those tools can be very effective in facilitating, supporting and monitoring group working, and potentially can be used as part of the assessment. Students can use the discussion groups for communications and Wiki for the preparation group reports.

Furthermore, as already mentioned, one of the main difficulties of assessing group work is to distinguish the individual contributions. Tools like Wiki and discussion groups provide an audit trail for an individual student's involvement. They could therefore be used as a reference point for tutors trying to assess individual effort and performance. In summary, the VLE use was quite limited and there is significant potential for its use, and that of other ICT systems, to be expanded in future.

## FUTURE CHALLENGES AND RECOMMENDATIONS

This interdisciplinary project case study largely revealed a positive experience among the students and staff through effective team working on a real world scenario. The majority of staff and students agreed it should be an important part of built environment degree programme. However, the case study also highlights several challenges:

- Simple, clear but challenging project brief should allow students to develop additional skills.
- Balanced assessment methods and weighting for both individual components and group components.
- Well coordinated and resourced module delivery team to ensure consistent communication with students and fair assessment.

In order to address above challenges, the following recommendations to improve the module for the next academic year were suggested:

- A clearer project scenario should be developed and tested well in advance of the beginning of the module.
- A core module delivery team consisting of 5-6 academic staff from different disciplines has to be formed and well briefed. A module leader needs to be appointed.
- Student groups need to be carefully organised to ensure the formation of multi-disciplinary teams.
- The assessment criteria need to reflect both individual and group performance.
- The module delivery team should have common understanding of the assessment criteria.
- Regular meetings for the module delivery team should ensure consistent communication with students.
- Detailed feedback should be given to the student through a final debriefing process.

It was found in the case study that, particularly at final level, students are often uncomfortable with group marks, or learning vehicles where their individual mark relies on the performance of other students. Hence further research is necessary to explore the broader body of literature surrounding the issues of group marking, and the means of accommodating weaker students without jeopardising the results of their more able teammates. Subsequently, it will be necessary to devise and test more sensitive assessment mechanisms to deal with this concern whilst still recognising the importance of collective responsibility.

In addition, a transparent internal moderation process is important to ensure consistency of marking where a number of different tutors are involved. Although this module included the use of a virtual learning environment, this was really only used as a communication and administrative tool. Given the advancing technologies within built environment design and construction processes, there is scope for much greater application of ICT within the delivery and learning on such interdisciplinary modules.

## CONCLUSION

This case study highlights the challenging nature of implementing interdisciplinary working within the built environment context. What is perhaps most pleasing (even surprising) is the generally positive attitude of students in terms of their own assessment of their learning outcomes. In this instance the module was at undergraduate final level, but the approach will be extended at Salford to all levels in the 2009-2010 academic year.

In terms of future research, it will be important to undertake a detailed review of staff and student experience and performance with a view to refining and improving this crucial aspect of built environment education. The student and staff experience needs to be compared across all levels of undergraduate programmes to assess whether the interdisciplinary projects and activities should follow the same format at each level or if, for example, there are different requirements in earlier and later years.

## REFERENCES


Andrews, J., & Derbyshire, A. (1993). *Crossing Boundaries*. London: Construction Industry Council.

Bailey, N. (2005). Professionalism and skills enhancement in the built environment: The role of higher education in a changing policy context in England. *CEBE Transactions*, *2*(2), 48–62.

Bennett, J., & Jayes, S. (1998). *The Seven Pillars of Partnering*. London: Thomas Telford.

Burton, R. (1992). *Steering Group on Architectural Education - Report and Recommendations*. London: Royal Institute of British Architects.

Canning, J. (Ed.). (2007). *Disciplines in dialogue: disciplinary perspectives on interdisciplinary teaching and learning*. Southampton, UK: The University of Southampton Interdisciplinary Teaching & Learning Group.

Cheng, N. Y. (2003). Approaches to design collaboration research. *Automation in Construction*, *12*, 715–723. doi:10.1016/S0926-5805(03)00059-1

Chettiparamb, A. (2007). *Interdisciplinarity: a literature review*. Southampton, UK: The University of Southampton Interdisciplinary Teaching & Learning Group.

Clelford, T., & Hopkins, A. (2009). *Working Together, Learning Together: Methods & Process in Multidisciplinary Group Projects*. Retrieved March 24, 2009, from http://www.cebe.heacademy.ac.uk/learning/casestudies/case_pdf/clelford.pdf

Collier, A., Bacon, J., Burns, D., & Muir, T. (1991). *Interdisciplinary studies for the Built Environment*. London: Council for National Academic Awards.

Construction Industry Board. (1996). *Educating the professional team*. London: Thomas Telford.

Cooper, I. (2002). Transgressing discipline boundaries: Is BEQUEST an example of the 'new production of knowledge'? *Building Research and Information, 30*(2), 116–129. doi:10.1080/096132102753436495

Dowden, M. J. (2007). Climate Change Adaptation: a Multi-Disciplinary Challenge. In RICS, *COBRA 2007 Conference Proceedings, Georgia Institute of Technology*. London: RICS.

Engineering & Physical Sciences Research Council. (2009). *The future of research: Joint statement from EPSRC and the Learned Societies*. Retrieved April 2, 2009, from http://www.epsrc.ac.uk/Content/News/futureofresearch.htm

Fox, W. (Ed.). (2000). *Ethics and the Built Environment*. London: Routledge.

Gann, D., & Salter, A. (1999). *Interdisciplinary Skills for Built Environment Professionals: a scoping study*. London: The Ove Arup Foundation.

Glaser, B. G., & Strauss, A. L. (1967). *The discovery of grounded theory: strategies for qualitative research*. Chicago: Aldien.

Haniff, A. M., Kaka, A. P., Barron, S. J., & Banfill, P. F. G. (2003). Changing industry culture through graduates: promoting knowledge sharing and interdisciplinary teamworking amongst students. In R. Newton, A. Bowden & M. Betts (Eds.), *CIB W89 Building Education & Research Symposium: Volume 3* (pp.1367-1377). Salford, UK: The University of Salford.

Higgins, M. (2004). Urban design project in a multidisciplinary European context: Collaboration between education and practice. *CEBE Transactions, 1*(2), 16–36.

Higher Education Academy. (2009). *Interdisciplinary Teaching and Learning Group*. Retrieved April 2, 2009, from http://www.heacademy.ac.uk/ourwork/networks/itlg

Howes, J. (2003). Working Together. In R. Newton, A. Bowden & M. Betts (Eds.), *CIB W89 Building Education & Research Symposium: Volume 3* (pp.1378-1389). Salford, UK: The University of Salford.

Kolarevic, B., Schmitt, G., Hirschberg, U., & Kurmann, D. (2000). An experiment in design collaboration. *Automation in Construction, 9*, 73–81. doi:10.1016/S0926-5805(99)00050-3

Latham, M. (1994). *Constructing the Team*. London: Her Majesty's Stationery Office.

Macrea, K. (2003). The Interact Student Competition. In R. Newton, A. Bowden & M. Betts (Eds.), *CIB W89 Building Education & Research Symposium: Volume 3* (pp.1418-1426). Salford, UK: The University of Salford.

Newell, S., & Swan, J. (2000). Trust and interorganisational networking. *Human Relations, 53*(10), 1287–1328.

Nunnington, N. (2009). *The European Challenge*. Retrieved March 24, 2009, from http://www.cebe.heacademy.ac.uk/learning/acbee/pdf/The%20European%20Challenge%20_Nick%20Nunnington.pdf

Oliver, G., Spence, R. & Kirby, P. (1994). Charting a New Course - Education for the Built Environment. *The Cambridge Review, May 1994*.

Robson, C. (2002). *Real world research: a resource for social scientists and practitioner –researchers* (2nd ed.). Oxford, UK: Blackwell Publishing.

Scruton, R. (1979). *The Aesthetics of Architecture*. London: Methuen & Co.

Spence, R. J. S., Macmillan, S., & Kirby, P. (2001). *Interdisciplinary design in practice*. London: Thomas Telford.

Wood, G. D. (1999). Interdisciplinary working in built environment education. *Education+Training, 41*(8&9), 373-380.

Wood, G. D. (2005). Partnering practice in the relationships between clients and main contractors, *RICS Research paper series, 5*(2), April.

Wotton, H. (1624). *Elements of Architecture Collected from the Best Sources and Examples. London: John Bull.* Farnborough, UK: Gregg Press. Reprinted 1969

# Section 5
# Future Possibilities

# Chapter 26
# The Use of Personal Digital Assistants in Nursing Education

**Nina Godson**
*Coventry University, UK*

**Adrian Bromage**
*University of Birmingham, UK*

## ABSTRACT

*The use of Personal Digital Assistants (PDAs) and smartphones (combined mobile telephone and PDA) in Nurse Education is a relatively new development, in its infancy. The use of mobile technologies by health care professionals is increasing, and it seems likely to accelerate, as mobile information and communication technologies become more ubiquitous in wider society. The chapter reports on a small-scale feasibility study to evaluate the practicalities of supporting student nurses on their first clinical placements with PDAs that have been pre-loaded with reusable e-learning objects. The student nurses generally found the PDA easy to use and carry on their person, valued the availability of the reusable e-learning object on their clinical placements and called for more of them to be made available to learners.*

## INTRODUCTION

Clinical Nurse Tutors face the challenge of managing large numbers of students in clinical skills sessions, when demonstrating and practising skills acquisition, including interprofessional working. This involves maintaining accurate and fair assessments of individual performance that may later serve as evidence of safe practice for their clinical placements. There needs to be a means of continuing this supportive learning environment, in the transition from simulation to reality, to ensure the bridge from theory to practice is strong.

Ideally, to enhance learning and support development of decision-making skills, students should have access to information at the moment and the physical location where it is needed. In relation to this, a major aim of pre registration nursing programs is to assist students to attain competencies for practice, for example, in the use of the tools and technology associated with nursing. Healthcare professionals need to continually update knowledge and skills, in order to enhance their clinical practice and professional develop-

DOI: 10.4018/978-1-61520-889-0.ch026

ment. Similarly, student nurses in any speciality must get into the habit of continually becoming familiar with the latest practice recommendations, such as guidelines for preventing the spread of infection and standards from regulatory agencies.

Using the world wide web can help meet these needs, given that anyone can publish web-pages that are then available instantly across the globe. Information published on line offers advantages compared to text books and journal articles, which are often outdated by the time they reach publication and distribution. However, up to 2003, connection to the internet was typically on either a desktop or laptop personal computer. Since that time, pocket-sized personal digital assistants (PDAs), also known as palmtop computers, with wireless access to the World Wide Web have become available.

The first PDA, the Apple Newton MessagePad, was launched in 1993. In 1996, Nokia introduced the first mobile telephone with full PDA functionality (a genre known as 'smartphones'). However, early PDAs relied on wired connection to networked desktop or laptop PCs, until the launch in 2003 of the Hewlett-Packard iPAQ H5450. This was the first PDA featuring built-in Wi-Fi for connecting to wireless computer networks and the World Wide Web. In 2009, the majority of PDAs sold are Wi-Fi enabled smartphones, for example, the RIM Blackberry, the Apple iPhone and the Nokia N-Series. It is argued in this chapter that the Personal digital assistants (PDAs) and smartphones can enable nurse educators can incorporate into the curriculum opportunities for students to develop critical approaches, and also introduce information technology as a tool in clinical decision making. This will serve as a foundation for using mobile computing technologies later in their careers.

This chapter describes a small-scale feasibility study in which the practicalities of using smartphones to support the undergraduate nursing curriculum programme at a UK University were evaluated. The students who participated were drawn from all branches of nursing, and were in their first year of nurse training.

## BACKGROUND

In the mid-1990's the World Wide Web radically changed the possibilities of our information landscape, and simultaneously became available to users of pocket-sized hand held computing devices such as Personal Digital Assistants (PDAs) (Murphy 2005). Even so, the use of PDAs in nursing was rare until comparatively recently. The Cumulative Index to Nursing and Allied Health Literature (CINAHL) has had a subject heading for PDAs ("Computers, Hand-Held") since 1997, while the National Library of Medicine first used the term "Computers, Handheld" in the Medical Subject Headings (MeSH) in 2003.

Discussions in the early articles focused on theories of informatics and technology. More recent articles indicate that PDAs are in wider use (De Groote, 2004), for example in nursing practice and student education. Interestingly, the Medical professions have written most of the literature on PDA use in health care, and it is clear that as a group they are very interested in this. Members of other health care disciplines also demonstrate interest in using PDAs (De Groote, 2004). The integration of personal digital assistants (PDAs) into health care delivery continues to grow; and health professionals are adopting PDAs faster than the general public (Stolworthy et al., 2000).

However, there has been little evidence that PDAs improve patient care (Fischer, 2003). PDA use by healthcare professionals tends to be in the context of on everyday routine. Many use their PDAs as a diary or address book, rather than as a knowledge base (Criswell & Parchman, 2002). Considering these facts, it might be argued that the use of PDA s has not been fully understood or its full potential recognised by health care professionals. However, the PDA can be considered a new learning tool, offering the potential to keep healthcare professionals up to date with developments in evidence based practice, and having potential to help nurse educators meet students' educational needs, as will be seen.

There is emerging evidence of PDA use in graduate nurse education as well as undergraduate nursing programs (Thomas et al., 2001). There have been accounts of how various institutions have used PDAs, for example, Huffstutler (2002). While university libraries have been known to provide information and support to students who use PDAs, many health care programs making use of PDAs seek help from institutional information technology department instead (Young, 2002).

The incorporation of information and communication technologies into nursing practice and education by student nurses and staff is potentially challenging, particularly on their clinical work placements. To meet the healthcare needs of today's society, student nurses are expected to demonstrate compassion when providing patient care. However, reduced time for patient contact may have eroded their ability to do so. It has become clear that increasing staff workloads have increased while staffing levels have decreased in the clinical setting (Aston & Molassiotis, 2003). Simultaneously, nursing curricula are forever evolving, introducing new ways of learning and new learning technologies, partly in response to the challenge of accommodating increasingly large student cohorts.

When student nurses go on clinical placements, teaching in the ward area is guided by both time and space limits. In light of nursing shortages, teaching that takes place at the patient's bedside is not always possible, and it is becoming more difficult to take students away from the ward area to train them. Thus the students' educational needs are not always reliably met within the current context of patient care settings. Nursing educators can usefully integrate technology as a means of support to student's learning and to prepare the new generation of nurses as independent learners.

Pre-registration nursing programmes aim to assist undergraduate students to attain competencies for practice, including interprofessional working, and the earlier these skills are introduced the better. Nurse tutors need to constantly review areas where information and training gaps exist, to highlight where nurse mentors at clinical placements and university-based tutors can play a larger part in training and preparation, both before and during students' clinical placements. In the case described here, a need was identified for first-year undergraduate students on their first clinical placements to have access to learning materials and resources at the moment they are needed, to enhance learning and support development of practical clinical skills (Huffstutler, 2002). In the feasibility study reported in this chapter, tutors responded to these pressures by developing a way to meet students' individual learning needs and styles.

## The Use of PDAs by Student Nurses

The evidence discussed in the preceding section of increasing adoption of PDAs by healthcare professionals suggests that PDAs can be considered as if any other piece of healthcare equipment. It is arguable that nurse educators should begin to introduce PDAs to student nurses, given the importance of providing opportunities for students to become familiar with PDAs as a tool for clinical skills development. Both tutors and students are more likely to learn to use PDAs effectively through consistent use in both clinical and classroom settings. However, it was seen in the preceding section that there is currently limited literature addressing use of PDAs in the nursing curriculum, and it appears that their educational potential in clinical settings is relatively unexplored.

The preparation of student nurses will always involve the development of clinical skills at the patient bedside, therefore quick and easy access to information is needed, not only in the skills laboratory but also in clinical practice. PDAs can operate as if electronic text books to help student nurses manage the information they need to learn. In the past, when students had queries about practical skills, they would turn to textbooks for answers, often having to find relevant volumes. This can be very time consuming, when there are so many clinical skills to learn, not to mention the problem of carrying heavy books to clinical placements. A PDA can enable student nurses to retrieve information in the clinical area quickly and easily, conserving precious learning time (Koeniger-Donohue, 2008).

PDAs can provide nurses with portable access to extensive reference materials as well as other organizational resources and time saving benefits. This can be in the form of reusable e-learning objects accessible from a central point by an entire cohort of students. The experience of learning from a PDA potentially offers students an opportunity to revisit at will any practical skill and reflect on their interpretation and understanding of it, so knowledge is accumulated and transferred into an individual's own understanding (White et al., 2005). This way of learning alongside other supportive mechanisms promises an enriching clinical experience for the student. This is especially relevant when mentors are busy with heavy workloads and staffing has decreased due to cut backs in the economy (Yonge et al., 2002), and ward areas are as a result struggling to meet the educational needs of student on clinical placements.

The ultimate goal of nurse educators is to develop competent and passionate student learners, who feel supported in transition from academia to clinical practice. PDAs can be a powerful tool that 'holds the hand' of trainee nurses when reassurance is needed. This could help reduce students' stress, which has been found to lead them to miss work shifts or abandon their studies (Timmins & Kaliszer, 2002). However, handheld devices must fit into the ward routine, and not be perceived as extra work, or as interfering with the nurse-patient interaction or nurse mentor relationships. The question of the practicalities of PDAs on clinical placement required investigation before tutors at the university in which the case study reported in this chapter took place could decide whether to invest time, effort and money in their adoption as learning tools.

## Feasibility Study: Knowledge in the Palm of your Hands

In 2008, a small-scale feasibility study was undertaken at Coventry University to examine the fundamental practicalities and issues surrounding the use of hand held devices (PDAs) by first year student nurses on their first clinical placement. It was thought that by creating reusable learning objects covering relevant skills and competencies and loading them onto a PDA, student nurses from any branch of nursing can practice the skills and competencies in the clinical room, were their mentor unavailable.

The student nurses can gain confidence and requisite background knowledge prior to the hands-on portion of their learning, saving the mentor time to work toward mastering these skills. The student nurse can then move to the patient bedside to gain the crucial supervised practice required to successfully and safely perform clinical skills. While the learning object used in the study was not explicitly designed to promote interprofessional learning, this was appropriate, give the aims of the feasibility study to explore the fundamental practicalities of smartphones on clinical placements. It is worth briefly outlining the nature of the learning object, however, and considerations that were taken into account to conduct a fair trial of the PDA.

The reusable e-learning object was designed to make the learning experience personally relevant and meaningful, related to unambiguous skills that students had previously learned in the skills laboratories run in face-to-face university-based teaching. This was so the nature of its content did not distort the students' smartphone use by being perceived as irrelevant or too challenging in its content matter. It comprised a "teachable moment" that was developed by nurse educators at the University into a reusable e-learning object that depicts a generic clinical skill for all branches of nursing, a hand-washing technique known as the 'Ayliff procedure' (Ayliffe, 1992). The resulting reusable e-learning object features a set of film clips that demonstrate and explain the Ayliff procedure (Gould, 2008), and was loaded on to each smartphone, enabling the participants to take it into their clinical practice placements.

The issue of using smartphones in a clinical area was discussed with an infection control nurse. It was decided that students should not use smartphones at patients' bedside and only in a study area. Thus it was explained to the participants that as an infection control precaution, smartphones were only to be used in the study room in the clinical area, and never at patients' bedsides. Ethical approval for the study was sought from the University ethics committee, and was duly granted.

## Methodology

### PDA Choice

To ensure that the PDA has a long service life, given the capital investment necessary for their purchase, the best choice of product needs to be considered carefully. Having a standard device was crucial, and one that is simple to use by all. Basically, there are two operating systems systems: the 'Palm' System and the 'Pocket PC' System. This study used a smartphone running the palm operating system. It was found by the course tutors that this operating system has the advantage of simplicity and ease of use. Smartphones are becoming as popular as ordinary mobile phones, which most students already possess, thus it was thought that the students would be more likely to adapt to a smartphone rather than a pure PDA.

It is worth noting that there is also a learning curve for the tutors involved. It was important that the research team had access to the log in information, so they themselves could not be 'locked out'of the smartphones. Preparing the smartphones for use is time consuming, in that reusable e-learning resources must be loaded onto each one. However once loaded the data are stored permanently.

### Participants

The sample comprised eighteen female students volunteers from a cohort of 215 students, as the researcher only had funding for twenty smartphones. The students were from the first year of a pre-registration nursing program that ran in January 2008, covering all four branches of nursing; adult, mental health, child and young persons, and learning disabilities. They were about to go on their first clinical placement, and they were to visit different types of placements, hospital based and community. It was considered important that the participants expressed an interest in the use of such technology, as should any technical hurdles become apparent while the students are on placement, they would have to solve them with only telephone or email support from the researchers.

## Ethical Briefing of Participants

The project was introduced to the students during one of their initial lectures, and a participant ethical briefing sheet was distributed. The briefing explained that each participant would be asked to complete both a pre- and post-placement questionnaire to evaluate the use of the smartphone, that they had the right to withdraw from the study at any time, and that their responses would be anonymised and pooled, then statistically evaluated and published. Students read the information and asked any questions, and those who were interested signed up to participate. A further date was arranged for those who signed up to meet with the researchers and learn how to use the smartphone and discuss the project further, after which they signed the necessary 'consent to participate' forms.

## Training for Smartphone Use

The smartphone is a tool that a student must learn to use effectively, just as they would a medical tool such as a tympanic thermometer or sphygmomanometer. Clinical Skills are developed through consistent practice in the clinical skills laboratory, and similar techniques can be used to familiarise individuals with a smartphone.

To these ends, the participants were invited to a workshop on how to use the smartphone. This session provided hands on experience, in a structured supportive environment. Each student was briefed about how to use the smartphone and provided with an instruction leaflet, and was then shown how to log into the device. They were briefed that after their twelve-week placement had finished, they were to return the smartphone to the researchers. It was also explained that as an infection control precaution, the devices were only to be used in the study room in the clinical area, and never at patients' bedsides.

Getting started with a PDA is like learning any other new skill. A positive attitude, practice and willingness to make mistakes will help to overcome initial unfamiliarity. The students were thus asked to keep the smartphone in their coat pocket or work bag while in placement, so as to have the reusable learning objects available whenever they felt they needed to review the handwashing technique. The students were reminded that they were not alone. A contact number for the researchers was provided, in case the participants needed technical support. They were also made aware that they were effectively in a networking group with other students in the placement using the same device.

## Pre-Placement Questionnaire

The students next completed a pre clinical placement questionnaire. Its purpose was to gather information about their general familiarity with information and communication technologies, and any learning resources they had used so far in their training for the hand washing technique. It was found that the majority of participants owned a mobile phone. Furthermore, nine of the eighteen participants posessed PDAs of their own, and were familiar with the function and layout of these devices. All students seemed keen to participate in the project, and did not appear worried about using the smartphones.

The main learning resources that the students had used prior to clinical placement were clinical skills laboratories, online learning, handouts/leaflets, Objective Structured Clinical Examinations and Clinical skills tutors. Objective structured clinical examinations are used as a form of assessment of clinical skills at the beginning of the nursing curriculum. Students found clinical skills laboratories and open laboratories the most effective way of learning the hand washing technique. The majority of participants found pictorial presentation, continual practice of the skill and memory aides the most effective ways of remembering the eight-step hand washing technique. Some felt that they needed reminders of these steps, not only in the skills laboratories but in clinical practice placements.

The participants felt that, pre-placement, they were able to demonstrate the correct hand washing technique, and would practice further to become proficient at this clinical skill. The participants were then asked to comment upon the least effective method of learning. The majority stated that the least effective was learning from textbooks and from previous practical experience before entering the nursing course. It could be argued that textbooks offer abstract experience, not 'hands on' concrete experience, and that previous care experience may not have taught the most up-to-date technique.

Finally the students were given the option to state their Objective Structured Clinical Examination (OSCE) score on the pre- placement questionnaire. All OSCEs include the assessment of hand washing. All of the participants in the

*Figure 1. Overall ease of use*

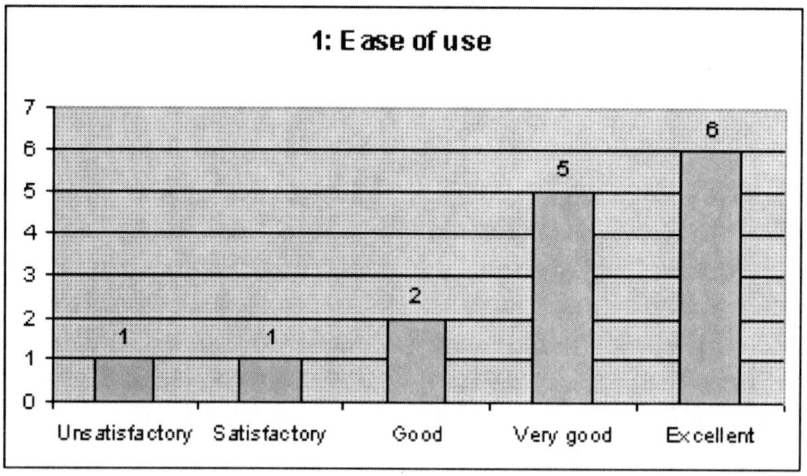


sample indicated that they had passed the year one OSCE, however, some chose not to record their actual score on the pre-placement questionnaire.

## Post-Placement Questionnaire

Each participant then took their smartphone into their first clinical placement, and twelve weeks later returned from their placements to the University for the theoretical component of their course. At this point, each participant completed a post- placement evaluation questionnaire, so that their thoughts and experiences on using the smartphones while on their placements could be evaluated.

The questionnaire comprised eighteen likert scale questions and answer options in three 'blocks' of questions. The first block covered the usability of the smartphone. The second covered the educational effectiveness of the reusable e-learning object. The final block covering the participants' feelings about the value of using the learning objects and smartphone on placement. The data of interest in this account are the first and third question blocks, which dealt with the usability of the smartphone, rather than the block covering the reusable e-learning object itself. This is partly because the latter was not explicitly designed to promote interprofessional learning, but mainly because this chapter seeks to examine the practicalities of mobile learning on clinical placements.

The scores on each question for each participant were entered onto a MS-Excel spreadsheet, and the number of participants choosing each response option for each question were summed and then entered onto bar charts that summarised the overall results for each question. The measure of central tendency is the modal response, the response option chosen by the greatest number of participants. The findings are presented below.

## Q1: Overall Ease of Use

Thirteen of the eighteen participants rated the overall usability of the smartphone 'good' 'very good' or 'excellent'. The central tendency, the mode, was six participants who agreed with the latter (see Figure 1).

## Q2: Smartphone Portability

Thirteen of the eighteen participants rated the portability of the smartphone 'good' 'very good'

*Figure 2. Smartphone portability*

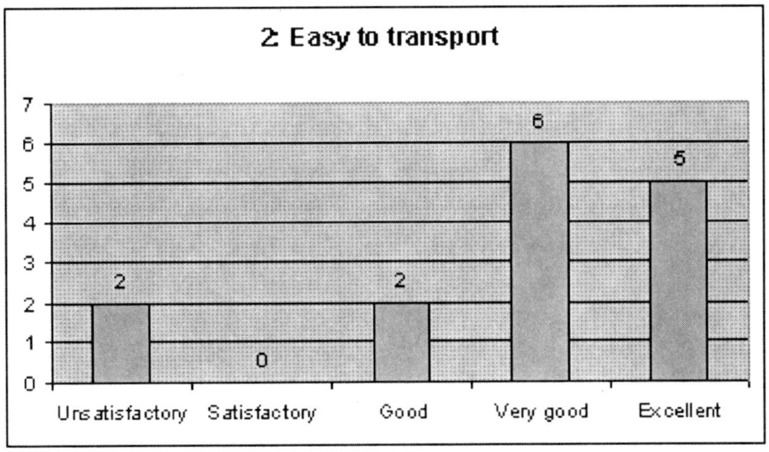


or 'excellent'. The central tendency, the mode, was six participants who agreed that the smartphone was 'very good' in this respect (see Figure 2)

## Q3: Comprehensibility of the Smartphone 'User Guide'

Thirteen of the eighteen participants rated the user guide to the smartphone that was provided by the researcher to the participants as 'good' 'very good' or 'excellent'. The central tendency, the mode, was eight participants who agreed that the user guide was 'very good' in this respect (see Figure 3).

## Q4: Device Interface Usability

Fourteen of the eighteen participants rated the ease-of-use of the smartphone user interface as 'good' 'very good' or 'excellent'. The central tendency, the mode, was six participants who agreed that the smartphone was 'good' in this respect (see Figure 4).

## Q5: Smartphone Battery Life

Ten of the eighteen participants rated battery life (the time between recharges) of the smartphone

*Figure 3. Comprehensibility of the smartphone 'user guide'*

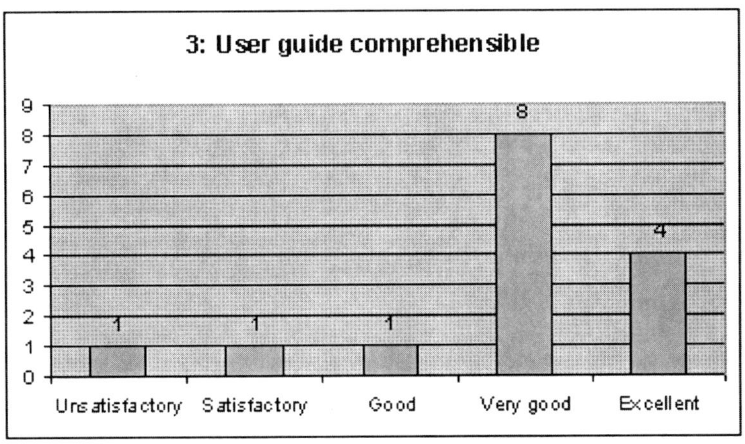

*Figure 4. Device interface easy to manipulate*

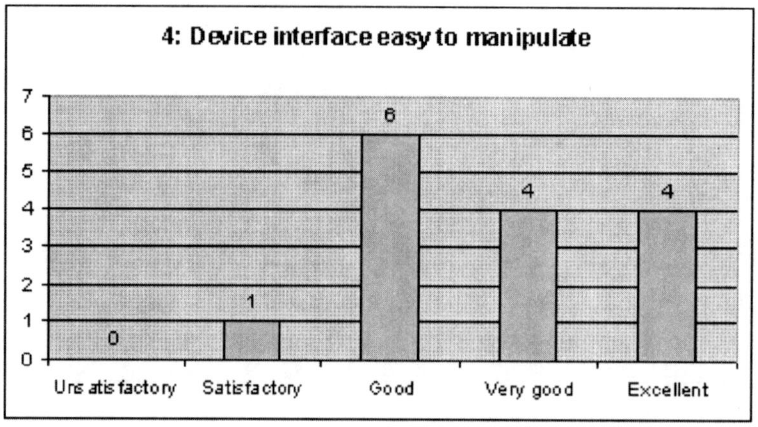


'satisfactory' or 'good'. The central tendency, the mode, was seven participants who agreed with the latter (see Figure 5).

## Q6: Smartphone Usefulness in the 'Clinical Area'

Ten of the eighteen participants rated the usefulness of the smartphone in the 'clinical area' as 'satisfactory' or 'very good'. The central tendency, the mode, was six participants who agreed with the latter (see Figure 6).

Questions eight to twelve focus on the learning object itself. As has already been discussed, it is felt more appropriate to focus on the delivery technology and the practicalities of its use. This is in part because the learning object was intended for cross-professional use, rather than one which explicitly promotes interprofessional learning. For these reasons, the next question to be examined is number 13.

## Q13: Reflection on Learning

All 18 participants tended to agree that using the smartphone on clinical placement enabled them to reflect on clinical skills, with no ratings of 'unsatisfactory' given. The central tendency, the

*Figure 5. Smartphone recharging frequency*

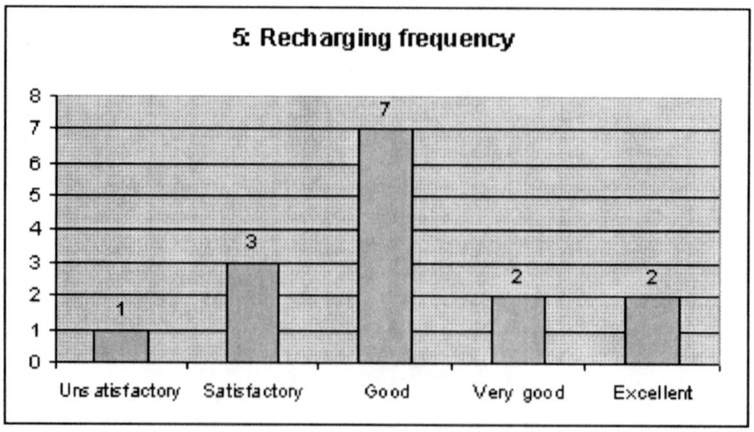

*Figure 6. Smartphone use in the 'clinical area*

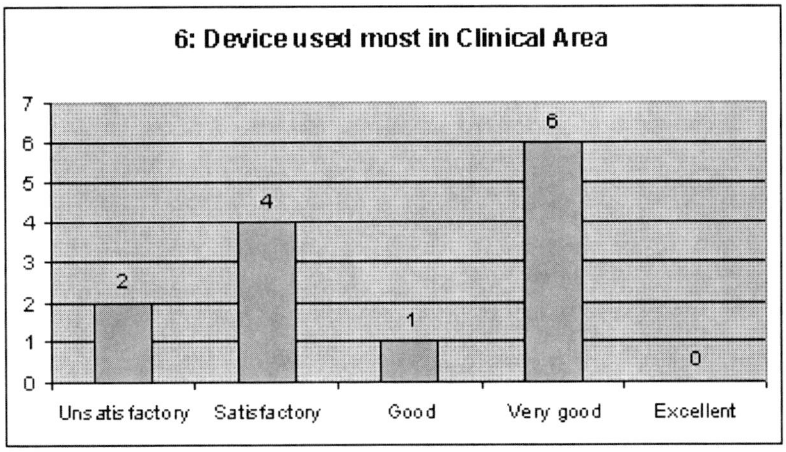


mode, was six participants rating the smartphone as 'excellent' in this respect (see Figure 7).

## Q14: Smartphone Enhancement of Learning

Fourteen of the eighteen participants rated the way the smartphone enhanced independent learning as 'satisfactory', 'very good' or 'excellent'. The central tendency, the mode, was eight participants agreeing with the latter (see Figure 8).

## Q15: Smartphone Facilitates Flexibility of Learning Place and Rime

Fourteen of the eighteen participants rated the ability of the smartphone to facilitate flexibility of learning place and time as 'satisfactory' to 'excellent'. The central tendency, the mode, was nine participants agreeing with the latter (see Figure 9).

*Figure 7. Reflection on learning materials*

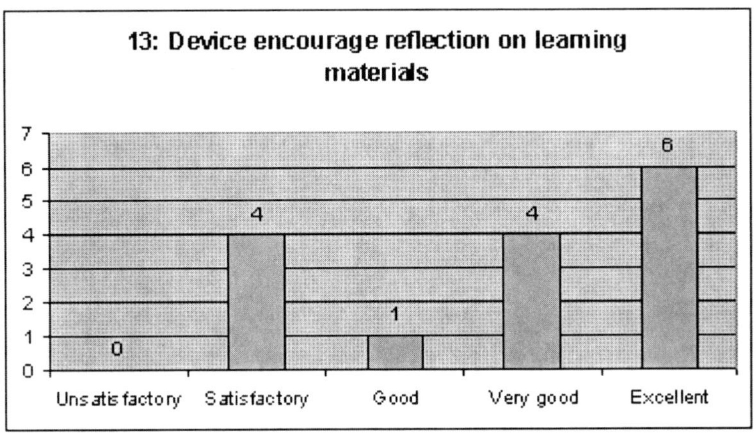

*Figure 8. Smartphone enhancement of independent learning*

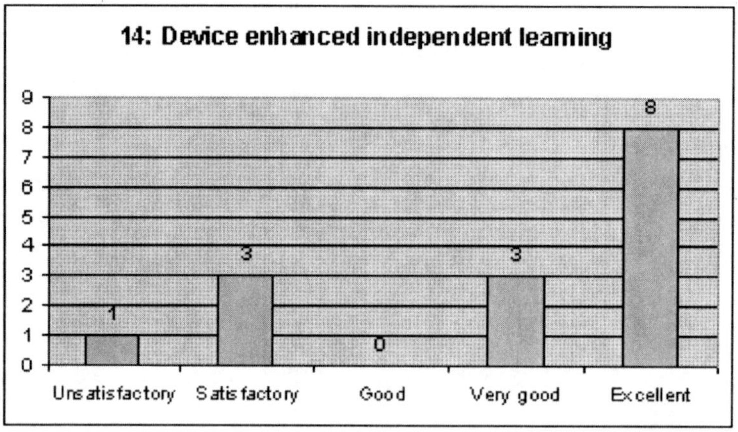


## Q16: Mobile Learning Accommodates Individual Learning Atyles

Thirteen of the eighteen participants rated the ability of mobile learning to accommodate individual learning styles as 'satisfactory' to 'excellent'. The central tendency, the mode, was five participants who agreed that mobile learning was 'very good' in this respect (see Figure 10).

## Q17: Mobile Learning Sustains Individual Learner Motivation

Thirteen of the eighteen participants rated the ability of mobile learning to sustain their motivation to learn as 'satisfactory' to 'excellent'. The central tendency, the mode, was seven participants who agreed that mobile learning was 'excellent' in this respect (see Figure 11).

*Figure 9. Smartphone facilitates flexibility of learning place and time*

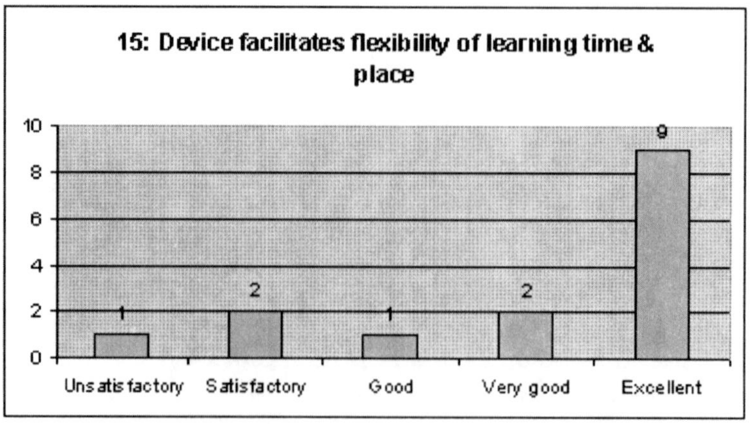

*Figure 10. Mobile learning accommodates individual learning styles*

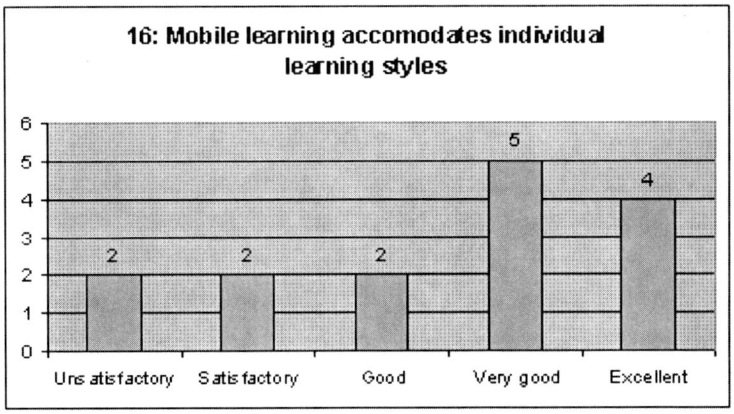


*Figure 11. Mobile learning sustains learner motivation*

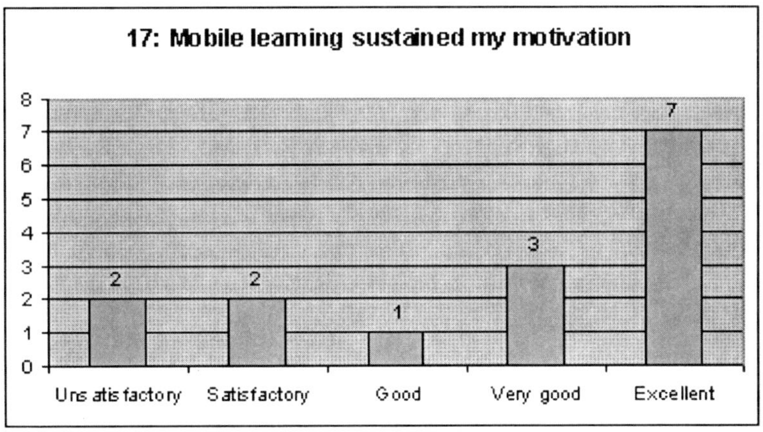


*Figure 12. Willingness to engage in mobile learning again*

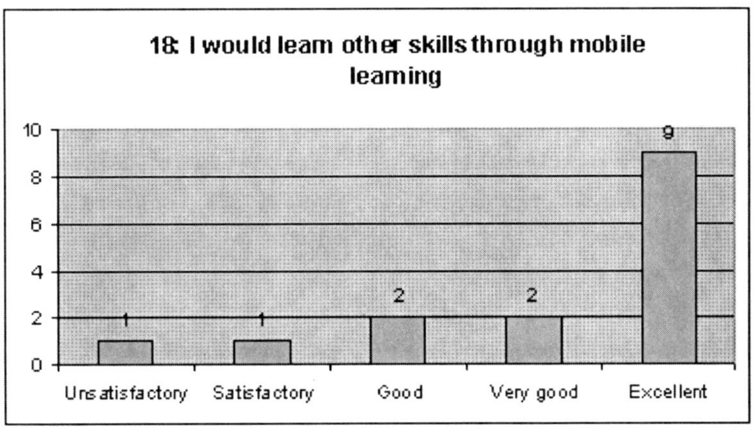

## Q18: Willingness to Engage in Mobile Learning Again

Fourteen of the eighteen participants rated their willingness to engage in mbile learning again as 'satisfactory' to 'excellent'. The central tendency, the mode, was nine participants who agreed that mobile learning was 'excellent' in this respect (see Figure 12).

It is interesting to note that many students made comments that indicated they wanted more smartphone-based learning and learning objects, as the selected quotes below clearly demonstrate:

*happy to do more learning supported by a mobile device in the future*

*It would have been good to have all the clinical skills sessions on a device, the Aseptic technique especially*

## DISCUSSION

The small-scale feasibility study reported here provides evidence that smartphones pre-loaded with reusable e-learning objects are a useful and practical means by which to support students nurses on their first clinical placements. Opportunities do not always arise in the clinical area to practice certain skills, this study strongly suggests that the smartphone can act as a back up mechanism. It thus seems reasonable to conclude that nurse academics and hospital trusts can use such mobile information and communication technologies to take a more proactive approach to bridge the theory-practice gap.

The students in the study felt that a smartphone pre-loaded with reusable e-learning objects facilitated independent learning. It seems plausible to imagine that their use may facilitate the phenomenon whereby students retain a higher percentage of knowledge when they learn at their own pace (Nelson 2003). In addition, Students who for whatever reason had no opportunity to attend the clinical skills sessions offered will not lose out. There are however organizational and technical barriers to the large-scale adoption of PDAs or smartphones, and they will now be discussed briefly.

## Change Management

Moving towards supporting collaborative learning with PDAs and smartphones requires consideration of not only the student's learning needs, but also those of their tutors and mentors, as well as the circumstances of their clinical placement. It must be remembered that clinical Staff are not always technical experts, and they are primarily concerned with providing patient care under tight time constraints. Imposed changes in work habits may be met with resistance from staff and students. It is important that these barriers are considered before any introduction of handheld devices in the clinical area.

One possible barrier to student nurses' use of PDAs is lack of knowledge of how to use them. While the majority of the participants in the feasibility study reported in this chapter owned a mobile telephone, not all individuals have grown up with mobile technology, for example, students returning to study aged over forty years. It could also be argued that the nursing culture has not encouraged nurses toward information and communication technology use either. Furthermore, training that enables students to use new technology takes time; time that is taken away from patients.

For PDAs to be successfully integrated into their education, it should be explained to student nurses why they should be using them, what they can offer, which models to buy, and how to operate their PDA applications. The value of introducing mobile computing devices to students early in their education has been recognised for some time (Thomas et al., 2001). Nurses are skilled in information management, routinely collecting and organizing data. Tutors can explain that by using a PDA that has relevant electronic resources uploaded onto it, nurses will find answers in their hands, not at the nurses' station. It can be explained that PDAs can bring benefits such as increased productivity, reduced errors, better patient care, and increased worker satisfaction (Huffstutler et al., 2002). All nurses need to network with other nurses to share information, to assist in research, provide and receive mentoring, emotional support, and friendship. PDAs can play a role in many of these areas, thus their adoption can be justified in these terms.

PDAs can also be a means by which to help health care staff stay abreast of new information about clinical skills and support mentors in their teaching role. It follows that the information on the PDA will need continually updating, as evidence based research changes as fast as technology. However, it is possible that, nurses can support PDA development by creating their own learning resources. The creation of learning objects that meet the needs of nursing students on placement and reflects and supports previous learning in clinical skills laboratories can itself be incorporated into the curriculum as a reflective learning exercise.

## FUTURE DEVELOPMENTS

The feasibility study reported in this chapter involved a small sample of eighteen students, due to only twenty PDA s being available. While sufficient to demonstrate the feasibility of PDAs on clinical placement, however, it is too few for a thorough evaluation to be made and firm conclusions to be reached. Furthermore, the participants put themselves forward on the basis that they had an interest in using information and communications technologies. It is quite possible that those who chose to participate were eager to have access to as much help as possible, whereas those who did not may have been too anxious about using new technology.

Further action research can shed light on these questions, which has implications for the possibilities of scaling-up the use of PDAs in nurse education. The next developmental step would be to develop and evaluate a core online repository of PDA-ready reusable e-learning objects to satisfy the needs of students on placement. Of particular interest would be the value of this approach as a means to support the development of interprofessional working in students on placement.

This would require a set of reusable e-learning objects that explicitly promote and explore interprofessionalism, rather than a genric skills object that is essentially cross-professional rather than interprofessional in nature. On the other hand, cross-professional e-learning objects can form the basis of interprofessional learning if used as discussion triggers between students from different professional groups, for example, nurses, medics, occupational therapists, dieticians and social workers. The inherent potential for communication through smartphones with Wi-Fi internet capability opens interesting possibilities in this regard. For example, online discussion boards are already available to the students of the university where the feasibility study was conducted through the interprofessional learning module that all healthcare students, regardless of their professional pathway, must undertake.

## CONCLUSION

Student and practicing nurses are likely become more familiar with handheld technology in their daily lives as these technologies become more ubiquitous in society, and such devices are likely to become more prevalent in the workplace. In the near future, it is thus likely that nurses will be increasingly open to the use of handheld technology to enhance nursing research and nursing practice.

PDAs open the possibilities for effectively delivering educational material that is always available at the point of need, regardless of an individual's physical location, as long as its use does not compromise infection control. The clinical mentor is supported during busy times, as students can use this as an intermediate measure. Learning can take place at any time and place, with one or multiple users, providing great flexibility and ensuring work coverage.

The results of the small feasibility study described in this chapter indicate that most of the students participating in the study were comfortable with this approach to mobile learning, finding it useful and calling for greater use of PDAs. The findings demonstrate that this approach is a feasible way to enable universities to meet students' individual needs to be on-demand while they are on clinical placement.

## ACKNOWLEDGMENT

The feasibility study that is reported in this chapter was funded as a small research project by the Centre of InterProfessional e-learning (CIPeL).

## REFERENCES

Aston, L., & Molassiotis, A. (2003). Supervising and supporting student nurses in clinical placements: the peer support initiative. *Nurse Education Today*, *3*(3), 202–210. doi:10.1016/S0260-6917(02)00215-0

Ayliffe, C. A. J., Lowbury, E. J. L., Geddes, A. M., & Williams, J. D. (1992). *Control of Hospital Infection: A practical handbook* (3rd ed.). London: Chapman, and Hall Medical.

Criswell, D. F., & Parchman, M. L. (2002). Handheld computer use in U.S. family practice residency programs. *Journal of the American Medical Informatics Association*, *9*(1), 80–86.

De Groote, S. (2004). The use of personal digital assistants in the health sciences: results of a survey. *Medical Library association, 92*(3), 341-349.

Fischer, S., Stewart, T. E., Mehta, S., Wax, R., & Lapinsky, S. E. (2003). Handheld computing in medicine. *Journal of the American Medical Informatics Association*, *10*(2), 139–149. doi:10.1197/jamia.M1180

Gould, D., & Drey, N. (2008). Hand hygiene technique. *Nursing Standard*, *22*(34), 42–46.

Huffstutler, S., Wyait, T. H., & Wright, C. R. (2002). The use of handheld technology in nursing education. *Nurse Educator*, *27*(6), 271–275. doi:10.1097/00006223-200211000-00008

Koeniger-Donohue, R. (2008). Hand held computers in nurse education: a PDA pilot project. *The Journal of Nursing Education*, *47*(2). doi:10.3928/01484834-20080201-01

Murphy, E. A. (2005). Handheld computers software for school nurses. *The Journal of School Nursing*, *21*(6), 56–360. doi:10.1177/10598405050210061101

Nelson, E. (2003). A Practical Solution for Training and Tracking in Patient-care Settings. *Nursing Administration Quarterly*, *27*(I), 29.

Stolworthy, Y. & Suszka-Hildebrandt, S. (2000). Mobile information technology at the point-of-care: the grass roots origin of mobile computing in nursing. *PDA Cortex 2000* [serial online].

Thotnas, B. A., Coppola, J. F., & Feldman, H. (2001). Adopting handheld computers for community-based curriculum: Case study. *The Journal of the New York State Nurses' Association*, *32*(1), 3–6.

Timmins, F., & Kaliszer, M. (2002). Aspects of nurse education programmes that frequently cause stress to nursing students -- fact-finding sample survey. *Nurse Education Today*, *22*(3), 203–211. doi:10.1054/nedt.2001.0698

White, A., Allen, P., Goodwin, L., Breckinridge, D., Dowell, J., & Garvy, R. (2005). Infusing PDA technology into nursing education. *Nurse Educator*, *30*(4), 150–154. doi:10.1097/00006223-200507000-00006

Yonge, O., Krahn, H., Trojan, L., Reid, D., & Haase, M. (2002). Being a preceptor is stressful! *Journal for Nurses in Staff Development*, *18*(1), 22–27. doi:10.1097/00124645-200201000-00005

Young, P. M., Leung, R. M., Ho, L. M., & McGhee, S. M. (2001). An evaluation of the use of hand-held computers for bedside nursing care. *International Journal of Medical Informatics*, *62*(2-3), 189–193. doi:10.1016/S1386-5056(01)00163-0

# Chapter 27
# Accessible Mobile Learning:
## Exploring the Concept of MobileLearning for All

**Christine Dearnley**
*University of Bradford, UK*

**Stuart Walker**
*University of Bradford, UK*

**John Fairhall**
*University of Bradford, UK*

## ABSTRACT

*This chapter explores the use of mobile devices in supporting practice based learning for health and social care students and practitioners, against a background of current UK disability legislation. The authors present the 'Mobile Enabled Disabled Students' case study in depth, accounting for its methodology and discussing the outcomes and implications. They will demonstrate how the features of mobile devices can support the organisational, memory and writing needs of dyslexic students specifically and how they can add value generally to learning and assessment processes. They make recommendations for implementing mobile assessment for practice learning based on a theory of mobile learning designed for accessibility.*

## INTRODUCTION

This chapter will consider the role of mobile devices in supporting practice based learning for disabled health and social care practitioners, against a background of current UK disability legislation. We draw briefly on our work as partners in the ALPS CETL (Assessment & Learning in Practice Settings - Centre for Excellence in Teaching & Learning) but focus specifically on the Mobile Enabled Disabled Students (MEDS) case study, undertaken at the University of Bradford (funded by the ALPS CETL Research Capacity Investment Fund) between October 2007 and 2008. We will then share the outcomes of MEDS and make recommendations for future applications of mobile learning designed for accessibility

DOI: 10.4018/978-1-61520-889-0.ch027

## BACKGROUND

The ALPS CETL, funded by the Higher Education Funding Council for England (HEFCE), is a collaborative programme between five UK Higher Education Institutions (Universities of Bradford, Huddersfield, Leeds, Leeds Metropolitan and York St John) and sixteen health and social care professions. Its overall aim is to improve the competence and confidence of health and social care students at the point of professional registration. To this end, the partners have been engaged in five years of collaborative activity to develop shared practice/work-based assessment processes. Distinctive elements of these processes are that they will be interprofessional and they will be delivered in electronic mobile format.

Interprofessional education (IPE) has been recognised internationally as an essential element of pre-registration health and social care provision. Colyer et al. (2005) report that this has required a paradigm shift in thinking about how learning is supported; it is no longer acceptable for students to be immersed in profession specific ideology without appreciation or examination of its origins, or how it relates to the beliefs, traditions and practices of other related professions. Today's health service requires professionals to collaborate and work together. IPE supports this by providing opportunities for students from different professions to gain an appreciation of each other's skills, values and knowledge base. Interprofessional assessment (IPA) of practice learning is perhaps the natural progression from IPE. IPA is not currently a requirement in many pre-registration professional programmes in the UK; its implementation however may potentially enhance the impact of IPE by requiring assessors to have a minimum understanding of the values and practices within the student's profession. The ALPS CETL aimed to introduce IPA across the partner sites and professions.

The practice assessment tools developed by ALPS were to be delivered on mobile digital devices (T-Mobile MDA Vario 1), funded by the CETL, carried and maintained by students. These were loaded with 'Intellisync' and 'Safeguard' software, allowing assessments to be 'pushed out' to students remotely whilst ensuring confidentiality and data security. Full training was given to students on using the devices and the assessment tools. Training of assessors was more complex due to their huge numbers and the wide geographical spread and variety of placements involved (hospitals, clinics & community settings). University staff liaised intensively with practice staff, opportunities were sought for awareness raising and training materials were developed to support students, who were often required to teach their mentors what the devices were for and how to enter data.

Several studies have demonstrated how mobile devices can be used to enhance interaction and learning, in different contexts. Naismith et al. (2004) reviewed the literature related to mobile technologies and learning and reports that these technologies are now employed in a wide range of educational activities and are fundamentally changing the nature of learning. The changes to the nature of learning and perceived pedagogic benefits of delivering assessment documents on mobile devices extend well beyond the change in presentation from paper to electronic format. The ALPS assessment processes, for example, required students to undertake a wide range of formative, work-based mini-assessments, which could facilitate the collection of evidence for student electronic portfolios. Mobile devices allow students to collect evidence in a wide range of media, such as audio files, images, videos, and reflections written or recorded in action (Schon, 1983) rather than relying on memory after work or a long journey home. This fundamentally changes how students interact with each other, learning materials and learning processes.

The ALPS devices allowed 24 hour access to the internet, thereby facilitating access to information anytime, anyplace, anywhere; they allowed access to emails and provided a calendar function (with reminder alarm system), a calculator, word processor, a spreadsheet ('Pocket' programmes align to those offered on a normal PC but with limited functionality), a note book function and games. They synchronised with PCs, allowing documents/material to be transferred from device to PC for storage or further development. They were not voice enabled and thus could not be used as mobile telephones.

Health and social care students frequently work in isolated practice placements without access to internet enabled PCs. It was anticipated that providing them with mobile devices could overcome subsequent problems; they could contact peers, colleagues and friends for social and academic support, collaborate interprofessionally with others and engage with learning processes, from remote locations. However, the specific issues faced by disabled students when using a mobile device for learning and assessment were still unknown. It was recognised that people cannot collaborate, or fully engage with the learning processes, if they cannot access all available resources. The MEDS study was developed to enable collaboration and engagement through accessibility.

## Mobile Enabled Disabled Students

The United Kingdom (UK) has legislation to protect disabled people from discrimination. The Disability Discrimination Act (DDA) became law within the UK in 1995. The Disability Rights Commission (DRC) was the key organisation to advance disabled people's rights within the UK, and has since been subsumed within the UK Equality and Human Rights Commission (EHRC). However, prior to this, the DRC began reviewing the legislation, regulations and statutory guidance within professional occupations.

The DDA makes it an offence to discriminate against a disabled person by treating him or her less favourably than others for a reason relating to their impairment. Part IV of the DDA directly applies to the provision of education services and is referred to as the Special Educational Needs and Disability Act (SENDA, 2001). Educational services covered under the act are deemed to include traditional activities such as lectures, as well as e-learning materials (Whillder, 2002).

Ruebain et al. (2006) report on the Disability Rights Commission's (DRC, 2006) review of legislation, regulations and statutory guidance within professional occupations, as part of their formal investigation into fitness standards in teaching, nursing and social work. Whilst this formal investigation focussed on these three areas, the legislation, regulations and statutory guidance for medicine, dentistry and the health professions, falling within the scope of the Health Professions Council, were also analysed for comparative purposes.

The final report "Maintaining Standards: Promoting Equality" (DCR, 2007), concluded that Disabled People often face barriers to entry and progression throughout large parts of the public sector. It also found that the mass of regulations and guidance often do little to protect the public, whilst deterring people with disabilities from applying or remaining in these professions (DRC, 2007, p. 24). If these barriers in the public sector were removed, it is not unreasonable to assume that more disabled people would enter Higher Education to become health and social care practitioners. Therefore, as in other areas of educational provision, accessibility must be ensured for all forms of teaching (Dearnley & Walker, 2009).

The aim of MEDS was to inform the ongoing work and development of ALPS in relation to the specific needs of disabled students when using mobile technologies for learning and assessment in practice settings. MEDS aimed to establish what worked well for disabled students who currently used mobile devices and to identify the challenges that mobile technologies presented to them. The MEDS team worked closely with ALPS, recruiting disabled users to trial the use of new assessment tools as they were developed, to assess their impact and identify changes that needed to be made for disabled users.

The principles of the SENDA (2001), directly apply to the provision of education services. The Act makes it an offence to discriminate against a disabled person by treating him or her less favourably than others for a reason relating to their disability. Pearson and Koppi (2006) suggest that e-learning, and we would suggest equally mobile learning, can offer disabled students opportunities in learning and participation that they might not otherwise have had.

Papadopoulos and Pearson (2007) note, however, that this can only happen if the learning activities and resources are designed to be accessible. Dunn (2003), investigating accessibility of e-learning materials found inaccessible elements both within VLE software itself, and within the content institutions put into them; they concluded that the lack of accessibility appeared to stem from a lack of knowledge about designing VLE systems that were accessible by disabled people.

The experience of VLE's demonstrated the need for accessible learning environments in order to ensure equality, good practice and reduce the risk of possible litigation. We therefore designed the MEDS research project to identify good practice when implementing mobile learning activities, and to inform the development of the ALPS mobile assessment tools.

## Design of the Study

An eclectic approach was agreed for this study within the phenomenological paradigm embracing a philosophy of collaboration based on feminist ontology (Letherby, 2003). This was appropriate because it acknowledges the complexity and diversity of societal structures and the people within those structures. This was reflected in what we were hoping to achieve in this project, not simply because our participants were from a marginalized group, but because they were human beings (Dearnley & Walker, 2009).

The study reflected the philosophy and structure of collaborative action research (AR), supporting the notions of Abbott and Sapsford (1998) that AR "arises out of practice and feeds back into practice". The study comprised of two distinct stages or cycles. The range of methods used reflected the nature of the study and the work of ALPS in that electronic devices were included to create innovative and powerful ranges of data. MEDS was conducted on a single site. It drew on participants from across the institution and utilised both quantitative and qualitative data. Specific objectives of the project were:

- To investigate the range of impairments that affect health and social care students. This could be used to form guidance for "reasonable expectations".
- To establish what works well for disabled students who currently use mobile devices and took part in this study.
- To identify the challenges that mobile technologies present to disabled students.
- To trial the use of new assessment methods as they were agreed by the ALPS Tools Working Group, among disabled users to assess their impact and identify changes that needed to be made for disabled users.

Inherent in the overall aim of this study was an obligation to ensure that accessible design principles were incorporated in the developing ALPS software. To this end the concept of an Accessibility Passport (Techdis UK, 2006), was built into the design of the study. This aims to encourage feedback and suggestions from users through an ongoing analysis of the learning materials. The model comprises four main elements which are:

- **Evaluation:** The User evaluates how the object works to deliver learning and offers feedback to the commissioner and designer.

*Figure 1. Selected words from the Microsoft Desirability Toolkit (Benedek & Miner, 2002)*

| | | | | |
|---|---|---|---|---|
| Accessible | Desirable | Gets in the way | Patronizing | Stressful |
| Appealing | Easy to use | Hard to use | Personal | Time-consuming |
| Attractive | Efficient | High quality | Predictable | Time-saving |
| Busy | Empowering | Inconsistent | Relevant | Too technical |
| Collaborative | Exciting | Intimidating | Reliable | Trustworthy |
| Complex | Familiar | Inviting | Rigid | Uncontrollable |
| Comprehensive | Fast | Motivating | Simplistic | Unconventional |
| Confusing | Flexible | Not valuable | Slow | Unpredictable |
| Connected | Fresh | Organized | Sophisticated | Usable |
| Consistent | Frustrating | Overbearing | Stimulating | Useful |
| Customizable | Fun | Overwhelming | Straight Forward | Valuable |

- **Commissioning:** Specification of the learning object or software accessibility requirements
- **Design:** Describes how the objectives will be accomplished.
- **Delivery:** Includes consultation with other stakeholders; for example the user, a student or tutor/support worker and ensures the accessibility of the materials by advising how they may be improved.

## Methods of Data Collection

### Stage 1

1:1: Quantitative Data from central data base:

Data was generated from the University central database to provide an overview of the frequency of impairments across the University and within the School of Health Studies.

1:2: Focus group to ascertain perceived benefits and barriers to using mobile devices among disabled students:

This was attended by 12 disabled student participants with a wide range of disabilities who had experience of using mobile devices in some form and representatives from ALPS partner software developers. Initially participants were asked to share their experiences of using mobile devices and their revelations are captured in the categories discussed below. Following this, participants were introduced to prototypes of the ALPS Assessment tools and encouraged to explore their use on an ALPS mobile device; the software developers were able to hear at first hand the participants' feedback on their design.

### Stage 2

2:1: Blogs/diaries:

Eight disabled students were given an ALPS mobile device and asked to keep a blog (or electronic diary) of how they used it generally and how it supported their learning. At a later date the ALPS assessment tools were uploaded to the devices and participants were again encouraged to record their experiences and thoughts about using these on the devices.

2:2: Focus group to ascertain the accessibility issues related to the ALPS mobile assessment tools:

This was attended by five participants who had used the devices and ALPS assessment tools.

2:3: Microsoft Desirability Toolkit (Benedek & Miner, 2002):

This is a process in which participants are supplied with a list of key words and asked to rank 5-10 of them in order of importance to them. This technique does not rely on memory, a questionnaire or rating scales and users do not have to generate words themselves (List of words used in Figure 1).

## Sample

Participants for stage one were recruited from across the University of Bradford as a whole to obtain a wider picture of disabled students in one HEI and to ensure a maximum range of disabilities were captured within the enquiry. Participants for stage two were recruited from Health and Social Care students at the University of Bradford, so that they could engage in realistic use of the ALPS Assessment tools. The prominent disability in this group was dyslexia.

## Data Analysis

The study examined the frequency of impairments of all disabled students who studied at the School of Health Studies 2007 / 8. Statistical information on the frequency of impairment for all disabled students at the University of Bradford for 2007 / 8 was then examined and comparisons were made.

Data from the focus groups was transcribed and entered in to NVivo computer software for initial open coding and the development of emerging categories, to which data from the electronic blogs and desirability tool kit was added, thereby increasing the depth and reliability of outcomes. Words selected by participants using the Microsoft desirability toolkit were used to generate word clouds that provide a visual overview of participant reactions to both the mobile devices and the ALPS mobile assessment tools.

## Implementation

A steering committee was convened and met three times throughout the duration of the one year project. The committee was made up of the project team, the Professor of Education Development, the ALPS mobile technologies project manager, a representative of TechDis, and a disabled student.

In order to allow participants to comment on both the ALPS device and software it was decided to use PebblePAD. This is a proprietary Macromedia Flash web-based Personal Learning Environment & EPortfolio system. PebblePAD provides a customisable graphic user environment to allow recordings of experiences, reflections, learning etc. It was anticipated that a mobile device would allow participants to record their thoughts and feelings as they experienced particular phenomena. It would also offer great flexibility and accessibility to disabled people who may find traditional paper based diary keeping a barrier to participation. A diary guide was produced so that all students would have the same guidance as to what was required of them.

## Ethical Issues

Approval to undertake this study was obtained from the University of Bradford Committee for Ethics in Research. Key issues to be considered in this study were (a) that participants were students of the University, (b) that they were disabled students and (c) there were issues related to the technology they were to be provided with. Safeguards included ensuring that participants understood the nature of the study and their right to withdraw at any time without repercussions to their status as a student of the University, and review of all correspondence by the disabled student member of the project steering committee.

The potential problem related to misuse of the devices in practice settings was a wider problem that was addressed by ALPS. Participants in the MEDS project were required to sign the same agreements and contracts, related to device use in practice, as any student who was allocated an ALPS mobile device. When planning how PebblePAD would be used we ensured that participants had access to their own 'gateway'. This meant that participants had to actively choose to submit data, e.g. a diary entry; it was not an automated process. Moreover, whilst participants were able to see their own contributions they did not have access to others; thus anonymity and privacy could be assured. If participants had a problem with the technology or the assessment tool they had a named member of the project team to contact in addition to the ALPS Help Desk (this is a shared service for all students using ALPS mobile devices).

*Figure 2. Percentage of student population with declared impairments in the school of health studies and general university.*

| Impairment | Percentage (School of Health Studies) | Percentage (General University) |
|---|---|---|
| Dyslexia | 57.9 | 50.4 |
| Visual Impairment | 2.1 | 3.0 |
| Hearing Impairment / Deaf | 9.7 | 5.7 |
| Mobility Impairment | 0.7 | 3.0 |
| Mental Health | 2.1 | 5.3 |
| Unseen Disability | 20.0 | 14.8 |
| Other Disability | 7.6 | 13.2 |
| Multiple Impairments | 0.0 | 3.4 |
| Autistic spectrum / Asperger's | 0.0 | 1.3 |

## Outputs and Results

Dyslexia is by far the largest group of impairments within the population of disabled students, both within the general University and the School of Health Studies. It is generally understood that Dyslexia accounts for between 60 – 70+ % of Disabled students at HEI's. This data can be seen in percentages, for comparative purposes in Figure 2.

We advise caution with these snap shot statistics as this data was only intended to provide a general overview. It is important to note that data collection occurred at the start of an academic year where many students from the previous year had left, and many new students were still to declare their impairment.

## Qualitative Outcomes

### Stage 1

In the first focus group, two participants had used a PDA (personal digital assistant), a handheld computer. First released in the 1980's, more recent PDAs have touch screens, audio, WiFi, and some incorporate a mobile phone ('Smartphones'). Software typically includes a calendar, a to-do list, an address book and E-mail and Web support. One participant, a part time student who worked as a nurse, had used a PDA since commencing her training six years previously. The other student used a wide variety of PDA functions including the built in camera to store images of diagrams, lecture slides etc for later use. The cost of PDAs appeared to be prohibitive for other students. However, they used a range of mobile devices, including lap-tops, hearing devices, audio recording devices, Dictaphones, MP3 players, mobile phones (with cameras), calculators and scanner pens. Three clear themes emerged:

*"Laptops are OK, but...." students reported that laptops offered mobility around the home but were too heavy to carry around university (or into practice settings).*

*"Benefits for learning & living" students were using a range of devices to support their learning activities but also to support their general day to day activities such as recording visits to the doctor and using the alarm system to get to lectures on time.*

*"Becoming mainstream" students reported that non disabled students were starting to observe them using devices, for example recording a lecture to aid memory, and realising they too could benefit from such technology.*

*Figure 3. Device desirability as identified by participants using Microsoft Desirability Toolkit (Benedek and Miner 2002)*

## Stage 2

In stage 2 of the study a wider range of data collection methods were employed and participants reported specifically on their experiences of using the ALPS mobile devices and the ALPS mobile assessment tools that had been developed throughout stage one. All participants were from students studying on health and social care programmes.

The following analysis presents a synthesis of data from participant blogs, focus group discussions and the Microsoft Desirability Toolkit. Word clouds presenting a visual overview of participant reactions to both the mobile devices and the ALPS mobile assessment tools can be seen in Figures 3 & 4. Many of the issues that arose during the stage one focus group emerge here too, adding to the reliability of those outcomes.

*Figure 4. ALPS Assessment tool as identified by participants using Microsoft Desirability Toolkit (Benedek and Miner 2002)*

## Benefits to Disabled Students in Using a Mobile Device for Assessment and Learning in Practice Settings

Many of the participants of this study were dyslexic. Difficulties with organisation, writing and spelling are common traits of dyslexia, it is not surprising therefore that key benefits reported involved how mobile devices supported these difficulties. The organisational capacities of the mobile devices were considered invaluable. The following quotes demonstrate participant views on this (NB: B=Blog; FG=Focus Group & AC=Accessibility Card):

*Essential with [learning difficulties] to be organised – this felt it could help (AC4).*

*I have found it invaluable in relation to keeping appointments and track of my days activities (B1).*

Another blog reported how the device became an aide memoire:

*Overall, I am very pleased with the PDA and I feel that with continued use will prove to be invaluable to future students (B1).*

Spell check was seen as extremely valuable by all participants. One participant who had used a similar device for a number of years recommended developing a "list of regular words that can just come up for a quick check", as she found this extremely helpful in practice settings. Another participant had worked in operating theatres and struggled when asked to label things with unfamiliar names/words and thought using the internet to look up these words for both their meaning and how to spell them would be very helpful.

The use of the audio function was useful to this group of students as a general function of a mobile device. Nursing handovers are regularly recorded and interestingly this student had found that it helped to provide practice for audio handovers, which she felt was 'quite a scary thing.' (FG1)

There was some indication that the participants found the devices motivational and fun. For example:

*Don't know how long it will take me to master this device. I'm sure it will be fun trying (B1).*

*I felt having a mobile device that I could use on placement and be relevant to me to be an improvement in my training (AC1).*

This participant used the accessibility cards to report that she found the device 'empowering'; she wrote:

*I believed having the extra help from the device would give me more confidence in my work and at university (AC1).*

The spell check, audio function and organisational aspects discussed above are specifically related to the use of a mobile device, but clearly they represent a significant added value to ALPS mobile assessment tools for this group of participants. They discussed this specifically. For example:

*Useful for memory, more than anything you could fill it in as you went along (FG2)*

*I found that if I could fill out the assessment straight away after the event it helped me to remember the interaction more accurately but as part of my disability relates to poor memory I found it difficult if I couldn't enter it straight away - I imagine this could be a problem when out on placements as time is limited (B5).*

Another participant added:

*If you sat me down and said 'here's a piece of paper analyse what you have just done' [I could not] but having the questions there that said what have you just done and how did that work, that was really helpful to me (FG4) (others agreed).*

When discussing the benefits of spell check one participant discussed how beneficial she had found the predictive text function and the key pad really helpful when filling in assessment documents. She said it was:

*brilliant... it tells you if your spelling is not right... I often get letters the wrong way round, so that was really helpful... it really speeded the whole process up (FG1).*

Several participants discussed how the audio function supported the ALPS mobile assessment tools. For example:

*Although I don't particularly like listening back to the audio recording I found it helpful to get all the information down without having to type it out (B5).*

And:

*I used it quite often, If I speak I don't lose my track as often, because if I'm typing I have to think about spelling and then I have lost what I was going to say (FG4).*

The participants of this study found the functions of the mobile devices supported them generally in learning and living activities. When assessment tools were made available on the mobile devices, participants reported that those supporting features transferred to their practice learning experience.

## Barriers to Disabled Students in Using a Mobile Device for Assessment and Learning in Practice Settings

It is interesting to note that the barriers this group of disabled participants reported to their use of mobile devices were not related to their impairments and reflected earlier ALPS work (Dearnley et al., 2008; Taylor et al., 2006). Barriers included problems with device functionality, e.g. it was slow when using the internet connection; lack of confidence/expertise in using the device; fear of losing it and the reaction of others whilst in professional settings. Here are some examples:

*As the device at times was slow & bulky, plus I had my phone as well, that it was too much to carry around all the time. If I had both mobile device and phone in one, I think I would use it more (AC1).*

*I'm still a little scared about how to use it without somehow managing to blow it up!! (B2)*

*While using my mobile device on placement I felt like I was doing something wrong because some of the nurses or employees looked at me as if I was using my phone and not working! (B2)*

This issue is intricately bound to how the ALPS mobile assessment tools are received in the professional settings and work related to this is ongoing within the partnership. One participant was a practice assessor and was able to give voice to the perspective of assessors generally and those with similar disabilities. She commented that it was good to see mobile devices being used in this way as she had previously used a device of her own to support her disability. She referred to herself as formerly being a 'closet user'. (AC4)

She expressed the view that the tools could be *"developmental for student and everybody – especially as a dyslexic mentor"* (FG3).

She went on to explain that current practices for assessor updates often involve being given a booklet to read. For dyslexic assessors this can clearly be problematic, however it was noted that *"the tool is much more interactive and supports mentor development*" (FG3).

It was noted by all participants however that there are time constraints in practice settings and it was problematic when the device wasn't working correctly as this confounded existing resistance to the devices. Participants reported that mentor reactions were *"very mixed"* and that *"some people dismissed it straight away"* (FG3)

This led to further discussions about student preparation and training. Students need to be confident in its use before they go out in practice where they will be required not only to show their mentors how the device and tools operate, but also to be champions and change agents. For example:

*Its difficult to show others when you are not confident yourself – doesn't portray it in its best light (FG3).*

A student participant added that it:

*Brings the student and the mentor together a bit more (FG1).*

The participants liked the ALPS mobile assessment tool and found it beneficial to their learning both because of the functionality of the mobile device such as spell check, diary and audio and because of the tools themselves which were organised and acted as prompts for recall and reflection; it was also felt that they enhanced student- assessor dialogue. The main challenges to using the tool in practice related primarily to those associated with resistance to the mode of delivery, i.e. the mobile device. Becoming familiar with the assessment tools was easier for students once they had become familiar with the device; assessors often were new to both and this was time consuming and problematic. It is therefore crucial that students are confident in the use of both the device and the assessment tool before going out into practice. Disabled students will need someone to show them how to use it, rather than being given instructions to read. It is anticipated that problems with the technology, such as slow connections etc, can only improve in time.

## Accessible Design

Participants indicated that what these users really want is something simple with extensive functionality. This applies both to the software and the hardware. Here are some of their suggestions:

*Small but big ...*

*I like the touch screen*

*Well I'd like it to be in my phone ... ... diary, internet*

*I just can't take it in at all so something simple and basic ...*

We would argue that many of these requirements are no different to those that all users, regardless of impairment would request and as such we support theories of total inclusivity, in particular the notion of Universal Design. Anders and Fechtner (1992) describe the key features of universal design as being that the design:

- Accommodates a wide range of individual preferences.

- Communicates information effectively, regardless of ambient conditions or the user's sensory abilities.
- Can be used efficiently and comfortably, and with a minimum of fatigue.
- Allows for appropriate reach, manipulation and use regardless of user's body size, posture or mobility.

Thus, the intention is that products are designed for people. Furthermore, it is recognised that individuals have a wide range of characteristics, be they height, age, race, gender or impairment. Therefore, accessible resources are more usable for every user, regardless of impairment. Accessible design does not only aid disabled people, there are potential benefits to everyone. In the physical world an example of this might be replacing steps at the entrance of a building with a ramp. Parents with push chairs or delivery people would also benefit from these changes.

However, Information Technology is quite different to the physical world, not only are there a myriad of access points, but protocols are often being renewed as technology advances. What is required is an understanding of users' needs, as well as an acceptance that these will change over time and must therefore be re-evaluated. This is particularly relevant to mobile learning as the environment may differ substantially from traditional classroom or home learning.

During the early stages of the MEDS study it became apparent that mobile devices and the software employed on them can have a substantial effect upon disabled users and their ability / willingness to engage with the learning process. Universal design was the model on which we initially based our approach. In its ideal form Universal design would suggest that both the device and software must be totally accessible to everyone. However, we recognise that producers of software and or learning materials have finite means. It is therefore highly desirable to minimise work and avoid duplication of effort in order to reduce both time and expenditure. We found that a pragmatic solution was to adopt a form of Accessibility Passport (Techdis UK, 2006), where the software developers were able to integrate users' feedback into the design process. This requires the recognition and inclusion of disabled people as important users.

### Implications

This study has demonstrated the importance of considering the rights and needs of disabled people when developing mobile learning systems. It is clear that when people face barriers, they will often use the tools at their disposal, in this case mobile technologies, in innovative and imaginative ways. We would argue that there are both considerable dangers and opportunities when designing mobile learning systems. Underestimating users' ability, can lead to lack of flexibility when designing learning environments. This in turn can lead to inflexible, inaccessible software which is both potentially exclusive and open to legal challenge.

It is vital that the needs of disabled people are included both during initial and subsequent stages of the design process, the Techdis Accessibility Passport (2006) is one example of this. Users of all forms of information technology have a diverse range of needs and study modes. Instead of viewing disabled people as sub group of special interest, we would suggest that considering their needs can offer all users improved access and flexibility. Whilst we do not propose that this is an easy solution we firmly believe that this can only improve access, engagement and inclusion of all users.

## FUTURE RESEARCH DIRECTIONS

There is clearly a need for further research into a number of issues that have been highlighted by this study. Some of these are currently being undertaken by the project team and /or their colleagues; these include a small follow up study with ultra mobile PCs (funded by TechDis) to identify how they can be used to support learning and assessment for disabled students. Of key interest here is the extent to which the benefits to learning and living, made possible by mobile devices and experienced and described by MEDS participants, can be achieved by ultra mobile PCs. These devices could potentially overcome the problematic issues of screen and keyboard size whilst retaining the benefits of lightweight mobility.

The MEDS study has also led directly to work currently being undertaken at the universities of Huddersfield and Bradford, funded by the Yorkshire & Humber Strategic Health Authority, into disabilities and fitness to practice. The MEDS study highlighted the opportunities afforded for technological solutions to support reasonable adjustments in health settings and added to our knowledge and understanding in this field, preparing us to take the work forward into this new project, which aims to identify and explore 'reasonable adjustment' as defined by the DDA (2005) in relation to preparation to practice for health professionals. Four key questions will guide this aim:

What does 'reasonable adjustment' mean in relation to NHS/SHA practice placements?
What is the perception of disabled students and staff in the NHS/ other health provision?
How can we help disabled students and prepare them to cope in practice?
How can the assessment process help us to identify and address fitness to practise issues?

Other related areas of interest ripe for investigation include the recommendations of the DCR report, that there is a need for further research into the provision of reasonable adjustments for students to ensure that disabled people were not disadvantaged in learning environments.

## RECOMMENDATIONS

### General Recommendations

- Students with dyslexia often find the functions of the mobile device extremely supportive for their learning, assessment and general organisation.
- Training for all students is crucial to the successful adoption of mobile learning and assessment processes.
- This includes training in how to use the device and all its functions in addition to any learning and assessment materials to be accessed on the device. As participants in this study have demonstrated, it is difficult for students to teach an assessor how to use the device and tools if they themselves are not sure how to use them.

### Technical Recommendations

- Adopting accessible design processes can help to ensure the needs of disabled people are considered at the design phase or any mobile learning processes.
- That mobile devices should be as preconfigured as possible before issuing to minimize impact and stress on the student.
- It should be possible for text and background colour to be customised
- Other recommendations related specifically to the ALPS mobile assessment tools and have informed their development.

## CONCLUSION

This study has identified the extent to which mobile technology can support some disabled learners generally and more specifically those with dyslexia and associated memory and organisational impairments. We have gained an early indication of how the ALPS mobile assessment tools can provide additional support for such learners because of the functions of the mobile devices; but also how the tools themselves appear to help students generally to organise their work, reflect on their activities and engage with their learning and practice educators.

The study has informed the development of the ALPS mobile assessment tools to date and further recommendations are currently being implemented. We have provided a list of general recommendations to arise from this study related to disabled students using the tools and also some very specific technological recommendations relating to how the tools could be improved for easier use by all students. We would argue that many of the requirements of disabled students are no different to those of all users, regardless of impairment and as such we have recommended total inclusivity.

## REFERENCES


Abbott, P., & Sapsford, R. (1998). *Research Methods for Nurses and the Caring Professions* (2nd ed.). Buckingham, UK: Open University Press.

Anders, R., & Fechtner, D. (1992). *Universal Design*. Brooklyn, NY: Pratt Institute Department of Industrial Design and Pratt Center for Advanced Design Research.

Benedek, J., & Miner, T. (2002). Measuring Desirability: New methods for evaluating desirability in a usability lab setting. In Proceedings of Usability Professionals Association 2002, Orlando, July 8-12.

Colyer, H., Helme, M., & Jones, I. (2005). Pragmatic approaches and the theory-practice relationship. In *Interprofessional Education*. Higher Education Academy Health Sciences and Practice Occasional Paper no. 7.

Dearnley, C. A., Haigh, J., & Fairhall, J. (2008). Using Mobile Technologies for Assessment and Learning in Practice. *Nurse Education in Practice*, *8*(3), 197–204. doi:10.1016/j.nepr.2007.07.003

Dearnley, C. A., & Walker, S. A. (2009). Mobile Enabled Research. In Vavoula, G. (Ed.), *Researching Mobile Learning: Frameworks, Methods and Research Designs*. Oxford, UK: Peter Lang.

*Disability Discrimination Act*. (2005). Retrieved August 27, 2009, from http://www.opsi.gov.uk/Acts/acts2005/ukpga_20050013_en_1

Disability Rights Commission. (2006). *Fitness Standards Formal Investigation*. Disability Rights Commission. Retrieved March 19, 2007, from http://www.drc-gb.org/Employers_and_Service_Provider/Fitness_Standards_Formal_Inves.aspx

Disability Rights Commission. (2007). *Maintaining Standards: Promoting Equality*. Report of a DRC (2007) Formal investigation Professional regulation within nursing, teaching and social work and disabled people's access to these professions. Retreived April 8, 2008, from http://www.maintainingstandards.org/files/Full%20 report%20_%20final.pdf

Dunn, S. (2003). *Return to SENDA? Implementing accessibility for disabled students in virtual learning environments in UK further and higher education*. London: City University of London. Retrieved March 19, 2007, from http://www.sara-dunn.net/VLEreport/documents/VLEreport.pdf

Letherby, G. (2003). *Feminist research in theory and practice. Milton Keynes*. UK: Open University Press.

Naismith, L., Lonsdale, P., Vavoula, G., & Sharples, M. (2004). *Mobile technologies and learning*. Retrieved April 7, 2008, from http://www.futurelab.org.uk/resources/publications_reports_articles/literature_reviews/Literature_Review203/

Papadopoulos, G., & Pearson, E. (2007). Accessibility awareness raising and continuing *professional development: The use of simulations as a motivational tool*. Association for Teaching and Learning, Online newsletter. Retrieved March 19, 2007, from http://newsletter.alt.ac.uk/e_article000735502.cfm

Pearson, E. J., & Koppi, T. (2006). Supporting Staff in Developing Inclusive Online Learning. In Adams, M., & Brown, S. (Eds.), *Towards Inclusive Learning in Higher Education*. London: Routledge Press.

Phipps, L., Sutherland, A., & Seale, J. (Eds.), *Access All Areas: disability, technology and learning* (pp. 6–9). Oxford, UK: JISC TechDis Service and ALT.

Ruebain, D., Honigmann, J., Mountfield, H., & Parker, C. (2006). *Analysis of the statutory and regulatory frameworks and cases relating to fitness standards in nursing, social work and teaching*. Disability Rights Commission. Retrieved August 4, 2009, from http://www.equalityhumanrights.com/Documents/DRC/Policy/fitness_regulatory_review_report.pdf

Schon, D. (1983). *The Reflective Practitioner*. San Francisco: Jossey-Bass.

*Special Educational Needs and Disability Act*. (2001). Retrieved July 20, 2009, from http://www.opsi.gov.uk/Acts/acts2001/en/ukpgaen_20010010_en_1

Taylor, J. D., Eastburn, S., Coates, C., & Ellis, I. (2006). Using mobile phones for critical incident assessment in health placement practice settings. *Third Biennial Northumbria/EARL SIG assessment Conference*.

Willder, B. (2002). Disability legislation: implications for learning technologists in the UK.

# Chapter 28
# Exploring Interprofessional Educational Possibilities:
## A Case Study from a Virtual World

**Elinor Clarke**
*Coventry University, UK*


## ABSTRACT

*This chapter reports a pilot research study investigating the possibility for Interprofessional education in a 3D multi user 'virtual world' known as Second life® (SL) (Linden Corporation). Following a brief introduction and context of interprofessional education (IPE) in health and social care, the pilot research project is reported. The goal was to gather insights into IPE and teaching and learning in virtual worlds. An action research approach enabled the author to collaborate with students and other experienced users of SL. Drawing on emerging interprofessional, constructivist and activity theories and insights from the research project, this chapter makes suggestions for utilising immersive worlds for interprofessional teaching and learning. It identifies the need for further research focusing on the vital ingredients necessary for interprofessional learning in virtual worlds, and a requirement to ensure that patient/client/service users remains the focus of the students' learning, which is the essence of successful interprofessional working.*


## INTRODUCTION

Headrick and Khaleel (2006) identify three key strategies for educating future health professionals to improve care. First, the integration or theory and practice, especially in the service of patients/clients, secondly, the assessment of learning, and finally, the creation of interprofessional experiences. It is the third of Headrick and Khaleel's strategies, creation of interprofessional e-learning experiences, that this chapter addresses. Multi user virtual environments (MUVE) and Multi user Dungeons (MUDS) have been used to facilitate multi-player games run over computer networks since the late 1970's (Dieterle & Clarke, 2009) The affordances of MUVE are well known amongst gamers but less familiar to educationalists.

DOI: 10.4018/978-1-61520-889-0.ch028

Teaching and learning in immersive worlds has interested many Higher Education (HE) establishments throughout the world and specifically the United Kingdom. Chandler et al. (2007) identified that "once in a virtual world such as Second Life academics are faced by a new environment with novel ways of teaching and a wealth of potential resources to facilitate their teaching". Conklin, (2007) suggested 101 uses for second Life in the college classroom. The evidence is that many Higher Education Institutions are using Immersive worlds for a variety of teaching and learning activities.

During 2008, teaching and learning in SL increased so rapidly that it is difficult to quantify the nature extent and variety of these activities. John Kirriemuir, (Eduserv 2007-2009) provides snapshots illustrating how higher education institutions in the United Kingdom (UK) are using SL. The January (2009) snapshot identified a record number of responses and the virtual world watch (VWW) has identified actual and potential needs of UK academic virtual world developers require. Examples of teaching and learning in SL range from research and composition skills (Robbins 2007/8), computer science (Hobbs et al., 2006) to archaeology (Wheeler et al., 2008). In depth peer reviewed studies of learning and virtual worlds are still rare (Carr, 2008). Small research projects (Carr, Oliver & Burn, 2008) have focused on the teaching and learning in immersive virtual worlds.

Larger studies (Preview JISC 2008) have focused specifically on problem based learning for specific health care professionals (paramedics and clinical managers). Savin-Badin (2008) explores the socio political impact of virtual world learning on higher education, a major research project funded by the Leverhulme Trust. Pedagogical issues and possibilities of second life are important not only because they might be different from other environments but because there are issues about power and control in games where avatars may be representative oneself or someone else (Savin-Badin, 2007): see http://cuba.coventry.ac.uk/leverhulme/category/research-themes/theme-3/. Some studies remain 'work in progress' (Hetherington, 2008; Chandler et al., 2008; (Savin-Badin, 2008), but for others findings are emerging (Toro-Troconis et al., 2008), (Carr et al., 2008).

## Interprofessional Education

The creation of IP learning experiences, has challenged educators to provide authentic, realistic and appropriate learning experiences (Wright 2009). There are many reports of IPE provision (Goosey and Barr, 1998; Scammell et al., 2008) and interprofessional (IP) experiences (Craddock et al, 2006). Clinical or service based interprofessional learning (IPL) experiences are common and a number of IPE initiatives such as ward based (Bray and Howkins, 2006) or placement learning (Armitage et al., 2009) based on short face to face interactive learning between the professions have been reported.

Emerging evidence of successful IPE experiences been demonstrated (Freeth et al., 2002), most evaluations focus on student satisfaction and successful module delivery and most evaluations of IPE programmes are often positive (Ireland et al., 2008). These evaluations, whilst important and informative do little to demonstrate that learners have changed their behaviour, acquired communication skills and all necessary attributes and competencies of interprofessional working (CUILU, 2006). An alternative measure of the success or otherwise of IPE could focus on evidence of collaboration, understanding of roles and acquisition and demonstration of team working skills.

Emerging theory suggests that IPE takes place when practitioners work together, learn together and learn about each other (Barr, 2000) and that this improves patient care. E-learning environments and online technology are often advocated as they support asynchronous learning and communication. Since 2005, the Centre for Interprofessional e-learning (CIPeL) a Higher Education

Funding Council (HEFCE) Centre for Excellence in Teaching and Learning (CETL) has worked with the Faculty of Health and Life Sciences at Coventry University to provide IP experiences for undergraduate health and social care students know as the interprofessional learning pathway (IPLP), as described and discussed in Bluteau and Jackson's chapter earlier in this volume. The intention of this chapter is to discuss a pilot study conducted within CIPeL to enable students to "learn with, from and about each other" (Barr et al 2005) utilising a multi user virtual environment (MUVE), Second Life® (SL).

## Second Life® Learning Environment

Second Life® (SL) is a 3D virtual world that can be downloaded free of charge to your personal computer from the following website http://www.secondlife.com. Jennings and Collins (2007) have considered how adopters of SL utilized it as a place of learning, research and socialisation. Jennings found two different types of virtual university campus: the first or "operative virtual campus" (Jennings, 2007) which resembled the real life version and provide activities based on traditional teaching (for example, virtual lectures, presentations and demonstrations) and the second a "reflective campus environment" (Jennings, 2007). Toro-Troconis, (2009) developed a virtual hospital in second life using game based learning activities with virtual patients around diagnostic, investigative and treatment skills for medical students. Increasingly, educationalists are recognising that MUVEs can contribute and shape how learners can communicate, participate, learn and identify themselves and that learning outside the classroom provides opportunities to change the traditional experiences of higher education.

Against the backdrop of IPE initiatives and experiences and a variety of virtual learning environments (VLE) there has also been interest in the transferability of computer acquired (virtual) skills to the real world. The most recent being the "virtual racing driver" transferring skills acquired using Gran Turismo©® to the real race track of the Dubai 24 hours (Meadon, 2009). This pilot study explores the possibility that health and social care students can acquire Interprofessional skills in a virtual world and transfer them to real life.

## Interprofessional Educational Possibilities in Virtual Worlds

It will be argued that the spectrum of IPE (Figure 1) may be extended further to immersive environments or virtual worlds. It was thought that a virtual world (simulation, games and communication) could expand the variety of opportunities for learners to experience and engage in interprofessional education meeting the needs of a variety of learning styles and the so called 'digital native' a term coined by Prensky (2001).

Four popular theories invoked to help understand IPE and e-learning are contact theory (Craddock et al., 2006), identity theory (Barr et al., 2005), constructivism (Craddock et al., 2006; Monty 2008) and activity theory (Engestrom, 1991). They have particular resonance with immersive worlds such as SL. For example, conditions associated with contact theory – interaction, egalitarian, co-operative and mutually supportive learning have also been found in SL (Savin-Badin, 2008; Carr et al., 2007). Identity theories are also used to underpin traditional IPE in practice to develop, trigger and argue issues around teams, roles and responsibilities and stereotyping. Activity theory has been associated with the location where learning takes place, learner activities, rules and roles. It is thought that by adopting an e-learning approach some of the barriers associated with interprofessional learning may be reduced, in particular, accommodating very large student cohorts that tend the be associated with interprofessional health and social care modules, and enabling students on clinical placement to have access to such modules.

*Figure 1. The potential spectrum of interprofessional education (IPE)*

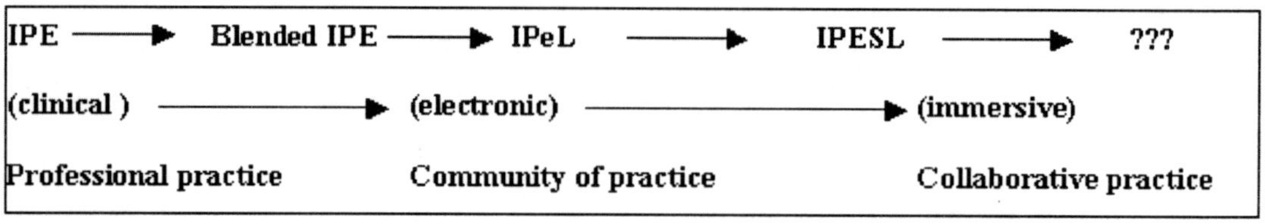


Key to Figure 1:
IPE; Interprofessional Education.
Blended IPE; a combination of teaching and learning techniques used to engage student in IPE.
IpeL; Interprofessional e-learning (using Information & Communication Technologies to engage students in IPE).
IPESL; Interprofessional Education using Second Life ®.

The learning environment chosen to provide IPE is important; it needs to support a variety of health and social care disciplines and learning styles. In addition, it must provide opportunities for students to acquire and demonstrate communication and reflective skills. It was necessary to identify whether a virtual world could support instructor facilitated (moderator/e-teacher) case-based stories, vignettes and other learning approaches. In addition, it was an opportunity for the author to transfer skills acquired as an e-facilitator on the IPLP to an alternative environment. It is, furthermore, timely to consider if immersive worlds such as SL can be used for interprofessional education.

## INTERPROFESSIONAL EDUCATION AND IMMERSIVE WORLD: CASE STUDY

Salmon (2000) identifies four generations of online learning environments which help to explain developments, differences and applications of e-learning environments. The IPLP is arguably a second generation learning environment as it supports web based, asynchronous discussions with linked resources through a virtual learning environment (VLE). Centres of excellence in teaching and learning (CETLS) such as CIPeL were intended not only to build on existing good practice, but to explore and develop that excellence through further research (Higher Education Academy HEA website). Emerging immersive technologies (Salmon's fourth generation learning environments) such as Xbox 360®, Playstation®, Ninentdo Wii® offer synchronous discussions, visual enhancement and possible alternative learning environments. Serious games such as flight simulators and simulation of emergency care also offer participants technologies derived from computer gaming for educational purposes. The author proposed a small pilot research project to ascertain if interprofessional education (IPE) could occur in different or other MUVEs to online discussion forums.

The aim of the study was to investigate whether IPE could take place in SL. The simplicity of the aim hides the complexity of the task. The objectives were to create a specific IP environment in SL, to engage undergraduate health and social care students in SL, to repurpose/reuse IPLO in SL and to see if IPE could take place in SL. The author first became active in SL in October 2007, and a small research proposal was supported by CIPeL in November 2007.

### The Research Procedure

By placing herself as the learner alongside the students the author planned to explore new ways for interprofessional teaching and learning. "The most appropriate way in which we can assess the impact of using SL in an educational context is by using a technique called Action Research" (Bignell, 2007). An action research approach was adopted involving undergraduate health and social care students. On a macro level, the author wished to identify if students would access a 3D virtual environment to support their learning and on a micro level would students be able to acquire IP knowledge and skills. If the results of the pilot study were favourable, it was intended to submit a bid for funding to support a full study. This is important to demonstrate that IPE in SL could enable acquisition of skills transferable to delivery of health and social care.

## The Second Life® Learning Environment

Second Life was chosen as the University had already purchased an island and built a virtual campus. Second life® (Linden Corporation) is a MUVE and a virtual world (VW) primarily intended as a social network. In preparation for students arriving in SL it was necessary to create an IP learning environment. Following visits to several other educational locations in SL, it became apparent that others had already created virtual hospitals (Imperial College, London), clinics (Plymouth) and zoo (Leicester University).

As the focus of this project was interprofessional learning, the intention was to work with existing students (from any course) who had experience of building in SL to find a suitable location and virtual building in SL. The Coventry University SL Island was already being used for teaching and learning in Art and Design courses, Problem Based Learning (PBL) and the Serious Games Institute (SGI). The author negotiated some space on the Island for the pilot project to take place.

It was also necessary to design and build a suitable environment for IP experiences. In partnership with a CIPeL student intern, the author created a multipurpose CIPeL Centre in SL. The Centre is a multi functional building in which rooms can be customised for a variety of health and social care learners' use. Capitalising on her experiences as an interprofessional facilitator, the author anticipated a ward based room, operating theatres, 'home from home' for health and social care students. In addition, a general reception area, café, shop and waiting area to allow social gathering of students and "patients". Several rooms would remain "empty" for future development for example, office space, bedsit or kitchen. The intention was that in future, other IP educators may wish to explore teaching potentials of this environment and could collaborate and develop the Centre further. The reuse or repurposing of the environment would potentially provide longevity for the research and be inclusive of a variety of health and social care professionals or other disciplines.

## Recruitment and Student Engagement

The next phase of the pilot study involved recruiting student participants in order for interprofessional teaching and learning to take place in SL. Recruitment was via an advert on the IPLP and on notice boards (see Figure 2). In an attempt to attract students, they were paid £10 per hour. Light refreshments would be provided for the first session, which took place during the early evening in a University Computer Workshop on PCs pre-loaded with the SL software.

It was envisaged that over three sessions, each group would consist of six to eight students, and the plan was that by session two the participants would comprise students from at least three different professions. However, recruitment proved sluggish. Finally a combination of word of mouth and personal invitation elicited some interest, and session one was arranged. Of the six students who participated in the initial workshop, five were paramedics and one was a medical student, and five were male and one female.

*Figure 2. Recruitment poster for the Pilot Study*

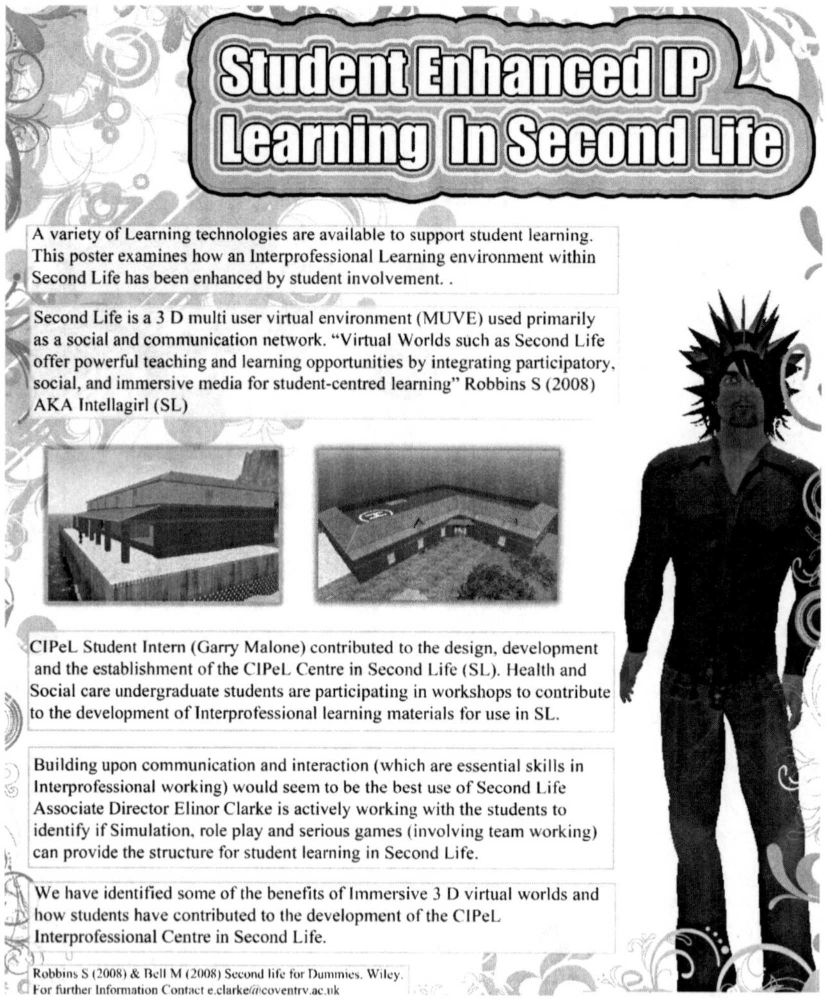


Student participation was voluntary and ethical approval was provided using the CIPeL ethical research approval application. Students participating in the study agreed to photographs being taken of their borrowed avatars during SL sessions and anonymity was ensured. In addition to evaluation of the SL experience, student opinions would be sought during debriefing after each SL Session. It was planned for students to provide verbal consent for each of the three workshops/sessions and there was no requirement to continue in the study in the future.

## Educational Tools

Each room or area of the CIPeL Centre could be customised for a variety of avatar activities. Littleton and Bayne (2008) reported the advantage of using SL for learning is that small, intimate and private learning spaces can be used so that students feel comfortable communicating with each other. The wholesale lifting of existing IPLO materials was considered inappropriate as they were designed and intended for use in asynchronous discussion forums, while SL supports synchro-

*Figure 3. An external view of the CIPeL Centre in SL*

nous discussions. Others (Savin-Badin, 2008; Robbins, 2008) have suggested that the value of transferring traditional materials to Second Life is also diminished.

Three SL sessions or workshops were planed:

- Session One: orientation to Second Life and acquisition of basic skills.
- Session Two: specific task (build a chair based on Conklin, 2007).
- Session Three: interprofessional working together to address a common dilemma or issue utilising an IPLO adapted for SL around a patient complaint.

Whilst others have created bespoke educational tools for example, Second Life PBL Game (Savin Badin, 2007) This study sought to repurpose patient journeys (Davidson & Courtney, 2004) and other IPLOs as a trigger to promote synchronous interprofessional discussions between the Avatars (students) within the virtual world. These IPE triggers were planned for session three, after the students had experienced shared learning in the first two sessions.

### Photo Timeline of the Research Project

In order to demonstrate the construction and development of the virtual learning environment (VLE) the author now provides a visual timeline of the pilot project. Figure 3 represents the early days –building the CIPeL Centre in Second Life (January – April 2008). Figure 4 shows the completed CIPeL Centre on Coventry University's SL island in use in July 2008.

## RESEARCH FINDINGS

This section addresses the key experiences and findings of the pilot project. The results will be considered using the previous themes of learning environment, recruitment and preparation, and educational tools. These are followed by discussion, reflections, commentary and recommendations. Readers should be mindful of the limitations evident in this pilot study and should be wary of making generalisations arising from such a small study.

*Figure 4. Session One: Second Life Learning Environment (Orientation and skills – Early July 2008)*

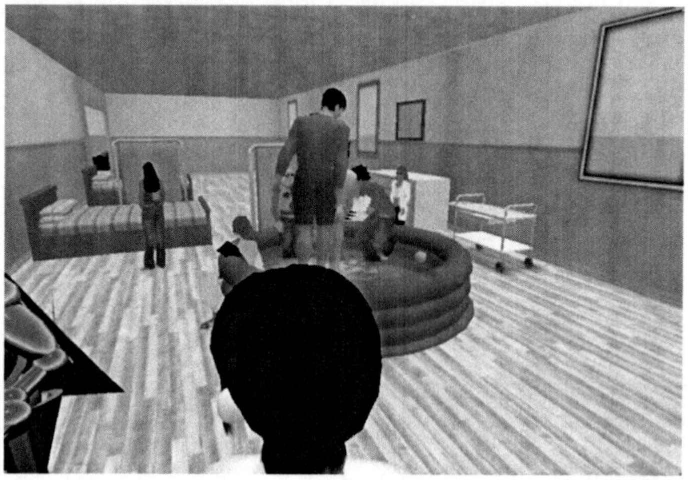

## The Second Life Learning Environment

The first workshop took place in the CIPeL resource room which has eight computers to which Second Life had already been downloaded. The aim of the first session was to expose students to this VLE. None of the student volunteers had accessed SL before and not all had access to a computer that could support this learning environment. The rationale being that SL is bandwidth hungry and students would be able to utilise avatars pre-created by Coventry University for the first workshop. It was not the intention to require students to acquire their own avatars or SL accounts until after the first workshop, when they would be able to choose if this environment was something they wished to use personally or further participation in the pilot study.

On the day of the first session scheduled in SL a message was sent to registered members suggesting that during this day SL would be down for scheduled maintenance. The information was received too late to cancel the session, and the author had to hope that the down load would be completed prior to the students arriving. Concerns about using an educational tool that is owned by a commercial provider and not by the university were raised by Harrison (2007), and this incident demonstrated the realities of such concerns.

## Student Engagement

a handbook or student guide were not provided as the main aim was to introduce the students to the environment and to enable them to identify if they wished to download the software and continue with the next stage (acquisition of own avatar and participate from home/other venue) Following on from the first visit to SL the intention was to direct students to (free) a guide and opportunity to engage in further SL skills using orientation places in SL.

The findings from workshop one are that the students found SL to be interesting and entertaining. They were also well able to move about, communicate and control their avatars. Despite positive comments about this environment, this group of students were reluctant to commit to downloading SL to their own computers and to contribute to workshop two. The students identified IT issues and time constraints as contributing to their decision to withdraw.

## Educational Tools

Six students completed an evaluation form for session one. Four students rated the session as "good" and two students "excellent". All the participants felt that the session met the objectives identified. Students had an opportunity to provide comments or recommendations, and the following are typical comments:

- Student 1: "might be a bit too distracting in real life, quite a bit going on"
- Student 2: "It's a great idea, and can see a good future with this."
- Student 3: "more structure"
- Student 4: "although a good idea, I would prefer a different style of PBL, steering away from online discussions"

The synchronous discussions enabled the participants to learn from each other some basic avatar skills for movement, expression and exchange of opinions during the workshop. Students were asked if they would like to take part in the next phase of the pilot study. Five of the six participants said that they would be interested in participation in session two. Unfortunately due to final exams, outstanding coursework and annual leave none of the students were able to commit to engaging in session two. At this stage it was clear that undergraduate students (especially those with requirements to attend clinical placements) felt unable to engage further in this pilot study, and the pilot study came to an abrupt end.

## Discussion and Analysis

On reflection, there were three main drivers for consideration of whether or not IPE could take place in SL. The first was an awareness that MUVE's and 3-d virtual worlds offered affordances for teaching and learning which were unavailable in other VLEs. Second life has the advantage of being a 3D virtual world, which some argue increases the fun and the experience of communication with friends. Advantages are that by creating an avatar, choosing where to visit and contributing to conversations where it is possible to instant message, see each others' avatar then the experience is enhanced. In short, the environment provides the users with a virtual 'life like' experience (visual, expressive, spontaneous and entertaining) in addition to the "immersive" attributes of this environment.

Secondly, a strong argument for supporting education in multi user virtual environments (MUVE) is that today's students grow up in a digital age and their thinking patterns and ways of processing information are different to those of their predecessors (Heatherington et al., 2008). However, whilst many visits to SL could be considered as educational, not all places and experiences can be considered to support academic or professional development. Educators choosing SL as their teaching environment need to consider health and safety issues as with any other learning environment. For example, Leicester University Media Zoo (Salmon, 2008) is a safe place for new avatars to explore SL and engages students in a variety of learning opportunities.

Thirdly, Students can access the materials at different times and places and can engage in learning twenty four seven (24/7):

*The asynchronous learning method can assist learning in a diverse and fragmented higher education world where students struggle with busy lives and increasingly face the pressure to work in paid employment while studying*

*the asynchronous learning method can promote greater reflection and higher levels of involvement from students who find that face to face seminar group contact is difficult and impersonal" (Bach et al., 2007, p. 186).*

Synchronous learning (for example audio or video conferencing) can be used for groups or individual students and benefits from organisation and chairing (Clarke, 2004, p.118).

Time constraints affected the development of a "virtual environment" for the sessions to take place and the number of SL sessions that could be achieved by one researcher. Effective IPE and training benefits from starting small (Freeth et al., 2005) and effectiveness is built on positive outcomes, acceptable cost and without unacceptable side effects. Freeth (2005) considers that getting a balance between these elements is an evolutionary and iterative process. The author planned a small initial project with the view of a subsequent larger study if the initial findings demonstrated positive aspects of this environment. The intention was not only to ascertain if existing interprofessional e-learning materials could be utilised, but also to identify if IPE skills could be acquired in a MUVE. MUVEs have been used in education as they enable multiple simultaneous participants to: (a) access virtual contexts, (b) interact with digital artefacts, (c) represent themselves through avatars, (d) communicate with other participants and (e) take part in experiences incorporating modelling and mentoring problems similar to those in real world contexts" (Dieterle & Clarke, 2009).

## The Second Life® Learning Environment

Salmon's (2000) five stage model suggests that the amount of interactivity of the learners increases as the learning increases. Salmon (2004, p. 30) suggests benefits from using the five stage model to "exploit the system at each stage" and "avoid the pitfalls". From access and motivation through to development the stages require good e-moderating as well as technical support.

Salmon is keen to point out that participants learn about ICT at the same time as learning through ICT; the integration of the two enables participants to learn along "with learning about a topic, and with and through other people" – fundamental to interprofessional learning. The self selected students all demonstrated a level of IT skills which enabled them to participate in SL. They also demonstrated acquisition of specific skills (instant messaging, flying and exploring their location). Dialogic learning, the cornerstone of personalised learning has also been suggested (Savin-Badin, 2008) as a theoretical underpinning for learning in MUVES. Effective communication is considered to be essential for effective interprofessional working and consequent delivery of quality health and social care to patients (Day, 2006, p. 81). Immersive worlds have potential as safe opportunities to acquire and demonstrate communication skills.

The findings from this pilot study demonstrated that interest in SL as an educational environment from undergraduate health and social care students does not match those of educators. SL may have the potential to support interprofessional learning but students do not necessarily share educators enthusiasm for this learning environment at this time. A particular limitation is the unwillingness of the students (due to time constraints) to continue with the pilot study. Given the current interest and fashion of this environment, however, it is likely that learners will in the future be exposed to it, and funded projects will provide the knowledge and understanding to develop new and engaging learning for health and social care students. In addition if patients (client, service user, carer, and consumer) also enter the world of SL new opportunities for interprofessional leaning will be afforded.

## Student Preparation and Engagement

The students in the pilot study demonstrated that they quickly masterd basic SL skills. Carr, Oliver and Burn (2008) and Toro-Troconis et al. (2008)

came to similar conclusions. Theoretically immersive worlds such as SL will be so engaging that students will wish to frequently return and meet and communicate with a wide variety of health and social care professionals. This study revealed that if this learning environment is not a core component of their course it is unlikely that students will utilise it for educational purposes. Interestingly, following-up the students also revealed that they were not intending to use SL for social networking. Undergraduates are preoccupied with coursework, assessments and clinical placement and are unlikely to wish to take on "additional learning" especially without a reduction in classroom contact or other reward. Following a period of reflection and follow up of the initial group, the author considers that a new pilot study would benefit from being embedded in a common core module to enable a reduction in the personal time and recognition of engagement and learning to be rewarded.

## Educational Tools

Access, immersiveness and engagement were considered a 'given' by the author, and these affordances were not demonstrated in this pilot study. Participants in SL require training on networking and skills development (Wheeler et al., 2008). Similarly, it is arguable that in order to "teach" in second life it is necessary to acquire skills and competence in using this environment. In keeping with other findings (Carr et al., 2008) the experience was busy, fast and chaotic. Teaching in e-learning environments have been reported by Moule (2007) and Salmon (2008). Little is known about the specific affordances of SL and in particular the transfer of skills and learning from a vitual world to the real world. More must be discovered about the nature of learning and the utilisation of educational tools in virtual worlds before widespread adoption in higher education.

## Recommendations

The function of a pilot study is to refine and focus the research study, test out data collection instruments and procedures and to confirm the feasibility and value of the chosen techniques. It was found that in order to engage in a full or definitive study the following recommendations should be implemented:

## The Second Life® Learning Environment

Education, according to Robbins (2008) does not have to be boring and teachers and institutions are showing the way to make it fun and interactive using SL. (Robbins, 2008, p. 284) suggests that SL "makes use of the flexibility of the space to allow students to have hands on experience with topics that might be difficult to teach in the classroom". The author recommends the use of "getting started in second life" Savin-Badin et al (2009) to enable teachers and learners to get the most out of the Second life experience.

## Recruitment and Preparation of Students

The participants need to be from a variety of health and social care backgrounds. Having one particular profession dominate does not enable experiences to be validated or generalisations to be drawn with regard to IPE. Students who wish to engage in the development of artefacts or encounters in SL need recognition and rewarding for their contributions. Students need to be supported with a reduction in coursework or be allocated time within their course to participate in SL. Undergraduate health and social care students are especially loaded with work as many courses have theoretical and clinical practice workloads. These constraints need to be addressed prior to embarking on a full study. The pilot study identified that students would like detailed information to be available prior to engaging in learning activities. Experienced Second Life Educators (Salmon, 2008; Savin-Badin, 2008; Carr, Oliver & Burn, 2008) all advocate acquisition of core technical skills which enable avatars to participate in this environment.

## Educators

Tutors should have prior experience of 'living' in SL and acquiring SL skills as well as being experienced in e-facilitation. One way for educators to prepare for effective use of virtual learning environments would be to use a training package such as InfoKit (JISC, 2008). Since the pilot project ended, a number of specific resources for educators in SL have emerged. Wheeler et al. (2008) produced a training programme using two .pdf guides (one for teachers and one for learners or participants). The guides use other Web 2.0 technologies (You Tube®) to support their training. Savin-Badin et al. (2009) have produced the JISC document "Getting started with Second life".

## Educational Tools

The tension between social network (informality and fun) and education (structure and outcomes) could be resolved with learning encounters using simulation or construction / development of a resource that is patient focused and student chosen. Robbins (2008) identified that simulations, role play, building and social events in SL can all contribute to learning. Specific spaces for education have been created in SL where different approaches to learning may be encountered. There are numerous locations and virtual campus which avatars can visit and engage in learning materials, demonstrating the truly immersive aspects of this environment. Immersive worlds have been identified as ideal for some types of learning. Specifically, simulation (whereby the students can participate in real life simulated experiences without fear of harming or killing real patients) Toro-Troconis et al., (2008).

## Conclusion

In summary, this chapter has identified some of the existing teaching and learning activity in an immersive 3D world namely Second Life ®, considered the benefits of this environment and identified the potential for SL to be used as a platform for IPeL. Readers should draw their own conclusions on account and the small scale of the pilot study and the limited experiences of the author in SL. While there is evidence that SL is a useful additional platform for learning activities and learner participation around team building and communication skills (Savin-Badin, 2008), this was not demonstrated in this small pilot study reported in this chapter. The social constructivist approach to IPE in second life would seem to be the way forward and contact with other learners will enable the social exchange of values, beliefs and attitudes of the health and social care professions.

Health and social care professionals may also find this open access, immersive environment suitable for interprofessional learning and whilst higher education has been quick to see the potential for second life the UK National Health Service (NHS) may also consider other applications for patient and staff learning and training. In the future, it is intended to focus specifically on using patient journeys and triggers to provide IP learning experiences for undergraduate health and social care students.

The use of 3D learning environments has immersive benefits for student learning but needs to be balanced against the high technological requirements to support this environment. In addition the utilisation of a social network for learning purposes demonstrates affordances and risks which are different to traditional learning environments. Teaching staff should be mindful of the dangers and difficulties of using an open source environment for professional courses.

In conclusion, the utilisation of a 3DVirtual World as an educational environment is a possibility for IPE but undergraduate health and social care practitioners may need SL to be a core element of their programmes before engagement is demonstrated. Further work is needed to consider the transferability of skills acquired in a virtual world to a real world. Future research is now required to explore the wider use of this MUVE and the contribution this type of e-learning can contribute to IPE.

## REFERENCES


Armitage, H., Pitt, R., & Jinks, A. (2009). Initial findings from TULIP (Trent Universities Interprofessional learning in practice) Project. *Journal of Interprofessional Care, 2391*, 101–103. doi:10.1080/13561820802379938

Barr, H. (2000). Working together to learn together: learning together to work together. *Journal of Interprofessional Care*, *14*(2), 177–179.

Barr, H., Koppel, I., Reeves, S., Hammick, M., & Freeth, D. (2005). *Effective Interprofessional Education: Argument, Assumption and Evidence (Promoting Partnership for Health)*. London: Wiley Blackwell.

Bignell, S. (2007). "Measuring" the impact of second life for educational purposes. In J. Kirriemuir, *Responses and second life meeting transcript*. Retrieved September 11, 2009, from http://eduserv.org.uk/foundation

Bray, J. & Howkins, E. (2006). Facilitating interprofessional learning in the workplace: a research project using the Delphi technique. *Work based Learning in Primary Care*, (4), 223-235.

Carr, D. (2008). Learning and virtual worlds. In Teaching and Learning Research Programme (TLRP) Education 2.0? Designing the web for teaching and learning. ESRC/ESPRC, (pp. 13-16).

Carr, D., Oliver, M. & Burn, A. (2007). *Learning from online worlds; Teaching in Second Life*. Edsurv project.

Chandler, J., Crellin, J., Collinson, T., & Duke-Williams, E. (2007). *Second Life Toolkits*. Higher Education Academy Development Fund Project. Retrieved September 11, 2009, from http://www.ic.heacademy.ac.uk?projects/development-fund/fund_details.php?id=106 on 01.08.08

Clarke, A. (2004). E-learning 'close to being a core skill in the 21st century'. Adults Learning, September 12-13. In G. Salmon (Ed.), E-moderating, the key to online teaching and learning. London: Routledge Falmer.

Conkin, M. S. (2007). *101 uses for Second Life in the College Classroom* (Version 2.0). Retrieved September 11, 2009, from http://trumpy.cs.elon.edu

Craddock, D., O'Halloran, C., Borthwick, A., & McPherson, K. (2006). Interprofessional education in health and social care: fashion or informed practice? [from http://www.blackwell-synergy.com/doi/pdf]. *Learning in Health and Social Care*, *5*(4), 220–242. Retrieved September 11, 2009. doi:10.1111/j.1473-6861.2006.00135.x

CUILU. (2006). *The Combined Universities Interprofessional Learning Unit, final report*. Retrieved September 11, 2009, from http://www.sheffield.ac.uk/cuilu

Davidson, A., & Courtney, K. (2004). *Using and reusing Electronic, Patient Case Studies to promote interprofessional learning*. Presented at Third Annual Web CT European Conference in Edinburgh.

Dieterle, E., & Clarke, J. (2009). Multi-user virtual environments for Teaching and Learning. In Pagani, M. (Ed.), *Encyclopaedia of multimedia technology and networking* (2nd ed., *Vol. 1*). Hershey, PA: IGI Global.

Engestrom, Y. (1991). Engestrom, Y. (2001). Expansive Learning at Work: toward an activity theoretical reconceptualization. *Journal of Education and Work, 14*(1), 133–156. doi:10.1080/13639080123238

Freeth, D., Hammick, M., Koppel, I., Reeves, S., & Barr, H. (2002). *A Critical review of Evaluations of Interprofessional Education*. London: The Learning and Teaching Support Network for Health Sciences and Practice.

Freeth, D., Hammick, M., Reeves, S., Koppel, I., & Barr, H. (2005). *Effective Interprofessional Education: Development, Delivery and Evaluation (Promoting Partnership for Health). CAIPE.* Wiley Blackwell Publishing.

Freeth, D., & Reeves, S. (2002). Evaluation of an Interprofessional training ward: pilot phase. In Glen, S., & Leiba, T. (Eds.), *Multiprofessional Learning for nurses – Breaking the Boundaries*. Basingstoke, UK: Macmillan.

Goosey, D., & Barr, H. (1998). Selected Case Studies of Interprofessional Education. Centre for the Advancement of Interprofessional Education, London. In Craddock, D. Borthwick, A. & McPherson, K. (2006). Interprofessional education in health and social care: fashion or informed practice? [from http://www.blackwell-synergy.com/doi/pdf]. *Learning in Health and Social Care, 5*(4), 220–242. Retrieved September 11, 2009.

Harrison, D. (2007). In Shepherd, J. *It's a world of possibilities*. The Guardian, May 8th.

Headrick, L. A., & Khaleel, N. I. (2008, August). Getting it right: Educating professionals to work together in improving health care. *Journal of Interprofessional Care, 22*(4), 364–374. doi:10.1080/13561820802227871

Heatherington, H., Bonar-Law, J., & Fleet, T. (2008). *Problem based learning in a Multi User Virtual Environment. Project description*. London: Higher Education Academy. Retrieved September 11, 2009, from http://www.ics.heacademy.ac.uk/projects/development-fund/fund_details.php?id=114 on 01.08.08

Hobbs, M., Gordon, M., & Brown, E. (2006). A virtual world environment for group work. *Current Developments in Technology-Assisted Education. Formatex, 1*, 1369–1373.

Ireland, J., Gibb, S., & West, B. (2008). Interprofessional education: reviewing the evidence. *British Journal of Midwifery, 16*(7), 446–453.

Jennings, N., & Collins, C. (2007). Virtual or virtually U: Educational Institutions in Second life. *International Journal of Social Sciences, 2*(3), 180–186.

JISC. (2008). *Effective use of Virtual Learning Environments (infoKit)*. Retrieved September 11, 2009, from http://www.jiscinfonet.ac.uk/InfoKits/effective-use-of-VLEs accessed on 08.12.08

Kirriemuir, J. (2007). *A July 2007 "snapshot" of UK Higher and Further Education Developments in Second Life*. Eduserv Foundation. Retrieved September 11, 2009, from www.eduserv.org.uk/foundation

Kirriemuir, J. (2007). *An update of the July 2007 "snapshot" of UK Higher and Further Education Development*. Eduserv Foundation. Retrieved September 11, 2009, from. www.eduserv.org.uk/foundation

Kirriemuir, J. (2008). *"Measuring" the impact of second life for Educational Purposes*. Eduserv Foundation. Retrieved September 11, 2009, from www.eduserv.org.uk/foundation

Littleton, F., & Bayne, S. (2008). Virtual worlds in education. *ESCalate News*, *10*(Spring), 26–28.

Meadon, R. (2009). Game On. Telegraph Motoring Supplement. Saturday, March 28th. M1-2.

Oliver, M., & Carr, D. (2008). Learning in virtual worlds: using Communities of Practice to explain how people learn from play. *BJET*. Abstract Retrieved September 11, 2009, from http://learningfromsocialworlds.worpress.com/

Prensky, M. (2001). Retrieved 13/02/09 from http://primaryblog.wordpress.com

Robbins, S. (2008). *Virtual worlds as Web2.0 Learning spaces*. Retrieved September 11, 2009, from http://hosted.mediasite.com/flash/ELI/Virtualworldsasweb2.0learningspaces/

Robbins, S., & Bell, M. (2008). *Second life for Dummies*. London: Wiley.

Salmon, G. (2002). *E-tivities: the key to Active online Learning*. London: Kogan Page.

Salmon, G. (2004). *E-moderating, the key to online teaching and learning* (2nd ed.). London: Routledge Falmer.

Salmon, G. (2007). *The SEAL: Second life: Advanced Learning project*. University of Leicester. Retrieved September 11, 2009, from www.le.ac.uk/beyonddistance/mediazoo

Salmon, G., Jones, S., & Armellini, A. (2008). Building institutional capability in e-learning design. *Alt-j. Research in Learning Technology*, *16*(2), 95–109.

Savin Baden, M. (2007). *A Practical Guide to Problem-Based Learning Online* (pp. 96–104). London: Routledge.

Savin-Baden, M. (2008a). *Learning Spaces. Creating Opportunities for Knowledge Creation in Academic Life* (pp. 136–150). The Society for Research into Higher Education.

Savin-Baden, M. (2008b). From cognitive capability to social reform? Shifting perceptions of learning in immersive worlds. *Alt-j. Research in Learning Technology*, *16*(3), 151–161.

Savin-Badin, M., Tombs, C., White, D., Poulton, T., Kavia, S., & Woodham, L. (2009). *Getting Started with Second life*. JISC. Retrieved September 11, 2009, from http://www.jisc.ac.uk

Scammell J., Hutchings M., Quinney, A., Hean, S., Dumbrell, A. & Pulman, A. (2008). *A virtual practice community for student learning and staff development in health and social work inter-professional education:changing practice through collaboration*. School of Health and Social Care, Bournemouth University. Higher Education Academy, A Health Sciences and Practice Subject Centre, Mini Project, 1/1/06-31/12/07.

Toro-Troconis, M., Partridge, M., Barrett, M., & Mellstöm, U. (2008). Game based learning for the delivery of virtual patients in Second Life. *The Academy Subject Centre for Medicine, Dentistry and Veterinary Medicine Newsletter01*, *17*, Summer, 3-4. Retrieved September 11, 2009, from http://www.medev.ac.uk/external_files/pdfs/01_newsletter/0117_lo_res.pdf

Wheeler, M., Nie, M., & Salmon, G. (2008). How avatars learn together: an immersive experience. *Online Newsletter*, Monday July 21st. Retrieved September 11, 2009, from http://newsletter.alt.ac.uk/e_article001151232.cfm

# Chapter 29
# Thinking Outside the Box

**Andrew Brooks**
*Centre for Interprofessional eLearning (CIPeL), UK*

## ABSTRACT

*Education is moving out of the classroom and into the real world, driven by both emerging Information and Communication Technologies (ICT) and new economic models. The boom of the web and social networking has revolutionised global communication and collaboration. A DIY culture and industries are emerging because of this. Mobile devices connecting to the digital superhighways are merging the real and digital worlds. This, coupled with the falling cost of the hardware as well as the free software movement may soon place a new model of education into the grasp of almost everyone. A model where education is no longer the process of being fed information, but rather a process of enquiry, exploration, discovery, expression and re-interpretation of the world around us on our own terms. The potential for learning from and with each other at this moment in time is unprecedented.*

## INTRODUCTION

It seems as if the traditional system of education taking place in the classroom has been mortally wounded by the possibilities afforded through modern information and communication Technologies (ICT), and this is not a bad thing. Another near fatal blow to the indoor education model is that of economics. If you can deliver the same information to your student via a cheaper method, then surely you will. Education has become a business in many cases, and even where it is not, money still talks. That is why there is now a growing trend towards e-learning (Dearing, 1997).

The advantages of delivering materials online become obvious when one realises that one's potential audience has suddenly become the whole world instead of just one's own city and also that one does not require a building with equipment, heating or lighting to house one's students in. For all the positives, there are also negatives. It is hard to beat face to face communication and instant

DOI: 10.4018/978-1-61520-889-0.ch029

feedback. Things like body language and the ability for someone to take a hands on approach also are potential casualties in this battle (Bromage, 2004), due to the reduced 'social bandwidth' afforded by information and communications technologies (Nelson-Kilger, 1993).

It seems that the most sensible approach is to move into the new world of virtual learning, while still retaining an element of the old face to face tradition (Bromage, 2004; Littlejohn & Pegler, 2007; Singh-Cooner, 2010, this volume). Hopefully that 'blended' sort of approach leaves you with the best of both worlds. It has been shown that children score higher grades in lessons taught outside the classroom (Klopfer, 2008), than those taught within. This is partly because in most cases the subject being taught, can be explored more clearly using reality, than through a representation or description of it. A representation or description lacks in the taste, touch, smell, look and feel and therefore is unable to get as secure a grip in the student's memory. (Fleming's 2001 Vark model)

## TECHNOLOGICAL ADVANCEMENTS ENABLING CHANGE

Mobile phones, GPS, Cameras, microphones, the World Wide Web, Shared processing, Virtual worlds, QR codes, Databases, cloud computing, and social networks. All these technologies are starting to merge and overlap with one another. All of this combined technology can now be found in a single device that fits into the palm of one's hand in the latest generation of Wi-Fi enabled 'smartphones'. It is as if they are becoming the organs and appendages of a great all knowing, all seeing beast, slowly coming to life. Yet all of this combined technology can now be found in a single device that fits into the palm of your hand. However, according to Moore's law, this great leap in the advancement and miniaturisation of digital technologies has its boundaries.

If the devices cannot become any smaller, then maybe their 'brains' can live outside of the devices themselves. This new great technological leap has begun and is already beaming over the airways via microwave, wifi and satellite. 'Cloud computing' is the process whereby users can access a software application, database of information or even processing power which lives on a larger computer, or even a collection of computers, somewhere else. This allows a small device with limited processing power and storage capacity to have all of the advantages of virtually infinite power and storage.

Advancements in sensor technology are also coming into their own. Portable devices such as mobile phones already come with devices like cameras, microphones GPS and wifi as standard. Add to this, devices such as accelerometers, lasers, projectors, night vision etc. and their potential for exploring and interacting with the world is further enhanced.

The mobile phone has been transformed from a brick into a slick multimedia device which tries to be all things to all people. One can watch movies, listen to music or film one's friends at a party. More than that however is the ability to share this information with the whole world via the World Wide Web. People can share their views and comment on your postings. Even people whom one has never met or who live on the other side of the world. They can share their perspective and hopefully enhance our understanding of the world we live in.

One of the most astounding things is how cheap this technology has become. It is no longer just in the hands of the elite few with the financial resources to afford it. This technology potentially belongs to us all. Mobile devices are still evolving and at some point they will be reduced down further to their vital components. These being a means of viewing and a means of interacting with the content. One of the most interesting things about this miniaturisation of the technology is the fact that one is no longer bound to a particular location. One is free to explore the world and take part in its ever-changing nature. It is in this place, where education, which is our understanding of the world, comes into its own (Klopfer, 2008).

Imagine wearing a pair of glasses that allow you to see both the real world and a virtual one simultaneously overlaid onto one another. This virtual world/could be a social one, a 2D world, 3D world, soundscape or even a combination of these. A practical application of something like this could be a tour of an actual building. As you walk around the physical space, you are presented with information about the building. A 3D animated structure may be overlaid onto the actual building so that the learner can see the building's inner architectural skeleton. Even video footage of actual builders constructing the building is layered seamlessly onto the real world, blurring the distinctions between the two. Inside the building, various painting and decorations could have an audio commentary which can be access if desired. Potentially virtual characters from various time periods could be wandering around the space getting on with their daily chores and even interacting and communing with the real visitors. You might choose to leave some comments or a video recording of something you noticed, for other visitors to find in a particular part of the building. Now, imagine this out in the big wide world. A tree that talks to you as you walk past it or a dinosaur grazing in an empty field. The possibilities are very exciting and have the potential to forever change our perception of the world we live in.

## Social DIY

The emergence of web2.0 'social software' (O'Reilly (2005) and the development of the open source movement has led to an emergence of a kind of DIY culture. Examples of these include the open source car project (www.theoscarproject.org) in which an online community of interested individuals are all working together to develop a car according to open source principles. Also the Arduino project (www.arduino.cc), which consists of a cheap electronic hardware circuit board and open source programming language. Arduino has multiple uses and can be used in conjunction with other bits of hardware to form virtually anything from electronic musical instruments to servo controlled robotic arms. Pockets of groups, interested in specific projects have sprung up all over the Internet. It is almost like the environment prior to the industrial revolution, where enthusiasts met each other in gentlemen's clubs to discuss and share their ideas, such as The Lunar Society in the midlands of England. Suddenly a new, democratised DIY revolution is taking place.

Websites like instructables (www.instructables.com) house a community of people who publish detailed instructions for building virtually anything from electronic circuit boards to home improvement techniques. Self publishing as also become a reality with Lulu (www.lulu.com), which enables individuals to sell their own digital works as books, CDs, or movies without any initial financial outlay.

One of the most interesting developments for me has been the advancements in rapid prototyping. This is the process whereby objects are designed in a 3D Computer Aided Design (CAD) environment, and then printed out as a physical object that one can hold in one's hand. There are a variety of ways of doing this, but ultimately most consist of building an object up slowly, layer on top of layer using a printing process that uses liquid plastic rather than ink.

The Fab@Home system (http://fabathome.org), developed by Massachusetts Institute of Technology (MIT), was the first open-source system that one could buy very cheaply and then build at home oneself. The Rep Rap (http://reprap.org), developed at Oxford University is similar. The idea is that this technology could empower developing countries, giving them the ability to manufacture complex machinery or specialised parts easily and affordably. This would allow them to catch up with the other industrialised nations and compete on an equal footing to some degree.

This new rapid processing technology has the potential to change the world in some very astounding ways. Printing technology has been refined in recent years. It is already possible to print electronic circuit boards using inkjet printing technology. There have also been recent developments in printing Liquid Crystal Display (LCD) and Organic Light-Emitting Diode (OLED) screens and even solar panels. All this combined has the potential for people to design and print out their own electronic devices, or virtually anything they could possibly need. One could print out a particular part that one needed to repair a car, or download and print out the latest iPod.

It is currently possible for a rapid prototyping machine to create a copy of itself (http://reprap.org) and if both the copy and the original were to copy themselves too, and so-forth, within 6 months this exponential self replicating process would be faster than any other known manufacturing method. Combining this Rapid Prototyping technology, with the open source and web 2.0 movements could truly democratise the world and bridge the wealth divide. Let me present you with a small glimpse of a possible future:

## The Potential

My friend prints me off a rapid processing machine, with his rapid processing machine and in exchange for this, I agree to print one off for someone else. Pretty soon, there is one available for everyone. I now proceed to download a copy of the portable device I would like to have, and print it out. It is a modular, open source design and I can customise it to my own personal requirements. Using this device, I can now connect to the Internet.

The World Wide Web is no longer controlled by the Internet Service Providers (ISPs). Instead, your computer talks directly to mine, or anyone else's nearby, and theirs in turn talks to others, ultimately connecting everyone to each other. I no longer have to pay for web space as I have as much space as my personal machine has disc space. When I connect, my computer's processor, becomes part of a world wide, shared processing monster. I share my processing power with everyone else, and in exchange for this, I get the power of everyone's combined processing.

The software applications that I use, live in the cloud and are extremely powerful because of the shared processing available to them through the network. The Internet houses a huge database of information that is available to me in an instant. Information that I gather about the world in the form of video, images, audio recordings etc., using my mobile device can be added to the pool of shared information. I can use this pool to find out what kind of tree I have just passed. A photographic analysis of the leaf shape could reveal what type of tree it is. Its GPS location is also stored in the image and someone else searching in the database, now has access to this new information which reflects a particular physical and psychological point-of-view.

At the heart of the forest stands a tree. The carpenter sees a table. The old man a sapling, the botanist a Salix, the herbalist some aspirin, the weaver a basket, the artist some charcoal and the child, a monster. In order to see the real tree, you have to shut your own eyes for a moment and look at it through the eyes of your companions. Then only, will you see the willow tree as it truly is, but more than this, you will also truly see your companions.

## The Reality

One of the most exciting things about being the resident Learning Technologist at CIPeL, is that, exploration and experimentation are seen as a key part of the role. So rather than creating a production line, an attempt has been made to be more like an artist's studio or scientist's laboratory. A conscious effort is made to keep abreast of new developments in media and multimedia

*Figure 1. Simulated human characters used in the project*

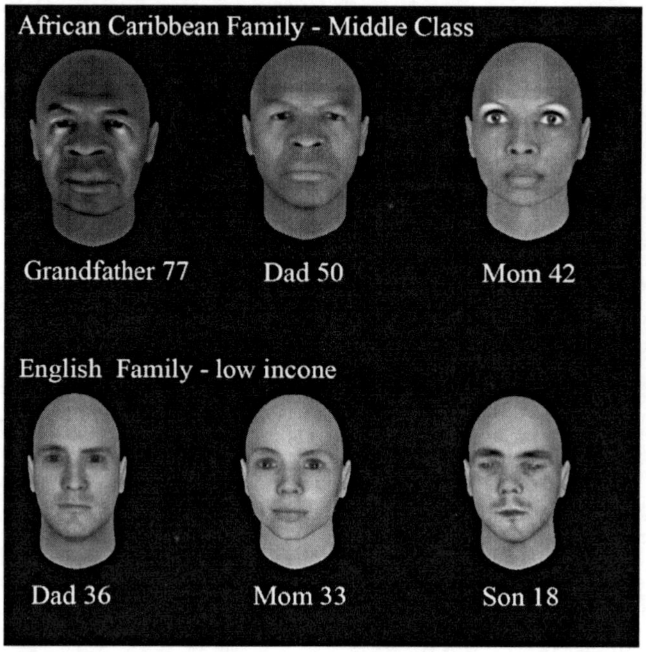


technologies in order to have an arsenal of tools and techniques at our disposal which have the potential to be re-purposed into our learning objects. A crossover of multiple disciplines takes place where these tools, techniques and technologies all merge, so it becomes essential to have a broad working knowledge in order to bring things together. The following case studies demonstrate this merging of various technologies into single learning objects:

## Reusable 3D Street: CIPeL Case Study 1

Whilst developing a Learning Object for the IPLP module, I needed to find a way of illustrating a sequence of scenarios which take place in a script. My options were as follows: I could use text to describe what was happening, potentially aided by some images (either photos or illustrations). A second option is to either film or photograph the scenarios. This however would be a Herculean task as the scripts were very specific in their requirements for specific locations and characters of varying age, physical condition and nationality. A third option would be to create this imagery using 3D characters and locations, and this ultimately was the path I chose to follow. It occurred to me very early on, that once I had created a set of virtual characters (see Figure 1) and locations (see Figure 3), these could be used and re-used on a virtually limitless set of projects. Characters or spaces could be easily tweaked to suit new situations.

I ended up exploring a wide variety of software: Endorphin™ (Natrualmotion), Poser™ (Smithmicro), Zbrush™ (Pixologic), Blender™ (Blender Foundation), Facegen™ (Singular Inversions), Kia's Supergoo™ (Scansoft), Crazytalk™ (Reallusion), Sketchup™ (Google), Photoshop™ (Adobe), Antics 3D™, Moviestorm (Short Fuse), iClone™ (Reallusion), Lively™ (Google), Second life™ (linden Research Inc), Comic Life™ (Plasq).

The final project came together using a combination of some of these software applications. The working methods I developed are still far from

*Figure 2. 'The Street' interface*

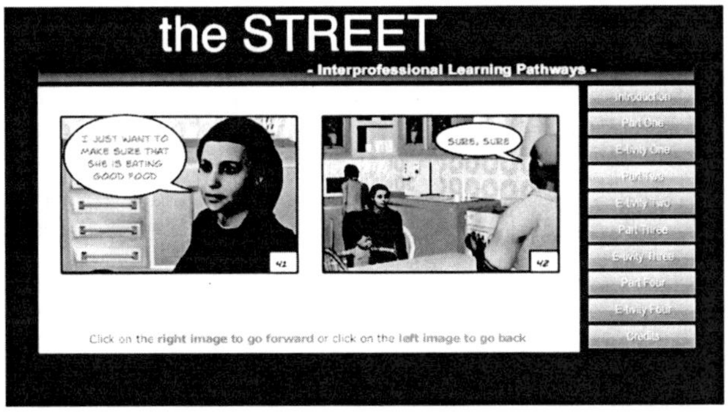


ideal and this is mainly due to the time constraints of the project. However, the Learning Object which has emerged from all of this, called 'The Street', clearly demonstrates the power of using a digital 3D environment in this way (see Figure 2).

I have tried to use mostly cheap or free software for this project as it ensures that the working methods can be utilised by others on a very limited budget. One of the most useful bits of software that I have been using is Sketchup. The thing that places this software into a league of its own is the fact that there is a huge community of people building and sharing 3D models with each other via the Google 3D Warehouse. It is possible to search this space and find a 3D model of virtually anything, that you can then download for free and modify to your personal requirements. On top of this, Sketchup is probably the most user friendly 3D modelling applications that I have ever come across and because of this, even someone with no

*Figure 3. Simulated city created in 'Antics 3D™'*

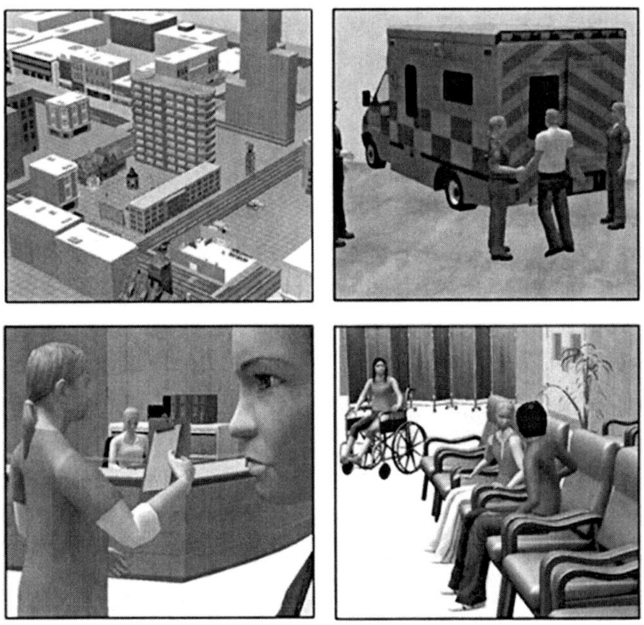

3D experience can get up and running very quickly. One of the major limitations with Sketchup for me though, is the fact that the animation capabilities are extremely limited. Organic shapes are also not one of its strengths. This led me to look in other directions for those missing features.

I looked at 3D Virtual environments such as Second Life and Google's Lively, but found things very difficult to control in terms of directing multiple characters to do specific things in a controlled and timed manner. The machinama (the use of real-time three-dimensional (3-D) graphics rendering engines to generate computer animation (Wikipedia) have done this with some success, but instead I turned to Pre-Viz (Pre-visualisation, mainly used in the film industry for visualising special effects sequences prior to filming. Usually generated in a 3-D computer environment). After trying a number of Pre-Viz applications I eventually settled with Antics 3D.

This was mainly to do with the fact that there was a free version available and also that it supported the import of Sketchup content. This meant that I had access to the Google 3D warehouse content and also to the power to tell stories with the kind of detail required by filmmakers. Antics 3D consists of a space where you can place objects, buildings, characters etc. There is a library of smart objects as well as characters. These smart objects each have a set of prebuilt animations built into them, and to some extent it is only a matter of clicking on characters and directing them to a particular object. They will then perform the actions themselves. These actions can be refined and multiple events orchestrated using a timeline. Cameras can be created, placed and also animated to replicate the look and feel of cinema productions.

Working in Antics 3D is a breath of fresh air, compared to traditional 3D animation, which is very time consuming and far more labour intensive. Antics 3D does unfortunately also have some severe limitations in terms of its characters. The free version of the software comes with a minimal set. One is also able to buy other characters or models, but none of these can be easily refined or adapted to a specific requirement. It would be great if one could easily personalise characters or even import ones created in other 3D applications.

## Theoretical Design Specification for an Interactive Video Simulation: CIPeL Case Study 2

Both video and 3D have their uses in the creation of learning objects. One is not necessarily inherently better than the other for representation of the physical world. Each project has to be evaluated individually and the most suitable tools should be selected appropriately.

I was asked to create a representation of an environment in which a nurse is presented with a new patient, whom she is required to assess and then treat. If the nurse follows the correct procedures, she will quickly realise that her patient needs to be treated for a particular condition. It however, the nurse follows incorrect procedures, or If a misdiagnosis of the patient's condition is made by the nurse and she then proceeds with an incorrect treatment, the patient could fall ill and ultimately die. I was very keen on retaining the realism of this situation, and also wanted to give the user a feeling of dealing with a real patient.

Although it is possible to create photorealistic 3D imagery and animation, this involves huge amounts of work and processing power. A more economical approach is to create a representation using video instead of 3D might be the best approach. This kind of situation can easily filmed as it requires very few actors and also, the whole scene takes place in a single environment that is easy to control in terms of lighting or accidental interruptions.

However, we can do more than just a linear representation with video. I drew some inspiration from www.subservientchicken.com, which was created for the fast food restaurant chain, Burger King by advertising and design agency "Crispin

Porter + Bogusky" and launched on April 8, 2004. It has achieved legendary status with 442 million hits in its first year. The website consists of what appears to be a live webcam image of a man dressed in a chicken suit. If you ask him to do something, by typing your request into a chat window, he usually will. At first glance, this feels fairly straightforward, but when one thinks about it, one realises that it is impossible for this one man to be obliging your every request. Are you the only person in the world asking him to jump up and down on the settee?

This website is in fact an interactive video simulation which is composed of many separate video clips of the chicken man character, performing all sorts of strange requests. At the end of each video clip, the chicken man always goes to stand back in the same place. This allows for the seamless cutting between the various video clips. Each separate video clip has a set of keyword metadata attached to it. Text requests by users are analysed by a computer program and then a video clip containing metadata most closely matching the text request is played. There is also an in-between video clip, looping as the waits for you to make your request.

In order to build this simulation, it is necessary to create a map of all the possible choices a user could have at each of the various stages of the simulation. Then each of these will have to be filmed in such a way as to allow the subject always to start and end from the same position on screen. Finally the video clips will be tagged with the appropriate metadata, and stored in a searchable database. Also, some form of error correction will be required for when the computer is unable to understand the user request very clearly, of if the user is asking the simulation to do something completely unanticipated. The overall effect hopefully is that of a user watching an actual patient being treated by a nurse. The onscreen nurse will follow the user's instructions completely until these lead to their eventual conclusion.

## CONCLUSION

Provide discussion of the overall coverage of the chapter and concluding remarks. The purpose of education is to enhance our understanding of the world we inhabit and hopefully to empower us, to make it a better place to live in. The relationship between education and the outside world, is a multifaceted one. On the one hand, education is the process whereby people are trained in the various skills required by industry and society. On the other hand, education is fed by the ideas that come from industry and society. It goes to follow then that as industry and society change, so does education, and visa versa. The two are mutually dependent.

It can be clearly seen that the e-learning approach to education, in particular Virtual learning Environments (VLEs), was born out of the development of the World Wide Web. Now, as the web has slowly mutated from a primarily information based medium into a more social environment, educational establishments are also shifting their attention to the new educational possibilities afforded by these technologies. The use of discussion forums, Voice over Internet protocol (VoIP), web 2.0 and social networking in education are now becoming the established way of doing things.

An interesting phenomenon is the fact that information has now become a commodity which is freely available and inexhaustible due to the international nature of the Internet. It is worth bearing in mind, that information is the primary commodity which most education establishments are selling (the other being tutors' advice and mentoring). So, why pay for something, when you can get it for free? Is it because free is inherently inferior to "paid for"? The Open Source software movement seems to be deflating that myth.

The world around us is changing at a rapid pace and we ignore these changes at our own peril. Adaption is the key to survival. It is not a formula taught verbatim, from a book. Adaption requires a leap of imagination. Without this leap, everything eventually becomes repetitive and stagnant. It is essential for the education process to nurture and ultimately facilitate this "Great Leap". If that does not occur, our education establishments themselves will eventually become irrelevant.

True education is a process, which leads to a state where the ability to think for oneself is attained and from here, the possibility of taking charge of one's own educational pathway. In this model, education is no longer the process of being fed information, but rather a process of enquiry, exploration, discovery, expression and re-interpretation of the world around us on our own terms.

So, what does the future hold for education? No one can say for certain, but the possibilities certainly look very exciting.

## REFERENCES


Bromage, A. (2004). Atavistic Avatars: Ontology, Education and 'Virtual Worlds. In Seth-Preston, D. (Ed.), *Virtual Learning and Higher Education* (pp. 133–150). Amsterdam: Rodopi.

Dearing, R. (1997) *Higher Education in the Learning Society: Report of the National Committee*. London: Her Majesty's Stationery Office. Retrieved September 11, 2009, from https://bei.leeds.ac.uk/Partners/NCIHE/

Fleming, N. D. (2001). *Teaching and Learning Styles: VARK Strategies*. Christchurch, New Zealand: Neil D. Fleming.

Klopfer, E. (2008). *Augmented Learning* (pp. 87–107). Cambridge, MA: The MIT Press.

Littlejohn, A., & Pegler, C. (2007). *Preparing for Blended e-Learning*. London: Routledge.

Nelson-Kilger, M. (1993). *The Digital individual.* The Third Conference on Computers Freedom and Privacy, March 9-12, 1993, Burlingame, CA. Retrieved May 14, 2009, from http://cpsr.org/prevsite/conferences/cfp93/nelson-kilger.html

O'Reilly, T. (2005). *What Is Web 2.0?* Retrieved May 14, 2009, from http://www.oreillynet.com/pub/a/oreilly/tim/news/2005/09/30/what-is-web-20.html?page=1

Singh-Cooner, T. (2010). Designing for enquiry: using Web 2.0 to enable mental health service user and carer involvement. In Bromage, A., Clouder, D., Thistlethwaite, J., & Gordon, F. (Eds.), *Interprofessional E-learning and Collaborative Work: Practices and Technologies*. Hershey, PA: IGI-Global.

# Compilation of References


Abbott, P., & Sapsford, R. (1998). *Research Methods for Nurses and the Caring Professions* (2nd ed.). Buckingham, UK: Open University Press.

Adams, A. (2005). Theorising Inter-professionalism. In The Higher Education Academy, Health Science and Practice Network, The Theory-Practice Relationship in Interprofessional Education. (pp. 31-37). London: Higher Education Academy.

Alavi, C. (Ed.). (1995). *Problem-based learning in a health science curriculum*. London: Routledge. doi:10.4324/9780203428023

Albion, P. R. (1999). *Heuristic evaluation of educational multimedia: from theory to practice. 16th Annual conference of the Australasian Society for Computers in Learning in Tertiary Education*. ASCILITE.

Alexander, S. (2001). E-learning developments and experiences. *Education + Training, 43*(4/5), 240-248.

Allen, D., & Hughes, D. (2002). *Nursing and the Division of Labour in Healthcare*. London: Palgrave MacMillan.

Allport, G. W., Vernon, P. G., & Lindzey, G. (1951). *Study of Values*. Boston: Houghton-Mifflin.

Aminzade, R., & Pescosolido, B. A. (1999). Introduction to the changing landscape of higher education . In Pescosolido, B. A., & Aminzade, R. (Eds.), *The social worlds of higher education: Handbook for teaching in a new century* (pp. 6–8). Thousand Oaks, CA: Pine Forge Press.

Anders, R., & Fechtner, D. (1992). *Universal Design*. Brooklyn, NY: Pratt Institute Department of Industrial Design and Pratt Center for Advanced Design Research.

Anderson, J. R. (1985). *Cognitive Psychology and its Implications*. Boston: Harvard University Press.

Anderson, T., Rourke, L., Garrison, D. R., & Archer, W. (2001). Assessing teaching presence in a computer conferencing context. *Journal of Asynchronous Learning Networks, 5*(2).

Andrews, J., & Derbyshire, A. (1993). *Crossing Boundaries*. London: Construction Industry Council.

Andrews, M., & Chilton, F. (2000). Student and mentor perceptions of mentoring effectiveness. *Nurse Education Today, 20*, 555–562. doi:10.1054/nedt.2000.0464

AODP. (2006). *Qualifications Framework for Mentors Supporting Learners in Practice: Standards and Guidance for Mentors and Practice Placements in Support of Pre-registration Diploma of Higher Education in Operating Department Practice Provision*. Cheshire, UK: Association of Operating Department Practitioners.

Arbaugh, J. B., & Hwang, A. (2006). Does "teaching presence" exist in online MBA courses? *The Internet and Higher Education, 9*(1), 9–21. doi:10.1016/j.iheduc.2005.12.001

Argyris, C., & Schön, D. (1974). *Theory and practice: increasing professional effectiveness*. San Francisco: Jossey-Bass.

Argyris, C., & Schön, D. (1978). *Organizational learning: a theory of action perspective*. Reading, MA: Addison-Wesley.

Armitage, H., Pitt, R., & Jinks, A. (2009). Initial findings from TULIP (Trent Universities Interprofessional learning in practice) Project. *Journal of Interprofessional Care, 2391*, 101–103. doi:10.1080/13561820802379938

Ashley, K., Davis, R., & Pinsent, E. (2008). *Significant properties of e-learning objects (SPeLOs)*. Draft Version, 0.92. A Report commissioned by the JISC Information Environment Committee under the Information Environment Strategic Theme. Retrieved February 16, 2009, from http://www.jisc.ac.uk/media/documents/programmes/preservation/spelos_report.pdf

Aston, L., & Molassiotis, A. (2003). Supervising and supporting student nurses in clinical placements: the peer support initiative. *Nurse Education Today*, *3*(3), 202–210. doi:10.1016/S0260-6917(02)00215-0

Atack, L. (2003). Becoming a Web-Based Learner: Registered Nurses' Experience. *Journal of Advanced Nursing*, *44*, 289–297. doi:10.1046/j.1365-2648.2003.02804.x

Atack, L., & Rankin, J. (2002). A descriptive study of registered nurses' experiences with web-based learning. *Journal of Advanced Nursing*, *40*(4), 457–465. doi:10.1046/j.1365-2648.2002.02394.x

Atkinson, R., & Flint, J. (2001). Accessing hidden and hard to reach populations: snowball research strategies. *Social Research update, 33*. Retrieved September 7, 2009, from http://www.soc.surrey.ac.uk/sru/SRU.HTML

Attwell, G. (2007). *Web 2.0 and the changing ways we are using computers for learning: what are the implications for pedagogy and curriculum?* Retrieved May 23, 2009, from http://www.e-learningeuropa.info/files/media/media13018.pdf

Atwal, A., & Caldwell, K. (2002). Do multidisciplinary integrated pathways improve inter-professional collaboration? *Scandinavian Journal of Caring Sciences*, 360–367. doi:10.1046/j.1471-6712.2002.00101.x

Australian Council of Professions. (2007). *About Professions Australia*. Retrieved April 15, 2008 from www.professions.com.au/define profession.html

Avery, A., & Pringle, M. (2005). Extended prescribing by UK nurses and pharmacists. *British Medical Journal*, *331*, 1154–1155. doi:10.1136/bmj.331.7526.1154

Ayliffe, C. A. J., Lowbury, E. J. L., Geddes, A. M., & Williams, J. D. (1992). *Control of Hospital Infection: A practical handbook* (3rd ed.). London: Chapman, and Hall Medical.

Babu, K. S., Law-Min, R., & Adlam, T., V. B. (2008). Involving service users and carers in psychiatric education: what do trainees think? *Psychiatric Bulletin*, *32*(1), 28–31. doi:10.1192/pb.bp.107.015354

Baikie, G., & Campbell, C. (2008a). *Conceptualizing a Critical Diversity Model*. Unpublished manuscript.

Baikie, G., & Campbell, C. (2008b). *Diversity awareness in Interprofessional Teams: A module prepared for the Tri-Faculty Interprofessional Academic Advisory Council*. Halifax, Canada: Dalhousie University.

Baikie, G., & Campbell, C. (2009). *Critical Diversity within Interprofessional Education: Constructing Professional Cultures of Belonging*. Collaborating Across Borders II Conference, May 2009, Halifax, Nova Scotia.

Bailey, N. (2005). Professionalism and skills enhancement in the built environment: The role of higher education in a changing policy context in England. *CEBE Transactions*, *2*(2), 48–62.

Baldwin, D. C. Jr, & Baldwin, M. A. (1979). Interdisciplinary education and health team training: A model for learning and service . In Hunt, A. D., & Weeks, L. E. (Eds.), *Medical education since 1960: Marching to a different drummer* (pp. 190–221). East Lansing, MI: Michigan State University Foundation.

Bandura, A. (1986). *Social foundations of thought and action: a social cognitive theory*. Englewood Cliffs, NJ: Prentice-Hall.

Banks, S., & Janke, K. (1998). Developing and implementing interprofessional learning in a Faculty of Health Professions. *Journal of Allied Health*, *27*, 132–136.

Banning, M. (2004). Nurse prescribing, nurse education and related research in the United Kingdom: a review of the literature. *Nurse Education Today*, *24*, 420–427. doi:10.1016/j.nedt.2004.05.002

Barberra, E. (2004). Quality in Virtual Learning Environment. *British Journal of Educational Technology*, *33*(1), 13–20. doi:10.1111/j.1467-8535.2004.00364.x

Barker, A. (2004). Using online learning technology to develop nurse leaders. *Nurse Leader*, *2*(5), 32–35. doi:10.1016/j.mnl.2004.07.011

Barker, E. (2005). Mentoring—A complex relationship . *Journal of the American Academy of Nurse Practitioners, 18*(2), 56–61. doi:10.1111/j.1745-7599.2006.00102.x

Barker, K. K., Bosco, C., & Oandasan, I. F. (2005). Factors in implementing interprofessional education and collaborative practice initiatives: Findings from key informant interviews. *Journal of Interprofessional Care, 19*(S1), 166–176. doi:10.1080/13561820500082974

Barnes, D., Carpenter, J., & Bailey, D. (2000). Partnerships with service users in interprofessional education for community mental health: a case study. *Journal of Interprofessional Care, 14*(2), 189–200. doi:10.1080/713678554

Barnes, D., Carpenter, J., & Dickinson, C. (2000). Interprofessional education for community mental health: attitudes to community care and professional stereotypes. *Social Work Education, 19*(6), 565–583. doi:10.1080/02615470020002308

Barnes, L. B., Christensen, C. R., & Hansen, A. J. (1994). *Teaching and the case method: Text, cases, and readings* (3rd ed.). Boston: Harvard Business School Publishing Division.

Barnett, R., & Coate, K. (2005). *Engaging the Curriculum in Higher Education.* Buckingham, UK: SRHE and Open University Press.

Barr, H. (1998). Competent to collaborate: towards a competency-based model for interprofessional education. *Journal of Interprofessional Care, 12*(2). doi:10.3109/13561829809014104

Barr, H. (2000). *Interprofessional Education: 1997-2000. A Review* (5:03, p. 9). United Kingdom Central Council of Nursing, Midwifery and Health Visiting.

Barr, H. (2000). Working together to learn together: learning together to work together. *Journal of Interprofessional Care, 14*(2), 177–179.

Barr, H. (2002). *Interprofessional education: today, yesterday and tomorrow.* Occasional Paper No1. London: Centre for Health Sciences and Practice, LTSN/CAIPE.

Barr, H. (2005). Learning together . In Meads, G., & Ashcroft, J. (Eds.), *The Case for Interprofessional Education in Health and Social Care*. Oxford, UK: Blackwell. doi:10.1002/9780470776445

Barr, H. (2007). *Interprofessional Education in the United Kingdom 1966-1997*. York, UK: Higher Education Academy.

Barr, H. (2009). Interprofessional education as an emerging concept . In Bluteau, P. A. S., & Jackson, J. A. (Eds.), *Interprofessional Education; making it happen.* London: Palgrave.

Barr, H., & Ross, F. (2006). Mainstreaming interprofessional education in the United Kingdom: A position paper. *Journal of Interprofessional Care, 20*(2), 96–104. doi:10.1080/13561820600649771

Barr, H., Koppel, I., & Reeves, S. (2005). *Effective interprofessional education: argument, assumption and evidence*. Oxford, UK: Blackwell. doi:10.1002/9780470776445

Barr, H., Koppel, I., Reeves, S., Hammick, M., & Freeth, D. (2005). *Effective Interprofessional Education: arguments, assumption & evidence*. Oxford, UK: Blackwell. doi:10.1002/9780470776445

Barrett, G., Greenwood, R., & Ross, K. (2003). Integrating interprofessional education into 10 health and social care programmes. *Journal of Interprofessional Care, 17*(3), 293–301. doi:10.1080/1356182031000122915

Barton, K. McKellar, P. & Maharg, P. (2005). *Authenticity in learning: transactional learning in virtual communities*. Draft paper submitted to the sixth International Clinical Conference. Retrieved August 7, 2009, from http://www.uwex.edu/disted/conference/Resource_library/handouts/03Info_H03.pdf

Basset, T., Campbell, P., & Anderson, J. (2006). Service User/Survivor Involvement in Mental Health Training and Education: *Overcoming the Barriers. Social Work Education, 25*(4), 393–402. doi:10.1080/02615470600593675

Becher, T., & Trowler, P. (2001). Academic Tribes and Territories, (2nd Ed.). Milton Keynes, UK: SRHE/Open University.

Beck, J., & Owens, M. (2007). Monitoring Student's Level's of Online Engagement in a Pre-registration, Interprofessional Education Module. *Proceedings of the EIPEN Conference*, September 2007, Poland.

Becker, B., & Mark, G. (2002). Social Conventions in Computer-mediated Communication: A Comparison of Three Online Shared Virtual Environments . In Schroeder, R. (Ed.), *The Social Life of Avatars* (pp. 19–39). London: Springer-Verlag.

Becta. (2007). *Virtual learning Environments (VLE's).* Retrieved July 2, 2009, from http://partners.becta.org.uk/index.php?section=rh&rid=13640

Beetham, H. (2007). An approach to learning activity design . In Beetham, H., & Sharpe, R. (Eds.), *Rethinking pedagogy for a digital age: designing and delivering e-learning* (pp. 26–40). Oxford, UK: Routledge.

Beetham, H., & Sharpe, R. (Eds.). (2007). *Rethinking pedagogy for a digital age*. London: Routledge.

Benedek, J., & Miner, T. (2002). Measuring Desirability: New methods for evaluating desirability in a usability lab setting. In Proceedings of Usability Professionals Association 2002, Orlando, July 8-12.

Benfield, G., Ramanau, R., & Sharpe, R. (2009). Student learning technology use: preferences for study and contact. *Brookes e-Journal of Learning and Teaching, 2*(4). Retrieved August 12, 2009, from http://bejlt.brookes.ac.uk/evaluating_learners_experiences_of_e-learning/

Benner, P. (1984). *From Novice to Expert: Excellence and Power in Clinical Nursing Practice*. Reading, MA: Addison-Wesley.

Bennett, J., & Jayes, S. (1998). *The Seven Pillars of Partnering*. London: Thomas Telford.

Bennett, S., Maton, K., & Kervin, L. (2008). The 'digital natives' debate: A critical review of the evidence. *British Journal of Educational Technology, 39*(5), 775–786. doi:10.1111/j.1467-8535.2007.00793.x

Bentley, G. W., Cook, P. P., Davis, K., Murphy, M. J., & Berding, C. B. (2003). RN to BSN program: transition from traditional to online delivery. *Nurse Educator, 28*(3), 121–126. doi:10.1097/00006223-200305000-00005

Bereiter, C., & Scardamalia, M. (1993). *Surpassing ourselves: An inquiry into the nature and implications of expertise*. Chicago: Open Court.

Berg, B., & Östergren, B. (1977). *Innovations and innovation processes in higher education*. Stockholm: UHÅ.

Berg, B., & Östergren, B. (1979). Innovation processes in higher education. *Studies in Higher Education, 4*(2), 261–268. doi:10.1080/03075077912331377027

Berman, N., Fall, L., Maloney, C., & Levine, D. (2008). Computer-Assisted Instruction in Clinical Education: a Roadmap to Increasing CAI Implementation. *Advances in Health Sciences Education : Theory and Practice, 13*(3), 373–383. doi:10.1007/s10459-006-9041-3

Bielaczyc, K., & Collins, A. (1999). Learning communities in classrooms: a reconceptualisation of educational practice . In Reigeluth, C. (Ed.), *Instructional design theories and models. A new paradigm of instructional theory* (*Vol. 2*, pp. 269–293). Mahwah, NJ: Lawrence Erlbaum Associates.

Biggs, J. (2003). *Teaching for Quality Learning at University* (2nd ed.). Berkshire, UK: Open University Press.

Bignell, S. (2007). "Measuring" the impact of second life for educational purposes. In J. Kirriemuir, *Responses and second life meeting transcript*. Retrieved September 11, 2009, from http://eduserv.org.uk/foundation

Bilinski, H. (2002). *Bates Guide to Physical Examination and History Taking* (8th ed.). Philadelphia: Lippincott.

Bines, H. (1992). Interprofessionalism. In H. Bines & D. Watson (Eds.), Developing Professional Education. (pp. 136-142). Milton Keynes, UK: Open University.

Biocca, F. (1997). The Cyborg's Dilemma: Progressive Embodiment in Virtual Environments. *Journal of Computer-Mediated Communication, 3*(2).

Bluteau, P. A. S., & Jackson, J. A. (2009). An e-learning model of interprofessional education . In Bluteau, P. A. S., & Jackson, J. A. (Eds.), *Interprofessional Education; making it happen*. London: Palgrave.

Bluteau, P. A. S., & Jackson, J. A. (2009). Interprofessional education unpacking the early challenges . In Bluteau, P. A. S., & Jackson, J. A. (Eds.), *Interprofessional Education; making it happen*. London: Palgrave.

Bluteau, P. A. S., & Jackson, J. A. (Eds.). (2009). *Interprofessional Education; making it happen*. London: Palgrave.

Boler, M., & Zembylas, M. (2003). Discomforting truths: The emotional terrain of understanding difference . In Trifonas, P. P. (Ed.), *Pedagogies of difference* (pp. 110–134). New York: RoutledgeFalmer.

Bond, M. (1997). A learning team in the making. *Journal of Interprofessional Care, 11*, 89–98. doi:10.3109/13561829709040247

Boud, D., Cohen, R., & Walker, D. (1993). *Using Experience for Learning*. Buckingham, UK: Society for Research into Higher Education and Open University Press.

Bowman, D. A. (2002). Principles for the Design of Performance-oriented Interaction Techniques . In Stanney, K. M. (Ed.), *Handbook of Virtual Environments; Design Implementation and Applications* (pp. 277–300). Mahwah, NJ: Lawrence Erlbaum Associates.

Boyce, B. A. (1992). Making the case for the case-method approach in physical education pedagogy classes. *Journal of Physical Education, Recreation & Dance, 63*(6), 17–20.

Boyle, T., Cook, J., Windle, R., Wharrad, H. J., Leeder, D., & Alton, R. (2006). An Agile Method for Developing Learning Objects. In L. Markauskaite, P. Goodyear & P. Reimann. (Eds.), *23rd annual conference of the Australasian Society for Computers in Learning in Tertiary Education: Who's learning, whose technology?* Retrieved April, 30, 2009, from http://papaya.edfac.usyd.edu.au/ascilite_papers/p64.pdf

Brabazon, T. (2008). *The University of Google: Education in the (Post) Information Age*. Aldershot, UK: Ashgate Publishing Limited.

Bradbury, H., Frost, N., Kilminster, S., & Zukas, M. (2009). *Beyond Reflective Practice: New Approaches to Professional Learning*. London: Routledge.

Bradley, E., Bradshaw, C., & Nolan, P. (2006). Nurse lecturers' observations on aspects of nurse prescribing training. *Nurse Education Today, 26*, 538–544. doi:10.1016/j.nedt.2006.01.008

Bradwell, P. (2009). *The Edgeless University Why Higher Education Must Embrace Technology*. London: DEMOS.

Brady, D. (1997). Further and higher education partnerships: the future for collaboration. *Journal of Vocational Education and Training, 49*(1), 167–175. doi:10.1080/13636829700200292

Bray, J. & Howkins, E. (2006). Facilitating interprofessional learning in the workplace: a research project using the Delphi technique. *Work based Learning in Primary Care*, (4), 223-235.

Brewer, M. B. (1979). In-Group Bias in the Minimal Intergroup Situation: A Cognitive-Motivational Analysis. *Psychological Bulletin, 86*(2), 307–324. doi:10.1037/0033-2909.86.2.307

Bristol Inquiry. (2001). *Final Report of the Bristol Royal Infirmary Inquiry*. Retrieved April 15, 2008 from www.bristol-inquiry.org.uk/final_report/report/index.htm

Bristol Royal Infirmary Inquiry. (2001). *Learning from Bristol: the report of the public inquiry into children's heart surgery at the Bristol Royal Infirmary 1984–1995*. Norwich, UK: The Stationery Office.

Britt, A. (2008). On Language: Avatar. *New York Times Magazine*, August 8th, 2008. Retrieved August 30, 2008, from http://www.nytimes.com/2008/08/10/magazine/10wwln-guest-t.html

Bromage, A. (2004). Atavistic Avatars: Ontology, Education and 'Virtual Worlds . In Seth-Preston, D. (Ed.), *Virtual Learning and Higher Education* (pp. 133–150). Amsterdam: Rodopi.

Bromage, A. (2006). The management of planned change: an interdisciplinary perspective . In Hunt, L., Bromage, A., & Tomkinson, B. (Eds.), *The Realities of Change in Higher Education: Interventions to Promote Learning and Teaching* (pp. 1–14). London: Routledge.

Brosnan, K. (2005). *Developing and sustaining a national learning object sharing network:A social capital theory perspective*. Retrieved October 28, 2008, from http://www.ascilite.org.au/conferences/brisbane05/blogs/proceedings/12_Brosnan.pdf

Brown, J., Collins, A., & Duguid, P. (1989). Situated Cognition and the Culture of Learning. *Educational Researcher, 18*(1), 32–42.

Bruner, J. S. (1966). *Towards a theory of instruction.* Cambridge, MA: Harvard University Press.

Bruner, J. S. (1990). *Acts of meaning.* Cambridge, MA: Harvard Univ. Press.

Bullock, A., & Trombley, S. (1999). *The New Fontana Dictionary of Modern Thought.* London: Harper-Collins.

Burton, R. (1992). *Steering Group on Architectural Education - Report and Recommendations.* London: Royal Institute of British Architects.

Butson, R. (2003). Learning objects: weapons of mass instruction. *British Journal of Educational Technology, 34,* 667–669. doi:10.1046/j.0007-1013.2003.00359.x

Cader, R., & McGovern, M. (2003). Introducing Blackboard: an electronic learning platform . *Nursing Times, 99*(32), 24–25.

CAIPE. (2002). *Defining IPE.* Retrieved July 26, 2009, from http://www.caipe.org.uk/about-us/defining-ipe/

CAIPE. (2002). Interprofessional Education- a Definition. CAIPE Bulletin. London: Centre for Advancement of Interprofessional Education (CAIPE).

Callon, M. (1986). Some elements of a sociology of translation: Domestication of the scallops and the fishermen of St Brieuc Bay . In Law, J. (Ed.), *Power, action and belief: A new sociology of knowledge?* (pp. 196–233). London: Routledge & Kegan Paul.

Camp, G. (1996). Problem-Based Learning: A paradigm shift or a passing fad? *Medical Education Online, 1*(2).

Campbell, M., Gibson, W., Hall, A., Richards, D., & Callery, P. (2005). Online vs. Face to Face Discussion in a Web-Based Research Methods Course for Postgraduate Nursing Students: A Quasi-Experimental Study. *International Journal of Nursing Studies, 45,* 750–759. doi:10.1016/j.ijnurstu.2006.12.011

Canning, J. (Ed.). (2007). *Disciplines in dialogue: disciplinary perspectives on interdisciplinary teaching and learning.* Southampton, UK: The University of Southampton Interdisciplinary Teaching & Learning Group.

Carlisle, C., Calman, L., & Ibbotson, T. (2009). Practice-based learning: The role of practice education facilitators in supporting mentors. *Nurse Education Today, 29*(7), 715–721. doi:10.1016/j.nedt.2009.02.018

Carpenter, J. (1995a). Doctors and nurses: stereotypes and stereotype change in interprofessional education. *Journal of Interprofessional Care, 9*(2), 151–161. doi:10.3109/13561829509047849

Carpenter, J. (1995b). Interprofessional Education for medical and nursing students: evaluation of a programme. *Medical Education, 29,* 265–275. doi:10.1111/j.1365-2923.1995.tb02847.x

Carr, D. (2008). Learning and virtual worlds. In Teaching and Learning Research Programme (TLRP) Education 2.0? Designing the web for teaching and learning. ESRC/ESPRC, (pp. 13-16).

Carr, D., Oliver, M. & Burn, A. (2007). *Learning from online worlds; Teaching in Second Life.* Edsurv project.

Carr, D., Oliver, M., & Burn, A. (2008). Learning, Teaching and Ambiguity in Virtual Worlds. In *Proceedings of the ReLIVE 08 conference,* November 20th and 21st, 2008, Open University, (pp. 83-93).

Carr, N. (2008). Is Google Making Us Stupid? *The Atlantic* (July/August 2008). Retrieved August 10, 2009, from http://www.theatlantic.com/doc/200807/google

Carter, L. M. (2008). A Study of Critical Thinking, Teacher- Student Interaction, and Discipline – Specific Writing in an Online Educational Setting for Registered Nurses. *Journal of Continuing Education in Nursing, 39*(3), 133–138. doi:10.3928/00220124-20080301-03

Center for Interprofessional e-learning (CIPEL). (2009). *Learning Objects.* Retrieved August 10, 32009, from http://www.cipel.ac.uk/learning_objects/learning_objects.htm.

Center for Technology in Learning. (2009). *Evaluation of Evidence-Based Practices in Online Learning: A Meta-Analysis and Review of Online Learning Studies.* Washington, DC: U.S. Department of Education.

Center for Activity Theory and Developmental Work Research. (2004). *Cultural-Historical Activity Theory*. Retrieved February 6, 2009, from http://www.edu.helsinki.fi/activity/pages/chatanddwr/chat/

Centre for the Advancement of Inter-professional Education (CAIPE). (2002). *Definition of Inter-professional Education*. Retrieved August 10, 2009, from http://www.caipe.org.uk

Centre for the Advancement of Interprofessional Education (CAIPE). (1997). Retrieved August 31, 2008, from http://www.caipe.org.uk/

Chambers, L. W., Conklin, J., Dalziel, W. B., MacDonald, C. J., & Stodel, E. J. (2008). E-learning education to promote interprofessional education with physicians, pharmacists, nurses and nurse practitioners in LTC facilities: promising potential solutions. *International Journal of Biomedical Engineering and Technology*, *1*, 233–249. doi:10.1504/IJBET.2008.016958

Chandler, J., Crellin, J., Collinson, T., & Duke-Williams, E. (2007). *Second Life Toolkits*. Higher Education Academy Development Fund Project. Retrieved September 11, 2009, from http://www.ic.heacademy.ac.uk?projects/development-fund/fund_details.php?id=106 on 01.08.08

Chapman, L. (2000). Distance learning for post-registered nursing: the facts . *Nursing Standard*, *14*(18), 33–36.

Chen, N., Wei, C., Uden, L., & Wu, K. (2008). *Effects of Reflective Teaching Strategies on Online Learners' Reflection Levels*. Eighth IEEE International Conference on Advanced Learning Technologies.

Cheng, L., Farnham, S., & Stone, L. (2002). Lessons Learned: Building and Deploying Shared Virtual Environments . In Schroeder, R. (Ed.), *The Social Life of Avatars* (pp. 90–112). London: Springer-Verlag.

Cheng, N. Y. (2003). Approaches to design collaboration research. *Automation in Construction*, *12*, 715–723. doi:10.1016/S0926-5805(03)00059-1

Chettiparamb, A. (2007). *Interdisciplinarity: a literature review*. Southampton, UK: The University of Southampton Interdisciplinary Teaching & Learning Group.

Chew, F. S., Stiles, R. S., Joseph, E., & Whitley, M. D. (1994). Computer-assisted instruction with interactive videodisc versus textbook for teaching radiology. *Academic Radiology*, *1*, 326–331.

Childs, M. (2007). Real Learning in Virtual Worlds. *Warwick Interactions Journal*, *30*(2).

Childs, S., Blenkinsopp, E., Hall, A., & Walton, G. (2005). Effective e-learning for health professionals and students – barriers and their solutions. A systematic review of the literature – findings from the HeXL project. *Health Information and Libraries Journal*, *22*(Suppl 2), 20–32. doi:10.1111/j.1470-3327.2005.00614.x

Choi, I., & Lee, K. (2009). Designing and implementing a case-based learning environment for enhancing ill-structured problem solving: Classroom management problems for prospective teachers. *Educational Technology Research and Development*, *57*(1), 99–129. doi:10.1007/s11423-008-9089-2

Chow, F. L. W., & Suen, L. K. P. (2001). Clinical staff as mentors in pre-registration undergraduate nursing education: students' perceptions of the mentors' roles and responsibilities. *Nurse Education Today*, *21*, 350–358. doi:10.1054/nedt.2001.0569

Chumley-Jones, H. A., Dobbie, A., & Alford, C. L. (2002)... *Academic Medicine*, *77*(10), S86–S93.

CICLE. (2009). *Higher Education in a Web 2.0 World. Committee of Inquiry into the Changing Learner Experience*. Retrieved May 18, 2009, from http://www.jisc.ac.uk/media/documents/publications/heweb20rptv1.pdf

Clark, P. G. (2002). Values and voices in teaching gerontology and geriatrics: Case studies as stories. *The Gerontologist*, *43*(3), 297–303.

Clarke, A. (2004). E-learning 'close to being a core skill in the 21st century'. Adults Learning, September 12-13. In G. Salmon (Ed.), E-moderating, the key to online teaching and learning. London: Routledge Falmer.

Clarke, A. C. (1982). *Profiles of the future: an inquiry into the limits of the possible* (2nd rev. ed.). London: Victor Gollancz.

Clarke, B.,A., Miers, M., Pollard, K. & Thomas, J. (2007). Complexities of learning together; students' experience of face to face interprofessional groups. *Learning in Health and Social Care, 6*(4), 201–212. doi:10.1111/j.1473-6861.2007.00162.x

Clarke, E. (2008). *Use and reuse of Inter-professional Learning objects in Second Life*. Paper presented at the 3rd Plymouth e-Learning Conference, Plymouth, 4th April, 2008.

Cleak, H., & Williamson, D. (2007). Preparing health science students for interdisciplinary professional practice. *Journal of Allied Health, 36*(3), 141–149.

Clelford, T., & Hopkins, A. (2009). *Working Together, Learning Together: Methods & Process in Multidisciplinary Group Projects*. Retrieved March 24, 2009, from http://www.cebe.heacademy.ac.uk/learning/casestudies/case_pdf/clelford.pdf

Clouder, D. (2008). Technology-enhanced learning: conquering barriers to interprofessional education. *The Clinical Teacher, 5*, 198–202. doi:10.1111/j.1743-498X.2008.00243.x

Collier, A., Bacon, J., Burns, D., & Muir, T. (1991). *Interdisciplinary studies for the Built Environment*. London: Council for National Academic Awards.

Collins, A., Brown, J. S., & Newman, S. E. (1989). Cognitive apprenticeship: teaching the craft of reading, writing and mathematics . In Resnick, L. B. (Ed.), *Knowing and learning: essays in honour of Robert Glaser* (pp. 453–494). Hillsdale, NJ: Erlbaum.

Collison, G., Elbaum, B., Haavind, S., & Tinker, R. (2000). *Facilitating Online Learning Effective Strategies for Moderators*. Madison, WI: Atwood Publishing.

Colyer, H., Helme, M., & Jones, I. (2005). Pragmatic approaches and the theory-practice relationship. In *Interprofessional Education*. Higher Education Academy Health Sciences and Practice Occasional Paper no. 7.

Combined Universities Interprofessional Learning Unit. (2004). *Interprofessional Capability Framework*. Retrieved August 10, 2009, from http://www.sheffield.ac.uk/cuilu

Comerford, S. A. (2003). Enriching classroom learning about diversity: Supports and strategies from a qualitative study. *Journal of Teaching in Social Work, 23*(3/4), 159–183.

Conkin, M. S. (2007). *101 uses for Second Life in the College Classroom* (Version 2.0). Retrieved September 11, 2009, from http://trumpy.cs.elon.edu

Connolly, M., Jones, C., & Jones, N. (2007). Managing collaboration across further and higher education: a case in practice. *Journal of Further and Higher Education, 31*(2), 159–169. doi:10.1080/03098770701267630

Connor, C. (2003). *Interprofessional Computer Assisted Learning*. Retrieved July 6, 2009, from http://www.health.heacademy.ac.uk/projects/miniprojects/connor.pdf

Conole, G. White, S. & Oliver, M. (2006). The impact of e-learning on organisational roles and structures. In G. Conole & M. Oliver, (Eds.), Contemporary perspectives in e-learning research: themes, methods and impact on practice. (Open and distance learning series). Oxford, UK: Routledge Falmer.

Conrad, D. (2005). Building and maintaining community in cohort-based online learning. *Journal of Distance Education, 20*(1), 1–20.

Construction Industry Board. (1996). *Educating the professional team*. London: Thomas Telford.

Cook, S. L. (1997). *Mindfulness and tolerance of ambiguity in online communication*. Unpublished PhD Thesis, University of Denver, Denver, CO.

Cooper, H., & Spencer-Dawe, E. (2006). Involving service users in interprofessional education: narrowing the gap between theory and practice. *Journal of Interprofessional Care, 20*(6), 603–617. doi:10.1080/13561820601029767

Cooper, H., Carlisle, C., Gibbs, T., & Watkins, C. (2001). Developing an evidence base for interdisciplinary learning: a systematic review. *Journal of Advanced Nursing, 35*(2), 228–237. doi:10.1046/j.1365-2648.2001.01840.x

Cooper, H., Spencer-Dawe, E., & Mclean, E. (2005). Beginning the process of teamwork: Design, implementation and evaluation of an inter-professional education intervention for first year undergraduate students. *Journal of Interprofessional Care, 19*(5), 492-508. Retrieved May 5, 2009, from http://www.informaworld.com/smpp/title~content=t713431856~db=all~tab=issueslist~branches=19 - v19

Cooper, H., Spencer-Dawe, E., & Mclean, E. (2006). Beginning the process of teamwork: design, implementation and evaluation of an inter-professional education intervention for first year undergraduate students. *Journal of Interprofessional Care, 19*(5), 492–508. doi:10.1080/13561820500215160

Cooper, I. (2002). Transgressing discipline boundaries: Is BEQUEST an example of the 'new production of knowledge'? *Building Research and Information, 30*(2), 116–129. doi:10.1080/096132102753436495

Cooperrider, D. L., & Srivastva, S. (1987). Appreciative Inquiry in Organizational Life. *Organizational Change and Development, 1*, 29–169.

Cope, P., Cuthbertson, P., & Stoddart, B. (2000). Situated learning in the practice placement. *Journal of Advanced Nursing, 31*(4), 850–856. doi:10.1046/j.1365-2648.2000.01343.x

Cornish, P., Church, E., Callanan, T., Bethune, C., & Curran, V. (2004). *From Multidisciplinary to Interdisciplinary: The Evolution of Shared Training, Research and Mental Health Care at Memorial University*. Paper presented at Shared Mental Health Care Conference, Vancouver, Canada.

Cornish, P., Church, E., Callanan, T., Bethune, C., Robbins, C., & Miller, R. (2003). Rural interdisciplinary mental health team building via satellite: a demonstration project. *Telemedicine Journal and e-Health*, (9): 63–71. doi:10.1089/153056203763317666

Courtney, K. (2008). Exploring virtual opportunities to enhance and promote an emergent community of practice. *Electronic journal of e-Learning, 5*(4), 261-270. Retrieved October 10, 2009, from http://www.ejel.org/Volume-5/v5-i4/Courtney.pdf

Craddock, D., O'Halloran, C., Borthwick, A., & McPherson, K. (2006). Interprofessional education in health and social care: fashion or informed practice? [from http://www.blackwell-synergy.com/doi/pdf]. *Learning in Health and Social Care, 5*(4), 220–242. Retrieved September 11, 2009. doi:10.1111/j.1473-6861.2006.00135.x

Criswell, D. F., & Parchman, M. L. (2002). Handheld computer use in U.S. family practice residency programs. *Journal of the American Medical Informatics Association, 9*(1), 80–86.

Cuban, L. (1988). *Teachers & machines: The classroom use of technology since 1920*. New York: Teachers College Press.

Cuban, L. (2002). *Oversold and Underused: Computers in the Classroom*. Cambridge, MA: Harvard University Press.

CUILU. (2006). *The Combined Universities Interprofessional Learning Unit, final report*. Retrieved September 11, 2009, from http://www.sheffield.ac.uk/cuilu

Currier, S., Barton, J., O'Beirne, R., & Ryan, B. (2004). Quality assurance for digital learning object repositories: issues for the metadata creation process. *ALT-J: Research in Learning Technology, 12*(1), 5–20.

Curtis, D. D., & Lawson, M. J. (2001). Exploring collaborative online learning. *JALN, 5*(1), 21–34.

D'Amour, D., Ferrada-Videla, M., San Martin Rodriguez, L., & Beaulieu, M. D. (2005). The conceptual basis for interprofessional collaboration: Core concepts and theoretical frameworks. *Journal of Interprofessional Care*, (Supplement 1), 116–131. doi:10.1080/13561820500082529

Dahlgren, L. O. (2005). Learning Conceptions and Outcomes. In F. Marton, D. Hounsell & N. Entwistle (Eds.), The Experience of Learning: Implications for teaching and studying in higher education. (3rd (Internet ed.). Edinburgh, UK: University of Edinburgh, Centre for Teaching, Learning and Assessment.

Dall'Alba, G. (2009). Learning professional ways of being: ambiguities of becoming. *Educational Philosophy and Theory, 41*(1), 34–45. doi:10.1111/j.1469-5812.2008.00475.x

Dark, G. G., & Perrett, R. (2007). Using e-learning as a tool for multidisciplinary education. *European Journal of Cancer Care*, (16): 90–93. doi:10.1111/j.1365-2354.2007.00777.x

Darken, R. P., & Peterson, B. (2002). Spatial Orientation, Wayfinding and Representation . In Stanney, K. M. (Ed.), *Handbook of Virtual Environments; Design Implementation and Applications* (pp. 379–402). Mahwah, NJ: Lawrence Erlbaum Associates.

Davidson, A., & Courtney, K. (2004). *Using and reusing Electronic, Patient Case Studies to promote interprofessional learning.* Presented at Third Annual Web CT European Conference in Edinburgh.

Davis, D., Mazmanian, P. E., Fordis, M., Harrison, V., Thorpe, K. E., & Perrier, L. (2006). Accuracy of Physician Self-assessment Compared with Observed Measures of Competence. A Systematic Review. *Journal of the American Medical Association, 296*(9), 1094–1102. doi:10.1001/jama.296.9.1094

Davis, L., Little, M., & Thornton, W. (1977). The art and angst of the mentoring relationship. *Academic Psychiatry, 21*, 61–67.

De Groote, S. (2004). The use of personal digital assistants in the health sciences: results of a survey. *Medical Library association, 92*(3), 341-349.

De Salas, K., & Ellis, L. (2006). The development and implementation of learning objects in a higher education setting. *Interdisciplinary Journal of E-learning and Learning Objects*, (2), 1-22. Retrieved July 10, 2009, from http://ijklo.org/volume2.html

Dearing, R. (1997) *Higher Education in the Learning Society: Report of the National Committee*. London: Her Majesty's Stationery Office. Retrieved September 11, 2009, from https://bei.leeds.ac.uk/Partners/NCIHE/

Dearnley, C. A., & Walker, S. A. (2009). Mobile Enabled Research . In Vavoula, G. (Ed.), *Researching Mobile Learning: Frameworks, Methods and Research Designs*. Oxford, UK: Peter Lang.

Dearnley, C. A., Haigh, J., & Fairhall, J. (2008). Using Mobile Technologies for Assessment and Learning in Practice. *Nurse Education in Practice, 8*(3), 197–204. doi:10.1016/j.nepr.2007.07.003

Deepwell, F., & Malik, S. (2008). On campus, but out of class: an invesitgation into students' experiences of learning technologies in their self-directed study. *Research in Learning Technology, 16*(1), 4–14.

deLottinville, C. (2006). *Team Communication Observation Guide*. Adapted from Participants in the Bayer-Fetzer Conference on Physician-Patient Communication in Medical Education. (2001). Essential Elements of Communication in Medical Encounters: The Kalamazoo Consensus Statement. *Academic Medicine, 76*, 390–393.

Demiris, G. (2006). The diffusion of virtual communities in health care: concepts and challenges. *Patient Education and Counseling, 62*(2), 178–188. doi:10.1016/j.pec.2005.10.003

Dennis, A. R., Valacich, J. S., Speier, C., & Morris, M. G. (1998). *Beyond Media Richness: An Empirical Test of Media Synchronicity Theory*. Paper presented at the Thirty-First Annual Hawaii International Conference on System Sciences, Volume 1. Retrieved April 16, 2009, from csdl2.computer.org/comp/proceedings/hicss/1998/8233/01/82330048.pdf

Department for Education and Skills. (2003). *Towards a Unified e-Learning Strategy*. Consultation Document. Retrieved March 4, 2009, from http://publications.teachernet.gov.uk/eOrderingDownload/DfES-0455-2003.pdf

Department of Families. Education & Schools. (2003). *Towards a Unified e-Learning Strategy Consultation Document.* Retrieved October 1, 2008, from http://www.dcsf.gov.uk/consultations/downloadableDocs/towards%20a%20unified%20e-learning%20strategy.doc

Department of Families. Education & Schools. (2005). *Harnessing Technology – Transforming Learning and Children's Services.* Retrieved October 1, 2008, from http://www.dcsf.gov.uk/publications/e-strategy/

Department of Health (DH). (2000a). *The NHS plan: a plan for investment, a plan for reform*. London: HMSO.

Department of Health (DH). (2000b). *Meeting the challenge: a strategy for allied health professionals*. London: HMSO.

Department of Health (DH). (2001). *Working together, learning together: a framework for lifelong learning for the NHS*. London: HMSO.

Department of Health (DOH). (1991). *The patients charter*. London: HMSO.

Department of Health (DOH). (1994). *Working in partnership*. London: HMSO.

Department of Health (DOH). (1998). *Modernising mental health services*. London: HMSO.

Department of Health (DOH). (2000). *The NHS Plan*. London: The Stationery Office.

Department of Health (DOH). (2000a). *NHS plan*. London: HMSO.

Department of Health (DOH). (2000b). *Expert patient*. London: HMSO.

Department of Health (DOH). (2001). *Involving patients and the public*. London: HMSO.

Department of Health (DOH). (2002). *Requirements for social work training. Department of Health*. London: HMSO.

Department of Health (DOH). (2004). *NHS Improvement Plan: Putting people at the heart of public services*. London: The Stationery Office.

Department of Health. (2000). *A Health Service of all the Talents: report on the review of workforce planning*. London: Department of Health.

Department of Health. (2001a). *Working Together-Learning Together: A Framework for Lifelong Learning in the NHS*. London: The Stationery Office.

Department of Health. (2001b). *A Service of all the Talents: developing the NHS workforce*. London: The Stationery Office.

Department of Health. (2002). *Learning from Bristol: The DH Response to the Report of the Public Inquiry into Children's Health Surgery at the Bristol Royal Infirmary 1984-1995*. London: Department of Health.

Department of Health. (2002). *Liberating the Talents*. London: The Stationery Office.

Department of Health. (2002). *Shifting the Balance of the Balance of Power: Securing Delivery*. London: Department of Health.

Department of Health. (2003). *Audit of Victoria Climbie Inquiry recommendations. Social Services Inspectorate*. London: Department of Health.

Department of Health. (2003). *The Victoria Climbie Inquiry: the Report of an Inquiry by Lord Laming*. London: The Stationery Office.

Department of Health. (2004). *Delivering HR in the NHS*. London: Department of Health.

Department of Health. (2006). *Our Health, Our Care, Our Say: A New Direction for Community Services*. London: Department of Health.

Department of Health. (2007). *National Stroke Strategy*. London: Department of Health.

Department of Health. (2008). *High Quality Care for all; NHS Next Stage Review Final Report*. London: The Stationary Office.

Department of Health. (2009). *The NHS Knowledge and Skills Framework (NHS KSF) and the Development Review Process*. Retrieved May 5, 2009, from http://www.dh.gov.uk/en/Publicationsandstatistics/Publications/PublicationsPolicyAndGuidance/DH_4090843

Department of Health/Government Strategy Unit. (2005). Independence, Well-being and Choice: Our vision for the Future of Adult Social Care in England. London: DoH.

DePoy, E., Wood, C., & Miller, M. (1997). Educating rural allied health professionals: an interdisciplinary effort. *Journal of Allied Health*, *26*(3), 127–132.

Dewey, J. (1916). *Democracy and education: an introduction to the philosophy of education*. New York: Simon and Schuster.

Dewey, J. (1929). *The Sources of a Science of Education*. New York: Horace Liveright.

Dewey, J. (1933). *How we think*. Boston, D.C.: Heath.

Dewey, J. (1938). *Experience and education*. New York: Simon & Schuster.

DeWitt, C., & Baldwin, J. R. (2007). Some historical notes on interdisciplinary and interprofessional education and practice in health care in the USA. *Journal of Interprofessional Care, 21*(S1), 23–37. doi:10.1080/13561820701594728

DfEE. (1993). *The learning age, Department for Education and Employment.* London: The Stationery Office.

DfES (Department for Education and Skills). (2003). Every Child Matters. Green Paper, Cm. 5860. London: The Stationery Office (TSO).

DfES. (2003). *The future of higher education. White Paper cm 5735.* London: The Stationery Office.

Dieterle, E., & Clarke, J. (2009). Multi-user virtual environments for Teaching and Learning . In Pagani, M. (Ed.), *Encyclopaedia of multimedia technology and networking* (2nd ed., *Vol. 1*). Hershey, PA: IGI Global.

*Disability Discrimination Act.* (2005). Retrieved August 27, 2009, from http://www.opsi.gov.uk/Acts/acts2005/ukpga_20050013_en_1

Disability Rights Commission. (2006). *Fitness Standards Formal Investigation.* Disability Rights Commission. Retrieved March 19, 2007, from http://www.drc-gb.org/Employers_and_Service_Provider/Fitness_Standards_Formal_Inves.aspx

Disability Rights Commission. (2007). *Maintaining Standards: Promoting Equality.* Report of a DRC (2007) Formal investigation Professional regulation within nursing, teaching and social work and disabled people's access to these professions. Retreived April 8, 2008, from http://www.maintainingstandards.org/files/Full%20report%20_%20final.pdf

DoH. (1999). *National service frameworks for mental health: modern standards and service models.* London: Department of Health. The Stationery Office.

DoH. (2000a). *A health service of all the talents: developing the NHS workforce.* London: Department of Health. The Stationery Office.

DoH. (2000b). *The NHS Plan.* London: Department of Health. The Stationery Office.

DoH. (2001). *Valuing people a new strategy for learning disability for the 21st century a white paper.* London: Department of Health. The Stationery Office.

DoH. (2002). *The expert patient: A new approach to chronic disease management for the 21st Century.* London: Department of Health. The Stationery Office.

DoH. (2004a). *The NHS Improvement Plan: putting people at the heart of public services. Cm 6268.* London: Department of Health. The Stationery Office.

DoH. (2004b). *Choosing Health: making healthy choices easier.* London: Department of Health The Stationery Office.

DoH. (July 2004c.) *National Standards, Local Action: health and social care standards and planning framework 2005/06-2007/08.* London: Department of Health. The Stationery Office.

Dornan, T., Boshuizen, H., King, N., & Scherpbier, A. (2007). Experience-based learning: a model linking the processes and outcomes of medical students' workplace learning. *Medical Education, 41*, 84–91. doi:10.1111/j.1365-2929.2006.02652.x

Dorsey, L., & Baker, C. (2004). Mentoring undergraduate nursing students. *Nurse Educator, 29*, 260–265. doi:10.1097/00006223-200411000-00013

Dowden, M. J. (2007). Climate Change Adaptation: a Multi-Disciplinary Challenge. In RICS, *COBRA 2007 Conference Proceedings, Georgia Institute of Technology.* London: RICS.

Downes, S. (2001). Learning Objects: resources for distance education worldwide. *The International Review of Research in Open and Distance Learning, 2*(1). Retrieved October 10, 2009, from http://www.irrodl.org/content/v2.1/downes.pdf

Dreyfus, H. (2000). Telepistemology: Descartes last stand . In Goldberg, K. (Ed.), *The Robot in the Garden; Telerobotics and Telepistemology in the Age of the Internet* (pp. 48–63). Cambridge, MA: MIT Press.

Dreyfus, H. L., & Dreyfus, S. E. (1986). *Mind over machine: The power of human intuition and expertise in the era of the computer.* New York: The Free Press.

Dreyfus, H., & Dreyfus, S. (2005). Expertise in real world contexts. *Organization Studies, 26*(5), 779–792. doi:10.1177/0170840605053102

Duffy, T. M., Lowyck, J., & Jonassen, D. (1993). *The Design of Constructivistic Learning Environments: Implications for Instructional Design and The Use of Technology*. Heidelburg, Germany: Springer-Verlag.

Dunn, S. (2003). *Return to SENDA? Implementing accessibility for disabled students in virtual learning environments in UK further and higher education*. London: City University of London. Retrieved March 19, 2007, from http://www.saradunn.net/VLEreport/documents/VLEreport.pdf

Dutfield, R., & Eling, C. (1994). Managing the individual's resistance to change . In Lovell, R. (Ed.), *Managing Change in the Public Sector*. Harlow, UK: Longman.

Dyke, M., & Conole, G. Ravenscroft, A. & de Freitas, S. (2006.) Learning theory and its application to e-Learning. In G. Conole & M. Oliver (Eds.), Contemporary Perspectives in E-Learning Research (pp. 80-95). London: RoutledgeFalmer.

Edelson, P. J., & Pittman, V. V. (2001). *E-Learning in the United States: New Directions and Opportunities for University Continuing Education*. Retrieved May 5, 2009, from http://www.sunysb.edu/spd/dean_papers/newdelhi.pdf

Eden, A. H. (2002). A Theory of Object-Oriented Design. *Information Systems Frontiers, 4*(4), 379–391. doi:10.1023/A:1020835709566

Eduserv Foundation. (2009). *Research and Innovation*. Retrieved March 4, 2009, from http://www.eduserv.org.uk/research

E-learning Facilitators (ELF). (2006). *The roles of E-learning facilitators: Skills, competences and training*. E-learning Facilitators (ELF). Retrieved July 11, 2009 from http://conseil-recherche-innovation.net/wp-content/download/ELF/ELF - facilitators roles report - Greta du Velay.pdf

Elliot, K., & Sweeney, K. (2008). Quantifying the reuse of learning objects. *Australasian Journal of Educational Technology, 24*(2), 137-142. Retrieved February 23, 2009, from http://www.ascilite.org.au/ajet/ajet24/elliott.pdf

Ellis, R. A., Goodyear, P., O'Hara, A., & Prosser, M. (2007). The university student experience of face-to-face and online discussions: coherence, reflection and meaning. *ALT-J . Research in Learning Technology, 15*(1), 83–97.

Elton. L. (1994). *Management of teaching and Learning: Towards change in universities*. London: CVCP.

Engestrom, Y. (1991). Engestrom, Y. (2001). Expansive Learning at Work: toward an activity theoretical reconceptualization. *Journal of Education and Work, 14*(1), 133–156. doi:10.1080/13639080123238

Engeström, Y. (1999). Activity Theory and individual and social transformation . In Engeström, Y., Miettinen, R., & Punamäki, R. L. (Eds.), *Perspectives on Activity Theory* (pp. 19–38). Cambridge, UK: Cambridge University Press.

Engineering & Physical Sciences Research Council. (2009). *The future of research: Joint statement from EPSRC and the Learned Societies*. Retrieved April 2, 2009, from http://www.epsrc.ac.uk/Content/News/futureofresearch.htm

English National Board/Department of Health. (2001). *Placements in Focus: Guidance for Education in Practice for Health Care Professions*. London: ENB/DoH.

Eraut, M. (1994). *Developing Professional Knowledge and Competence*. London: Falmer.

Eraut, M. (1999). *Theoretical and methodological perspectives on researching workplace learning*. Paper for AERA Annual Conference, Montreal, April 1999.

Eraut, M., Alderton, J., Cole, G., & Senker, P. (2000). Development of Knowledge and Skills at work . In Coffield, F. (Ed.), *Differing Visions of a Learning Society* (*Vol. 1*, pp. 231–262). Bristol, UK: The Policy Press.

Ericson, D. L. (1991). Guest Editorial: Territorialism in the health care professions. *The International Journal of Childbirth Education, 6*(3), 30–31.

Ericsson, K. A. (2008). Deliberate practice and acquisition of expert performance: a general overview. *Academic Emergency Medicine*, *15*(11), 988–994. doi:10.1111/j.1553-2712.2008.00227.x

Espeland, V., & Indrehus, O. (2003). Evaluation of students' satisfaction with nursing education in Norway. *Journal of Advanced Nursing*, *42*(3), 226–236. doi:10.1046/j.1365-2648.2003.02611.x

Etzioni, A. (1969). *The Semi-Professions and their Organization*. New York: Free Press.

Evers, H. (1981). Multidisciplinary teams in geriatric wards: myth or reality? *Journal of Advanced Nursing*, (6): 205–214. doi:10.1111/j.1365-2648.1981.tb03214.x

Falconer, I., Conole, G., Jeffrey, A., & Douglas, P. (2006). *Pedagogical Learning Activity Reference Guide*. Retrieved February 5, 2009, from http://athena.cs.man.ac.uk/apache2-default/r/LADIE/www.elframework.org/refmodels/ladie/guides/LARM_Pedagogy30-03-06.doc

Falconer, J. A., Roth, E. J., Sutin, J. A., Strasser, D. C., & Chang, R. W. (1993). The critical path method in stroke rehabilitation: lessons from an experiment in cost containment and outcome improvement. *Quality Review Bulletin*, *19*(1), 8–16.

Farrell, M. (2005). Enabling interprofessional education: the potential of e-learning. In Carlisle, C., Cooper, H., & Mercer, D. (Eds.), *Interprofessional Education: an Agenda for Healthcare Professionals*. Salisbury, UK: Quay Books.

Felton, A., & Stickley, T. (2004). Pedagogy, power and service user involvement. *Journal of Psychiatric and Mental Health Nursing*, *11*, 89–98. doi:10.1111/j.1365-2850.2004.00693.x

Fill, K., Leung, S., DiBiase, D., & Nelson, A. (2006). Repurposing a learning activity on academic integrity: the experience of three universities. *Journal of Interactive Media in Education, 1*. Retrieved February 15, 2009, from http://jime.open.ac.uk/2006/01/

Finch, J. (2000). Interprofessional education and teamworking: a view from the education providers. *British Medical Journal*, *321*, 1138–1140. doi:10.1136/bmj.321.7269.1138

Fink, L. D. (2003). *Creating Significant Learning Experiences: An Integrated Approach to Designing College Courses*. San Francisco: Jossey- Bass.

First Wave Team. (2005). *Meeting the Challenge Project Final Report*. Coventry, UK: Coventry University.

Fischer, S., Stewart, T. E., Mehta, S., Wax, R., & Lapinsky, S. E. (2003). Handheld computing in medicine. *Journal of the American Medical Informatics Association*, *10*(2), 139–149. doi:10.1197/jamia.M1180

Fleming, N. D. (2001). *Teaching and Learning Styles: VARK Strategies*. Christchurch, New Zealand: Neil D. Fleming.

Fleming, T., Tammone, W., & Wahl, M. (2002). *E-Learning: Addressing the Challenges via Collaboration*. Washington, DC: US Department of Education. Retrieved February 14, 2009, from http://eric.ed.gov/ERICDocs/data/ericdocs2sql/content_storage_01/0000019b/80/1a/52/e9.pdf

Forman, D. (2005). Recognising the Importance of the Interpersonal. In *The Theory Practice Relationship in Inter-professional Education* (pp. 55-60). Higher Education Academy Health Sciences and Practice Occasional Paper No. 7.

Forman, D. (2008). Networking IPE Networks. *The Network: Towards Unity for Health Newsletter, 26.*

Forman, D., Nyatanga, L., & Rich, T. (2002). E-learning and Educational Diversity. *Nurse Education Today*, *22*, 1–7. doi:10.1054/nedt.2001.0740

Fox, N., O'Rourke, A., Roberts, C., & Walker, J. (2001). Change management in primary care: design and evaluation of an internet-delivered course . *Medical Education*, *35*(8), 803–805. doi:10.1046/j.1365-2923.2001.00974.x

Fox, W. (Ed.). (2000). *Ethics and the Built Environment*. London: Routledge.

Franks, D., Dale, P., Hindmarsh, R., Fellows, C., Buckridge, M., & Cybinski, P. (2007). Interdisciplinary Foundations: Reflecting on interdisciplinarity and three decades of teaching and research at Griffith University, Australia. *Studies in Higher Education*, *32*(2), 167–185. doi:10.1080/03075070701267228

Freeth, D. (2001). Sustaining Interprofessional Collaboration. *Journal of Interprofessional Care*, *15*(1), 37–46. doi:10.1080/13561820020022864

Freeth, D., & Reeves, S. (2002). Evaluation of an Interprofessional training ward: pilot phase . In Glen, S., & Leiba, T. (Eds.), *Multiprofessional Learning for nurses – Breaking the Boundaries*. Basingstoke, UK: Macmillan.

Freeth, D., & Reeves, S. (2004). Learning to work together: using the presage, process, product (3P) model to highlight decisions and possibilities. *Journal of Interprofessional Care*, *18*(1), 43–56. doi:10.1080/13561820310001608221

Freeth, D., Hammick, M., Koppel, I., Reeves, S., & Barr, H. (2002). *A Critical review of Evaluations of Interprofessional Education*. London: The Learning and Teaching Support Network for Health Sciences and Practice.

Freeth, D., Hammick, M., Reeves, S., Koppel, I., & Barr, H. (2005). *Effective Interprofessional Education: Development, Delivery and Evaluation (Promoting Partnership for Health). CAIPE*. Wiley Blackwell Publishing.

Friesen, N. (2004). Three objections to learning objects . In McGreal, R. (Ed.), *Online Education Using Learning Objects* (pp. 59–70). London: Routledge/Falmer.

Fry, G. L. A. (2001). Multifunctional landscapes – towards transdisciplinary research. *Landscape and Urban Planning*, *57*, 159–168. doi:10.1016/S0169-2046(01)00201-8

Funnell, P. (1995). Exploring the value of interprofessional shared learning . In Soothill, K., Mackay, L., & Webb, C. (Eds.), *Interprofessional relations in health care* (pp. 163–171). London: Edward Arnold.

Gallew, H. A. (2005). Students Teaching Students: Learning Through Doing, Being, and Becoming. *Occupational Therapy in Health Care*, *19*(3), 105–117. doi:10.1300/J003v19n03_08

Gann, D., & Salter, A. (1999). *Interdisciplinary Skills for Built Environment Professionals: a scoping study*. London: The Ove Arup Foundation.

Garg, A. X., Norman, G. R., Eva, K. W., Spero, L., & Sharan, S. (2002). Is there any real virtue of virtual reality?: the minor role of multiple orientations in learning anatomy from computers. *Academic Medicine*, *77*(10Suppl), S97–S99.

Garrison, D. R., & Anderson, T. (2003). *E-learning in the 21st century: a framework for research and practice*. New York: RoutledgeFalmer. doi:10.4324/9780203166093

Garrison, D. R., & Arbaugh, J. B. (2007). Researching the community of inquiry framework: review, issues, and future directions. *The Internet and Higher Education*, *10*(3), 157–172. doi:10.1016/j.iheduc.2007.04.001

Garrison, D. R., Anderson, T., & Archer, W. (2000). Critical Inquiry in a Text-Based Environment: Computer Conferencing in Higher Education. *The Internet and Higher Education*, *2*(2-3), 87–105. doi:10.1016/S1096-7516(00)00016-6

Garrison, D. R., Anderson, T., & Archer, W. (2001). Critical thinking, cognitive presence and computer conferencing in distance education. *American Journal of Distance Education*, *15*(1), 7–23. doi:10.1080/08923640109527071

Gee, J. P. (2003). *What video games have to teach us about learning and literacy*. New York: Palgrave Macmillan.

Geertz, C. (1983). *Local knowledge: Further essays in interpretative anthropology*. New York: Basic Books.

Geissler, C., Freeth, D., Hammick, M., Koppel, I., Reeves, S., & Barr, H. (2002). *A critical review of evaluations of interprofessional education*. Retrieved August 28, 2008, from http://www.health.ltsn.ac.uk/publications/occasionalpaper/ occasionalpaper02.pdf

Gell, C. (2003). *Involving service users in the training of graduate primary care mental health workers*. Nottingham, UK: Trent Workforce Development Confederation.

Geller, Z., Rhyne, R., Hansbarger, L., Borrego, M., Van-Leit, B., & Scaletti, J. (2002). Interdisciplinary health professional education in rural New Mexico: A 10 year experience. *Learning in Health and Social Care*, *1*, 33–46. doi:10.1046/j.1473-6861.2002.00003.x

Gibbs, C. (2003). Explaining effective teaching: self-efficacy and thought control of action. *Journal of Educational Enquiry*, *4*(2), 1–14.

Gibson, J. (2009). The five 'Es'of an excellent teacher. *The Clinical Teacher*, *6*(1), 3–5. doi:10.1111/j.1743-498X.2008.00239.x

Gibson, J. M. (2003). Arrogance - A Dangerous Weapon of the Physics Trade? *Physics Today*, *56*(2), 54. doi:10.1063/1.1564350

Gijbeis, D., Gochy, F., Van den Bossche, P., & Segers, M. (2005). Effects of problem-based learning: A meta-analysis from the angle of assessment. *Review of Educational Research*, *75*(1), 27–61. doi:10.3102/00346543075001027

Gilbert, J. H. V. (2005). Interprofessional learning and higher education structural barriers. *Journal of Interprofessional Care*, *19*(S1), 87–106. doi:10.1080/13561820500067132

Gilbert, J. H. V. (2005). Interprofessional learning and higher education structural barriers. *Journal of Interprofessional Care*, *19*(S1), 87–106. doi:10.1080/13561820500067132

Gilbert, S. W. (2000). *A New Vision Worth Working Toward – Connected Education and Collaborative Change*. Retrieved February 27, 2009, from http://www.tltgroup.org/gilbert/NewVwwt2000--2-14-00.htm

Gilchrist, M. (2008). Issues around creating a reusable learning object to support statistics teaching. *Electronic journal of e-Learning, 5*(4), 271-278. Retrieved October 10, 2009, from http://www.ejel.org/Volume-5/v5-i4/Gilchrist.pdf

Gill, D., Parker, C., Spooner, M., Thomas, M., Ambrose, K., & Richardson, J. (2006). Tomorrow's Doctors and Nurses: Peer Assisted Learning. *The Clinical Teacher*, *3*(1), 13–18. doi:10.1111/j.1743-498X.2006.00087.x

Girvan, C., FitzGibbon, A., & Savage, T. (2009). *Identification and analysis of a pedagogy in action within Multi-User Virtual Environments*. Paper presented at the Virtual Worlds Best Practice in Education Conference, March 27-29, 2009, Second Life.

Glaser, B. G., & Strauss, A. L. (1967). *The discovery of grounded theory: strategies for qualitative research.* Chicago: Aldien.

Glen, S. (1999). Educating for interprofessional collaboration: Teaching about values. *Nursing Ethics*, *6*, 202–213.

GMC. (2003). *Tomorrow's doctors. Recommendations on undergraduate medical education.* London: GMC. Retrieved 03/04/09 from www.gmc-uk.org/education/undergraduate/tomdoc.pdf

GMC. (2009). *Tomorrow's doctors. Outcomes and Standards for undergraduate medical education.* London: GMC. Retrieved October 22, 2009, from http://www.gmc-uk.org/education/undergraduate/tomdoc.pdf

Golder, S. A., Wilkinson, D., & Huberman, B. A. (2007). *Rhythms of Social Interaction: Messaging within a Massive Online Network.* Paper presented at the 3rd International Conference on Communities and Technologies (CT2007).

Goodlad, S., & Hirst, B. (1989). *Peer tutoring: A Guide to Learning by Teaching.* London: Kogan Page.

Goodman, S., Lillis, T., Maybin, J., & Mercer, N. (2003). *Language, Literacy and Education: A Reader. Stoke on Trent*. UK: Trentham Books.

Goosey, D., & Barr, H. (1998). Selected Case Studies of Interprofessional Education. Centre for the Advancement of Interprofessional Education, London. In Craddock, D. Borthwick, A. & McPherson, K. (2006). Interprofessional education in health and social care: fashion or informed practice? [from http://www.blackwell-synergy.com/doi/pdf]. *Learning in Health and Social Care*, *5*(4), 220–242. Retrieved September 11, 2009.

Gordon, F., & Walsh, C. (2005). A Framework for Interprofessional Capability: developing students of health and social care as collaborative workers. *Journal of Integrated Care*, *13*(3), 26–33.

Gordon, F., Walsh, C., Marshall, M., Wilson, F., & Hunt, T. (2004). Involving patients and service users in student learning: developing practice and principles. *Journal of Integrated Care*, *12*(6), 28–35.

Gordon, F., Walsh, C., Marshall, M., Wilson, F., & Hunt, T. (2005). Good practice guidelines for patient and service user participation in teaching and assessing. Sheffield, UK: The Combined Universities Interprofessional Learning Unit: Sheffield Hallam University and University of Sheffield.

Gordon, F., Walsh, C., Marshall, M., Wilson, F., & Hunt, T. (2006). *The Combined Universities Interprofessional Learning Unit: Final report*. Sheffield, UK: Sheffield Hallam University and University of Sheffield.

Gosser, D. K., Cracolice, M. S., Kampmeier, J. A., Roth, V., Strozak, V. S., & Varma-Nelson, P. (2001). *Peer-led Team Learning: A guidebook*. Upper Saddle River, New Jersey: Pretice Hall.

Gould, D., & Drey, N. (2008). Hand hygiene technique. *Nursing Standard*, *22*(34), 42–46.

Gregson, B., Cartlidge, A., & Bond, J. (1991). *Interprofessional collaboration in primary health care organisations*. London: Royal College of General Practitioners.

Gustafson, P., & Gibbs, G. (2000). Guiding or Hiding? The Role of the Facilitator in Online Teaching and Learning. *Teaching Education*, *11*(2), 195–210. doi:10.1080/713698967

Hall, H. (2003). Borrowed theory: Applying exchange theories in information science research. *Library & Information Science Research*, *25*, 287–306. doi:10.1016/S0740-8188(03)00031-8

Hall, P. (2005). Interprofessional teamwork: professional cultures as barriers. *Journal of Interprofessional Care*, *19*(S1), 188–196. doi:10.1080/13561820500081745

Hall, P., & Weaver, L. (2001). Interdisciplinary education and teamwork: a long and winding road. *Medical Education*, *35*(9), 867–875. doi:10.1046/j.1365-2923.2001.00919.x

Hallawell, B., Windle, R., Wharrad, H. J., Laverty, H., & Williams, A. (2009). *Student generated e-learning within Learning/Intellectual Disabilities*. Paper presented at 4th International conference on e-learning. Toronto, Canada.

Hammersley, M. (1996). The relationship between qualitative and quantitative research: Paradigm loyalty versus methodological eclecticism . In Richardson, J. T. E. (Ed.), *Handbook of qualitative research methods for psychology and the social sciences*. Leicester, UK: BPS Books.

Hammick, M., Freeth, D., Koppel, I., Reeves, S., & Barr, H. (2007). A best evidence systematic review of interprofessional education: BEME Guide no. 9. *Medical Teacher*, *29*(8), 735–751. doi:10.1080/01421590701682576

Haniff, A. M., Kaka, A. P., Barron, S. J., & Banfill, P. F. G. (2003). Changing industry culture through graduates: promoting knowledge sharing and interdisciplinary teamworking amongst students. In R. Newton, A. Bowden & M. Betts (Eds.), *CIB W89 Building Education & Research Symposium: Volume 3* (pp.1367-1377). Salford, UK: The University of Salford.

Hanna, D. E. (1998). Higher education in an era of digital competition: Emerging organizational models. *Journal of Asynchronous Learning Networks*, *2*(1), 66–96.

Happell, B., & Roper, C. (2002). Promoting consumer participation through the implementation of a consumer academic position. *Nurse Education in Practice*, *2*(2), 73–79. doi:10.1054/nepr.2002.0068

Harden, R. M., & Hart, I. R. (2002). An international virtual medical school (IVIMEDS): the future for medical education? *Medical Teacher*, *24*(3), 261–267. doi:10.1080/01421590220141008

Harrison, D. (2007). In Shepherd, J. *It's a world of possibilities*. The Guardian, May 8th.

Hartnell-Young, E., Smallwood, A., Kingston, S., & Harley, P. (2006). Joining up the episodes of lifelong learning: A regional transition project. *British Journal of Educational Technology*, *37*(6), 853–866. doi:10.1111/j.1467-8535.2006.00668.x

Harvey, B. (2005). Learning Objects and instructional Design. *The International Review of Research in Open and Distance Learning*, *6*(2). Retrieved February 25, 2009, from http://www.irrodl.org/index.php/irrodl/article/view/227/310

Harvey, B., & Beards, D. (2004). E-learning in Scottish further and higher education. *Education + Training, 46*(6/7), 353–360. doi:10.1108/00400910410555268

Hawkins, J. M., & Allen, R. (1991). *Oxford Encyclopaedic English Dictionary*. Oxford: Clarendon.

Hayler, R. W. (1999). Assessing Proctors: solving problems in grading student performance in a peer support system. *Mentoring & Tutoring, 7*(1), 35–39. doi:10.1080/0968465990070103

Headrick, L. A., & Khaleel, N. I. (2008, August). Getting it right: Educating professionals to work together in improving health care. *Journal of Interprofessional Care, 22*(4), 364–374. doi:10.1080/13561820802227871

Heatherington, H., Bonar-Law, J., & Fleet, T. (2008). *Problem based learning in a Multi User Virtual Environment. Project description*. London: Higher Education Academy. Retrieved September 11, 2009, from http://www.ics.heacademy.ac.uk/projects/development-fund/fund_details.php?id=114 on 01.08.08

HEFCE. (2005). *Strategy for e-learning*. Retrieved July 5, 2009, from www.hefce.ac.uk

HEFCE. (2009). *Enhancing learning and teaching through the use of technology: A revised approach to HEFCE's strategy for e-learning*. Retrieved July 5, 2009, from www.hefce.ac.uk

Heidari, F., & Galvin, K. (2002). The role of open learning in nurse education. Does it have a place? *Nurse Education Today, 22*, 617–623. doi:10.1016/S0260-6917(02)00133-8

Heitzmann, R. (2008). Case study instruction in teacher education: Opportunity to develop students' critical thinking, school smarts and decision making. *Education, 128*(4), 523–542.

Hermsen, M. A., & Ten Have, A. M. J. (2005). Palliative care teams: Effective through moral reflection . *Journal of Interprofessional Care, 19*(6), 561–568. doi:10.1080/13561820500404871

Herreid, C. F. (2006). *Start with a story: The case study method of teaching college science*. Arlington, VA: NSTA Press.

Herrington, J., Reeves, T. C., Oliver, R., & Woo, Y. (2004). Designing authentic activities in web-based courses. *Journal of Computing in Higher Education, 16*(1), 3–29. doi:10.1007/BF02960280

Hewstone, M., & Brown, R. J. (1986). Contact is not Enough: an Inter-group Perspective on the "Contact Hypothesis" . In Hewstone, M., & Brown, R. J. (Eds.), *Contact and Conflict in Intergroup Encounters*. Oxford, UK: Blackwell.

Hewstone, M., Carpenter, J., Rought, D., & Franklyn-Stokes, A. (1994). Intergroup contact between professional groups: two evaluation studies. *Journal of Community & Applied Social Psychology, 4*, 347. doi:10.1002/casp.2450040504

Hickey, G., & Kipping, C. (1998). Exploring the concept of user involvement in mental health through a participation continuum. *Journal of Clinical Nursing, 7*(1), 83–88. doi:10.1046/j.1365-2702.1998.00122.x

Higgins, M. (2004). Urban design project in a multidisciplinary European context: Collaboration between education and practice. *CEBE Transactions, 1*(2), 16–36.

Higher Education Academy. (2008). *Case Study: Professional development through interdisciplinary and interprofessional learning activities*. Retrieved April 15, 2008 from http://cebe.cf.ac.uk/BPBN/casestudy/leeds_cdcli.htm

Higher Education Academy. (2009). *Interdisciplinary Teaching and Learning Group*. Retrieved April 2, 2009, from http://www.heacademy.ac.uk/ourwork/networks/itlg

Higher Education Funding Council for England (HEFCE). (2009). *Enhancing learning and teaching through the use of technology: a revised approach to HEFCE's strategy for e-learning*. London: HEFCE.

Hiltz, S. R., Turoff, M., & Harasim, L. (2007). Development and philosophy of the field of asynchronous learning networks . In Andrews, R., & Haythornthwaite, C. (Eds.), *The Sage Handbook of E-Learning Research*. London: Sage.

Hobbs, M., Gordon, M., & Brown, E. (2006). A virtual world environment for group work. *Current Developments in Technology-Assisted Education . Formatex, 1*, 1369–1373.

Hodgkinson-Williams, C., & Gray, E. (2008). *Degrees of Openness: The emergence of Open Educational Resources at the University of Cape Town*. Paper presented at EMERGE2008 virtual conference. Retrieved October 10, 2009, from http://emerge2008.net/access/content/group/emerge2008/PresentationFiles/Hodgkinson-Williams/DegreesOfOpeness.pdf

Hoffman, S. J., Rosenfield, D., Gilbert, J. H. V., & Oandasan, I. (2008). Student leadership in interprofessional education: benefits, challenges, and implications for educators, researchers and policy makers. *Medical Education, 42*, 654–661. doi:10.1111/j.1365-2923.2008.03042.x

Hope, R. (2004). *The ten essential shared capabilities: a framework for the whole of the mental health workforce*. London: Department of Health.

Horsburgh, M., Lamdin, R., & Williamson, E. (2001). Multiprofessional learning: the attitudes of medical, nursing and pharmacy students to shared learning. *Medical Education, 35*(9), 876–883. doi:10.1046/j.1365-2923.2001.00959.x

Howes, J. (2003). Working Together. In R. Newton, A. Bowden & M. Betts (Eds.), *CIB W89 Building Education & Research Symposium: Volume 3* (pp.1378-1389). Salford, UK: The University of Salford.

Howkins, E., & Bray, J. (2008). *Preparing for inter-professional teaching: theory and practice*. Oxford, UK: Radcliffe Medical Press.

Hoyle, E., & John, P. (1995). *Professional Knowledge and Professional Practice*. London: Cassell.

Hubbard, G., & Themessl-Huber, M. (2005). Professional perceptions of joint working in primary care and social care services for older people in Scotland. *Journal of Interprofessional Care, 19*(4), 371–385. doi:10.1080/13561820500165167

Huberman, B. A., Romero, D. M., & Fang, W. (2008). Social networks that matter: Twitter under the microscope. *First Monday, 14*(1), 1-5. Retreived October 13, 2009, from http://uk.arxiv.org/abs/0812.1045

Huckstadt, A., & Hayes, K. (2005). Evaluation of interactive online courses for advanced practice nurses . *Journal of the American Academy of Nurse Practitioners, 17*(3), 85–89. doi:10.1111/j.1041-2972.2005.0015.x

Huffstutler, S., Wyait, T. H., & Wright, C. R. (2002). The use of handheld technology in nursing education. *Nurse Educator, 27*(6), 271–275. doi:10.1097/00006223-200211000-00008

Hughes, M., Ventura, S., & Dando, M. (2004). On-line interprofessional learning: introducing constructivism through enquiry-based learning and peer review. *Journal of Interprofessional Care, 18*(3), 263–268. doi:10.1080/13561820410001731304

Humphris, D., & Hean, S. (2004). Educating the future workforce: building the evidence about interprofessional learning. *Journal of Health Services Research & Policy, 9*(1), 24–27. doi:10.1258/135581904322724103

Hutchings, A., Williamson, G. R., & Humphreys, A. (2005). Supporting learners in clinical practice: capacity issues. *Journal of Clinical Nursing, 14*, 945–955. doi:10.1111/j.1365-2702.2005.01239.x

Hutchings, M. (2002). Enhancing learning with computer mediated communication: teacher intentions and student conceptions of online learning . In Rust, C. (Ed.), *Improving Student Learning Using Learning Technologies* (pp. 99–111). Oxford, UK: Oxford Centre for Staff Development.

Hutchings, M., Hadfield, M., Howarth, G., & Lewarne, S. (2007). Meeting the challenges of active learning in Web-based case studies for sustainable development. *Innovations in Education and Teaching International, 44*(3), 331–343. doi:10.1080/14703290701486779

Hutchins, E. (1995). *Cognition in the wild*. Cambridge, MA: MIT Press.

Huws, Jagger, & Bates. (2001). *Where the Butterfly Alights: the Global Location of E- work*. IES Report 480.

Hyde, A., & Murray, M. (2005). Nurses' Experiences of Distance Education Programmes. *Journal of Advanced Nursing, 49*, 87–95. doi:10.1111/j.1365-2648.2004.03267.x

Ikkos, G. (2003). Engaging patients as teachers of clinical interview skills. *Psychiatric Bulletin, 27*(8), 312–315. doi:10.1192/pb.27.8.312

IMS Global Learning Consortium. (2003). IMS *Learning Design information model version 1.0 final specification.* Retrieved March 14, 2009, from http://www.imsglobal.org/learningdesign/ldv1p0/imsld_infov1p0.html

International Council of Nurses. (2000). *The ICN Code of Ethics for Nurses.* Geneva: ICN.

Ireland, J., Gibb, S., & West, B. (2008). Interprofessional education: reviewing the evidence. *British Journal of Midwifery, 16*(7), 446–453.

Irvine, R., Kerridge, I., McPhee, J., & Freeman, S. (2002). Interprofessionalism and ethics: consensus or clash of cultures? *Journal of Interprofessional Care, 16*(3), 2002. doi:10.1080/13561820220146649

Jackson, J. A., & Bluteau, P. A. S. (2007). At first it's like shifting sands: Setting up interprofessional learning within a secondary care setting. *Journal of Interprofessional Care, 21*(3), 351–353. doi:10.1080/13561820701233905

Jackson, J. A., & Bluteau, P. A. S. (2009). Creating a model: Overcoming the challenges of implementing Interprofessional Education . In Bluteau, P. A. S., & Jackson, J. A. (Eds.), *Interprofessional Education; making it happen.* London: Palgrave.

Jairath, N., & Mills, M. E. (2005). Online Health Science Education- Development and Implementation (pp. 4-6, 85, 106-114). Philadelphia, PA: Lippincott Williams & Wilkins.

Jakobsson, M. (2002). Rest in Peace, Bill the Bot: Death and Life in Virtual Worlds . In Schroeder, R. (Ed.), *The Social Life of Avatars* (pp. 63–76). London: Springer-Verlag.

Jameson, J., & DeFreitas, S. (2006). Editorial: Collaborative e-support for lifelong learning. *British Journal of Educational Technology, 37*(6), 817–824. doi:10.1111/j.1467-8535.2006.00665.x

Jameson, J., Ferrell, G., Kelly, J., Walker, S., & Ryan, M. (2006). Building trust and shared knowledge in communities of e-learning practice: collaborative leadership in the JISC eLISA and CAMEL lifelong learning projects. *British Journal of Educational Technology, 37*(6), 949–967. doi:10.1111/j.1467-8535.2006.00669.x

Jamieson, D. W., & Thomas, K. W. (1974). Power and Conflict in the Student-Teacher Relationship. *The Journal of Applied Behavioral Science, 10*, 321–336. doi:10.1177/002188637401000304

Jennings, N., & Collins, C. (2007). Virtual or virtually U: Educational Institutions in Second life. *International Journal of Social Sciences, 2*(3), 180–186.

Jentsch, F., & Bowers, C. A. (1998). Evidence for the Validity of PC-based Simulations in Studying Aircrew Coordination. *The International Journal of Aviation Psychology, 8*(3), 243–260. doi:10.1207/s15327108ijap0803_5

JISC. (2008). *Effective practice with e-portfolios.* Bristol, UK: Joint Information Systems Council.

JISC. (2008). *Effective use of Virtual Learning Environments (infoKit).* Retrieved September 11, 2009, from http://www.jiscinfonet.ac.uk/InfoKits/effective-use-of-VLEs accessed on 08.12.08

Johns, C. (1995). Framing Learning Through Reflection within Carper's Fundamental Ways of Knowing In Nursing. *Advance for Nurse Practitioners, 22*(2), 226–234.

Johnsen, K., Dickerson, R., Raij, A., Lok, B., Jackson, J., Min, S., et al. (2005). *Experiences in using immersive virtual characters to educate medical communication skills.* Proceedings of IEEE Virtual Reality 2005 (VR 2005), 179-186. doi: 10.1109/VR.2005.1492772

Johnson, L., Levine, A., & Smith, R. (2009). *The 2009 Horizon Report.* Austin, TX: The New Media Consortium.

Jonassen, D. H. (2000). Transforming learning with technology: Beyond modernism and postmodernism or whoever controls the technology creates the reality. *Educational Technology, 40*(2), 21–25.

Jonassen, D. H., & Land, S. M. (Eds.). (2000). *Theoretical foundations of learning environments*. Mahwah, NJ: Lawrence Erlbaum.

Jones, E. M. (2000). *Shaping team practice in the context of health reform: Challenges, tensions, benefits*. PhD Thesis, Flinders University of South Australia.

Jones, M. L., Walters, S., & Akehurst, R. (2001). The implications of contact with the mentor for pre-registration nursing and midwifery students. *Journal of Advanced Nursing*, *35*, 151–160. doi:10.1046/j.1365-2648.2001.01832.x

Jones, R. A. (2005). Prioritising to safeguard the integrity of nurse education. *Nursing Times*, *101*(34), 26–27.

Jonsson, P., Edelbring, S., Sennerö, A., Fors, U., Kumlien, S., Olt, H., et al. (2006). *MICS, Interprofessional case simulation*. Paper presented at the Karolinska Institutet's Educational Congress, Stockholm.

Joshua-Amadi, M. (2002). Recruitment and retention: a study in motivation. *Nursing Management*, *9*, 7–21.

Juntunen, A., & Heikkinen, E. (2004). Lessons from interprofessional e-learning: Piloting a care of the elderly module. *Journal of Interprofessional Care*, *18*(3), 269–278. doi:10.1080/13561820410001731313

Kamin, C., O'Sullivan, P., Deterding, R., Younge, M., & Wade, T. (2006). A case study of teaching presence in virtual problem-based learning groups. *Medical Teacher*, *28*(5), 425–428. doi:10.1080/01421590600628241

Kay, R., & Knaack, L. (2005). Developing Learning objects for secondary school students: A multi-component model. *Interdisciplinary Journal of Knowledge and Learning Objects*, *1*(1), 229–254. Retrieved March 30, 2009, from http://ijklo.org/Volume1/v1p229-254Kay_Knaack.pdf

Keen, A. (2008). *The Cult of the Amateur: How blogs, MySpace, YouTube and the rest of today's user-generated media are killing our culture and economy*. London: Nicholas Brealey Publishing Ltd.

Keen, S., & Todres, L. (2007). Strategies for disseminating qualitative research findings: three exemplars. *Forum Qualitative Sozial Forschung*, *8*(3). Retrieved from http://www.qualitative-research.net/fqs-texte/3-07/07-3-17-e.htm.

Kehrwald, B. (2007). *Social Presence and Learner Support: Understanding learners' experiences with mediated social processes in text-based online learning environments*. Unpublished doctoral thesis, University of Southern Queensland.

Kelly, A. (2004). *The curriculum. Theory and practice* (5th ed.). London: Sage.

Kemshal-Bell, G. (2001). *The online teacher*. Report prepared for the Project Steering Committee of the VET Teachers and Online Learning Project, ITAM, ESD, TAFE NSW. Sydney: New South Wales Dept. of Education and Training. Retrieved March 29, 2009 from http://cyberteacher.onestop.net/final report.pdf

Kennard, J. (2002). Illuminating the relationship between shared learning and the workplace. *Medical Teacher*, *24*, 379–384. doi:10.1080/01421590220145752

Kennedy, I. (2001). *The Bristol Royal Infirmary Inquiry, CM5207* (1). London: HMSO.

Kirriemuir, J. (2007). *A July 2007 "snapshot" of UK Higher and Further Education Developments in Second Life*. Eduserv Foundation. Retrieved September 11, 2009, from www.eduserv.org.uk/foundation

Kirriemuir, J. (2008). *"Measuring" the impact of second life for Educational Purposes*. Eduserv Foundation. Retrieved September 11, 2009, from www.eduserv.org.uk/foundation

Klopfer, E. (2008). *Augmented Learning* (pp. 87–107). Cambridge, MA: The MIT Press.

Knight, P. T. (2002). *Being a Teacher in Higher Education*. Buckingham, UK: SRHE and Open University.

Knowles, M. (1984). *Andagogoy in action: applying modern principles to adult education*. San Francisco: Jossey Bass.

Knowles, M. S. (1980). *The Modern Practice of Adult Education: from Pedagogy to Andragogy* (2nd ed.). Upper Saddle River, NJ: Prentice Hall.

Knowlton, D. S. (2000). A theoretical framework for the online classroom: A defense and delineation of a student-centered pedagogy. *New Directions for Teaching and Learning*, *84*, 5–13. doi:10.1002/tl.841

Koeler, M. J., & Mishra, P. (2005). What happens when teachers design educational technology? The development of technological pedagogical content knowledge. *Journal of Educational Computing Research, 32*(2), 131–152. doi:10.2190/0EW7-01WB-BKHL-QDYV

Koeniger-Donohue, R. (2008). Hand held computers in nurse education: a PDA pilot project. *The Journal of Nursing Education, 47*(2). doi:10.3928/01484834-20080201-01

Kohn, L. T., Corrigan, J. M., & Donaldson, M. S. (2005). *To Err is Human: Building a Safer Health System*. Washington, DC: Institute for Health.

Kolarevic, B., Schmitt, G., Hirschberg, U., & Kurmann, D. (2000). An experiment in design collaboration. *Automation in Construction, 9*, 73–81. doi:10.1016/S0926-5805(99)00050-3

Kolb, D. A. (1984). *Experiential Learning: Experience as a source of learning and development*. Upper Saddle River, NJ: Prentice Hall.

Kolodner, J., & Guzdial, M. (1996). Effects with and of CSCL: Tracking learning in a new paradigm . In Koschmann, T. (Ed.), *CSCL: Theory and Practice of an Emerging Paradigm* (p. 353). Hillsdale, NJ: Lawrence Erlbaum and Associates.

Koschmann, T. D. (1996). *CSCL: theory and practice of an emerging paradigm*. Hillsdale, NJ: Lawrence Erlbaum and Associates.

Kotter, J. P. (1990). *A force for change: How leadership differs from management*. New York: Free Press.

Kozlowski, D. (2002). Using online learning in a traditional face-to-face environment . *Computers in Nursing, 20*(1), 23–30.

Kristjanson, L., Dudgeon, D., Nelson, F., Henteleff, P., & Balneaves, L. (1997). Evaluation of an interdisciplinary training program in palliative care: addressing the needs of rural and northern communities. *Journal of Palliative Care, 13*(3), 5–12.

Kumaş-Tan, & Zofia Ö. (2005). *Beyond cultural competence. Taking difference into account in occupational therapy*. Unpublished Master's thesis, Dalhousie University, Halifax, Canada.

Kunter, M., Tsai, Y., Klusmann, U., Brunner, M., Krauss, S., & Baumert, J. (2008). Students' and mathematics teachers' perceptions of teacher enthusiasm and instruction. *Learning and Instruction, 18*, 468–482. doi:10.1016/j.learninstruc.2008.06.008

Kurz, T. L., Llama, G., & Savenye, W. (2005). Issues and challenges of creating video cases to be used with preservice teachers. *TechTrends: Linking Research and Practice to Improve Learning, 49*(4), 67–73.

Kwan, M. M., & Balasubramanian, P. (2003). KnowledgeScope: Managing knowledge in context. *Decision Support Systems, 35*(4), 467. doi:10.1016/S0167-9236(02)00126-4

Laher, A., & Penny, J. (2005). Service improvement. In E. Pack (Ed.), Organisational Development In Healthcare: Approaches, innovations, achievements. Oxford, UK: Radcliffe publishing.

Lai, T., & Land, S. M. (2009). *Supporting Reflection in Online Learning Environments* (pp. 141–154). Educational Media and Technology Yearbook.

Laming, L. (2003). *The Victoria Climbié Inquiry*. Norwich, UK: HMSO.

Lane, H. C. (2006). *Intelligent Tutoring Systems: Prospects for Guided Practice and Efficient Learning*. Whitepaper for the Army's Science of Learning Workshop, Hampton, VA. Aug 1-3, 2006. Retrieved October 10, 2009, from http://people.ict.usc.edu/~lane/papers/ITSProspectsLane-Aug06.pdf

Langley, G. L., Nolan, K. M., Nolan, T. W., Norman, C. L., & Provost, L. P. (2009). *The Improvement Guide: A Practical Approach to Enhancing Organizational Performance* (2nd ed.). San Francisco, CA: Jossey-Bass Publishers.

Latham, M. (1994). *Constructing the Team*. London: Her Majesty's Stationery Office.

Latour, B. (1993). *We have never been modern*. New York: Harvester.

Latter, S., Maben, J., Myall, M., & Young, A. (2007). Evaluating nurse prescribers' education and continuing professional development for independent prescribing practice: Findings from a national survey in England. *Nurse Education Today*, *27*, 685–696. doi:10.1016/j.nedt.2006.10.002

Latter, S., Rycroft-Malone, J., Yerrell, P., & Shaw, D. (2001). Nurses' educational preparation for a medication education role: Findings from a national survey. *Nurse Education Today*, *21*(2), 143–154. doi:10.1054/nedt.2000.0528

Lattuca, L. R. (2002). Learning interdisciplinarity: sociocultural perspectives on academic work. *The Journal of Higher Education*, *73*(6), 711–739. doi:10.1353/jhe.2002.0054

Lattuca, L. R. (2003). Creating Interdisciplinarity: grounded definitions from college and university faculty. *History of Intellectual Culture*, *3*(1), 1–20.

Laurillard, D. (2002). *Rethinking University Teaching: A Conversational Framework for the Effective Use of Learning Technologies*. London: RoutledgeFalmer. doi:10.4324/9780203304846

Laurillard, D. (2002). *Rethinking university teaching: a framework for the effective use of educational technology* (2nd ed.). London: Routledge. doi:10.4324/9780203304846

Lave, J., & Wenger, E. (1991). *Situated learning: legitimate peripheral participation*. Cambridge, UK: Cambridge University Press.

Law, J. (1987). Technology and heterogeneous engineering: The case of Portuguese expansion . In Bijker, W. E., Hughes, T. P., & Pinch, T. J. (Eds.), *The social construction of technological systems: New directions in the sociology and history of technology* (pp. 111–134). Cambridge, MA: MIT Press.

Lawless, C., & Kirkwood, A. (1976). Training the Educational Technologist. *British Journal of Educational Technology*, *1*(7), 54–60. doi:10.1111/j.1467-8535.1976.tb00188.x

Le Var, R. M. H. (2002). Patient involvement in education for enhanced quality of care. *International Nursing Review*, *49*(4), 219–225. doi:10.1046/j.1466-7657.2002.00128.x

Leeder, D., McLachlan, J. C., Rodrigues, V., Stephens, N., Wharrad, H. J., & McElduff, P. (2004). Universities' Collaboration in e-learning (UCeL): a virtual community of practice in health professional education. In P. Kommers, P. Isaias & M. B. Nunes (Eds.), Lisbon, Portugal IADIS Web-based communities, (pp. 386-393).

Letherby, G. (2003). *Feminist research in theory and practice. Milton Keynes*. UK: Open University Press.

Levin, E. (2004). *Involving service users and carers in social work education. Social Care Institute for Excellence*. London: Policy Press.

Levison, D. A. (2003). *CHMS Position Paper: Interprofessional Education*. Council of Heads of Medical Schools and Deans of UK Faculties of Medicine.

Limburg, F., & Clark, C. (2006). Diversity initiatives in higher education: Teaching multicultural education online. *Multicultural Education*, *13*(3), 49–55.

Lindqvist, S., Duncan, A., Shepstone, L., Watts, F., & Pearce, S. (2005). Case-based learning in cross-professional groups: The development of a pre-registration interprofessional learning programme. *Journal of Interprofessional Care*, *19*(5), 509–520. doi:10.1080/13561820500126854

Lipsett, A. (2008). UK universities should take online lead - Calls for revival of e-learning to secure future of higher education. *The Guardian*. Retrieved May, 5, 2009, from http://www.guardian.co.uk/education/2008/nov/11/e-learning-university

Littlefield, M. B., & Roberson, K. C. (2005). Computer technology for the feminist classroom. *Affilia*, *20*(2), 186–202. doi:10.1177/0886109905274676

Littlejohn, A., & Peglar, C. (2007). *Preparing for blended e-learning*. Abingdon, UK: Routledge.

Littleton, F., & Bayne, S. (2008). Virtual worlds in education. *ESCalate News*, *10*(Spring), 26–28.

Littleton, K., Toates, F., & Braisby, N. (2007). Three Approaches to Learning . In Miell, D., Phoenix, A., & Thomas, K. (Eds.), *DSE212 Mapping Psychology Book 1. Milton Keynes*. UK: The Open University.

Lloyd-Jones, N., Hutchings, S., & Hobson, S. (2007). Interprofessional learning in practice for pre-registration health care: Interprofessional learning occurs in practice – is it articulated or celebrated? *Nurse Education in Practice*, *7*, 11–17. doi:10.1016/j.nepr.2006.03.002

Loar, B. (1981). *Mind and meaning*. London: Cambridge University Press.

Lobel, M., Neubauer, M., & Swedburg, R. (2005). Comparing how student collaborate to learn about the self and relationships in a real-time non-turn taking online and turn-taking face-face environment. *Journal of Computer-Mediated Communication*, *10*(4), 18.

Lockee, B., Burton, J., & Cross, L. (1999). No comparison: Distance education finds a new use for 'No significant difference'. *Educational Technology Research and Development*, *47*(3), 33–42. doi:10.1007/BF02299632

Lopes, T., & Thomas, B. (2006). *Dancing on live embers: Challenging racism in organizations*. Toronto: Between the Lines.

Lord Laming. (2003). *The Victoria Climbié Inquiry, report of an inquiry.* Retrieved April 3, 2009, from www.victoria-climbie-inquiry.org.uk

Lord Laming. (2009). *The protection of Children in England: a progress report*. London: The Stationery Office.

Luecht, R., Madsen, M., Taugher, M., & Petterson, B. (1990). Assessing professional perceptions: Design and validation of an interdisciplinary education scale. *Journal of Allied Health*, *19*(2), 181–191.

Lumague, M., Morgan, A., Mak, D., Hanna, M., Kwong, J., & Cameron, C. (2006). Interprofessional education: The student perspective. *Journal of Interprofessional Care*, *20*(3), 246–253. doi:10.1080/13561820600717891

Lusted, D. (1986). Why pedagogy? An introduction to this issue. *Screen*, *26*(5), 2–14.

Lymn, J., Bath-Hextall, F., & Wharrad, H. J. (2008). Pharmacology education for nurse prescribing students - a lesson in reusable learning objects. *Biomed Central Nursing*, *7*(1), 2.

MacDonald, C. J., Stodel, E. J., & Chambers, L. W. (2008). An Online inter-professional Learning Resource for Physicians, Pharmacists, Nurse Practitioners, and Nurses in Long Term Care: Benefits, Barriers and Lessons Learned. *Informatics for Health & Social Care*, *33*(1), 21–38. doi:10.1080/14639230801886824

Macome, E. (2008). On implementation of an information system in the Mozambican context: the EDM case viewed through ANT lenses. *Information Technology for Development*, *14*(2), 154–170. doi:10.1002/itdj.20063

Macrea, K. (2003). The Interact Student Competition. In R. Newton, A. Bowden & M. Betts (Eds.), *CIB W89 Building Education & Research Symposium: Volume 3* (pp.1418-1426). Salford, UK: The University of Salford.

Magnussen, L. (2006). Applying the Principles of Significant Learning in the E–Learning Environment. *The Journal of Nursing Education*, 82–86.

Mallik, M., & McGowan, B. (2007). Issues in practice based learning in nursing in the United Kingdom and the Republic of Ireland: Results from a multi professional scoping exercise. *Nurse Education Today*, *27*(1), 52–59. doi:10.1016/j.nedt.2006.02.007

Mallow, G. E., & Gilje, F. (1999). Technology-based nursing education: overview and call for further dialogue. *The Journal of Nursing Education*, *38*(6), 248–251.

Mancuso-Murphy, J. (2006). Distance Education in Nursing: An Integrated Review of Online Nursing Students' Experiences with Technology – Delivered Instruction. *The Journal of Nursing Education*, 252–260.

Manochehr, N.-N. (2006). The influence of learning styles on learners in e-learning environments: An empirical study. *Computers in Higher Education Economics Review*, *18*, 10–14.

Margaryan, A., Currier, S., Littlejohn, A., & Nicol, D. (2006). *Report on Learning Communities and Repositories. Community Dimensions of Learning Object Repositories project Deliverable 1*. Retrieved October 10, 2009, from http://academy.gcal.ac.uk/cd-lor/learning-communitiesreport.pdf

Margison, F. (2001). Practice based evidence in psychotherapy . In Mace, C., Mooray, S., & Roberts, B. (Eds.), *Evidence in psychological therapies. A critical guide for practitioners*. Hove, UK: Brunner-Routledge.

Maslow, A. H. (1943). A theory of human motivation. *Psychological Review*, *50*(4), 370–396. doi:10.1037/h0054346

Mason, J., & Lefrere, P. (2003). Trust, collaboration, e-learning and organisational transformation. *International Journal of Training and Development*, *7*(4), 259–270. doi:10.1046/j.1360-3736.2003.00185.x

Mason, R., & Rennie, F. (2008). *E-learning and social networking handbook: resources for higher education*. London: Routledge.

Mason, R., Pegler, C., & Weller, M. (2005). A learning object success story. *Journal of Asynchronous Learning Networks*, *9*(1), 97–105.

Masterman, E. (2008). Activity Theory and the Design of Pedagogic Planning Tools. In L. Lockyer, S. Bennett, S. Agostinho and B. Harper (Eds.), Handbook of Research on Learning Design and Learning Objects: Issues, Applications and Technologies (2 Vol., pp. 209-227). Hershey, PA: Information Science Reference.

Masters, H., Forrest, S., Harley, A., Hunter, M., Brown, N., & Risk, I. (2002). Involving mental health service users and carers in curriculum development: moving beyond 'classroom' involvement. *Journal of Psychiatric and Mental Health Nursing*, *9*, 309–316. doi:10.1046/j.1365-2850.2002.00493.x

Mattick, K., & Bligh, J. (2009). *Readiness for Interprofessional learning Scale*. In P. A. S.

Maxwell, N. (2003). Two Great Problems of Learning. *Teaching in Higher Education*, *8*(1), 129–134. doi:10.1080/1356251032000052375

Mayes, J. T. (1995). Learning Technology and Groundhog Day . In Strang, W., Simpson, V. B., & Slater, D. (Eds.), *Hypermedia at Work: Practice and Theory in Higher Education*. Canterbury, UK: University of Kent Press.

Mayes, T., & de Freitas, S. (2004). *Review of e-learning theories, frameworks and models. JISC project report*. Retrieved October 10, 2009, from http://www.e-learning.ac.uk/resources/modelsdeskreview/view

Mayes, T., & de Freitas, S. (2007). Learning and e-learning: the role of theory . In Beetham, H., & Sharpe, R. (Eds.), *Rethinking pedagogy for a digital age* (pp. 13–25). London: Routledge.

Mayo, J. A. (2004). Using case-based instruction to bridge the gap between theory and practice. *Journal of Constructivist Psychology*, *17*(2), 137–146. doi:10.1080/10720530490273917

McAleese, R. (1996). Technology in education to technology of education: concepts, conflicts and compromises . In Armstrong, S., Thompson, G., & Brown, S. (Eds.), *Facing up to radical changes in universities and colleges* (pp. 51–63). London: Kogan Page.

McCall, B. (2007). Cry for help from tutor 'counsellors'. *Times Higher Education Supplement*, *1785*, 4–5.

McConville, S. (2007). *Widening Participation: A Virtual Approach to F.E. Collaboration*. Wolverhampton, UK: University of Wolverhampton. Retrieved February 14, 2009, from http://wlv.openrepository.com/wlv/handle/2436/18076

McCroskey, J., & Robertson, P. J. (1999). Challenges and Benefits of Interprofessional Education: Evaluation of the Inter-Professional Initiative at the University of Southern California. *Teacher Education Quarterly*, 69–87.

McDonald, J. (2006). Learning object: A new definition, a case study and an argument for change. *Who's learning? Whose technology?: Proceedings of the 23rd Annual ASCILITE Conference,* Sydney, Australia. Retrieved December 30, 2008, from http://www.ascilite.org.au/conferences/sydney06/proceeding/pdf_papers/p99.pdf

McKee, M. (2003, July). Excavating our frames of mind: The key to dialogue and collaboration. *Social Work, 48*(3), 401–408.

McKey, P. (2000). *Total Student Experience Model.* White Paper, NextEd.

McLuhan, M., & Fiore, Q. (1967). *The medium is the massage*. Harmondsworth, UK: Penguin.

Meadon, R. (2009). Game On. Telegraph Motoring Supplement. Saturday, March 28th. M1-2.

Meads, G., Jones, I., Harrison, R., Forman, D. & Turner, W. (2009). Cultural Congruence between Higher Education and Health and Social Care Agencies. *Journal of Education and Work.*

Mellor, M. J., Hyer, K., & Howe, J. (2002). The geriatric interdisciplinary team approach: challenges and opportunities in educating trainees together from a variety of disciplines. *Educational Gerontology, 28*, 867–880. doi:10.1080/03601270290099868

Merleau-Ponty, M. (2002). *Phenomenology of perception.* London: Routledge.

Merryfield, M. (2003). Like a veil: Cross-cultural experiential learning online. *Contemporary Issues in Technology & Teacher Education, 3*(2), 146–171.

Merryfield, M. M. (2001). The paradoxes of teaching a multicultural education course online. *Journal of Teacher Education, 52*(4), 283–299. doi:10.1177/0022487101052004003

Meyer, J. H. F., & Land, R. (2003). Threshold concepts and troublesome knowledge: linkages to ways of thinking and practising, In C. Rust (Ed.), Improving Student Learning: Theory and Practice Ten Years On (pp. 412-424). Oxford, UK: Oxford Centre for Staff and Learning Development (OCSLD).

Mezirow, J. (2000). *Learning to think like an adult: core concepts of transformation theory. Learning as transformation: critical perspectives on a theory in progress.* San Francisco: Jossey Bass.

Micari, M., Streitwieser, B., & Light, G. (2006). Undergraduates Leading Undergraduates: Peer Facilitation in a Science Workshop Program. *Innovative Higher Education, 30*(4), 269–288. doi:10.1007/s10755-005-8348-y

Miccolo, M. A., & Spanier, A. H. (1993). Critical care management in the 1990s: making collaborative practice work. *Critical Care Clinics, 9*, 443–453.

Miers, M. E., Clarke, B. A., Pollard, K. C., Rickaby, C. E., Thomas, J., & Turtle, A. (2007). Online Inter-professional Learning: The Student Experience. *Journal of Interprofessional Care, 21*(5), 529–542. doi:10.1080/13561820701585296

Millen, J. (1997). Par for the Course: designing course outlines and feminist freedoms. *Curriculum Studies, 5*(1), 9–27.

Miller, C., Ross, N., & Freeman, M. (2001). *Interprofessional education in health and social care*. London: Arnold Publications.

Miller, C., Woolfe, C., & Mackintosh, N. (2006). Evaluation of Common Learning Pilots and Allied Health Professions First Wave Sites FINAL REPORT. Retrieved April 3, 2009, from http://www.smd-edu.qmul.ac.uk/student_info/undergrad/ipe/docs/CLAM_Final_Report_Revised_Sept_06.pdf

Milner, M. (1994). *Status and Sacredness: A General Theory of Status Relations and an Analysis of Indian Culture*. New York: Oxford University.

Mishra, P., & Koehler, M. J. (2006). Technological Pedagogical Content Knowledge: A framework for integrating technology in teacher education. *Teachers College Record, 108*(6), 1017. doi:10.1111/j.1467-9620.2006.00684.x

Mitchell, B. (2008). Chaired debate, Anna Peachey, Roo Reynolds, Edward Castronova, Claudia L'Amoreaux, Bill Thompson, Ren Reynolds. *ReLIVE08*, November 21st, 2008, Open University, Retrieved August 10, 2009, from http://stadium.open.ac.uk/stadia/preview.php?whichevent=1248&s=31&schedule=1468

Monahan, T., McArdle, G., & Bertolotto, M. (2008). Virtual reality for collaborative e-learning. *Computers & Education*, *50*, 1339–1353. doi:10.1016/j.compedu.2006.12.008

Monteith, M., & Smith, J. (2001). Learning in a virtual campus: the pedagogical implications of students' experiences. *Innovations in Education and Teaching International*, *38*(2), 119–125. doi:10.1080/13558000010030176

Mooney Melvin, P. (1987). In Quest of the Professional Historian: The Introduction to Public History Course. *The Public Historian*, *9*(3), 67–79.

Moore, M. G. (1993). Theory of Transactional Distance . In Keegan, D. (Ed.), *Theoretical Principles of Distance Education* (pp. 22–38). London: Routledge.

Morales, R., Carmichael, P., Wharrad, H. J., Bradley, C., & Windle, R. (2006). *Developing a Multi-Method Evaluation Strategy for Reusable Learning Objects: an approach informed by Cultural-Historical Activity Theory.* Presented at the 1st European Practice-based and Practitioner Research Conference - Improving quality in teaching and learning: Developmental work and Implementation challenges, University of Leuven, Belgium.

Morgan, J., Rawlinson, M., & Weaver, M. (2006). Facilitating online reflective learning for health and social care professionals. *Open Learning*, *21*(2), 167–176. doi:10.1080/02680510600715594

Morison, S., Boohan, M., Jenkins, J., & Mountray, M. (2003). Facilitating undergraduate interprofessional learning in healthcare: Comparing classroom and clinical learning for nursing and medical students. *Learning in Health and Social Care*, *2*(2), 93–104. doi:10.1046/j.1473-6861.2003.00043.x

Mu, K., & Royeen, C. B. (2004). Interprofessional vs. interdisciplinary services in school-based occupational therapy. *Occupational Therapy International*, *11*(4), 244–247. doi:10.1002/oti.214

Muirhead, R. J. (2007). E-learning: is this teaching at students or teaching with students? *Nursing Forum*, *42*(4), 178–184. doi:10.1111/j.1744-6198.2007.00085.x

Mulholland, J., Mallik, M., & Moran, P. (2005). *An Overview of the Nature of the Preparation of Practice Educators in Five Health Care Disciplines.* 6th Occasional Paper. Higher Education Academy, London: Higher Education Academy: Health Sciences and Practice. Accessed August 10, 2009, from http://www.health.heacademy.ac.uk/publications/occasionalpaper/occp6.pdf

Mulholland, J., Mallik, M., Moran, P., Scammell, J., & Turnock, C. (2005). *Making practice-based learning work. An overview of the nature of the preparation of practice educators in five health care disciplines.* London: Higher Education Academy, Health Science and Practice Network.

Munro, L. (2000). Non-Disciplinary Power and the Network Society. *Organization*, *7*(4), 679–695. doi:10.1177/135050840074010

Murphy, E. A. (2005). Handheld computers software for school nurses. *The Journal of School Nursing*, *21*(6), 56–360. doi:10.1177/10598405050210061101

Myall, M., Levett-Jones, T., & Lathlean, J. (2008). Mentorship in contemporary practice: the experiences of nursing students and practice mentors . *Journal of Clinical Nursing*, *17*(14), 1834–1842. doi:10.1111/j.1365-2702.2007.02233.x

Naismith, L., Lonsdale, P., Vavoula, G., & Sharples, M. (2004). *Mobile technologies and learning.* Retrieved April 7, 2008, from http://www.futurelab.org.uk/resources/publications_reports_articles/literature_reviews/Literature_Review203/

Nardi, B., Whittaker, S., & Bradner, E. (2000). Interaction and Outeraction: Instant Messaging in Action. *Proceedings of ACM Conference on Computer Supported Cooperative Work (CSCW '00)*, (pp. 79-88). New York: ACM Press.

Nash, S. S. (2005). Learning objects, learning object repositories, and learning theory: Preliminary best practices for online courses. *Interdisciplinary Journal of Knowledge and Learning Objects*, *1*, 217–228.

NCIHE - National Committee of Inquiry into Higher Education. (1997). *Higher Education in the Learning Society.* London: HMSO.

Nelson, E. (2003). A Practical Solution for Training and Tracking in Patient-care Settings. *Nursing Administration Quarterly*, *27*(I), 29.

Nelson-Kilger, M. (1993). *The Digital individual.* The Third Conference on Computers Freedom and Privacy, March 9-12, 1993, Burlingame, CA. Retrieved May 14, 2009, from http://cpsr.org/prevsite/conferences/cfp93/nelson-kilger.html

Neville, M. W. (2003). Testing techniques. Blackboard allows students to take tests on the go . *Nurse Educator*, *28*(5), 207–209. doi:10.1097/00006223-200309000-00005

Newell, S., & Swan, J. (2000). Trust and inter-organisational networking. *Human Relations*, *53*(10), 1287–1328.

NHS Institute for Innovation & Improvement. (2008). *Improvement in Pre-Registration Education for better, safer healthcare: An NHS Institute initiative to introduce pre-registration health and social care students to improvement*. Retrieved May 7, 2008, from http://www.institute.nhs.uk/images//documents/BuildingCapability/Building%20improvement%20capability/Preregistration%20training/51427%20Pre-registration%20leaflet.pdf

Nicol, D., & Sinclair, C. (2003). The Social Dimensions of Online Learning. *Innovations in Education and Teaching International*, *40*(3), 270–280. doi:10.1080/1470329032000103807

NIMHE. (2003). *Cases for Change: User involvement*. Leeds, UK: NIMHE.

Nisbet, G., Hendry, G. D., Rolls, G., & Field, M. J. (2008). Interprofessional learning for prequalification health care students: An outcomes-based evaluation. *Journal of Interprofessional Care*, *22*(1), 57–68. doi:10.1080/13561820701722386

Noble, D. (1997). Digital Diploma Mills – Part 1: The automation of Higher Education. *FirstMonday, 3*(1). Retrieved October 10, 2009, from http://www.firstmonday.dk/issues/issue 3_1/noble/

Norman, I. (2005). Inter-professional education for pre-registration students in the health professions: recent developments in the UK and emerging lessons. *International Journal of Nursing Studies*, *42*, 119–123. doi:10.1016/j.ijnurstu.2004.11.003

Nunnington, N. (2009). *The European Challenge*. Retrieved March 24, 2009, from http://www.cebe.heacademy.ac.uk/learning/acbee/pdf/The%20European%20Challenge%20_Nick%20Nunnington.pdf

Nursing and Midwifery Council (NMC). (2007). *Standards for the preparation of teachers of nursing and midwifery*. London: Nursing and Midwifery Council.

Nyatanga, L. (2005.) The archetypal roots of ethnocentrism. In *Interprofessional Education*. Higher Education Academy Health Sciences and Practice Occasional Paper no. 7.

O'Keeffe, M., O'Regan, L., & Cashman, D. (2008). *Supporting the development of Communities of Practice: informal versus formal communities*. Paper presented at the Association for Learning Technology conference 'Rethinking the Digital Divide' University of Leeds, UK.

O'Malley, J., & McGraw, H. (1999). Students perceptions of distance learning, online learning and the traditional classroom. *Online Journal of Distance Learning Administration*, *2*(4).

O'Neil, C., & Fisher, C. (2008). Should I take this Course Online? *The Journal of Nursing Education*, *47*(2), 53–58. doi:10.3928/01484834-20080201-04

O'Neil, E.Pew Health Professions Commission. (1993). *Recreating Health Professional Practice for a New Century*. San Francisco, CA: Pew Health.

O'Neill, B., Wyness, A., McKinnon, S., & Granger, P. (2000). *Partnership, Collaboration and Course Design: An emerging model of interprofessional education*. Retrieved November 8, 2008, from http://www.cstudies.ubc.ca/facdev/services/registry/pcacdaemoie.html

O'Reilly, T. (2005). *What Is Web 2.0?* Retrieved May 14, 2009, from http://www.oreillynet.com/pub/a/oreilly/tim/news/2005/09/30/what-is-web-20.html?page=1

Oandasan, I., & Reeves, S. (2005). factors, processes and outcomes in IPE. *Journal of Interprofessional Care*, *19*(1), 39–48. doi:10.1080/13561820500081703

Oandasan, I., & Reeves, S. (2005). Key elements for interprofessional education: Part 1: The learner, the educator and the learning context. *Journal of Interprofessional Care*, *19*(Suppl. 1), 21–38. doi:10.1080/13561820500083550

Ohio State University. (1999). *Assumptions for Interprofessional Education and Practice*. Ohio: The Interprofessional Commission of Ohio. Retrieved August 30, 2008, from http://www.osu.edu/ico/assumptn.html

Oliver, G., Spence, R. & Kirby, P. (1994). Charting a New Course - Education for the Built Environment. *The Cambridge Review, May 1994.*

Oliver, M. & Carr, D. (2009). Learning in Virtual Worlds: Using Communities of Practice to Explain How People Learn From Play. *British Journal and Education and Technology, Special issue, 'Out of This World: 3-D MUVEs in Higher Education'*, *40*(3), April.

Oliver, M. (2002). What do Learning Technologists Do? *Innovations in Education and Training International*, *39*(4), 245–252.

Oliver, M. (2003). *Curriculum Design as acquired social practice: a case study*. Paper presented at the 84th Annual Meeting of the American Educational Research Association, Chicago.

Oliver, M. (2004). Metadata vs. educational culture: roles, power and standardisation . In Land, R., & Bayne, S. (Eds.), *Education in Cyberspace* (pp. 112–138). London: RoutledgeFalmer.

Oliver, M., & Carr, D. (2008). Learning in virtual worlds: using Communities of Practice to explain how people learn from play. *BJET*. Abstract Retrieved September 11, 2009, from http://learningfromsocialworlds.worpress.com/

Oliver, M., & Conole, G. (2003). Evidence-based practice and E-Learning in Higher Education: Can We and Should We? *Research Papers in Education*, *18*(4), 385–397. doi:10.1080/0267152032000176873

Oliver, M., Price, S., Boycheva, S., Dugstad Wake, J., Jones, C., Mjelstad, S., Kemp, B., Nikolov, R., & van der Meij, H. (2005). *Empirical studies of the impact of technology-enhanced learning on roles and practices in Higher Education*. Kaleidoscope project deliverable 30-03-01-F.

Oliver, M., Sharpe, R., Duggleby, J., Jennings, D., & Kay, D. (2004). *Accrediting learning technologists: a review of the literature, schemes and programmes*. ALT accreditation project report No. 1. Retrieved October 10, 2009, from http://www.ucl.ac.uk/epd/alt-accreditation/Initial_review.doc

Oliver, R. (1999). Exploring strategies for on-line teaching and learning . *Distance Education*, *20*(2), 240–254. doi:10.1080/0158791990200205

Oliver, R., & Herrington, J. (2001). Online learning design for dummies: professional development strategies for beginning online designers. In P. Barker & S. Rebelsky (Eds.), *Proceedings of ED-MEDIA 2002, World Conference on Educational Multimedia, Hypermedia and Telecommunications*. Norfolk, VA: AACE (p. 1500-1505). Retrieved September 20, 2009, from http://elrond.scam.ecu.edu.au/oliver/

Oliver, R., & McMahon, M. (2006). *Trials of Learning Objects (Series 7).* Retrieved February 16, 2009, from http://toolboxes.flexible-learning.net.au/documents/pdfs/lotrials_report06.pdf

Orick, J. (2000). The Virtual Library: Changing Roles and Ethical Challenges for Librarians. *The International Information & Library Review*, *32*(3-4), 313–324. doi:10.1006/iilr.2000.0138

Ousey, K. J., & White, S. A. (2009). Developing a Multi-professional e-Learning Mentor Update Package. In Proceedings *of the International Conference on Mobile, Hybrid, and On-line Learning, elml,* (pp.103-107).

Owens, M., Allen, L., Hutchinson, J. & Pearson, D. (2007). *Final report on: Promoting interprofessional learning across West Yorkshire*. Report presented to the West Yorkshire Strategic Health Authority.

Palloff, R. M., & Pratt, K. (2005). Collaborating Online-Learning Together in Community (pp. 3-18, 57-97). San Francisco, CA: Jossey-Bass.

Papadopoulos, G., & Pearson, E. (2007). Accessibility awareness raising and continuing *professional development: The use of simulations as a motivational tool.* Association for Teaching and Learning, Online newsletter. Retrieved March 19, 2007, from http://newsletter.alt.ac.uk/e_article000735502.cfm

Papert, S. (1980). *Mindstorms: Children, Computers and Powerful Ideas*. New York: Basic Books.

Paris, D. C. (2000). Is there a professor in the class? In R. A. Cole, (Ed.), (2000), Issues in web-based pedagogy: A critical primer. London: Greenwood Press.

Paris, M. (2003). Reuse in Practice: Learning objects and software development. *Interact, Integrate, Impact: Proceedings of the 20th Annual ASCLITE Conference*. Adelaide, Australia. Retrieved March 12, 2009, from http://www.ascilite.org.au/conferences/adelaide03/docs/pdf/679.pdf

Parsell, G., & Bligh, H. (1999). Interprofessional learning. *Postgraduate Medical Journal, 74*, 89–95. doi:10.1136/pgmj.74.868.89

Parsell, G., Spalding, R., & Bligh, J. (1998). Shared goals, shared learning: evaluations of a multiprofessional course for undergraduate students. *Medical Education, 32*, 304–311. doi:10.1046/j.1365-2923.1998.00213.x

Paulus, T. M., & Phipps, G. (2008). Approaches to case analyses in synchronous and asynchronous environments. *Journal of Computer-Mediated Communication,* (13): 459–484. doi:10.1111/j.1083-6101.2008.00405.x

Payler, J., Meyer, E., & Humphries, D. (2008). Pedagogy for Interprofessional education – what do we know and how can we evaluate it? *Learning in Health and Social Care, 7*(2), 64–68. doi:10.1111/j.1473-6861.2008.00175.x

Payne, M. (2006). Intentity Politics in Multiprofessional teams: Palliative care social work. *Journal of Social Work, 6*(2), 137–150. doi:10.1177/1468017306066741

Pea, R. (1996). Seeing what we build together: Distributed multimedia learning environments for transformative communications . In Koschmann, T. (Ed.), *CSCL: Theory and Practice of an Emerging Paradigm* (p. 353). Hillsdale, NJ: Lawrence Erlbaum and Associates.

Peacock, S., & Hooper, J. (2006). *How Do Diverse Groups of Learners in the Health Sciences Respond To A New Virtual Learning Experience?* Retrieved July 6, 2009, from http://www.health.heacademy.ac.uk/projects/miniprojects/completeproj/peacock.pdf

Peacock, S., & Hooper, J. (2007). E-learning in physiotherapy education. *Physiotherapy, 93*(3), 218–228. doi:10.1016/j.physio.2006.11.009

Pearson, E. J., & Koppi, T. (2006). Supporting Staff in Developing Inclusive Online Learning . In Adams, M., & Brown, S. (Eds.), *Towards Inclusive Learning in Higher Education*. London: Routledge Press.

Pecukonis, E., Doyle, O., & Bliss, D. L. (2008). Reducing barriers to interprofessional training: Promoting interprofessional cultural competence. *Journal of Interprofessional Care, 22*(4), 417–428. doi:10.1080/13561820802190442

Pellatt, G. C. (2006). The role of mentors in supporting preregistration nursing students. *British Journal of Nursing (Mark Allen Publishing), 15*(6), 336–340.

Perry, W. (1970). *Forms of intellectual and academic developments in the college years*. New York: Holt.

Peters, M. (2000). Does constructivist epistemology have a place in nursing education? *The Journal of Nursing Education, 39*, 166–172.

Pethybridge, J. (2004). How team working influences discharge planning from hospital: a study of four multi-disciplinary teams in an acute hospital in England. *Journal of Interprofessional Care*, (18): 29–41. doi:10.1080/13561820410001639334

*Pew Internet & American Life Project*. (2009). Retreived August 10, 2009, from www.pewinternet.org Washington, DC: Pew Internet & American Life Project.

Phipps, L., Sutherland, A., & Seale, J. (Eds.), *Access All Areas: disability, technology and learning* (pp. 6–9). Oxford, UK: JISC TechDis Service and ALT.

Piaget, J. (1929). *The Child's Conception of the World.* New York: Harcourt, Brace Jovanovich.

Piertroni, P. (1991). Stereotypes or archetypes? A study of perceptions of health-care students. *Journal of Social Work Practice, 5*(1), 61–69.

Pollard, & Hillage. (2000). *Exploring E-learning*. IES Report 376.

Polsani, R. P. (2003). Use and Abuse of Reusable Learning Objects. *Journal of Digital Information, 3*(4). Retrieved January 15, 2009, from http://jodi.tamu.edu/Articles/v03/i04/Polsani/

Pomeroy, V. M., & Philip, I. (1994). Health-care teams: An interdisciplinary workshop for undergraduates. *Medical Teacher, 16*(4), 341–346. doi:10.3109/01421599409008271

Posey, L., & Pintz, C. (2006). Online teaching strategies to improve collaboration among nursing students. *Nurse Education in Practice, 6*(2), 372–379. doi:10.1016/j.nepr.2006.07.018

Prensky, M. (2001). Digital Natives, Digital Immigrants. *On the Horizon. NCB University Press, 9* (5). Retrieved October 10, 2009, from http://www.marcprensky.com/writing/

Prensky, M. (2001). Do They Really Think Differently? *Horizon, 9*(6).

Prensky, M. (2001). Retrieved 13/02/09 from http://primaryblog.wordpress.com

Price, J. (2005). Complexity and Interprofessional Education. In The Higher Education Academy, Health Science and Practice Network, The Theory-Practice Relationship in Interprofessional Education (pp. 76-87). London: Higher Education Academy.

Price, S., & Oliver, M. (2007). A Framework for Conceptualising the Impact of Technology on Teaching and Learning. *Educational Technology & Society, 10*(1), 16–27.

PriceWaterhouseCoopers. (2000). *Annex 3: Learning products and services for the e-U*. Retrieved October 10, 2009, from http://www.hefce.ac.uk/pubs/hefce/2000/0044/00_44a3.pdf

Priest, H., Sawyer, A., Roberts, P., & Rhodes, S. (2005). A survey of interprofessional education in communication skills in health care programmes in the UK. *Journal of Interprofessional Care, 19*(3), 236–250. doi:10.1080/13561820500053447

Pulman, A. (2007). On the road to Virtual Europe – Redux. *The Electronic Journal of E-learning, 5*(4), 297–304.

Pulman, A., Scammell, J., & Martin, M. (2008). Enabling interprofessional education: the role of technology to enhance learning. *Nurse Education Today, 29*(2), 232–239. doi:10.1016/j.nedt.2008.08.012

Pulsford, D., Boit, K., & Owen, S. (2002). Are mentors ready to make a difference? A survey of mentors' attitudes towards nurse education. *Nurse Education Today, 22*(6), 439–446.

Quinn, C. N. (1996). *Pragmatic evaluation: Lessons from usability*. Paper presented at the 13th Annual Conference of the Australasian Society for Computers in Learning in Tertiary Education.

Quinney, A. (2005). Placements online; student experiences of a website to support learning in practice settings. *Social Work Education, 24*(4), 439–450. doi:10.1080/02615470500096951

Quinney, A. (2006). *Collaborative Social Work Practice*. Exeter, UK: Learning Matters.

Quinney, A., Hutchings, M., & Scammell, J. (2008). Student and staff experiences of using a virtual community, Wessex Bay, to support interprofessional learning: messages for collaborative practice. *Social Work Education, 27*(6), 658–664. doi:10.1080/02615470802201754

Rafter, M. E., Pesun, I. J., Herren, M., Linfante, J. C., Mina, M., & Wu, C. D. (2006). A preliminary survey of interprofessional education. *Journal of Dental Education, 70*(4), 417–427.

Ragan, L. C. (1999). Good Teaching is Good Teaching: An Emerging Set of Guiding Principles and Practice for the Design and Development of Distance Education. *Cause/Effect, 22*, 20-24.

Ramsden, P. (1991). A performance indicator of teaching quality in higher education: the course experience questionnaire. *Studies in Higher Education, 16*, 129–150. doi:10.1080/03075079112331382944

Ramsden, P. (1992). *Learning to teach in higher education*. London: Routledge. doi:10.4324/9780203413937

Rawson, D. (1994). Models of Inter-professional work. Likley theories and possibilities . In Leathard, A. (Ed.), *Going Inter-professional: Working Together for Health and Welfare* (pp. 38–63). London: Routledge.

Reason, P. (1996). Power and conflict in multi-disciplinary collaboration . In Piertroni, P., & Piertroni, C. (Eds.), *Innovation in community care and primary health: the Marylebone experiment*. New York: Churchill Livingstone.

Recker, M., Dorward, J., & Dawson, D. Xin Mao, Ye Liu, Palmer, B. et al. (2005). Teaching, Designing, and Sharing: A Context for Learning Objects. *Interdisciplinary Journal of Knowledge and Learning Objects,* (1), 197-216. Retrieved July 10, 2009, from http://ijklo.org/Volume1/v1p197-216Recker.pdf

Rees, D., & Johnson, R. (2007). All together now? Staff views and experiences of a pre-qualifying interprofessional curriculum. *Journal of Interprofessional Care, 21*(5), 543–555. doi:10.1080/13561820701507878

Reeves, S., & Freeth, D. (2003). New forms of information technology, new forms of collaboration? In Leathard, A. (Ed.), *Interprofessional Collaboration: From Policy to Practice in Health and Social Care*. London: Routledge.

Reeves, S., Zwarenstein, M., Goldman, J., Barr, H., Freeth, D., Hammick, M. & Koppel, I. (2008). Interprofessional education: effects on professional practice and health care outcomes. *Cochrane Database of Systematic Reviews,* Issue 1.

Reid, I. C. (1999). Beyond models: Developing a university strategy for online instruction. *Journal of Asynchronous Learning Networks, 3*(1), 19–31.

Richardson, B., & Cooper, N. (2003). Developing a virtual interdisciplinary research community in higher education. *Journal of Interprofessional Care, 17*, 173–182. doi:10.1080/1356182031000081777

Richardson, W. (2009). *Blogs, wikis, podcasts, and other powerful web tools for classrooms* (2nd ed.). Thousand Oaks, CA: Corwin Press.

Robbins, S. (2008). *Virtual worlds as Web2.0 Learning spaces*. Retrieved September 11, 2009, from http://hosted.mediasite.com/flash/ELI/Virtualworldsasweb2.0learningspaces/

Robbins, S., & Bell, M. (2008). *Second life for Dummies*. London: Wiley.

Roberts, G. (2008) Enhancing Second Life as an Educational Tool: Resolving Tensions Between Anonymity and Trust, poster presentation, ReLIVE 08 conference, November, 20th and 21st 2008, Open University, http://www.open.ac.uk/relive08/documents/Detailed-final-programme.pdf

Roberts, J., O'Sullivan, J., & Howard, J. (2005). The roles of emerging and conventional technologies in serving children and adolescents with special needs in rural and Northern communities. *Journal of Distance Education, 20*(1), 84–103.

Roberts, K. K. (1998). *A Naturalistic Study of Students' Experience in a Computer-Based Nursing Course*. Doctoral Dissertation University of Kansas, Dissertation Abstracts International 59 (12) 4352A (UMI No 9914 105).

Robley, L. R., Farnsworth, B. J., Flynn, J. B., & Horne, C. D. (2004). This New House: Building Knowledge Through Online Learning. *Journal of Professional Learning, 20*, 333–343. doi:10.1016/j.profnurs.2004.07.012

Robson, C. (2002). *Real world research: a resource for social scientists and practitioner –researchers* (2nd ed.). Oxford, UK: Blackwell Publishing.

Rogers, E. M. (2003). *Diffusion of Innovations* (5th ed.). New York: Free Press.

Rogoff, B. (2003). *The cultural nature of human development*. New York: Oxford University Press.

Rose, D. (2001). *Users' Voices: the Perspectives of Mental Health Service Users on Community and Hospital Care*. London: Sainsbury Centre for Mental Health.

Rosenberg, M. J. (2001). *E-Learning – Strategies for Delivering Knowledge in the Digital Age*.

Rosenlund, C. H., & Damark-Bembenek, B. (1999). Assessing the effectiveness of an online program. *Nurse Educator*, *24*(1), 5–6. doi:10.1097/00006223-199901000-00001

Ross, F., & Southgate, L. (2000). Learning together in medical and nursing training: aspirations and activity. *Medical Education*, *34*(9), 739–743. doi:10.1046/j.1365-2923.2000.00626.x

Rourke, L., Anderson, T., Garrison, D. R., & Archer, W. (1999). Assessing social presence in asynchronous text-based computer conferencing. *Journal of Distance Education*, *14*(2), 50–71.

Rouse, D. P. (2000). The effectiveness of computer-assisted instruction in teaching nursing students about congenital health disease. *Computers in Nursing*, *18*(6), 282–287.

Rowley, J. B., & Hart, P. M. (1996). How video case studies can promote reflective dialogue. *Educational Leadership*, *53*, 28–29.

Royal College of Physicians. (2008). *National Clinical Guidelines for Stroke* (3rd ed.). London: RCP.

Ruddle, K., & Feeny, D. (1997). *Transforming the Organisation: New Approaches to Management Measurement and Leadership. Oxford Executive Research Briefing*. Oxford, UK: Templeton College.

Ruebain, D., Honigmann, J., Mountfield, H., & Parker, C. (2006). *Analysis of the statutory and regulatory frameworks and cases relating to fitness standards in nursing, social work and teaching*. Disability Rights Commission. Retrieved August 4, 2009, from http://www.equalityhumanrights.com/Documents/DRC/Policy/fitness_regulatory_review_report.pdf

Russell, J. E. A., & Adams, D. M. (1997). The changing nature of mentoring in organizations: An introduction to the special issue on mentoring in organizations. *Journal of Vocational Behavior*, *51*, 1–14. doi:10.1006/jvbe.1997.1602

Ryan, D. P., & Marlow, B. (2004). Build-a-case: A brand new continuing medical education technique that is peculiarly familiar. *The Journal of Continuing Education in the Health Professions*, *24*(2), 112–118. doi:10.1002/chp.1340240208

Sainsbury Centre for Mental Health. (1997). *Pulling together*. London: Sainsbury Centre Publications.

Salas, E., Bowers, C. A., & Rhodenizer, L. (1998). It Is Not How Much You Have but How You Use It: Toward a Rational Use of Simulation to Support Aviation Training. *The International Journal of Aviation Psychology*, *8*(3), 197–208. doi:10.1207/s15327108ijap0803_2

Salmon, G. (2000). *E-moderating: the key to teaching and learning on-line*. London: Kogan Page.

Salmon, G. (2002). *E-tivities: the key to active online learning*. London: Routledge Falmer.

Salmon, G. (2002). Mirror, mirror, on my screen... Exploring online reflections. *British Journal of Educational Technology*, *33*(4), 379–391. doi:10.1111/1467-8535.00275

Salmon, G. (2004). *E-moderating, the key to online teaching and learning* (2nd ed.). London: Routledge Falmer.

Salmon, G. (2007). *The SEAL: Second life: Advanced Learning project*. University of Leicester. Retrieved September 11, 2009, from www.le.ac.uk/beyonddistance/mediazoo

Salmon, G., Jones, S., & Armellini, A. (2008). Building institutional capability in e-learning design. *Alt-j . Research in Learning Technology*, *16*(2), 95–109.

San Diego, J., Laurillard, D., Boyle, T., Bradley, C., Ljubojevic, D., Neumann, T., & Pearce, D. (2008). Towards a user-oriented analytical approach to learning design. *ALT-J: Research in Learning Technology*, *16*(1), 15–29.

San Martin-Rodriguez, L., Beaulieu, M. D., D'Amour, D., & Ferrada-Videla, M. (2005). The determinants of successful collaboration: A review of theoretical and empirical studies. *Journal of Interprofessional Care*, (Supplement 1), 132–147. doi:10.1080/13561820500082677

Santy, J., Beadle, M. & Needham, Y. (2009). Using an online case conference to facilitate interprofessional learning. *Nurse Education in Practice*, doi:10.1016/j.nepr. 2008.11.04

Saunders, D. (1992). Peer Tutoring in Higher Education. *Studies in Higher Education, 17*(2), 211–218. doi:10.1080/03075079212331382677

Savin Baden, M. (2007). *A Practical Guide to Problem-Based Learning Online* (pp. 96–104). London: Routledge.

Savin-Baden, M. (2000). *Problem-based learning in higher education: untold stories*. Buckingham, UK: SRHE and Open University Press.

Savin-Baden, M. (2008a). *Learning Spaces. Creating Opportunities for Knowledge Creation in Academic Life* (pp. 136–150). The Society for Research into Higher Education.

Savin-Baden, M. (2008b). From cognitive capability to social reform? Shifting perceptions of learning in immersive worlds. *Alt-j . Research in Learning Technology, 16*(3), 151–161.

Savin-Badin, M., Tombs, C., White, D., Poulton, T., Kavia, S., & Woodham, L. (2009). *Getting Started with Second life*. JISC. Retrieved September 11, 2009, from http://www.jisc.ac.uk

Scammell J., Hutchings M., Quinney, A., Hean, S., Dumbrell, A. & Pulman, A. (2008). *A virtual practice community for student learning and staff development in health and social work inter-professional education:changing practice through collaboration*. School of Health and Social Care, Bournemouth University. Higher Education Academy, A Health Sciences and Practice Subject Centre, Mini Project, 1/1/06-31/12/07.

Schon, D. (1983). *The Reflective Practitioner*. London: Temple Smith.

Schön, D. A. (1987). *Educating the Reflective Practitioner*. San Francisco, CA: Jossey-Bass.

Schön, D. A. (1992). The crises of professional knowledge and the pursuit of an epistemology of practice. *Journal of Interprofessional Care, 6*(1), 49–63. doi:10.3109/13561829209049595

Scollon, R. (2001). *Mediated Discourse: The Nexus of Practice*. London: Routledge.

Scottish Centre for Information on Language Teaching and Research. (2008). *Cross sector collaborative activities to promote modern languages in Scotland*. Report to the Subject Centre for Languages, Linguistics and Area Studies.

Scruton, R. (1979). *The Aesthetics of Architecture*. London: Methuen & Co.

Senge, P. (1990). *The Fifth Discipline: The Art and Practice of the Learning Organisation*. New York: Doubleday.

Sfard, A. (1998). On Two Metaphors for Learning and the Dangers of Choosing Just One. *Educational Researcher, 27*(2), 4–13.

Sharpe, R., & Oliver, M. (2007). Supporting practitioners' design for learning: principles of effective resources and interventions . In Beetham, H., & Sharpe, R. (Eds.), *Rethinking Pedagogy for a Digital Age* (pp. 117–128). London: Routledge.

Sharpe, R., Benfield, G., Roberts, G., & Francis, R. (2006). *The undergraduate experience of blended e-learning: a review of UK literature and practice*. York, UK: Higher Education Academy. Retrieved August 20, 2009, from http://www.heacademy.ac.uk/assets/York/documents/ourwork/research/literature_reviews/blended_e-learning_full_review.pdf

Shulman, L. S. (1992). Toward a pedagogy of cases . In Shulman, J. H. (Ed.), *Case methods in teacher education* (pp. 1–30). New York: Teachers College Press.

Sigulem, D. M. (2001). A Web-Based distance education course in nutrition in public health: case study. *Journal of Medical Internet Research, 3*(2), E16. doi:10.2196/jmir.3.2.e16

Sijstermans, R., Jaspers, M., Bloemendaal, P., & Schoonderwaldt, E. (2007). Training inter-physician communication using the dynamic patient simulator. *International Journal of Medical Informatics, 76*, 336–343. doi:10.1016/j.ijmedinf.2007.01.007

Silen, C. (2001). *Between chaos and cosmos-a driving force for responsibility and independence in learning, The Power of Problem Based Learning. PROBLARC, The 3rd Asia Pacific conference on PBL, 2001*. Australia: The University of Newcastle.

Silverman, R., Welty, W. M., & Lyon, S. (1996). *Case Studies for Teacher Problem Solving* (2nd ed.). New York: McGraw-Hill.

Simpson, V. (1996). The Hypertext Campus project. In TLTSN Case Studies 2: Managing the Adoption of Technology for Learning. Teaching and Learning Technology Support Network.

Sinclair, L. (2004). Interprofessional education (IPE): New knowledge in need of translation? *Canadian Physiotherapy Association: Research News and Views Newsletter*, Spring, 11-13.

Singh-Cooner, T. (2010). Designing for enquiry: using Web 2.0 to enable mental health service user and carer involvement . In Bromage, A., Clouder, D., Thistlethwaite, J., & Gordon, F. (Eds.), *Interprofessional E-learning and Collaborative Work: Practices and Technologies*. Hershey, PA: IGI-Global.

Sit, J. W. H., Chung, J. W. Y., Chow, M. C. M., & Wong, T. K. S. (2005). Experiences of online learning: students' perspective. *Nurse Education Today*, *25*, 140–147. doi:10.1016/j.nedt.2004.11.004

Skinner, J. (1958). Teaching Machines. [from http://www.bfskinner.org/f/teaching_machines.pdf]. *Science*, *128*(3330), 969–977. Retrieved October 10, 2009. doi:10.1126/science.128.3330.969

Skjørshammer, M. (2001). Co-operation and conflict in a hospital: interprofessional differences in perception and management of conflicts. *Journal of Interprofessional Care*, (15): 7–18. doi:10.1080/13561820020022837

Skorga, P. (2002). Interdisciplinary and distance education in the Delta: the Delta Health Education Partnership. *Journal of Interprofessional Care*, *16*(2), 149–157. doi:10.1080/13561820220124166

Smith, H., & Oliver, M. (2002). University teachers' attitudes to the impact of innovations in ICT on their practice. In C. Rust (Ed.), *Proceedings of the 9th International Improving Student Learning Symposium* (pp. 237-246). Oxford, UK: Oxford Centre for Staff and Learning Development.

Smith, J. (2005). *The Shipman Enquiry: The Sixth Report. Shipman: the final report*. London: The Shipman Enquiry.

Soetendorp, R., & Byles, L. (2000, January). *Law for non-lawyers: facilitating inter-professional dialogue*. Paper presented at 2nd HEA Annual Conference on Learning in Law Initiative, Warwick University, UK. Retrieved April 15, 2008 from http://www.ukcle.ac.uk/interact/lili/2000/soetendorp.html

Solomon, P., & Geddes, E. L. (in press). An interprofessional e-learning module on health care ethics. *Journal of Interprofessional Care*.

Soong, B. M. H., Chan, H. C., Chua, B. C., & Loh, K. F. (2001). Critical success factors for on-line course resources. *Computers & Education*, *36*(2).

*Special Educational Needs and Disability Act*. (2001). Retrieved July 20, 2009, from http://www.opsi.gov.uk/Acts/acts2001/en/ukpgaen_20010010_en_1

Spence, R. J. S., Macmillan, S., & Kirby, P. (2001). *Interdisciplinary design in practice*. London: Thomas Telford.

Spencer, J., Blackmore, D., Heard, S., McCrorie, P., McHaffie, D., & Scherpbier, A. (2000). Patient-oriented learning: a review of the role of the patient in the education of medical students. *Medical Education*, *34*(10), 851–857. doi:10.1046/j.1365-2923.2000.00779.x

Spouse, J. (2000). Supervision of Clinical Practice: The Nature of Professional Development . In Spouse, J., & Redfern, L. (Eds.), *Successful Supervision in Health Care Practice* (pp. 126–154). Oxford, UK: Blackwell Science.

Squires, D., & Preece, J. (1999). Predicting quality in educational software: Evaluating for learning, usability and the synergy between them. *Interacting with Computers*, *11*, 467–483. doi:10.1016/S0953-5438(98)00063-0

Stanton, G. (2008). Access to higher education in England: its position within the wider education ecology . In *Unfinished Business in Widening Participation* (pp. 9–26). London: The Learning and Skills Network.

Stepans, M. B., Thompson, C. L., & Buchanan, M. L. (2002). The role of the nurse on a transdisciplinary early intervention assessment team. *Public Health Nursing (Boston, Mass.)*, *19*, 238–245. doi:10.1046/j.1525-1446.2002.19403.x

Steuer, J. (1995). Defining virtual reality: Dimensions determining telepresence . In Biocca, F., & Levy, M. R. (Eds.), *Communication in the Age of Virtual Reality* (pp. 33–56). Mahwah, NJ: Lawrence Erlbaum Associates.

Stevens, A., Hernandez, J., Johnsen, K., Dickerson, R., Raij, A., & Harrison, C. (2006). The use of virtual patients to teach medical students history taking and communication skills. *American Journal of Surgery*, *191*(6), 806–811. doi:10.1016/j.amjsurg.2006.03.002

Stew, G. (2000). Interprofessional Learning – An Evaluation . In Bourner, T., Katz, T., & Watson, D. (Eds.), *New Directions in Professional Higher Education* (pp. 52–59). Buckingham, UK: SRHE and Open University.

Stew, G. (2005). Learning together in practice: a survey of Interprofessional education in clinical settings in south-east London. *Journal of Interprofessional Care*, *19*, 223–235. doi:10.1080/13561820500138685

Stolworthy, Y. & Suszka-Hildebrandt, S. (2000). Mobile information technology at the point-of-care: the grass roots origin of mobile computing in nursing. *PDA Cortex 2000* [serial online].

Stumpf, S. H., & Clark, J. Z. (1999). The Promise And Pragmatism Of Inter-Disciplinary Education. *Journal of Allied Health*, *28*, 30–32.

Suchan, J. & Hayzak, G. (2001). The communication characteristics of virtual teams: a case study. *Professional Communication, IEEE transactions*, *44*(3), 174-186.

Sullivan, T. J. (1998). *Collaboration – A Health Care Imperative*. London: McGraw Hill Company.

Swan, K. P., Richardson, J. C., Ice, P., Garrison, D. R., Cleveland-Innes, M. & Arbaugh, J. B. (2008). Validating a Measurement Tool of Presence in Online Communities of Inquiry. *E-mentor*, *2*(24).

Tait, M., Tait, D., Thorton, F., & Edwards, M. (2008). Development and evaluation of a critical care e-learning scenario. *Nurse Education Today*, *28*, 970–980. doi:10.1016/j.nedt.2008.05.016

Taradi, S., Taradi, M., Radic, K., & Pokrajac, N. (2005). Blending problem-based learning with Web technology positively impacts student learning outcomes in acid-based physiology. *Advances in Physiology Education*, *29*, 35–39. doi:10.1152/advan.00026.2004

Tarvainen, M. (2004). *Engineering Education and Interdisciplinary Studies*. Paper presented at 2004 SEF Annual Congress: The XXI Century: The Golden Opportunity for Engineering Education. Finland.

Tate, M., & Hoshek, D. (2009). A model for the effective management of reusable learning objects (RLOs): lessons from a case study. *Interdisciplinary Journal of E-learning & Learning Objects*, *5*, 51–72.

Tatnall, A., & Davey, B. (2005). *An actor-network approach to informing clients through portals*. Informing Science and Information Technology Education Joint Conference, Flagstaff, Arizona, USA, 16-19, 2005. Retrieved August 12, 2009, from http://informingscience.org/proceedings/InSITE2005/I60f75Tatn.pdf

Taylor, I. (1997). *Developing learning in interprofessional education: Partnerships for practice*. Buckingham, UK: Open University Press.

Taylor, J. D., Eastburn, S., Coates, C., & Ellis, I. (2006). Using mobile phones for critical incident assessment in health placement practice settings. *Third Biennial Northumbria/EARL SIG assessment Conference.*

Taylor, T. L. (2002). Living Digitally: Embodiment in Virtual Worlds . In Schroeder, R. (Ed.), *The Social Life of Avatars* (pp. 40–62). London: Springer-Verlag.

Teare, R., Davies, D., & Sandelands, E. (1998). *The Virtual University An Action Paradigm and Process for Workplace Learning*. London: Cassell.

TechDis. (2002). *The accessible curricula: good practice for all*. TechDis, LTSN Generic Centre and UWIC.

Tew, J., Gell, C., & Foster, S. (2004). *Learning from Experience: Involving Service Users and Carers in Mental Health Education and Training: A Good Practice Guide*. Higher Education Academy/NIMHE/Trent Workforce Development Confederation.

Thatcher, J. D. (2006). Computer Animation and Improved Student Comprehension of Basic Science Concepts. *The Journal of the American Osteopathic Association*, *106*(1), 9–14.

The Audit Commission. (1999). *Critical to success*. London: The Stationery Office.

The Bristol Royal Infirmary Inquiry. (2001). *The report of the public inquiry into children's heart surgery at the Bristol Royal Infirmary 1984-1995*. Retrieved April 3, 2009, from www.bristol-inquiry.org.uk/finalreport/the_report.pdf

The Combined Universities Interprofessional Learning Unit. (2006). *The Combined Universities Interprofessional Learning Unit final report*. Retrieved July 10, 2008, from http://www.sheffield.ac.uk/cuilu

Thistlethwaite, J., & Nesbit, G. (2007). Interprofessional Education: what's the point and where we're at. *The Clinical Teacher*, *4*, 67–72. doi:10.1111/j.1743-498X.2007.00148.x

Thompson, L., & Cupples, J. (2008). Seen and not heard? Text messaging and digital sociality. *Social & Cultural Geography*, *9*(1), 95–108. doi:10.1080/14649360701789634

Thompson, L., & Kimberley, S. (2006). *Audit of Toolbox usage* (Final Report). Australian Flexible Learning Framework: Department of Education, Science and Training, Commonwealth of Australia. Retrieved March 16, 2009, from http://toolboxes.flexible-learning.net.au/documents/pdfs/NationalAuditOfToolboxes05.pdf

Thorpe, M. (2002). From independent learning to collaborative learning: new communities of practice in open, distance and distributed learning . In Lea, M., & Nicoll, K. (Eds.), *Distributed Learning: Social and Cultural Approaches to Practice* (pp. 131–151). London: RoutledgeFalmer.

Thotnas, B. A., Coppola, J. F., & Feldman, H. (2001). Adopting handheld computers for community-based curriculum: Case study. *The Journal of the New York State Nurses' Association*, *32*(1), 3–6.

Thralls, C., & Blyler, N. R. (1993). The Social Perspective and Pedagogy in Technical Communication. *Technical Communication Quarterly*, *2*(3), 249–270.

Timmins, F., & Kaliszer, M. (2002). Aspects of nurse education programmes that frequently cause stress to nursing students -- fact-finding sample survey. *Nurse Education Today*, *22*(3), 203–211. doi:10.1054/nedt.2001.0698

TLRP. (2008). *Education 2.0? Designing the web for teaching and learning. Teaching and Learning Research Programme*. Retrieved May 18, 2009, from http://www.tlrp.org/tel/files/2008/11/tel_comm_final.pdf

Todres, L. (2008). Being with that; the relevance of embodied understanding for practice. *Qualitative Health Research*, *18*(11), 1566–1573. doi:10.1177/1049732308324249

Todres, L., Galvin, K., & Dahlberg, K. (2007). Lifeworld-led healthcare: revisiting a humanising philosophy that integrates emerging trends. *Medicine, Health Care, and Philosophy*, *10*(1), 53–63. doi:10.1007/s11019-006-9012-8

Toro-Troconis, M., Partridge, M., Barrett, M., & Mellstöm, U. (2008). Game based learning for the delivery of virtual patients in Second Life. *The Academy Subject Centre for Medicine, Dentistry and Veterinary Medicine Newsletter01, 17*, Summer, 3-4. Retrieved September 11, 2009, from http://www.medev.ac.uk/external_files/pdfs/01_newsletter/0117_lo_res.pdf

Towell, J., & Towell, E. (1997). Presence in Text-Based Networked Virtual Environments or "MUDS". *Presence (Cambridge, Mass.)*, *6*(5), 590–595.

Trigwell, K., Prosser, M., & Taylor, P. (1994). Qualitative Differences in Approaches to Teaching 1st Year University Science. *Higher Education*, *27*(1), 75–84. doi:10.1007/BF01383761

Trim, P. (2001). An Analysis of a Partnership Arrangement between an Institution of Further Education and an Institution of Higher Education. *Journal of Further and Higher Education*, *25*(1), 107–116. doi:10.1080/03098770020030542

Tulloch, J. (2006). *One day in July: experiencing 7/7*. London: Little, Brown.

Tunstall-Pedoe, S., Rink, E., & Hilton, S. (2003). Student Attitudes To Undergraduate Interprofessional Education. *Journal of Interprofessional Care*, *17*(2), 161–172. doi:10.1080/1356182031000081768

Turner, P., Sheldon, F., Coles, C., Mountford, B., Hiller, R., Radway, P., & Wee, B. (2000). Listening to and learning from the family carer's story: an innovative approach in interprofessional education. *Journal of Interprofessional Education, 15*(4), 387–395.

Tyler, G. (2006). Addressing Barriers to Participation: Service User Involvement in Social Work Training. *Social Work Education, 25*(4), 385–392. doi:10.1080/02615470600593394

UKCC. (2000). *Registrar's Letter 17/2000 Requirements for Pre-Registration Nursing Programmes*. London: United Kingdom Central Council for Nursing, Midwifery and Health Visiting.

Universities Collaboration in E-Learning (UCEL). (2009). *UCEL Homepage*. Retrieved February 14, 2009, from http://www.ucel.ac.uk/

Van den Besselaar, P. & Heimeriks (2001). *Disciplinary, Multidisciplinary, Interdisciplinary: Concepts and Indicators*. Paper presented at 8th Cnference on Scientometrics and Infometrics, Syndney, Australia.

Vince, R. (1996). *Managing Change: Reflections on equality and management learning*. Bristol, UK: Polity.

Vygotsky, L. (1978). *Mind in society: the development of higher psychology processes*. Cambridge, MA: Harvard University Press.

Vygotsky, L. S., & Cole, M. (1978). *Mind in Society: The Development of Higher Psychological Processes*. Cambridge, MA: Harvard University Press.

Wackerhausen, S. (2007). *Reflection as transformation Collaboration and reflection across boundaries*. Key note speech, Proceedings of the First International Conference of the European Interprofessional Network (EIPEN), Krakow, Poland.

Wagner, T. (2008). RIGOR redefined. *Educational Leadership, 66*(2), 20–24.

Waldegrave, W. (1996). *Chief Secretary William Waldegrave speech at the Chartered Institute of Public Finance and Accountancy's conference*: Building Partnerships - Working Together. Held in Brighton 11-13 June 1996.

Wall, S., & Shankar, I. (2008). Adventures in transdisciplinary learning. *Studies in Higher Education, 33*(5), 551–565. doi:10.1080/03075070802373008

Walsh, K. (2007). Inter-professional Education Online: The BMJ Learning Experience. *Journal of Interprofessional Care, 21*(6), 691–693. doi:10.1080/13561820701436912

Wanless, D. (2004). *Securing Good Health for the Whole Population: final report. Commissioned by the Department of Health*. London: Department of Health.

Warkentin, M., & Beranek, P. M. (1999). Training to improve virtual team communication. *Information Systems Journal, 9*(4), 271–289. doi:10.1046/j.1365-2575.1999.00065.x

Watson, S. (2000). The support that mentors receive in the clinical setting. *Nurse Education Today, 20*(7), 585–592. doi:10.1054/nedt.2000.0463

Weiss, J. (1987). Pathways to cooperation among public agencies. *Journal of Policy Analysis and Management, 7*(1), 94–117. doi:10.2307/3323353

Weiss, L., Capozzi, M. & Prusak, L. (2004). Learning from the internet giants. *MIT Sloan Management Review, Summer,* 79-84.

Wenger, E. (1999). *Communities of Practice; Learning, Meaning and Identity*. Cambridge, MA: Cambridge University Press.

Wesch, M. (2008). *An anthropological introduction to YouTube*. Presentation at the Library of Congress. Retrieved September 29, 2009, from http://www.youtube.com/watch?v=TPAO-lZ4_hU&feature=fvw

West, M. A., Borrill, C., Dawson, J., Scully, J., Carter, M., & Anelay, S. (2002). The link between management of employees and patient mortality in acute hospitals. *International Journal of Human Resource Management, 13*(8), 1299–1310. doi:10.1080/09585190210156521

Wharrad, H. J., Allcock, N., & Chapple, M. (1994). A survey of the teaching of biological sciences on undergraduate nursing courses. *Nurse Education Today, 14*, 436–442. doi:10.1016/0260-6917(94)90004-3

Wharrad, H. J., Herman, L., Timmons, S., Randle, J., Lymn, J., Shaw, W., et al. (2008). *How easy is cross sector collaboration? A case study of the collaborative development of e-learning in infection control, medicines and prescribing*. Proceedings of International Conference Education Research & Innovation, Madrid, Spain.

Wharrad, H. J., Kent, C., Allcock, N., & Wood, B. (2001). A comparison of CAL with a conventional method of delivery of cell biology to undergraduate nursing students using an experimental design. *Nurse Education Today, 21*(7), 579–588. doi:10.1054/nedt.2001.0602

Wheeler, M., Nie, M., & Salmon, G. (2008). How avatars learn together: an immersive experience. *Online Newsletter*, Monday July 21st. Retrieved September 11, 2009, from http://newsletter.alt.ac.uk/e_article001151232.cfm

Wheeler, S. (2007). The influence of communication technologies and approaches to study on transactional distance in blended learning. *ALT-J, 15*(2), 103–117. doi:10.1080/09687760701470924

White, A., Allen, P., Goodwin, L., Breckinridge, D., Dowell, J., & Garvy, R. (2005). Infusing PDA technology into nursing education. *Nurse Educator, 30*(4), 150–154. doi:10.1097/00006223-200507000-00006

Whitehead, C. (2007). The doctor dilemma in interprofessional education and care: how and why will physicians collaborate? *Medical Education, 41*, 1010–1016. doi:10.1111/j.1365-2923.2007.02893.x

Whittaker, K. A., & Taylor, J. (2004). Learning from the experience of working with consumers in educational development. *Nurse Education Today, 24*, 530–537. doi:10.1016/j.nedt.2004.06.005

Whittington, C. (2003). *Learning for collaborative practice with other professions and agencies*. London: Department of Health.

Wilcock, P. M., & Headrick, L. (2000). Interprofessional learning for the improvement of healthcare: why bother? *Journal of Interprofessional Care, 14*, 111–117. doi:10.1080/713678552

Wilde, R. (2007). Europeanhistory.about.com/library/glossaryhist/ bldefdiscipl.htm [Retrieved 2nd October 2007].

Wiley, D. (2000). Connecting learning objects to instructional design theory: A definition, a metaphor, and a taxonomy. In D. Wiley (Ed.), *The Instructional Use of Learning Objects* (online version). Retrieved October 10, 2009, from http://www.reusability.org/read/chapters/wiley.doc

Wilkes, Z. (2006). The student-mentor relationship: a review of the literature. *Nursing Standard, 20*(37), 42–47.

Wilkinson, A., Forbes, A., Bloomfield, J., & Gee, C. F. (2004). An exploration of four web-based open and flexible learning modules in post-registration nurse education. *International Journal of Nursing Studies, 41*, 411–424. doi:10.1016/j.ijnurstu.2003.11.001

Willder, B. (2002). Disability legislation: implications for learning technologists in the UK.

Wilson, V., & Pirie, A. (2000). *Multidisciplinary Teamworking: Indicators of Good Practice. Spotlights, Scottish Council for Research in Education.* Retrieved 15th April 2008 from http://www.scre.ac.uk/spotlight/spotlight77.html

Wilson-Barnett, J., Butterworth, T., & White, E. (1995). Clinical support and the Project 2000 nursing student: Factors influencing this process. *Journal of Advanced Nursing, 21*, 1152–1158. doi:10.1046/j.1365-2648.1995.21061152.x

Windle, R., McCormick, D., Dandrea, J. & Wharrad, H.J. (2009). The use and reuse of reusable learning objects for addressing challenging areas of the curriculum. *Nurse Education Today* (submitted for publication).

Windle, R., Wharrad, H. J., Laverty, H., Keay, C., Hallawell, B., & Williams, A. (2008). *Student generated content. Permanent footprints or shifting sands*. Paper presented at Online Educa Berlin: ICWA.

Windle, R., Wharrad, H. J., Leeder, D., & Morales, R. (2007b). *Analysis of the Pedagogical Attributes of Learning Objects in an attempt to identify Reusable Designs.* In C. Montgomerie & J. Seale (Eds.), *Proceedings of World Conference on Educational Multimedia, Hypermedia and Telecommunications* (pp. 2676-2685). Chesapeake, VA: AACE.

Windle, R., Wharrad, H. J., McCormick, D., Dandrea, J., Bath-Hextall, F., Leighton, B., et al. (2007a). Does A Community Of Practice Approach To The Development of Learning Objects Support Reuse of E-Learning Materials In Health Science Education. In L. Gómez Chova, D. Martí Belenguer, & I. Candel Torres (Eds), *International Technology, Education and Development Conference*. Valencia, Spain: INTED Proceedings.

Wolfson, G. K., Marsom, G., & Magnuson, C. W. (2005). Changing the nature of the discourse: Teaching field seminars online. *Journal of Social Work Education, 41*(2), 335–361. doi:10.5175/JSWE.2005.200300346

Woo, K., Gosper, M., McNeill, M., Preston, G., Green, D., & Philips, R. (2008). Web-based lecture technologies: blurring the boundaries between face-to-face and distance learning. *Research in Learning Technology, 16*(2), 81–93.

Wood, G. (1999). Interdisciplinary working in built environment education. *Education + Training, 41*(8), 373–380. doi:10.1108/00400919910298648

Wood, G. D. (2005). Partnering practice in the relationships between clients and main contractors, *RICS Research paper series, 5*(2), April.

Wood, J., & Wilson Barnett, J. (1999). The influence of user involvement on the learning of mental health nursing students. *Nursing Times Research, 4*(4), 257–270. doi:10.1177/136140969900400403

Woods, C. (2007). Researching and developing interdisciplinary teaching: towards a conceptual framework for classroom communication. *Higher Education, 54*, 853–866. doi:10.1007/s10734-006-9027-3

World Health Organisation. (in press). *A Framework for Action on Inter-professional Education and Collaborative Practice*. Belgium . *World Health Organisation.*

World Health Organisation. (in press). World Health Organization Framework for Action on Interprofessional Education & Collaborative Practice. *WHO Library Cataloguing.*

World Health Organization (WHO). (1988). *Learning Together to Work Together for Health*. Technical Report Series 769. Geneva: World Health Organization.

World Health Organization. (1988). *Learning together to Work Together*. Geneva, Switzerland: World Health Organization.

World Health Organization. (2009). *Chronic Diseases and Health Promotion.* Retrieved September 30, 2009, from http://www.who.int/chp/chronic_disease_report/part2_ch2/en/index1.html

Wotton, H. (1624). *Elements of Architecture Collected from the Best Sources and Examples. London: John Bull.* Farnborough, UK: Gregg Press. Reprinted 1969

Wright, A., & Lindqvist, S. (2008). The development, outline and evaluation of the second level of an interprofessional learning programme – listening to the students. *Journal of Interprofessional Care, 22*(5), 475–487. doi:10.1080/13561820802210968

Wright, C., & Rowe, N. (2005). Protecting Professional Identities: Service User Involvement and Occupational Therapy. *British Journal of Occupational Therapy, 68*(1), 45–47.

Wykurz, G., & Kelly, D. (2002). Developing the role of patients as teachers: literature review. *BMJ (Clinical Research Ed.), 325*(7368), 818–821. doi:10.1136/bmj.325.7368.818

Yadav, A., & Koehler, M. J. (2007). The role of epistemological beliefs in preservice teachers' interpretation of video cases of early-grade literacy instruction. *Journal of Technology and Teacher Education, 15*(3), 335–361.

Yan, J., Gilbert, J., & Hoffman, S. (2007). World Health Organization study group on interprofessional education and collaborative practice. *Journal of Interprofessional Care, 2*(6), 587.

Yanosky, R., Harris, M., & Zastrocky, M. (2004*). Hype cycle for Higher education Gartner report G00121162.* Retrieved September 28, 2008, from http://www.gartner.com/it/section.jsp?type=press_releases&format=archive&year=2004

Yonge, O., Krahn, H., Trojan, L., Reid, D., & Haase, M. (2002). Being a preceptor is stressful! *Journal for Nurses in Staff Development, 18*(1), 22–27. doi:10.1097/00124645-200201000-00005

Young, P. M., Leung, R. M., Ho, L. M., & McGhee, S. M. (2001). An evaluation of the use of hand-held computers for bedside nursing care. *International Journal of Medical Informatics*, *62*(2-3), 189–193. doi:10.1016/S1386-5056(01)00163-0

Zhao, S. (2003). Toward a taxonomy of copresence. *Presence (Cambridge, Mass.)*, *12*(5), 445–455. doi:10.1162/105474603322761261

# About the Contributors

**Adrian Bromage's** academic background is in Psychology and Cognitive Science. Between 2002-2007 he worked at Coventry University's education development unit, then between 2007-2009 at the Centre for Interprofessional eLearning (CIPeL). At CIPeL, Adrian was involved in evaluating Reusable Interprofessional learning Objects and interprofessional learning in general. His most recent research is an unfunded private action research study, 'the College Of Online Liberal Arts (COOLA)', which explores the intersection of free and open-source software and academics' professional autonomy. In 2010 his work at The University of Birmingham Learning Development Unit drew in part on the latter project.

**Lynn Clouder** is Director of the Centre for Interprofessional eLearning (CIPeL). Lynn's professional background is in physiotherapy and she has worked in higher education since joining Coventry University in 1991. The focus of her PhD was on identity development and professional socialization. A secondment to the Centre for the Study of Higher Education, as a pedagogical research fellow, allowed Lynn to further develop her interests in pedagogical research. She was awarded a National Teaching Fellowship in 2007.

**Jill Thistlethwaite** is Director of the Institute of Clinical Education at Warwick Medical School and a practising GP. She has been involved with interprofessional education since working at Leeds University from 1996-2003. From 2003 to 2008 she was a medical educator in Australia, first at James Cook University in North Queensland where she developed and ran a Graduate Certificate in Health Professional Education, and then at the University of Sydney. In Sydney she received an Australian Learning and Teaching Council grant with colleagues for a project relating to interprofessional education and practice within higher education. At Warwick she continues as associate editor of the Journal of Interprofessional Care. She is president of InterEd, a board member of CAIPE and was a contributor to the World Health Organisation's 'Framework for Action on IPE and Collaborative Practice (2009).

**Frances Gordon** joined Sheffield Hallam University in 2002 and is a nurse by professional background. She completed her master of Education degree in 1990 at the University of Aberdeen and was awarded a research fellowship by the National Board for Scotland for Nursing, Midwifery and Health Visiting to undertake doctoral studies, completing her PhD, again from the University of Aberdeen, in 2000. She was awarded a personal chair in Interprofessional Education by Sheffield Hallam University in 2007 in recognition of her national and international reputation in the field of interprofessional education in health and social care. Presently, at Sheffield Hallam University she is Head of Interprofessional and Multidisciplinary Learning and Co-Director of the Centre for Interprofessional eLearning (CIPeL).

***

**Sue Baptiste,** OT, MHSc, is a tenured professor at McMaster University, Hamilton, Ontario in the School of Rehabilitation Science and is a past Assistant Dean of the occupational therapy program. She has focused much of her academic work on faculty development, organizational cultural change and ongoing professional competence plus curriculum design and development. Current research includes the development of interprofessional teaching and learning resources as well as the creation and delivery of a program for internationally educated occupational therapists to enable their integration into Canadian professional practice

**Hugh Barr** holds visiting chairs in the interprofessional field at the University of Greenwich and Kingston University with St George's University of London, and previously at the University of Nottingham and King's College London, with honorary doctorates from the universities of Southampton and East Anglia. His extensive overseas commitments include academic appointments in Japan (current) and Finland and New Zealand (former). He is Editor of the 'Promoting Partnership for Health' series for Wylie and consultant editor to of the Journal of Interprofessional Care. His background is in probation, prison aftercare, criminology and social work education.

**Fiona Bath-Hextall** is a Reader in Evidence Based Health Care. She conducts systematic reviews that often are used to inform clinical trial design. She is a member of the teaching staff of the Centre of Excellence in Teaching and Learning for Reusable Learning Objects (RLO-CETL). She also works with colleagues in multidisciplinary teams to author and produce learning objects, often involving working with colleagues from across the partner institutions in developing communities of practice.

**Patricia (Pat) A. S. Bluteau** originally trained as a mental health nurse. Patricia's work in interprofessional student learning has informed the development of Interprofessional Learning Pathway, now underway for thirteen professional groups at Coventry University. From 2005-2006, she combined Associate Director of CIPeL and Interprofessional Learning Pathway Co-ordinator, both within the Faculty of Health & Life Sciences. In October 2006, she became full time in CIPeL. Patricia currently undertakes collaborative research into students' perceptions of an interprofessional learning Pathway and is exploring historical perspectives of IPL through a case study approach.

**Karen Booth's** professional background is in medical imaging. Her involvement with interprofessional learning began in 1995 with leading the planning and development of the IPL programme, which now spans all eight health and social care courses in the Faculty of Health and Wellbeing of Sheffield Hallam University. Her interest in e-learning started with the introduction of Blackboard as a VLE, and she has undertaken a number of studies into its use in enhancing and facilitating learning. Her current research interests are related to interprofessional learning, particularly in using technology to support learning, and in building learner autonomy.

**Helen Bradbury** is a Senior Lecturer at the University of Leeds working in both education and healthcare. Helen worked in the NHS for 20 years as a hospital pharmacist before commencing her academic career. Her research interests include inter-professional education and teaching and learning in clinical settings.

**Andrew Brooks** did his MA in Media Art at Coventry University. His work explores the interactive use of webcam technologies as well as audience participation with the moving image. He has lectured in multimedia and has over 15 years filmmaking experience. He recently spent a year living in Southern China, where he wrote and directed a feature length film. He is currently Employed as the Learning Technologist at CIPeL, supporting staff in the development and creation of a variety of learning objects. Email: words@hotmail.com

**Marion Brown**, RSW. Since 2006, Marion Brown has worked as Assistant professor at Dalhousie University. Marion coordinates research based and International field placements for the School of Social Work. Her research focuses on social work education requirements across national borders, the migration of social workers both within Canada and internationally, and experiences of social workers and social work students who complete social work education in one national jurisdiction and practice in another. Marion's other research includes critical girlhood studies, specifically, constructions of femininity and use of violence by girls, distance learning in social work education, and pedagogies for interprofessional education and practice.

**Helen Bywater** qualified as a Registered General Nurse / Registered Sick Children's Nurse in 1987 and her practical experience has been focussed in paediatric nursing. In 1994 she joined the education team at Sheffield Children's Hospital to address the personal and professional development needs for all staff including the development and delivery of in-house training programmes. Helen joined Sheffield Hallam University in 2002 as a training coordinator in the Forging Ahead project team. This was a UK Department of Health First Wave site for modernising interprofessional education in undergraduate health and social care programmes, and for the last five years she has been a member of the core team of the Centre for Interprofessional eLearning (CIPeL).

**Mark Childs**, MScEcon, MA, BSc, PGCE, joined Coventry University in 2008 as a Teaching Development Fellow in e-learning and is now located in Coventry's Faculty of Engineering and Computing. Between 2001 and 2008 he worked at the University of Warwick on e-learning research, previously exploring national and international collaborative e-learning at the University of Wolverhampton. Mark currently studies Immersive Virtual Worlds as a doctoral student at the Warwick Institute of Education.

**Elinor Clarke** was seconded from the nursing and midwifery department to the Centre for Interprofessional e-learning (CIPeL). She maintains her professional Midwifery registration, and continues to teach on the undergraduate midwifery programs. Elinor is experienced in curriculum design and delivery, and gained experience in the development of a patient journey (Betty, 2005). Following completion of the e-spire course in 2005, Elinor works as an e-facilitator on the interprofessional learning pathway (IPL) at Coventry University. CIPeL responsibilities are student interns, special interest group, repository and service user involvement. Elinor is actively engaged in the development of new IP learning objects, researching IP attributes and interested the use of technology to support learning.

**Tarsem Singh Cooner** worked as a qualified senior social work practitioner and practice teacher prior to joining the University of Birmingham in 2001. As Associate Director of CEIMH, Tarsem's primary role is to promote the development of innovative and flexible methods of teaching and learning in interdisciplinary mental health within the University. He takes a lead role in assisting the contributing colleges in overcoming the traditional barriers to interdisciplinary teaching and learning. Tarsem's research interests include the role the Internets communication and collaboration properties offer in developing interdisciplinary training for social care and health professionals; mobile e-Learning in practice settings; students experiences of enquiry-based blended learning designs.

**Bernadette Davies** is a registered nurse, and is module leader for pre and post registration modules at Coventry university including continence care and wound management. She is involved with the inter-professional learning pathway (IPLP) as both an e-facilitator and a year lead. Students from several health care professional courses at Coventry University and Warwick medical school undertake IPLP in the first second and final year of their course. She has worked with CIPeL to develop learning objects used within IPLP and to support teaching for long term conditions and theoretical aspects of manual handling. These learning objects can be used by a range of health and social care professions and students.

**Christine Dearnley** is the University of Bradford's Site lead for the Assessment and Learning in Practice Settings (ALPS) Centre of Excellence in Teaching and Learning, a collaborative interprofessional programme between five Higher Education Institutions and a wide range of partners from the health and social care professions. Chris has a keen interest and a comprehensive research portfolio in learning and assessment for health care practitioners. Her early work was applied to open and distance learning and her interest in the potential of e-Learning and more recently mobile learning, in helping students learn how to learn has built on this work. Her main area of research is Pedagogy for Health & Social Care Practice, specifically: Components of Independent learning & the implications for students, tutors and the NHS; Independent learning & its relationship with personal and professional development; Learning technologies that promote inclusivity and support independent learning for all students.

**Samuel Edelbring** works at the Centre for Medical Education (CME) at the Karolinska Institutet, Sweden. His research concerns learning with a particular kind of technology called "Virtual Patients" (VPs). His PhD project, "The nature and function of Virtual Patients in medical education", aims at articulating the character and clarifying the functions of VPs in medical education. The research methods used are combinations of qualitative and quantitative approaches. At CME, Samuel teaches on e-learning, teacher training and communication courses.

**John Fairhall** is the Mobile Technology Adviser at the University of Bradford – a role that is part technical specialist, part project manager and part learning technologist. He is currently working on the University's E-Strategy to deliver a 'Web enabled campus supported by wireless and mobile computing'. As part of this work, John is particularly interested in using mobile technology in teaching and learning, and so represents Bradford on the IT group of the Assessment and Learning in Practice Settings (ALPS) Centre of Excellence in Teaching and Learning (CETL), and continues to play a key role in supporting students and staff who are using mobile devices. His work in this regard involves researching and developing uses of mobile technology in teaching, learning and assessment in practice- and work-based settings, in particular, the Mobile Enabled Disabled Students (MEDS) project.

**Dawn Forman** held the post of Dean and Professor for 12 years whilst she worked at the University of Derby and Sheffield Hallam University. She is a Key Associate of the Leadership Foundation for Higher Education, and has particular expertise in governance, strategic leadership, interprofessional education, team development, executive coaching and change management with extensive practical experience within Higher Education and Health Service fields. Dawn is also a European representative for the Network: Towards Unity for Health, which is closely linked with the World Health Organisation and has been a board member of the Centre for the Advancement of Interprofessional Education (CAIPE) for 15 years.

**Marit Fougner** has collaborated with Laurence Habib to test the utility of tools selected from Soft Systems Methodology to support action research in IPE development in Oslo to shed light on a rich picture from multiple perspectives. Marit is engaged in networks developing models of interprofessional learning in practice within healthcare education. Within the realm of her teaching practice, she has been developing pedagogical frameworks using art as an approach to get insights into the use of observation as a scientific method in health work. Her research interests include also learning models in physiotherapy education as well as strategies aiming at developing intercultural competence among physiotherapy students.

**Nina Godson** is a Senior lecturer and lead for clinical skills at Coventry University. She leads on the management and development of the clinical skills team and develops and co-ordinate clinical skills across the pre-registration nursing curriculum with an applied research focus. Part of here remit is to review and develop interprofessional moving and handling policies through the interprofessional moving and handling working group. Her CIPeL secondment was related to these aims, and involved developing reusable learning objects for clinical skills training, to be delivered on hand-held mobile devices.

**Pete Greasley** is a Lecturer at the School of Health Studies, University of Bradford. His academic interests include educational research and psychosocial aspects of health & social care.

**Laurence Habib** has collaborated with Marit Fougner to test the appropriateness of tools selected from Soft Systems Methodology (SSM) to support action research in IPE development in Higher Education. Her work with SSM includes using rich pictures to gain insights into problem situations from multiple perspectives. Laurence's research interests include issues of adoption, appropriation and domestication of technology, as well as action-oriented methods and methodologies for academic and educational development.

**Elizabeth Hanna,** MHSc, Reg. CASLPO, coordinates clinical education at Bridgepoint Hospital, a mid-size complex chronic care and rehabiliation hospital with a community affiliation with University of Toronto. Elizabeth has a hand in the clinical placements, internships, practicums and associated work of more than 500 students in roughly 15 different disciplines a year. In doing so, Elizabeth has the opportunity to work at multiple levels with multiple partners: academic institutions, staff at Bridgepoint from the clinical frontlines up to the executive corridor, and the students themselves. Her back ground training is in Speech-Language Pathology.

**Karen Harrison** has departmental management responsibility for Physiotherapy and Dietetics at Coventry University, and is Professor of Physiotherapy Education. Karen is involved in teaching on health inequalities, health science management, clinical teaching and research supervision at undergraduate and postgraduate level. Her research interests include curriculum evaluation and quality management in higher education, exploration of the patient perspective on the experience of living with chronic disease and Consumer perspective and empowerment in health and education.

**Melissa Highton** is head of the Learning Technologies Group at the University of Oxford. Previously she was a senior staff development officer at University of Leeds. She has many years experience in teaching blended learning courses for professional development. She works closely with academics and supports innovation in the use of technology in teaching. At Oxford her group provide IT services, teach skills to thousands of members of the University, publish open code and carry out pedagogical research. Projects at Oxford include the virtual learning environment for research led teaching, podcasting, community created collections, virtual worlds, green IT and computer modelling.

**Maggie Hutchings'** work as Senior Lecturer in Education in the School of Health & Social Care focuses on leading and supporting the development of learning, teaching and assessment praxis. Her research interests include interprofessional education, lifeworld-led education and e-learning pedagogy. Her particular areas of expertise are in developing and supporting the pedagogical design of new and innovative learning opportunities using a variety of e-learning tools. Projects have included collaborating with colleagues on Wessex Bay, a virtual practice community to enhance interprofessional education, implementation of the School's blended learning strategy, and the development of e-learning modules in information literacy and research awareness for practice. Current activities are focused on developing lifeworld-led online case studies for a transprofessional unit and implementing computer based assessment and e-submission and e-feedback initiatives within the School.

**Ann Jackson** works at Warwick Medical School. Graduate entry medical students share the Interprofessional Learning Pathway with those of Coventry University. Her background prior to higher education was nursing, midwifery and health visiting. She is a critical Reader for CIPeL, reviewing secondees' learning objects during their development, both formally using peer review tools and informally. Critical Readers offer advice in relation to the input of their particular discipline and from an e-learning viewpoint. Ann with colleagues at Warwick and Coventry University has developed e learning service user case studies as a support to the more usual service user interviews. Ann is particularly interested in learning objects that can be used on line but also stand alone allowing them to be used in a variety of different settings.

**Marion Jones** RN; BA; MEdAdmin (HONS); PhD is the Associate Dean Postgraduate in the Faculty of Health and Environmental Sciences at Auckland University of Technology in New Zealand. She is a Registered Nurse with experience in acute care, perioperative nursing, pain management, leadership, team practice and interprofessional practice. She is also visiting Professor at Derby University in the U.K. Marion has a strong commitment to interprofessional practice, nursing and health practice nationally and internationally. Marion's research and teaching areas are in theory practice debate, interprofessional team practice, perioperative nursing, pain management and research methodology. Her present areas of research are Interprofessional Practice, Client Centred Care and Team Practice.

**Marie Krumins** MA, PgDip, CIM joined CIPeL in 2005 and is responsible for the operational running of the CETL, including developing policy, financial handling, project management, and exploiting promotional and dissemination opportunities. Marie is also involved in researching aspects of CIPeL's work including online peer facilitation and benefits to academics in engagement with the CETL.

**Joanne S. Lymn** has a background in education developments for teaching pharmacology across a range of student groups, using techniques including problem-based learning and reusable learning object (RLOs). At the University of Nottingham, Joanne was responsible for moving the non-medical prescribing course to a virtual learning Environment (VLE), the use of which resource has increased year-on-year. The VLE now houses all the lecture information, pre-reading, links to relevant websites and papers, as well as links to relevant RLOs and lecture recordings. The VLE is also used for formative on-line assessments and students use discussion forums for easy communication and group work.

**Helen M. Lynch** has a background as an educator and flexible learning enthusiast in the vocational education and training and university sectors in Australia. She is the ACT (Australian Capital Territory) Toolbox Champion for the Australian Flexible Learning Framework and a committed advocate of e-learning and the use of the world wide web for education. Helen is also an online educational designer with Canberra Institute of Technology and holds post graduate qualifications in educational management, online education and Indigenous education. Helen is currently undertaking a doctorate at the University of Southern Queensland researching professional development and e-learning object reuse.

**Lorraine McFarland** has been involved in applied research in the self-management of long-term health conditions, Emotional disclosure, emotional experience of living with a long-term health condition and health interventions. She has also contributed to research to explore participants', tutors' and health professionals' views of the Expert Patient Programme (EPP) in the Sandwell area. The EPP focuses on assisting patients to feel confident in their own ability to self-manage their medical condition. Her academic background is in psychology and education.

**Heather MacNeill,** MD, BSc (PT) FRCPC, is a lecturer in the Division of Physiatry at the University of Toronto. Heather is project lead for the Collaborative Online Interprofessional Learning (C.O.I.L.) program. The project seeks to create online educational tools that foster real-life collaboration between different professional groups. Bringing together multiple health professionals requires a common focus on which to base the development of community and collaboration. The common denominator of health professionals is the patient- which in a learning environment can be translated into case based learning. To learn more about the project visit www.bridgepointhealth.ca under research and teaching-->COIL

**Martin Oliver** is a Reader in the Faculty of Culture and Pedagogy at the London knowledge lab, based at the University of London. His research interests include impacts of new technology on roles and practices within Higher Education (including how this changes what students learn and do), evaluating ICT use and the development of theory and methodologies in e-learning. Martin teaches on the MA in ICT in Education, and is seconded part-time to the Higher Education Academy, where he is involved with work on EvidenceNet, a national service to support evidence-informed practice.

**Rebecca O'Rourke's** academic background is in Cultural Studies, where she has focused on writing as a social practice. Most of her career has been spent as an adult educator, in both formal and informal settings, and she returned to work in higher education in 1991. Rebecca's research and teaching interests at the University of Leeds include creative writing, academic writing and professional learning. She is interested in the pedagogic potential of new technology in relation to writing development and was

PI on the JISC-funded AWESOME Dissertation Environment project. She is a member of the Clinical Education Course team at the University of Leeds.

**Karen Ousey** clinical background is within orthopaedics and tissue viability. Karen moved into education in 1997 as a lecturer/practitioner at the University of Salford and moved to the University of Huddersfield in 2005, and is currently the Divisional Head for Acute and Critical Care. Her educational research interests include exploration of the theory-practice gap, the socialisation of student nurses into clinical practice and advancing clinical practice. She has authored numerous publications relating to education and wound care and is the author of 'Pressure Area Care' and co author of Lower Extremity Wounds: A problem based approach.

**Melissa Owens** is interprofessional learning co-ordinator at the University of Bradford and is also a board member of CAIPE (Centre for the Advancement of Interprofessional Education). She has a particular interest in interprofessional learning in the primary care setting.

**Caroline Plews** works as the School Lead for Service User and Carer Involvement at the University of Bradford's School of Health Studies. Her work focuses on service user and carer involvement in the training and education of health professional students. The school is working to increase the number of service users and carers who collaborate with academic staff to develop the service user and carers' perspectives within all aspects of their educational programmes. The idea is inspired by the UK Government Department of Health initiative to have patient centred care, where, service users and carers help to shape the UK National Health Service.

**Anne Quinney** is committed to delivering high quality social work education. Recent research projects include exploring student self-efficacy in research skills for practice as part of the national Evaluating the Outcomes of Social Work Education 3 year project, funded by the Social Care Institute for Excellence and the Social Policy and Social Work subject centre of the Higher Education Academy, and a project funded by the Health Sciences and Practice subject centre of the Higher Education Academy (with Janet Scammell and Maggie Hutchings of Bournemouth University) to evaluate 'Wessex bay', a virtual practice community for student learning and staff development in inter-professional education. Along with Judith Thomas (UWE) and Colin Whittington (Independent Consultant) Anne recently developed a series of open access multimedia interactive learning resources on Interprofessional and Interagency Collaboration for the Social Care Institute for Excellence. Anne is also the Editor of the journal Practice; Social Work in Action, and the author of the textbook Collaborative Social Work Practice.

**Jacqueline Randle's** specialist areas of teaching are clinical skills and infection prevention and control. She has developed a range of reusable learning objects, e-learning resources aimed at clinical audiences, so clinical skills can be improved for the safety of patient care. The majority of Jacqueline's work has been clinically focussed and aimed at making a significant contribution to clinical practice by working in teams consisting of academics and clinicians, with studies primarily focussed on long-term behaviour change in order to reduce healthcare associated infections.

**Steve Rankin.** Since 2005, Steve Rankin has developed software solutions to make inter-professional collaboration and medical image access an integrated and seamless part of care delivery. As part of an

EHR or as a stand-alone application, eUnity provides secure access to full-fidelity DICOM images with unprecedented performance using standard web-browsers (no client software to manage.) This allows Medical professionals to access medical images across multiple sites and multiple PACS using eUnity as a universal viewer.

**Scott Reeves'** research work is focused on advancing scholarship in interprofessional education and practice. His main interests are exploring and evaluating the processes and outcomes related to interprofessional education and practice by use of ethnographic, mixed methods and systematic review. Scott also has appointments in the Wilson Centre for Research in Education, University of Toronto, the Centre for Faculty Development at St Michael's Hospital and the Keenan Research Centre, Li Ka Shing Knowledge Institute of St Michael's Hospital, Toronto.

**Janet Scammell** works as Academic and Professional Lead for Adult and Child Health Nursing at the School of Health and Social Care at Bournemouth University. Her work experience is varied most recently involving practice learning and education management. She has been involved in a number of funded projects concerned with uni-professional and interprofessional learning in practice. Her recent doctoral studies concern race and the mentor-student relationship in nursing. Janet has collaborated on a project funded by the Health Sciences and Practice subject centre of the Higher Education Academy (with Ann Quinney and Maggie Hutchings of Bournemouth University) to evaluate 'Wessex Bay', a virtual practice community for student learning and staff development in inter-professional education.

**Steve Smith** is currently Associate Head of Quality Management and Enhancement in the Faculty of Business, Environment and Society at Coventry University. His background is in environmental biology and he has a PhD in plant ecology. Following has interest in interdisciplinary education, Steve undertook a two year 0.5 secondment to CIPeL. Steve has lectured on undergraduate and postgraduate courses in environmental biology and wine studies. He also has extensive experience of programme development, course management and quality assurance. Steve's research interests are in environmental biology but he has undertaken some educational research through two HEA small grants.

**Patricia (Patty) Solomon** is Professor and Director of the Program for Interprofessional Practice, Education and Research (PIPER) in the Faculty of Health Sciences at McMaster University. She is a well known educator and has published and consulted widely in the area of curricular innovation in the health sciences. Currently she is the Director of Online IPE a virtual learning center devoted to e-learning in an interprofessional context.

**Stephen Timmons** worked in the UK National Health Service at the Royal London Hospital and the Hammersmith Hospital, and then went to the University of Glasgow to do an M.Sc. in Health Informatics. After that, he worked in the School of Health Care Practice at Anglia Ruskin University until 2000. Since that time he has worked at the University of Nottingham. His interests in using technology to support student learning includes developing and evaluating an e-learning resource to support the post-graduate Philosophy of Social Science module, and researching the acceptability and usefulness of an online statistics module and assessment using a Virtual Learning Environment (VLE).

**Kerry Trabinger, M.Ed.** has been involved in online Learning since 1998, initially as an Online tutor for Qantas College and then as an Online Lecturer at the Canberra Institute of Technology. In 2001 Kerry moved into online course design, and was awarded the prestigious Australian Flexible Learning Leader in 2004. Between 2006 and 2007 Kerry was the manager/facilitator of the Australian E-Learning Networks Community forum. Kerry currently works for the Centre for Excellence at the Canberra Institute of Technology and her current research passion is investigating the use of Virtual classrooms, and she is studying for a Doctorate in Education.

**Paraskevas Vezyridis** graduated in Nursing from the Technological Educational Institute of Athens in 2001. He has completed a MSc (Dist) in Health Informatics at City University of London and a BSc (Hons) in Nursing Studies at King's College, London. Since 2007, he has been undertaking a PhD studying the implementation of a clinical information system in an emergency department. As a tutor for the Graduate School at the University of Nottingham he has delivered training on the use of NVivo for qualitative data analysis and has worked as a researcher in the qualitative evaluation of a Virtual Portfolio Tool for first year nursing students and a pressure sensor-pager alert system for reducing falls in in-patient elderly. Has a particular research interest in science and technology studies.

**Stuart Walker** is a Disability Advisor and the IT Officer for Disabled people at the University of Bradford. His role involves offering advice to enable disabled people in many areas e.g.: Estates, Teaching and Learning and has extensive experience in the use of IT / Assistive technologies. Stuart has both aided and led in successfully gaining grants to research areas such as Netbooks, Speech Recognition Software and mobile communications. He is also involved in the Assessment and Learning in Practice Settings (ALPS) Centre of Excellence in Teaching and Learning, and his work in this regard involves researching and developing uses of mobile technology to enable disabled people in teaching, learning and assessment in practice- and work-based settings, in particular, the Mobile Enabled Disabled Students (MEDS) project. He has a keen interest in many forms of technology especially Mobile, Speech Recognition and open source software such as Linux. Stuart is currently in the process of completing a PhD into Speech Recognition Systems and user experiences.

**Heather Wharrad** is Reader & Associate Professor in Education and Health Informatics at University of Nottingham. She is academic lead for the SONET Applied Research group and the research group in Education & Technology for Health in the School of Nursing, Midwifery and Physiotherapy and has been principal investigator on many externally funded national and international projects on e-learning pedagogy, evaluation and research in health. She wrote the successful bid for the University of Nottingham partner of the HEFCE £3.3 million funded Centre for Excellence in Teaching & Learning (CETL) for Reusable Learning Objects. She is a Fellow of the Higher Education Academy. She has published widely in the fields of medical physiology, health informatics and e-learning and is a regular presenter and workshop facilitator at national and international e-learning conferences.

**Stephen White** trained as an Operating Theatre Technician in the Royal Air Force. He served in the UK and overseas before becoming an instructor in operating department practice. On leaving the Service he began this post at the University of Huddersfield. Stephen's interests are clearly focussed on the use of technology in teaching and learning, and he is presently completing a PhD on the Infection Potential of Mobile Technology Used in Healthcare Education. He is also the Project Lead for the

Yorkshire and the Humber SHA-funded development of a Multi-professional Online Mentor Update tool and is a member of the University of Huddersfield's core team for the Assessment and Learning in Practice Settings (ALPS) Centre of Excellence in Teaching and Learning. Stephen's latest project is a computer-based tool that utilises the principles of novice to expert competency development in guiding the learners' cognition within problem-based learning scenarios.

**Dr. Richard Windle** has been involved in developing and deploying reusable learning objects for the last 7 years. His most recent role was as the local academic coordinator for the Reusable Learning Object Centre for Excellence in Teaching and Learning (RLO-CETL) within the University of Nottingham, and as a member of the management group for this initiative. He has responsibility for developing, evaluating and disseminating reusable learning objects in health care related subjects. This involves working with colleagues in multidisciplinary teams to produce learning objects and working with colleagues from across partner institutions in developing communities of practice. Richard is exploring ways in which reusable learning objects can contribute to, and inform the emerging open educational resources (OER) agenda. His research activities focus around the community of practice approach to e-learning development and representations of pedagogical attributes within the materials produced, and around the engagement of students and other stakeholder groups in developing e-learning resources within health sciences. Richard.windle@nottingham.ac.uk

**Gerard Wood** is the Associate Head of School (Teaching) in the School of the Built Environment at the University of Salford, specialising in procurement, construction economics and project management. He worked for several years as a chartered quantity surveyor in both private practice and management contracting before pursuing an academic career. His most recent research interests include dispute avoidance, partnering arrangements and the practices of risk and value management within the construction industry, together with a parallel focus on built environment education and interdisciplinary working.

**Song Wu** is the Programme Director for BSc (Hons) in Quantity Surveying at the University of Salford. He has worked on several EPSRC funded project including '3D to nD modelling', 'Process Protocol', 'Salford: China research tour', 'nD game'. He is also an academic investigator for EPSRC funded Salford Centre for Research and Innovation. His research interests include product and process modelling, data modelling, and computer simulation. Dr Wu completed MSc IT in Construction and PhD at the University of Salford, and in 2009 was awarded UK China Fellowship for Excellence for his collaboration research with leading Chinese research institutions.

# Index

## D



## E